The Categories of Mathematics

" The Many Fields of Study "

Edited by Paul F. Kisak

Contents

1 **Mathematics**   1

    1.1 History 2

       1.1.1 Evolution 2

       1.1.2 Etymology 5

    1.2 Definitions of mathematics 5

       1.2.1 Mathematics as science 7

    1.3 Inspiration, pure and applied mathematics, and aesthetics 9

    1.4 Notation, language, and rigor 10

    1.5 Fields of mathematics 10

       1.5.1 Foundations and philosophy 11

       1.5.2 Pure mathematics 12

       1.5.3 Applied mathematics 14

    1.6 Mathematical awards 15

    1.7 See also 15

    1.8 Notes 15

    1.9 References 18

    1.10 Further reading 18

    1.11 External links 19

2 **Definitions of mathematics**   21

    2.1 Early definitions 21

    2.2 Greater abstraction and competing philosophical schools 21

    2.3 General, nonspecialist perspectives 22

    2.4 Playful, metaphorical, and poetic definitions 22

    2.5 See also 23

    2.6 References 23

    2.7 Further reading 24

3 **Mathematical beauty**   25

    3.1 Beauty in method 25

    3.2   Beauty in results   26

    3.3   Beauty in experience   28

    3.4   Beauty and philosophy   28

    3.5   Beauty and mathematical information theory   30

    3.6   Mathematics and the arts   30

          3.6.1   Music   30

          3.6.2   Visual arts   30

    3.7   See also   31

    3.8   Notes   31

    3.9   References   32

    3.10 External links   33

4   Mathematical notation   34

    4.1   Definition   34

    4.2   Expressions   34

    4.3   Precise semantic meaning   35

    4.4   History   35

          4.4.1   Counting   35

          4.4.2   Geometry becomes analytic   35

          4.4.3   Counting is mechanized   35

          4.4.4   Modern notation   36

          4.4.5   Computerized notation   36

          4.4.6   Ideographic notation   36

    4.5   Non-Latin-based mathematical notation   36

    4.6   See also   36

    4.7   Notes   37

    4.8   References   37

    4.9   External links   37

5   Areas of mathematics   38

    5.1   Classification systems   38

    5.2   Major divisions of mathematics   39

          5.2.1   Foundations   39

          5.2.2   Arithmetic   39

          5.2.3   Algebra   40

          5.2.4   Analysis   40

          5.2.5   Combinatorics   40

          5.2.6   Geometry and topology   41

          5.2.7   Applied mathematics   41

   5.3 See also 42

   5.4 Notes 43

   5.5 External links 43

6 Philosophy of mathematics   44

   6.1 Recurrent themes 44

   6.2 History 45

        6.2.1 20th century 45

   6.3 Major themes 46

        6.3.1 Mathematical realism 46

        6.3.2 Mathematical anti-realism 46

   6.4 Contemporary schools of thought 46

        6.4.1 Platonism 46

        6.4.2 Empiricism 48

        6.4.3 Mathematical monism 48

        6.4.4 Logicism 48

        6.4.5 Formalism 49

        6.4.6 Conventionalism 51

        6.4.7 Psychologism 51

        6.4.8 Intuitionism 51

        6.4.9 Structuralism 52

        6.4.10 Embodied mind theories 53

        6.4.11 Fictionalism 54

        6.4.12 Social constructivism or social realism 55

        6.4.13 Beyond the traditional schools 56

   6.5 Arguments 57

        6.5.1 Indispensability argument for realism 57

        6.5.2 Epistemic argument against realism 57

   6.6 Aesthetics 58

   6.7 See also 58

        6.7.1 Related works 58

        6.7.2 Historical topics 58

   6.8 Notes 58

   6.9 Further reading 60

   6.10 External links 62

        6.10.1 Journals 62

7 Glossary of areas of mathematics   63

   7.1 A 63

7.2   B   65

7.3   C   65

7.4   D   68

7.5   E   69

7.6   F   70

7.7   G   71

7.8   H   72

7.9   I   73

7.10 J   74

7.11 K   74

7.12 L   74

7.13 M   75

7.14 N   76

7.15 O   76

7.16 P   77

7.17 Q   78

7.18 R   78

7.19 S   79

7.20 T   80

7.21 U   81

7.22 V   81

7.23 W   81

7.24 X   82

7.25 Y   82

7.26 Z   82

7.27 See also   82

8   Arithmetic   83

8.1   History   83

8.2   Arithmetic operations   84

    8.2.1   Addition (+)   84

    8.2.2   Subtraction (−)   85

    8.2.3   Multiplication (× or · or *)   85

    8.2.4   Division (÷ or /)   85

8.3   Decimal arithmetic   86

8.4   Compound unit arithmetic   87

    8.4.1   Basic arithmetic operations   87

    8.4.2   Principles of compound unit arithmetic   87

    8.4.3   Operations in practice   87

8.5 Number theory 88

8.6 Arithmetic in education 88

8.7 See also 88

    8.7.1 Related topics 88

8.8 Notes 89

8.9 References 90

8.10 External links 90

9 Order theory   **94**

9.1 Background and motivation 94

9.2 Basic definitions 94

    9.2.1 Partially ordered sets 95

    9.2.2 Visualizing a poset 95

    9.2.3 Special elements within an order 95

    9.2.4 Duality 97

    9.2.5 Constructing new orders 97

9.3 Functions between orders 97

9.4 Special types of orders 98

9.5 Subsets of ordered sets 99

9.6 Related mathematical areas 99

    9.6.1 Universal algebra 99

    9.6.2 Topology 99

    9.6.3 Category theory 100

9.7 History 100

9.8 See also 100

9.9 Notes 101

9.10 References 101

9.11 External links 101

10 Algebraic structure   **102**

10.1 Introduction 102

10.2 Examples 103

    10.2.1 One set with operations 103

    10.2.2 Two sets with operations 104

10.3 Hybrid structures 105

10.4 Universal algebra 106

10.5 Category theory 106

10.6 See also 107

10.7 References 107

   10.8 External links 107

11 Number theory   **108**

   11.1 History 108

      11.1.1 Origins 108

      11.1.2 Early modern number theory 111

      11.1.3 Maturity and division into subfields 112

   11.2 Main subdivisions 113

      11.2.1 Elementary tools 113

      11.2.2 Analytic number theory 113

      11.2.3 Algebraic number theory 114

      11.2.4 Diophantinegeometry 114

   11.3 Recent approaches and subfields 115

      11.3.1 Probabilistic number theory 115

      11.3.2 Arithmetic combinatorics 116

      11.3.3 Computations in number theory 116

   11.4 Applications 117

   11.5 Literature 117

   11.6 Prizes 117

   11.7 See also 117

   11.8 Notes 118

   11.9 References 119

   11.10 Sources 122

   11.11 External links 125

12 Field (mathematics)   **138**

   12.1 Definition and illustration 138

      12.1.1 First example: rational numbers 139

      12.1.2 Second example: a field with four elements 140

      12.1.3 Alternative axiomatizations 140

   12.2 Related algebraic structures 140

      12.2.1 Remarks 140

   12.3 History 141

   12.4 Examples 141

      12.4.1 Rationals and algebraic numbers 141

      12.4.2 Reals, complex numbers, and p-adic numbers 141

      12.4.3 Constructible numbers 142

      12.4.4 Finite fields 143

      12.4.5 Archimedean fields 143

        12.4.6 Field of functions   143

        12.4.7 Local and global fields   144

    12.5 Some first theorems   144

    12.6 Constructing fields   144

        12.6.1 Closure operations   144

        12.6.2 Subfields and field extensions   144

        12.6.3 Rings vs fields   145

        12.6.4 Ultraproducts   145

    12.7 Galois theory   146

    12.8 Generalizations   146

        12.8.1 Exponentiation   146

    12.9 Applications   147

    12.10 See also   147

    12.11 Notes   147

    12.12 References   147

    12.13 Sources   148

    12.14 External links   148

13 Commutative ring   149

    13.1 Definition and first examples   149

        13.1.1 Definition   149

        13.1.2 First examples   149

    13.2 Ideals and the spectrum   150

        13.2.1 Ideals and factor rings   150

        13.2.2 Localizations   151

        13.2.3 Prime ideals and the spectrum   151

    13.3 Ring homomorphisms   152

    13.4 Modules   152

    13.5 Noetherian rings   153

    13.6 Dimension   153

    13.7 Constructing commutative rings   154

        13.7.1 Completions   154

    13.8 Properties   154

    13.9 See also   154

    13.10 Notes   154

        13.10.1 Citations   155

    13.11 References   155

14 Commutative algebra   156

14.1 Overview   156

14.2 History   156

14.3 Main tools and results   158

   14.3.1 Noetherian rings   158

   14.3.2 Hilbert's basis theorem   158

   14.3.3 Primary decomposition   159

   14.3.4 Localization   159

   14.3.5 Completion   160

   14.3.6 Zariski topology on prime ideals   160

14.4 Examples   160

14.5 Connections with algebraic geometry   160

14.6 See also   161

14.7 References   161

15 Mathematical analysis   **163**

15.1 History   164

15.2 Important concepts   165

   15.2.1 Metric spaces   165

   15.2.2 Sequences and limits   165

15.3 Main branches   166

   15.3.1 Real analysis   166

   15.3.2 Complex analysis   166

   15.3.3 Functional analysis   166

   15.3.4 Differential equations   166

   15.3.5 Measure theory   167

   15.3.6 Numerical analysis   167

15.4 Other topics in mathematical analysis   167

15.5 Applications   168

   15.5.1 Physical sciences   168

   15.5.2 Signal processing   168

   15.5.3 Other areas of mathematics   168

15.6 See also   169

15.7 Notes   169

15.8 References   170

15.9 External links   170

16 Cryptography   **171**

16.1 Terminology   172

16.2 History of cryptography and cryptanalysis   173

     16.2.1 Classic cryptography   173

     16.2.2 Computer era   176

  16.3 Modern cryptography   177

     16.3.1 Symmetric-key cryptography   177

     16.3.2 Public-key cryptography   179

     16.3.3 Cryptanalysis   182

     16.3.4 Cryptographic primitives   183

     16.3.5 Cryptosystems   183

  16.4 Legal issues   183

     16.4.1 Prohibitions   184

     16.4.2 Export controls   184

     16.4.3 NSA involvement   184

     16.4.4 Digital rights management   185

     16.4.5 Forced disclosure of encryption keys   185

  16.5 See also   186

  16.6 References   186

  16.7 Further reading   188

  16.8 External links   189

17 Abstract algebra   **192**

  17.1 History   192

     17.1.1 Early group theory   193

     17.1.2 Modern algebra   194

  17.2 Basic concepts   195

  17.3 Applications   196

  17.4 See also   196

  17.5 References   196

  17.6 Sources   196

  17.7 External links   197

18 Combinatorics   **198**

  18.1 History   198

  18.2 Approaches and subfields of combinatorics   199

     18.2.1 Enumerative combinatorics   199

     18.2.2 Analytic combinatorics   199

     18.2.3 Partition theory   199

     18.2.4 Graph theory   199

     18.2.5 Design theory   200

     18.2.6 Finite geometry   200

18.2.7   Order theory   200

18.2.8   Matroid theory   200

18.2.9   Extremal combinatorics   200

18.2.10 Probabilistic combinatorics   201

18.2.11 Algebraic combinatorics   201

18.2.12 Combinatorics on words   201

18.2.13 Geometric combinatorics   201

18.2.14 Topological combinatorics   202

18.2.15 Arithmetic combinatorics   202

18.2.16 Infinitary combinatorics   202

18.3 Related fields   202

18.3.1   Combinatorial optimization   202

18.3.2   Coding theory   202

18.3.3   Discrete and computational geometry   202

18.3.4   Combinatorics and dynamical systems   203

18.3.5   Combinatorics and physics   203

18.4 See also   203

18.5 Notes   203

18.6 References   204

18.7 External links   204

19 **Geometry**   **214**

19.1 Overview   215

19.1.1   Practical geometry   216

19.1.2   Axiomatic geometry   216

19.1.3   Geometric constructions   218

19.1.4   Numbers in geometry   218

19.1.5   Geometry of position   219

19.1.6   Geometry beyond Euclid   219

19.1.7   Dimension   219

19.1.8   Symmetry   220

19.2 History   220

19.3 Contemporary geometry   225

19.3.1   Euclidean geometry   225

19.3.2   Differential geometry   225

19.3.3   Topology and geometry   225

19.3.4   Algebraic geometry   225

19.4 See also   226

19.4.1   Lists   226

      19.4.2 Related topics   227

      19.4.3 Other fields   227

   19.5 Notes   228

   19.6 Sources   230

   19.7 Further reading   230

   19.8 External links   230

20 Convex geometry   **232**

   20.1 Classification   232

   20.2 Historical note   233

   20.3 See also   233

   20.4 Notes   233

   20.5 References   233

21 Discrete geometry   **235**

   21.1 History   235

   21.2 Topics in discrete geometry   235

      21.2.1 Polyhedra and polytopes   235

      21.2.2 Packings, coverings and tilings   236

      21.2.3 Structural rigidity and flexibility   237

      21.2.4 Incidence structures   237

      21.2.5 Oriented matroids   238

      21.2.6 Geometric graph theory   239

      21.2.7 Simplicial complexes   240

      21.2.8 Topological combinatorics   240

      21.2.9 Lattices and discrete groups   240

      21.2.10 Digital geometry   240

      21.2.11 Discrete differential geometry   241

   21.3 See also   241

   21.4 Notes   241

   21.5 References   242

22 Differential geometry   **243**

   22.1 Branches of differential geometry   244

      22.1.1 Riemannian geometry   244

      22.1.2 Pseudo-Riemannian geometry   244

      22.1.3 Finsler geometry   244

      22.1.4 Symplectic geometry   244

      22.1.5 Contact geometry   245

         22.1.6 Complex and Kähler geometry   245

         22.1.7 CR geometry   246

         22.1.8 Differential topology   246

         22.1.9 Lie groups   246

    22.2 Bundles and connections   246

    22.3 Intrinsic versus extrinsic   246

    22.4 Applications   247

    22.5 See also   248

    22.6 References   248

    22.7 Further reading   249

    22.8 External links   249

23 Algebraic geometry   250

    23.1 Basic notions   251

         23.1.1 Zeros of simultaneous polynomials   252

         23.1.2 Affine varieties   253

         23.1.3 Regular functions   254

         23.1.4 Morphism of affine varieties   254

         23.1.5 Rational function and birational equivalence   254

         23.1.6 Projective variety   255

    23.2 Real algebraic geometry   256

    23.3 Computational algebraic geometry   257

         23.3.1 Gröbner basis   257

         23.3.2 Cylindrical Algebraic Decomposition (CAD)   258

         23.3.3 Asymptotic complexity vs. practical efficiency   258

    23.4 Abstract modern viewpoint   259

    23.5 History   259

         23.5.1 Prehistory: before the 16th century   259

         23.5.2 Renaissance   260

         23.5.3 19th and early 20th century   260

         23.5.4 20th century   260

    23.6 Analytic geometry   261

    23.7 Applications   261

    23.8 See also   261

    23.9 Notes   262

    23.10 Further reading   262

    23.11 External links   264

24 Topology   265

24.1 History 267

24.2 Introduction 268

24.3 Concepts 270

    24.3.1 Topologies on Sets 270

    24.3.2 Continuous functions and homeomorphisms 270

    24.3.3 Manifolds 271

24.4 Topics 271

    24.4.1 General topology 271

    24.4.2 Algebraic topology 271

    24.4.3 Differential topology 271

    24.4.4 Geometric topology 272

    24.4.5 Generalizations 272

24.5 Applications 272

    24.5.1 Biology 272

    24.5.2 Computer science 272

    24.5.3 Physics 273

    24.5.4 Robotics 273

24.6 See also 273

24.7 References 273

24.8 Further reading 274

24.9 External links 275

25 General topology   276

25.1 History 277

25.2 A topology on a set 277

    25.2.1 Basis for a topology 277

    25.2.2 Subspace and quotient 278

    25.2.3 Examples of topological spaces 278

25.3 Continuous functions 279

    25.3.1 Alternative definitions 279

    25.3.2 Properties 280

    25.3.3 Homeomorphisms 281

    25.3.4 Defining topologies via continuous functions 281

25.4 Compact sets 282

25.5 Connected sets 282

    25.5.1 Connected components 283

    25.5.2 Disconnected spaces 283

    25.5.3 Path-connected sets 283

25.6 Products of spaces 283

25.7 Separation axioms 284

25.8 Countability axioms 285

25.9 Metric spaces 286

25.10 Baire category theory 286

25.11 Main areas of research 287

     25.11.1 Continuum theory 287

     25.11.2 Pointless topology 287

     25.11.3 Dimension theory 287

     25.11.4 Topological algebras 287

     25.11.5 Metrizability theory 288

     25.11.6 Set-theoretic topology 288

25.12 See also 288

25.13 References 288

25.14 Further reading 289

26 Algebraic topology   290

26.1 Main branches of algebraic topology 291

     26.1.1 Homotopy groups 291

     26.1.2 Homology 291

     26.1.3 Cohomology 291

     26.1.4 Manifolds 291

     26.1.5 Knot theory 291

     26.1.6 Complexes 292

26.2 Method of algebraic invariants 292

26.3 Setting in category theory 292

26.4 Applications of algebraic topology 293

26.5 Notable algebraic topologists 293

26.6 Important theorems in algebraic topology 295

26.7 See also 295

26.8 Notes 295

26.9 References 295

26.10 Further reading 296

27 Manifold   297

27.1 Motivational examples 297

     27.1.1 Circle 297

     27.1.2 Enriched circle 300

     27.1.3 Sphere 301

     27.1.4 Other curves 302

27.2 History   303

   27.2.1 Early development   303

   27.2.2 Synthesis   304

   27.2.3 Poincaré's definition   304

   27.2.4 Topology of manifolds: highlights   305

27.3 Mathematical definition   305

   27.3.1 Broad definition   306

27.4 Charts, atlases, and transition maps   306

   27.4.1 Charts   306

   27.4.2 Atlases   306

   27.4.3 Transition maps   307

   27.4.4 Additional structure   307

27.5 Manifold with boundary   307

   27.5.1 Boundary and interior   307

27.6 Construction   308

   27.6.1 Charts   308

   27.6.2 Patchwork   308

   27.6.3 Identifying points of a manifold   309

   27.6.4 Gluing along boundaries   309

   27.6.5 Cartesian products   310

27.7 Manifolds with additional structure   310

   27.7.1 Topological manifolds   310

   27.7.2 Differentiable manifolds   310

   27.7.3 Riemannian manifolds   311

   27.7.4 Finsler manifolds   311

   27.7.5 Lie groups   311

   27.7.6 Other types of manifolds   312

27.8 Classification and invariants   312

27.9 Examples of surfaces   313

   27.9.1 Orientability   313

   27.9.2 Genus and the Euler characteristic   314

27.10 Maps of manifolds   314

   27.10.1 Scalar-valued functions   314

27.11 Generalizations of manifolds   314

27.12 See also   315

   27.12.1 By dimension   316

27.13 Notes   316

27.14 References   317

    27.15External links   317

28 Probability theory   **324**

    28.1 History   324

    28.2 Treatment   324

        28.2.1 Motivation   325

        28.2.2 Discrete probability distributions   325

        28.2.3 Continuous probability distributions   326

        28.2.4 Measure-theoretic probability theory   327

    28.3 Classical probability distributions   328

    28.4 Convergence of random variables   328

        28.4.1 Law of large numbers   329

        28.4.2 Central limit theorem   329

    28.5 See also   330

    28.6 Notes   330

    28.7 References   330

    28.8 External links   330

29 Statistics   **331**

    29.1 Scope   333

        29.1.1 Mathematical statistics   333

    29.2 Overview   333

    29.3 Data collection   334

        29.3.1 Sampling   334

        29.3.2 Experimental and observational studies   334

    29.4 Types of data   335

    29.5 Terminology and theory of inferential statistics   336

        29.5.1 Statistics, estimators and pivotal quantities   336

        29.5.2 Null hypothesis and alternative hypothesis   336

        29.5.3 Error   336

        29.5.4 Interval estimation   338

        29.5.5 Significance   338

        29.5.6 Examples   340

    29.6 Misuse of statistics   341

        29.6.1 Misinterpretation: correlation   341

    29.7 History of statistical science   341

    29.8 Applications   344

        29.8.1 Applied statistics, theoretical statistics and mathematical statistics   344

        29.8.2 Machine learning and data mining   345

29.8.3   Statistics in society 347

29.8.4   Statistical computing 347

29.8.5   Statistics applied to mathematics or the arts 348

29.9   Specialized disciplines 348

29.10 See also 349

29.11 References 350

30 Numerical analysis   **352**

30.1   General introduction 352

30.1.1   History 353

30.1.2   Direct and iterative methods 354

30.1.3   Discretization 354

30.2   Generation and propagation of errors 354

30.2.1   Round-off 354

30.2.2   Truncation and discretization error 354

30.2.3   Numerical stability and well-posed problems 355

30.3   Areas of study 356

30.3.1   Computing values of functions 356

30.3.2   Interpolation, extrapolation, and regression 356

30.3.3   Solving equations and systems of equations 356

30.3.4   Solving eigenvalue or singular value problems 357

30.3.5   Optimization 357

30.3.6   Evaluating integrals 357

30.3.7   Differential equations 357

30.4   Software 358

30.5   See also 358

30.6   Notes 358

30.7   References 358

30.8   External links 359

31 Symbolic computation   **360**

31.1   Terminology 360

31.2   Scientific community 361

31.3   Computer science aspects 361

31.3.1   Data representation 361

31.3.2   Simplification 362

31.4   Mathematical aspects 362

31.4.1   Equality 363

31.5   See also 363

31.6 References   363

31.7 Further reading   364

32 Mechanics   **365**

32.1 Classical versus quantum   365

32.2 Relativistic versus Newtonian mechanics   366

32.3 General relativistic versus quantum   366

32.4 History   366

    32.4.1 Antiquity   366

    32.4.2 Medieval age   366

    32.4.3 Early modern age   366

    32.4.4 Modern age   368

32.5 Types of mechanical bodies   368

32.6 Sub-disciplines in mechanics   368

    32.6.1 Classical mechanics   368

    32.6.2 Quantum mechanics   370

32.7 Professional organizations   370

32.8 See also   370

32.9 References   371

32.10 Further reading   371

32.11 External links   371

33 Fluid mechanics   **372**

33.1 Brief history   372

33.2 Main branches   372

    33.2.1 Fluid statics   372

    33.2.2 Fluid dynamics   373

33.3 Relationship to continuum mechanics   373

33.4 Assumptions   373

    33.4.1 Continuum hypothesis   374

33.5 Navier–Stokes equations   375

    33.5.1 General form of the equation   375

33.6 Newtonian versus non-Newtonian fluids   376

    33.6.1 Equations for a Newtonian fluid   377

33.7 See also   377

33.8 Notes   378

33.9 References   378

33.10 Further reading   378

33.11 External links   378

34 Operations research 379

  34.1 Overview 379

  34.2 History 380

    34.2.1 Historical origins 380

    34.2.2 Second World War 380

    34.2.3 After World War II 383

  34.3 Problems addressed 383

  34.4 Management science 384

    34.4.1 Related fields 384

    34.4.2 Applications 384

  34.5 Societies and journals 385

  34.6 See also 386

  34.7 References 386

  34.8 Notes 388

  34.9 Further reading 388

  34.10 External links 389

35 Mathematical optimization 391

  35.1 Optimization problems 392

  35.2 Notation 392

    35.2.1 Minimum and maximum value of a function 393

    35.2.2 Optimal input arguments 393

  35.3 History 394

  35.4 Major subfields 394

    35.4.1 Multi-objective optimization 395

    35.4.2 Multi-modal optimization 396

  35.5 Classification of critical points and extrema 396

    35.5.1 Feasibility problem 396

    35.5.2 Existence 396

    35.5.3 Necessary conditions for optimality 396

    35.5.4 Sufficient conditions for optimality 396

    35.5.5 Sensitivity and continuity of optima 397

    35.5.6 Calculus of optimization 397

  35.6 Computational optimization techniques 397

    35.6.1 Optimization algorithms 397

    35.6.2 Iterative methods 397

    35.6.3 Global convergence 399

    35.6.4 Heuristics 399

  35.7 Applications 399

35.7.1 Mechanics and engineering   399

35.7.2 Economics   400

35.7.3 Operations research   400

35.7.4 Control engineering   400

35.7.5 Petroleum engineering   400

35.7.6 Molecular modeling   400

35.8 Solvers   400

35.9 See also   401

35.10 Notes   401

35.11 Further reading   401

35.11.1 Comprehensive   401

35.11.2 Continuous optimization   402

35.11.3 Combinatorial optimization   403

35.11.4 Relaxation (extension method)   403

35.12 Journals   403

35.13 External links   404

35.14 Text and image sources, contributors, and licenses   405

35.14.1 Text   405

35.14.2 Images   418

35.14.3 Content license   426

Chapter 1

Mathematics

This book is about the study of mathematical topics such as quantity and structure.

Euclid (holding calipers), Greek mathematician, 3rd century BC, as imagined by Raphael in this detail from The School of Athens.[1]

Mathematics (from Greek μάθημα *máthēma*, "knowledge, study, learning") is the study of topics such as quantity (numbers),[2] structure,[3] space,[2] and change.[4][5][6] There is a range of views among mathematicians and philosophers as to the exact scope and definition of mathematics.[7][8]

Mathematicians seek out patterns[9][10] and use them to formulate new conjectures. Mathematicians resolve the truth or falsity of conjectures by mathematical proof. When mathematical structures are good models of real phenomena, then mathematical reasoning can provide insight or predictions about nature. Through the use of abstraction and logic, mathematics developed from counting, calculation, measurement, and the systematic study of the shapes and motions of physical objects. Practical mathematics has been a human activity for as far back as written records exist. The research required to solve mathematical problems can take years or even centuries of sustained inquiry.

Rigorous arguments first appeared in Greek mathematics, most notably in Euclid's *Elements*. Since the pioneering work of Giuseppe Peano (1858–1932), David Hilbert (1862–1943), and others on axiomatic systems in the late 19th century, it has become customary to view mathematical research as establishing truth by rigorous deduction from appropriately chosen axioms and definitions. Mathematics developed at a relatively slow pace until the Renaissance, when mathematical innovations interacting with new scientific discoveries led to a rapid increase in the rate of mathematical discovery that has continued to the present day.[11]

Galileo Galilei (1564–1642) said, "The universe cannot be read until we have learned the language and become familiar with the characters in which it is written. It is written in mathematical language, and the letters are triangles, circles and other geometrical figures, without which means it is humanly impossible to comprehend a single word. Without these, one is wandering about in a dark labyrinth."[12] Carl Friedrich Gauss (1777–1855) referred to mathematics as "the Queen of the Sciences".[13] Benjamin Peirce (1809–1880) called mathematics "the science that draws necessary conclusions".[14] David Hilbert said of mathematics: "We are not speaking here of arbitrariness in any sense. Mathematics is not like a game whose tasks are determined by arbitrarily stipulated rules. Rather, it is a conceptual system possessing internal necessity that can only be so and by no means otherwise."[15] Albert Einstein (1879–1955) stated that "as far as the laws of mathematics refer to reality, they are not certain; and as far as they are certain, they do not refer to reality."[16] French mathematician Claire Voisin states "There is creative drive in mathematics, it's all about movement trying to express itself." [17]

Mathematics is used throughout the world as an essential tool in many fields, including natural science, engineering, medicine, finance and the social sciences. Applied mathematics, the branch of mathematics concerned with application of mathematical knowledge to other fields, inspires and makes use of new mathematical discoveries, which has led to the development of entirely new mathematical disciplines, such as statistics and game theory. Mathematicians also engage in pure mathematics, or mathematics for its own sake, without having any application in mind. There is no clear line separating pure and applied mathematics, and practical applications for what began as pure mathematics are often discovered.[18]

1.1 History

1.1.1 Evolution

Main article: History of mathematics

The evolution of mathematics can be seen as an ever-increasing series of abstractions. The first abstraction, which is shared by many animals,[19] was probably that of numbers: the realization that a collection of two apples and a collection of two oranges (for example) have something in common, namely quantity of their members.

As evidenced by tallies found on bone, in addition to recognizing how to count physical objects, prehistoric peoples may have also recognized how to count abstract quantities, like time – days, seasons, years.[20]

More complex mathematics did not appear until around 3000 BC, when the Babylonians and Egyptians began using arithmetic, algebra and geometry for taxation and other financial calculations, for building and construction, and for astronomy.[21] The earliest uses of mathematics were in trading, land measurement, painting and weaving patterns and the recording of time.

Greek mathematician Pythagoras (c. 570 – c. 495 BC), commonly credited with discovering the Pythagorean theorem

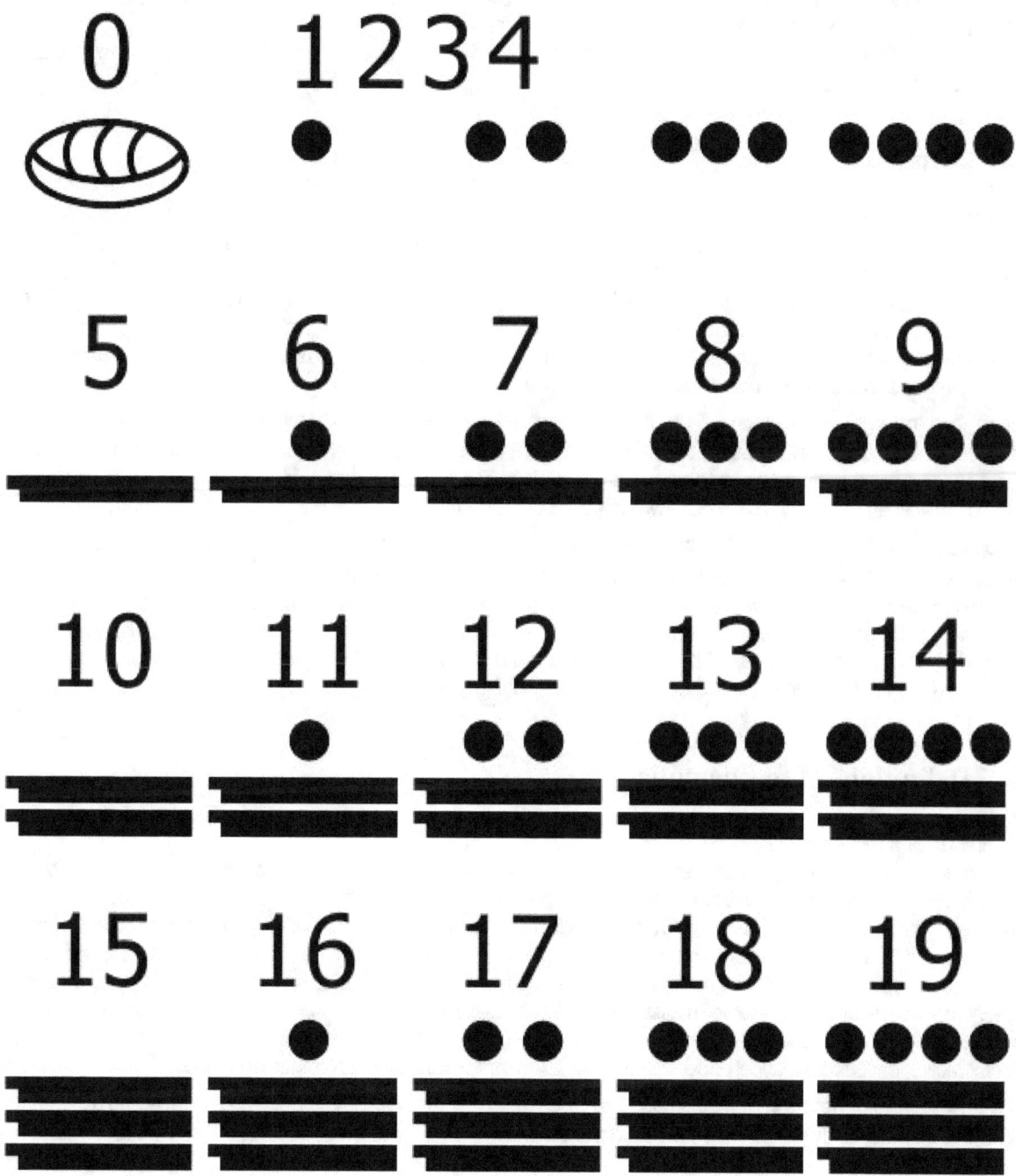

Mayan numerals

In Babylonian mathematics elementary arithmetic (addition, subtraction, multiplication and division) first appears in the archaeological record. Numeracy pre-dated writing and numeral systems have been many and diverse, with the first known written numerals created by Egyptians in Middle Kingdom texts such as the Rhind Mathematical Papyrus.

Between 600 and 300 BC the Ancient Greeks began a systematic study of mathematics in its own right with Greek mathematics.[22]

Mathematics has since been greatly extended, and there has been a fruitful interaction between mathematics and science, to the benefit of both. Mathematical discoveries continue to be made today. According to Mikhail B. Sevryuk, in the

January 2006 issue of the *Bulletin of the American Mathematical Society*, "The number of papers and books included in the *Mathematical Reviews* database since 1940 (the first year of operation of MR) is now more than 1.9 million, and more than 75 thousand items are added to the database each year. The overwhelming majority of works in this ocean contain new mathematical theorems and their proofs."[23]

1.1.2 Etymology

The word *mathematics* comes from the Greek μάθημα (*máthēma*), which, in the ancient Greek language, means "that which is learnt",[24] "what one gets to know", hence also "study" and "science", and in modern Greek just "lesson". The word *máthēma* is derived from μανθάνω (*manthano*), while the modern Greek equivalent is μαθαίνω (*mathaino*), both of which mean "to learn". In Greece, the word for "mathematics" came to have the narrower and more technical meaning "mathematical study" even in Classical times.[25] Its adjective is μαθηματικός (*mathēmatikós*), meaning "related to learning" or "studious", which likewise further came to mean "mathematical". In particular, μαθηματική τέχνη (*mathēmatikḗ tékhnē*), Latin: *ars mathematica*, meant "the mathematical art".

In Latin, and in English until around 1700, the term *mathematics* more commonly meant "astrology" (or sometimes "astronomy") rather than "mathematics"; the meaning gradually changed to its present one from about 1500 to 1800. This has resulted in several mistranslations: a particularly notorious one is Saint Augustine's warning that Christians should beware of *mathematici* meaning astrologers, which is sometimes mistranslated as a condemnation of mathematicians.[26]

The apparent plural form in English, like the French plural form *les mathématiques* (and the less commonly used singular derivative *la mathématique*), goes back to the Latin neuter plural *mathematica* (Cicero), based on the Greek plural τα μαθηματικά (*ta mathēmatiká*), used by Aristotle (384–322 BC), and meaning roughly "all things mathematical"; although it is plausible that English borrowed only the adjective *mathematic(al)* and formed the noun *mathematics* anew, after the pattern of physics and metaphysics, which were inherited from the Greek.[27] In English, the noun *mathematics* takes singular verb forms. It is often shortened to *maths* or, in English-speaking North America, *math*.[28]

1.2 Definitions of mathematics

Main article: Definitions of mathematics

Aristotle defined mathematics as "the science of quantity", and this definition prevailed until the 18th century.[29] Starting in the 19th century, when the study of mathematics increased in rigor and began to address abstract topics such as group theory and projective geometry, which have no clear-cut relation to quantity and measurement, mathematicians and philosophers began to propose a variety of new definitions.[30] Some of these definitions emphasize the deductive character of much of mathematics, some emphasize its abstractness, some emphasize certain topics within mathematics. Today, no consensus on the definition of mathematics prevails, even among professionals.[7] There is not even consensus on whether mathematics is an art or a science.[8] A great many professional mathematicians take no interest in a definition of mathematics, or consider it undefinable.[7] Some just say, "Mathematics is what mathematicians do."[7]

Three leading types of definition of mathematics are called logicist, intuitionist, and formalist, each reflecting a different philosophical school of thought.[31] All have severe problems, none has widespread acceptance, and no reconciliation seems possible.[31]

An early definition of mathematics in terms of logic was Benjamin Peirce's "the science that draws necessary conclusions" (1870).[32] In the *Principia Mathematica*, Bertrand Russell and Alfred North Whitehead advanced the philosophical program known as logicism, and attempted to prove that all mathematical concepts, statements, and principles can be defined and proven entirely in terms of symbolic logic. A logicist definition of mathematics is Russell's "All Mathematics is Symbolic Logic" (1903).[33]

Intuitionist definitions, developing from the philosophy of mathematician L.E.J. Brouwer, identify mathematics with certain mental phenomena. An example of an intuitionist definition is "Mathematics is the mental activity which consists in carrying out constructs one after the other."[31] A peculiarity of intuitionism is that it rejects some mathematical ideas considered valid according to other definitions. In particular, while other philosophies of mathematics allow objects that can be proven to exist even though they cannot be constructed, intuitionism allows only mathematical objects that one can actually construct.

Leonardo Fibonacci, the Italian mathematician who established the Hindu–Arabic numeral system to the Western World

Formalist definitions identify mathematics with its symbols and the rules for operating on them. Haskell Curry defined mathematics simply as "the science of formal systems".[34] A formal system is a set of symbols, or *tokens*, and some *rules* telling how the tokens may be combined into *formulas*. In formal systems, the word *axiom* has a special meaning, different from the ordinary meaning of "a self-evident truth". In formal systems, an axiom is a combination of tokens that is included in a given formal system without needing to be derived using the rules of the system.

1.2.1 Mathematics as science

Gauss referred to mathematics as "the Queen of the Sciences".[13] In the original Latin *Regina Scientiarum*, as well as in German *Königin der Wissenschaften*, the word corresponding to *science* means a "field of knowledge", and this was the original meaning of "science" in English, also; mathematics is in this sense a field of knowledge. The specialization restricting the meaning of "science" to *natural science* follows the rise of Baconian science, which contrasted "natural science" to scholasticism, the Aristotelean method of inquiring from first principles. The role of empirical experimentation and observation is negligible in mathematics, compared to natural sciences such as psychology, biology, or physics. Albert Einstein stated that "as far as the laws of mathematics refer to reality, they are not certain; and as far as they are certain, they do not refer to reality."[16] More recently, Marcus du Sautoy has called mathematics "the Queen of Science ... the main driving force behind scientific discovery".[35]

Many philosophers believe that mathematics is not experimentally falsifiable, and thus not a science according to the definition of Karl Popper.[36] However, in the 1930s Gödel's incompleteness theorems convinced many mathematicians that mathematics cannot be reduced to logic alone, and Karl Popper concluded that "most mathematical theories are, like those of physics and biology, hypothetico-deductive: pure mathematics therefore turns out to be much closer to the natural sciences whose hypotheses are conjectures, than it seemed even recently."[37] Other thinkers, notably Imre Lakatos, have applied a version of falsificationism to mathematics itself.

An alternative view is that certain scientific fields (such as theoretical physics) are mathematics with axioms that are intended to correspond to reality. The theoretical physicist J.M. Ziman proposed that science is *public knowledge*, and thus includes mathematics.[38] Mathematics shares much in common with many fields in the physical sciences, notably the exploration of the logical consequences of assumptions. Intuition and experimentation also play a role in the formulation of conjectures in both mathematics and the (other) sciences. Experimental mathematics continues to grow in importance within mathematics, and computation and simulation are playing an increasing role in both the sciences and mathematics.

The opinions of mathematicians on this matter are varied. Many mathematicians feel that to call their area a science is to downplay the importance of its aesthetic side, and its history in the traditional seven liberal arts; others feel that to ignore its connection to the sciences is to turn a blind eye to the fact that the interface between mathematics and its applications in science and engineering has driven much development in mathematics. One way this difference of viewpoint plays out is in the philosophical debate as to whether mathematics is *created* (as in art) or *discovered* (as in science). It is common to see universities divided into sections that include a division of *Science and Mathematics*, indicating that the fields are seen as being allied but that they do not coincide. In practice, mathematicians are typically grouped with scientists at the gross level but separated at finer levels. This is one of many issues considered in the philosophy of mathematics.

Carl Friedrich Gauss, known as the prince of mathematicians

1.3 Inspiration, pure and applied mathematics, and aesthetics

Main article: Mathematical beauty

Isaac Newton (left) and Gottfried Wilhelm Leibniz (right), developers of infinitesimal calculus

Mathematics arises from many different kinds of problems. At first these were found in commerce, land measurement, architecture and later astronomy; today, all sciences suggest problems studied by mathematicians, and many problems arise within mathematics itself. For example, the physicist Richard Feynman invented the path integral formulation of quantum mechanics using a combination of mathematical reasoning and physical insight, and today's string theory, a still-developing scientific theory which attempts to unify the four fundamental forces of nature, continues to inspire new mathematics.[39]

Some mathematics is relevant only in the area that inspired it, and is applied to solve further problems in that area. But often mathematics inspired by one area proves useful in many areas, and joins the general stock of mathematical concepts. A distinction is often made between pure mathematics and applied mathematics. However pure mathematics topics often turn out to have applications, e.g. number theory in cryptography. This remarkable fact, that even the "purest" mathematics often turns out to have practical applications, is what Eugene Wigner has called "the unreasonable effectiveness of mathematics".[40] As in most areas of study, the explosion of knowledge in the scientific age has led to specialization: there are now hundreds of specialized areas in mathematics and the latest Mathematics Subject Classification runs to 46 pages.[41] Several areas of applied mathematics have merged with related traditions outside of mathematics and become disciplines in their own right, including statistics, operations research, and computer science.

For those who are mathematically inclined, there is often a definite aesthetic aspect to much of mathematics. Many mathematicians talk about the *elegance* of mathematics, its intrinsic aesthetics and inner beauty. Simplicity and generality are valued. There is beauty in a simple and elegant proof, such as Euclid's proof that there are infinitely many prime numbers, and in an elegant numerical method that speeds calculation, such as the fast Fourier transform. G.H. Hardy in *A Mathematician's Apology* expressed the belief that these aesthetic considerations are, in themselves, sufficient to justify the study of pure mathematics. He identified criteria such as significance, unexpectedness, inevitability, and economy as factors that contribute to a mathematical aesthetic.[42] Mathematicians often strive to find proofs that are particularly elegant, proofs from "The Book" of God according to Paul Erdős.[43][44] The popularity of recreational mathematics is another sign of the pleasure many find in solving mathematical questions.

1.4   Notation, language, and rigor

Main article: Mathematical notation

Most of the mathematical notation in use today was not invented until the 16th century.[45] Before that, mathematics was written out in words, a painstaking process that limited mathematical discovery.[46] Euler (1707–1783) was responsible for many of the notations in use today. Modern notation makes mathematics much easier for the professional, but beginners often find it daunting. It is extremely compressed: a few symbols contain a great deal of information. Like musical notation, modern mathematical notation has a strict syntax (which to a limited extent varies from author to author and from discipline to discipline) and encodes information that would be difficult to write in any other way.

Mathematical language can be difficult to understand for beginners. Words such as *or* and *only* have more precise meanings than in everyday speech. Moreover, words such as *open* and *field* have been given specialized mathematical meanings. Technical terms such as *homeomorphism* and *integrable* have precise meanings in mathematics. Additionally, shorthand phrases such as *iff* for "if and only if" belong to mathematical jargon. There is a reason for special notation and technical vocabulary: mathematics requires more precision than everyday speech. Mathematicians refer to this precision of language and logic as "rigor".

Mathematical proof is fundamentally a matter of rigor. Mathematicians want their theorems to follow from axioms by means of systematic reasoning. This is to avoid mistaken "theorems", based on fallible intuitions, of which many instances have occurred in the history of the subject.[47] The level of rigor expected in mathematics has varied over time: the Greeks expected detailed arguments, but at the time of Isaac Newton the methods employed were less rigorous. Problems inherent in the definitions used by Newton would lead to a resurgence of careful analysis and formal proof in the 19th century. Misunderstanding the rigor is a cause for some of the common misconceptions of mathematics. Today, mathematicians continue to argue among themselves about computer-assisted proofs. Since large computations are hard to verify, such proofs may not be sufficiently rigorous.[48]

Axioms in traditional thought were "self-evident truths", but that conception is problematic.[49] At a formal level, an axiom is just a string of symbols, which has an intrinsic meaning only in the context of all derivable formulas of an axiomatic system. It was the goal of Hilbert's program to put all of mathematics on a firm axiomatic basis, but according to Gödel's incompleteness theorem every (sufficiently powerful) axiomatic system has undecidable formulas; and so a final axiomatization of mathematics is impossible. Nonetheless mathematics is often imagined to be (as far as its formal content) nothing but set theory in some axiomatization, in the sense that every mathematical statement or proof could be cast into formulas within set theory.[50]

1.5   Fields of mathematics

See also: Areas of mathematics and Glossary of areas of mathematics

Mathematics can, broadly speaking, be subdivided into the study of quantity, structure, space, and change (i.e. arithmetic, algebra, geometry, and analysis). In addition to these main concerns, there are also subdivisions dedicated to exploring links from the heart of mathematics to other fields: to logic, to set theory (foundations), to the empirical mathematics of the various sciences (applied mathematics), and more recently to the rigorous study of uncertainty.

Leonhard Euler, who created and popularized much of the mathematical notation used today

1.5.1   Foundations and philosophy

In order to clarify the foundations of mathematics, the fields of mathematical logic and set theory were developed. Mathematical logic includes the mathematical study of logic and the applications of formal logic to other areas of mathematics;

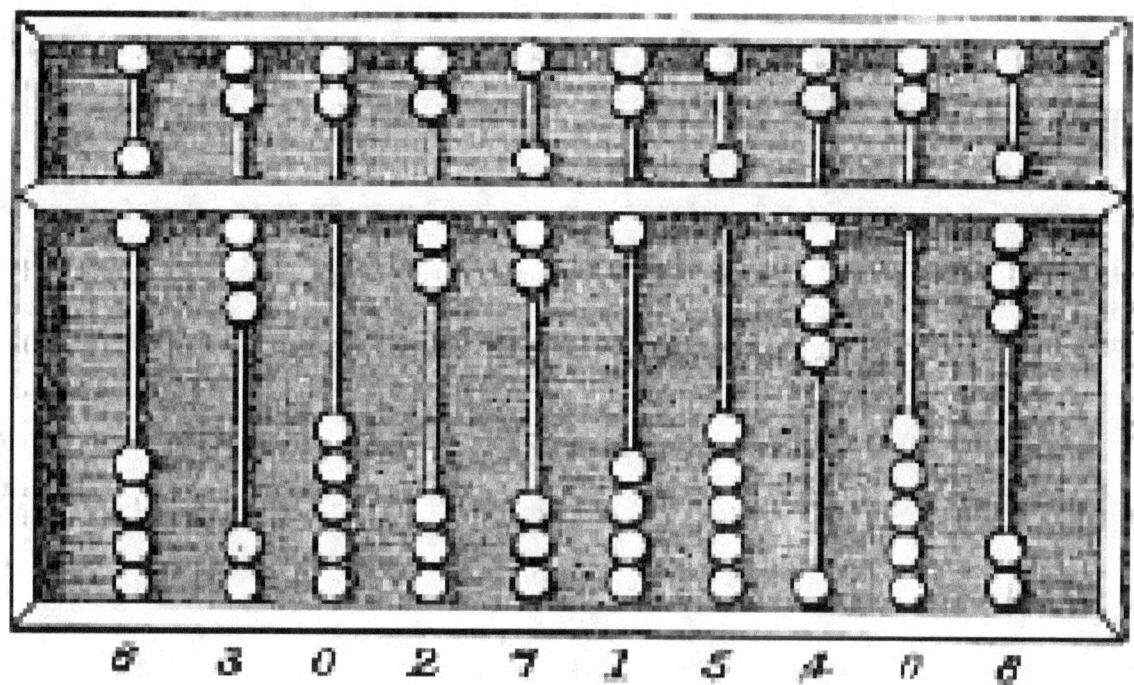

An abacus, a simple calculating tool used since ancient times

set theory is the branch of mathematics that studies sets or collections of objects. Category theory, which deals in an abstract way with mathematical structures and relationships between them, is still in development. The phrase "crisis of foundations" describes the search for a rigorous foundation for mathematics that took place from approximately 1900 to 1930.[51] Some disagreement about the foundations of mathematics continues to the present day. The crisis of foundations was stimulated by a number of controversies at the time, including the controversy over Cantor's set theory and the Brouwer–Hilbert controversy.

Mathematical logic is concerned with setting mathematics within a rigorous axiomatic framework, and studying the implications of such a framework. As such, it is home to Gödel's incompleteness theorems which (informally) imply that any effective formal system that contains basic arithmetic, if *sound* (meaning that all theorems that can be proven are true), is necessarily *incomplete* (meaning that there are true theorems which cannot be proved *in that system*). Whatever finite collection of number-theoretical axioms is taken as a foundation, Gödel showed how to construct a formal statement that is a true number-theoretical fact, but which does not follow from those axioms. Therefore, no formal system is a complete axiomatization of full number theory. Modern logic is divided into recursion theory, model theory, and proof theory, and is closely linked to theoretical computer science, as well as to category theory.

Theoretical computer science includes computability theory, computational complexity theory, and information theory. Computability theory examines the limitations of various theoretical models of the computer, including the most well-known model – the Turing machine. Complexity theory is the study of tractability by computer; some problems, although theoretically solvable by computer, are so expensive in terms of time or space that solving them is likely to remain practically unfeasible, even with the rapid advancement of computer hardware. A famous problem is the "P = NP?" problem, one of the Millennium Prize Problems.[52] Finally, information theory is concerned with the amount of data that can be stored on a given medium, and hence deals with concepts such as compression and entropy.

1.5.2   Pure mathematics

Quantity

The study of quantity starts with numbers, first the familiar natural numbers and integers ("whole numbers") and arithmetical operations on them, which are characterized in arithmetic. The deeper properties of integers are studied in number theory, from which come such popular results as Fermat's Last Theorem. The twin prime conjecture and Goldbach's conjecture are two unsolved problems in number theory.

As the number system is further developed, the integers are recognized as a subset of the rational numbers ("fractions"). These, in turn, are contained within the real numbers, which are used to represent continuous quantities. Real numbers are generalized to complex numbers. These are the first steps of a hierarchy of numbers that goes on to include quaternions and octonions. Consideration of the natural numbers also leads to the transfinite numbers, which formalize the concept of "infinity". Another area of study is size, which leads to the cardinal numbers and then to another conception of infinity: the aleph numbers, which allow meaningful comparison of the size of infinitely large sets.

Structure

Many mathematical objects, such as sets of numbers and functions, exhibit internal structure as a consequence of operations or relations that are defined on the set. Mathematics then studies properties of those sets that can be expressed in terms of that structure; for instance number theory studies properties of the set of integers that can be expressed in terms of arithmetic operations. Moreover, it frequently happens that different such structured sets (or structures) exhibit similar properties, which makes it possible, by a further step of abstraction, to state axioms for a class of structures, and then study at once the whole class of structures satisfying these axioms. Thus one can study groups, rings, fields and other abstract systems; together such studies (for structures defined by algebraic operations) constitute the domain of abstract algebra.

By its great generality, abstract algebra can often be applied to seemingly unrelated problems; for instance a number of ancient problems concerning compass and straightedge constructions were finally solved using Galois theory, which involves field theory and group theory. Another example of an algebraic theory is linear algebra, which is the general study of vector spaces, whose elements called vectors have both quantity and direction, and can be used to model (relations between) points in space. This is one example of the phenomenon that the originally unrelated areas of geometry and algebra have very strong interactions in modern mathematics. Combinatorics studies ways of enumerating the number of objects that fit a given structure.

Space

The study of space originates with geometry – in particular, Euclidean geometry. Trigonometry is the branch of mathematics that deals with relationships between the sides and the angles of triangles and with the trigonometric functions; it combines space and numbers, and encompasses the well-known Pythagorean theorem. The modern study of space generalizes these ideas to include higher-dimensional geometry, non-Euclidean geometries (which play a central role in general relativity) and topology. Quantity and space both play a role in analytic geometry, differential geometry, and algebraic geometry. Convex and discrete geometry were developed to solve problems in number theory and functional analysis but now are pursued with an eye on applications in optimization and computer science. Within differential geometry are the concepts of fiber bundles and calculus on manifolds, in particular, vector and tensor calculus. Within algebraic geometry is the description of geometric objects as solution sets of polynomial equations, combining the concepts of quantity and space, and also the study of topological groups, which combine structure and space. Lie groups are used to study space, structure, and change. Topology in all its many ramifications may have been the greatest growth area in 20th-century mathematics; it includes point-set topology, set-theoretic topology, algebraic topology and differential topology. In particular, instances of modern day topology are metrizability theory, axiomatic set theory, homotopy theory, and Morse theory. Topology also includes the now solved Poincaré conjecture, and the still unsolved areas of the Hodge conjecture.

Other results in geometry and topology, including the four color theorem and Kepler conjecture, have been proved only with the help of computers.

Change

Understanding and describing change is a common theme in the natural sciences, and calculus was developed as a powerful tool to investigate it. Functions arise here, as a central concept describing a changing quantity. The rigorous study of real numbers and functions of a real variable is known as real analysis, with complex analysis the equivalent field for the complex numbers. Functional analysis focuses attention on (typically infinite-dimensional) spaces of functions. One of many applications of functional analysis is quantum mechanics. Many problems lead naturally to relationships between a quantity and its rate of change, and these are studied as differential equations. Many phenomena in nature can be described by dynamical systems; chaos theory makes precise the ways in which many of these systems exhibit unpredictable yet still deterministic behavior.

1.5.3   Applied mathematics

Applied mathematics concerns itself with mathematical methods that are typically used in science, engineering, business, and industry. Thus, "applied mathematics" is a mathematical science with specialized knowledge. The term *applied mathematics* also describes the professional specialty in which mathematicians work on practical problems; as a profession focused on practical problems, *applied mathematics* focuses on the "formulation, study, and use of mathematical models" in science, engineering, and other areas of mathematical practice.

In the past, practical applications have motivated the development of mathematical theories, which then became the subject of study in pure mathematics, where mathematics is developed primarily for its own sake. Thus, the activity of applied mathematics is vitally connected with research in pure mathematics.

Statistics and other decision sciences

Applied mathematics has significant overlap with the discipline of statistics, whose theory is formulated mathematically, especially with probability theory. Statisticians (working as part of a research project) "create data that makes sense" with random sampling and with randomized experiments;[53] the design of a statistical sample or experiment specifies the analysis of the data (before the data be available). When reconsidering data from experiments and samples or when analyzing data from observational studies, statisticians "make sense of the data" using the art of modelling and the theory of inference – with model selection and estimation; the estimated models and consequential predictions should be tested on new data.[54]

Statistical theory studies decision problems such as minimizing the risk (expected loss) of a statistical action, such as using a procedure in, for example, parameter estimation, hypothesis testing, and selecting the best. In these traditional areas of mathematical statistics, a statistical-decision problem is formulated by minimizing an objective function, like expected loss or cost, under specific constraints: For example, designing a survey often involves minimizing the cost of estimating a population mean with a given level of confidence.[55] Because of its use of optimization, the mathematical theory of statistics shares concerns with other decision sciences, such as operations research, control theory, and mathematical economics.[56]

Computational mathematics

Computational mathematics proposes and studies methods for solving mathematical problems that are typically too large for human numerical capacity. Numerical analysis studies methods for problems in analysis using functional analysis and approximation theory; numerical analysis includes the study of approximation and discretization broadly with special concern for rounding errors. Numerical analysis and, more broadly, scientific computing also study non-analytic topics of mathematical science, especially algorithmic matrix and graph theory. Other areas of computational mathematics include computer algebra and symbolic computation.

1.6 Mathematical awards

Arguably the most prestigious award in mathematics is the Fields Medal,[57][58] established in 1936 and now awarded every four years. The Fields Medal is often considered a mathematical equivalent to the Nobel Prize.

The Wolf Prize in Mathematics, instituted in 1978, recognizes lifetime achievement, and another major international award, the Abel Prize, was introduced in 2003. The Chern Medal was introduced in 2010 to recognize lifetime achievement. These accolades are awarded in recognition of a particular body of work, which may be innovational, or provide a solution to an outstanding problem in an established field.

A famous list of 23 open problems, called "Hilbert's problems", was compiled in 1900 by German mathematician David Hilbert. This list achieved great celebrity among mathematicians, and at least nine of the problems have now been solved. A new list of seven important problems, titled the "Millennium Prize Problems", was published in 2000. A solution to each of these problems carries a $1 million reward, and only one (the Riemann hypothesis) is duplicated in Hilbert's problems.

1.7 See also

Main article: Lists of mathematics topics

- Mathematics and art
- Mathematics education
- Relationship between mathematics and physics
- STEM fields

1.8 Notes

[1] No likeness or description of Euclid's physical appearance made during his lifetime survived antiquity. Therefore, Euclid's depiction in works of art depends on the artist's imagination (see *Euclid*).

[2] "mathematics, n.". *Oxford English Dictionary*. Oxford University Press. 2012. Retrieved June 16, 2012. The science of space, number, quantity, and arrangement, whose methods involve logical reasoning and usually the use of symbolic notation, and which includes geometry, arithmetic, algebra, and analysis.

[3] Kneebone, G.T. (1963). *Mathematical Logic and the Foundations of Mathematics: An Introductory Survey*. Dover. pp. 4. ISBN 0-486-41712-3. Mathematics ... is simply the study of abstract structures, or formal patterns of connectedness.

[4] LaTorre, Donald R., John W. Kenelly, Iris B. Reed, Laurel R. Carpenter, and Cynthia R Harris (2011). *Calculus Concepts: An Informal Approach to the Mathematics of Change*. Cengage Learning. pp. 2. ISBN 1-4390-4957-2. Calculus is the study of change—how things change, and how quickly they change.

[5] Ramana (2007). *Applied Mathematics*. Tata McGraw–Hill Education. p. 2.10. ISBN 0-07-066753-5. The mathematical study of change, motion, growth or decay is calculus.

[6] Ziegler, Günter M. (2011). "What Is Mathematics?". *An Invitation to Mathematics: From Competitions to Research*. Springer. pp. 7. ISBN 3-642-19532-6.

[7] Mura, Roberta (Dec 1993). "Images of Mathematics Held by University Teachers of Mathematical Sciences". *Educational Studies in Mathematics* 25 (4): 375–385.

[8] Tobies, Renate and Helmut Neunzert (2012). *Iris Runge: A Life at the Crossroads of Mathematics, Science, and Industry.* Springer. pp. 9. ISBN 3-0348-0229-3. It is first necessary to ask what is meant by *mathematics* in general. Illustrious scholars have debated this matter until they were blue in the face, and yet no consensus has been reached about whether mathematics is a natural science, a branch of the humanities, or an art form.

[9] Steen, L.A. (April 29, 1988). *The Science of Patterns* Science, 240: 611–616. And summarized at Association for Supervision and Curriculum Development, www.ascd.org.

[10] Devlin, Keith, *Mathematics: The Science of Patterns: The Search for Order in Life, Mind and the Universe* (Scientific American Paperback Library) 1996, ISBN 978-0-7167-5047-5

[11] Eves

[12] Marcus du Sautoy, *A Brief History of Mathematics: 1. Newton and Leibniz*, BBC Radio 4, September 27, 2010.

[13] Waltershausen

[14] Peirce, p. 97.

[15] Hilbert, D. (1919–20), Natur und Mathematisches Erkennen: Vorlesungen, gehalten 1919–1920 in Göttingen. Nach der Ausarbeitung von Paul Bernays (Edited and with an English introduction by David E. Rowe), Basel, Birkhäuser (1992).

[16] Einstein, p. 28. The quote is Einstein's answer to the question: "how can it be that mathematics, being after all a product of human thought which is independent of experience, is so admirably appropriate to the objects of reality?" He, too, is concerned with *The Unreasonable Effectiveness of Mathematics in the Natural Sciences.*

[17] "Claire Voisin, Artist of the Abstract". .cnrs.fr. Retrieved October 13, 2013.

[18] Peterson

[19] Dehaene, Stanislas; Dehaene-Lambertz, Ghislaine; Cohen, Laurent (Aug 1998). "Abstract representations of numbers in the animal and human brain". *Trends in Neuroscience* 21 (8): 355–361. doi:10.1016/S0166-2236(98)01263-6. PMID 9720604.

[20] See, for example, Raymond L. Wilder, *Evolution of Mathematical Concepts; an Elementary Study, passim*

[21] Kline 1990, Chapter 1.

[22] "*A History of Greek Mathematics: From Thales to Euclid*". Thomas Little Heath (1981). ISBN 0-486-24073-8

[23] Sevryuk

[24] "mathematic". Online Etymology Dictionary.

[25] Both senses can be found in Plato. μαθηματική. Liddell, Henry George; Scott, Robert; *A Greek–English Lexicon* at the Perseus Project

[26] Cipra, Barry (1982). "St. Augustine v. The Mathematicians". *osu.edu.* Ohio State University Mathematics department. Retrieved July 14, 2014.

[27] *The Oxford Dictionary of English Etymology, Oxford English Dictionary, sub* "mathematics", "mathematic", "mathematics"

[28] "maths, n." and "math, n.3". *Oxford English Dictionary,* on-line version (2012).

[29] James Franklin, "Aristotelian Realism" in *Philosophy of Mathematics*", ed. A.D. Irvine, *p. 104. Elsevier (2009).*

[30] Cajori, Florian (1893). *A History of Mathematics.* American Mathematical Society (1991 reprint). pp. 285–6. ISBN 0-8218-2102-4.

[31] Snapper, Ernst (September 1979). "The Three Crises in Mathematics: Logicism, Intuitionism, and Formalism". *Mathematics Magazine* 52 (4): 207–16. doi:10.2307/2689412. JSTOR 2689412.

[32] Peirce, Benjamin (1882). *Linear Associative Algebra.* p. 1.

[33] Bertrand Russell, *The Principles of Mathematics,* p. 5. University Press, Cambridge (1903)

[34] Curry, Haskell (1951). *Outlines of a Formalist Philosophy of Mathematics.* Elsevier. pp. 56. ISBN 0-444-53368-0.

[35] Marcus du Sautoy, *A Brief History of Mathematics: 10. Nicolas Bourbaki*, BBC Radio 4, October 1, 2010.

[36] Shasha, Dennis Elliot; Lazere, Cathy A. (1998). *Out of Their Minds: The Lives and Discoveries of 15 Great Computer Scientists*. Springer. p. 228.

[37] Popper 1995, p. 56

[38] Ziman

[39] Johnson, Gerald W.; Lapidus, Michel L. (2002). *The Feynman Integral and Feynman's Operational Calculus*. Oxford University Press. ISBN 0-8218-2413-9.

[40] Wigner, Eugene (1960). "The Unreasonable Effectiveness of Mathematics in the Natural Sciences". *Communications on Pure and Applied Mathematics* 13 (1): 1–14. doi:10.1002/cpa.3160130102.

[41] "Mathematics Subject Classification 2010" (PDF). Retrieved November 9, 2010.

[42] Hardy, G.H. (1940). *A Mathematician's Apology*. Cambridge University Press. ISBN 0-521-42706-1.

[43] Gold, Bonnie; Simons, Rogers A. (2008). *Proof and Other Dilemmas: Mathematics and Philosophy*. MAA.

[44] Aigner, Martin; Ziegler, Günter M. (2001). *Proofs from* The Book. Springer. ISBN 3-540-40460-0.

[45] "Earliest Uses of Various Mathematical Symbols". Retrieved September 14, 2014.

[46] Kline, p. 140, on Diophantus; p. 261, on Vieta.

[47] See *false proof* for simple examples of what can go wrong in a formal proof.

[48] Ivars Peterson, *The Mathematical Tourist*, Freeman, 1988, ISBN 0-7167-1953-3. p. 4 "A few complain that the computer program can't be verified properly", (in reference to the Haken–Apple proof of the Four Color Theorem).

[49] " The method of "postulating" what we want has many advantages; they are the same as the advantages of theft over honest toil." Bertrand Russell (1919), *Introduction to Mathematical Philosophy*, New York and London, p 71.

[50] Patrick Suppes, *Axiomatic Set Theory*, Dover, 1972, ISBN 0-486-61630-4. p. 1, "Among the many branches of modern mathematics set theory occupies a unique place: with a few rare exceptions the entities which are studied and analyzed in mathematics may be regarded as certain particular sets or classes of objects."

[51] Luke Howard Hodgkin & Luke Hodgkin, *A History of Mathematics*, Oxford University Press, 2005.

[52] Clay Mathematics Institute, P=NP, claymath.org

[53] Rao, C.R. (1997) *Statistics and Truth: Putting Chance to Work*, World Scientific. ISBN 981-02-3111-3

[54] Like other mathematical sciences such as physics and computer science, statistics is an autonomous discipline rather than a branch of applied mathematics. Like research physicists and computer scientists, research statisticians are mathematical scientists. Many statisticians have a degree in mathematics, and some statisticians are also mathematicians.

[55] Rao, C.R. (1981). "Foreword". In Arthanari, T.S.; Dodge, Yadolah. *Mathematical programming in statistics*. Wiley Series in Probability and Mathematical Statistics. New York: Wiley. pp. vii–viii. ISBN 0-471-08073-X. MR 607328.

[56] Whittle (1994, pp. 10–11 and 14–18): Whittle, Peter (1994). "Almost home". In Kelly, F.P. *Probability, statistics and optimisation: A Tribute to Peter Whittle* (previously "A realised path: The Cambridge Statistical Laboratory upto 1993 (revised 2002)" ed.). Chichester: John Wiley. pp. 1–28. ISBN 0-471-94829-2.

[57] "*The Fields Medal is now indisputably the best known and most influential award in mathematics*." Monastyrsky

[58] Riehm

1.9   References

- Courant, Richard and H. Robbins, *What Is Mathematics? : An Elementary Approach to Ideas and Methods*, Oxford University Press, USA; 2 edition (July 18, 1996). ISBN 0-19-510519-2.

- Einstein, Albert (1923). *Sidelights on Relativity: I. Ether and relativity. II. Geometry and experience (translated by G.B. Jeffery, D.Sc., and W. Perrett, Ph.D).* E.P. Dutton & Co., New York.

- du Sautoy, Marcus, *A Brief History of Mathematics*, BBC Radio 4 (2010).

- Eves, Howard, *An Introduction to the History of Mathematics*, Sixth Edition, Saunders, 1990, ISBN 0-03-029558-0.

- Kline, Morris, *Mathematical Thought from Ancient to Modern Times*, Oxford University Press, USA; Paperback edition (March 1, 1990). ISBN 0-19-506135-7.

- Monastyrsky, Michael (2001). "Some Trends in Modern Mathematics and the Fields Medal" (PDF). Canadian Mathematical Society. Retrieved July 28, 2006.

- Oxford English Dictionary, second edition, ed. John Simpson and Edmund Weiner, Clarendon Press, 1989, ISBN 0-19-861186-2.

- *The Oxford Dictionary of English Etymology*, 1983 reprint. ISBN 0-19-861112-9.

- Pappas, Theoni, *The Joy Of Mathematics*, Wide World Publishing; Revised edition (June 1989). ISBN 0-933174-65-9.

- Peirce, Benjamin (1881). Peirce, Charles Sanders, ed. "Linear associative algebra". *American Journal of Mathematics* (Corrected, expanded, and annotated revision with an 1875 paper by B. Peirce and annotations by his son, C.S. Peirce, of the 1872 lithograph ed.) (Johns Hopkins University) 4 (1–4): 97–229. doi:10.2307/2369153. JSTOR 2369153. Corrected, expanded, and annotated revision with an 1875 paper by B. Peirce and annotations by his son, C. S. Peirce, of the 1872 lithograph ed. *Google* Eprint and as an extract, D. Van Nostrand, 1882, *Google* Eprint..

- Peterson, Ivars, *Mathematical Tourist, New and Updated Snapshots of Modern Mathematics*, Owl Books, 2001, ISBN 0-8050-7159-8.

- Popper, Karl R. (1995). "On knowledge". *In Search of a Better World: Lectures and Essays from Thirty Years*. Routledge. ISBN 0-415-13548-6.

- Riehm, Carl (August 2002). "The Early History of the Fields Medal" (PDF). *Notices of the AMS* (AMS) 49 (7): 778–782.

- Sevryuk, Mikhail B. (January 2006). "Book Reviews" (PDF). *Bulletin of the American Mathematical Society* 43 (1): 101–109. doi:10.1090/S0273-0979-05-01069-4. Retrieved June 24, 2006.

- Waltershausen, Wolfgang Sartorius von (1965) [first published 1856]. *Gauss zum Gedächtniss*. Sändig Reprint Verlag H. R. Wohlwend. ASIN B0000BN5SQ. ISBN 3-253-01702-8. ASIN 3253017028.

1.10   Further reading

- Benson, Donald C., *The Moment of Proof: Mathematical Epiphanies*, Oxford University Press, USA; New Ed edition (December 14, 2000). ISBN 0-19-513919-4.

- Boyer, Carl B., *A History of Mathematics*, Wiley; 2nd edition, revised by Uta C. Merzbach, (March 6, 1991). ISBN 0-471-54397-7.—A concise history of mathematics from the Concept of Number to contemporary Mathematics.

- Davis, Philip J. and Hersh, Reuben, *The Mathematical Experience*. Mariner Books; Reprint edition (January 14, 1999). ISBN 0-395-92968-7.

- Gullberg, Jan, *Mathematics – From the Birth of Numbers*. W. W. Norton & Company; 1st edition (October 1997). ISBN 0-393-04002-X.

- Hazewinkel, Michiel (ed.), *Encyclopaedia of Mathematics*. Kluwer Academic Publishers 2000. – A translated and expanded version of a Soviet mathematics encyclopedia, in ten (expensive) volumes, the most complete and authoritative work available. Also in paperback and on CD-ROM, and online.

- Jourdain, Philip E. B., *The Nature of Mathematics*, in *The World of Mathematics*, James R. Newman, editor, Dover Publications, 2003, ISBN 0-486-43268-8.

- Maier, Annaliese, *At the Threshold of Exact Science: Selected Writings of Annaliese Maier on Late Medieval Natural Philosophy*, edited by Steven Sargent, Philadelphia: University of Pennsylvania Press, 1982.

1.11 External links

- Mathematics at *Encyclopaedia Britannica*

- Mathematics on *In Our Time* at the BBC. (listen now)

- Free Mathematics books Free Mathematics books collection.

- Encyclopaedia of Mathematics online encyclopaedia from Springer, Graduate-level reference work with over 8,000 entries, illuminating nearly 50,000 notions in mathematics.

- HyperMath site at Georgia State University

- FreeScience Library The mathematics section of FreeScience library

- Rusin, Dave: *The Mathematical Atlas*. A guided tour through the various branches of modern mathematics. (Can also be found at NIU.edu.)

- Polyanin, Andrei: *EqWorld: The World of Mathematical Equations*. An online resource focusing on algebraic, ordinary differential, partial differential (mathematical physics), integral, and other mathematical equations.

- Cain, George: Online Mathematics Textbooks available free online.

- Tricki, Wiki-style site that is intended to develop into a large store of useful mathematical problem-solving techniques.

- Mathematical Structures, list information about classes of mathematical structures.

- Mathematician Biographies. The MacTutor History of Mathematics archive Extensive history and quotes from all famous mathematicians.

- *Metamath*. A site and a language, that formalize mathematics from its foundations.

- Nrich, a prize-winning site for students from age five from Cambridge University

- Open Problem Garden, a wiki of open problems in mathematics

- *Planet Math*. An online mathematics encyclopedia under construction, focusing on modern mathematics. Uses the Attribution-ShareAlike license, allowing article exchange with Wikipedia. Uses TeX markup.

- Some mathematics applets, at MIT

- Weisstein, Eric et al.: *MathWorld: World of Mathematics*. An online encyclopedia of mathematics.

- Patrick Jones' Video Tutorials on Mathematics

- Citizendium: Theory (mathematics).

- du Sautoy, Marcus, *A Brief History of Mathematics*, BBC Radio 4 (2010).

- MathOverflow A Q&A site for research-level mathematics

- Math – Khan Academy

- National Museum of Mathematics, located in New York City

Chapter 2

Definitions of mathematics

Definitions of mathematics vary widely and different schools of thought, particularly in philosophy, have suggested radically different and controversial accounts.[1][2]

2.1   Early definitions

Aristotle defined mathematics as:

> The science of quantity.

In Aristotle's classification of the sciences, discrete quantities were studied by arithmetic, continuous quantities by geometry.[3] Auguste Comte's definition tried to explain the role of mathematics in coordinating phenomena in all other fields:[4]

> The science of indirect measurement.[5] Auguste Comte 1851

The "indirectness" in Comte's definition refers to determining quantities that cannot be measured directly, such as the distance to planets or the size of atoms, by means of their relations to quantities that can be measured directly.[6]

2.2   Greater abstraction and competing philosophical schools

The preceding kinds of definitions, which had prevailed since Aristotle's time,[3] were abandoned in the 19th century as new branches of mathematics were developed, which bore no obvious relation to measurement or the physical world, such as group theory, projective geometry,[5] and non-Euclidean geometry.[7] As mathematicians pursued greater rigor and more-abstract foundations, some proposed definitions purely in terms of logic:

> Mathematics is the science that draws necessary conclusions.[8] Benjamin Peirce 1870

> All Mathematics is Symbolic Logic.[7] Bertrand Russell 1903

Peirce did not think that mathematics is the same as logic, since he thought mathematics makes only hypothetical assertions, not categorical ones.[9] Russell's definition, on the other hand, expresses the logicist philosophy of mathematics[10] without reservation. Competing philosophies of mathematics put forth different definitions.

Opposing the completely deductive character of logicism, intuitionism emphasizes the construction of ideas in the mind. Here is an intuitionist definition:[10]

> Mathematics is mental activity which consists in carrying out, one after the other, those mental constructions which are inductive and effective.

meaning that by combining fundamental ideas, one reaches a definite result.

Formalism denies both physical and mental meaning to mathematics, making the symbols and rules themselves the object of study.[10] A formalist definition:

> Mathematics is the manipulation of the meaningless symbols of a first-order language according to explicit, syntactical rules.

Still other approaches emphasize pattern, order, or structure. For example:

> Mathematics is the classification and study of all possible patterns. Walter Warwick Sawyer, 1955

Yet another approach makes abstraction the defining criterion:

> Mathematics is a broad-ranging field of study in which the properties and interactions of idealized objects are examined. Wolfram MathWorld

2.3   General, nonspecialist perspectives

Most contemporary reference works define mathematics mainly by summarizing its main topics and methods and referencing its history:

> The abstract science which investigates deductively the conclusions implicit in the elementary conceptions of spatial and numerical relations, and which includes as its main divisions geometry, arithmetic, and algebra. Oxford English Dictionary, 1933

> I believe maths is concerned with the development of language for expression, validation, falsification, deduction, calculation. This also involves the development of concepts for expression and description of structure and patterns. Ronald Brown

> The study of the measurement, properties, and relationships of quantities and sets, using numbers and symbols. American Heritage Dictionary, 2000

> The science of structure, order, and relation that has evolved from elemental practices of counting, measuring, and describing the shapes of objects.[11] Encyclopaedia Britannica

2.4   Playful, metaphorical, and poetic definitions

Bertrand Russell wrote this famous tongue-in-cheek definition, describing the way all terms in mathematics are ultimately defined by reference to undefined terms:

> The subject in which we never know what we are talking about, nor whether what we are saying is true.[12] Bertrand Russell 1901

Many other attempts to characterize mathematics have led to humor or poetic prose:

> "Mathematics is about making up rules and seeing what happens."[13] Vi Hart

A mathematician is a blind man in a dark room looking for a black cat which isn't there.[14] Charles Darwin

A mathematician, like a painter or poet, is a maker of patterns. If his patterns are more permanent than theirs, it is because they are made with ideas. G. H. Hardy, 1940

Mathematics is the art of giving the same name to different things.[8] Henri Poincaré

Mathematics is the science of skilful operations with concepts and rules invented just for this purpose. [this purpose being the skilful operation ....][15] Eugene Wigner

Mathematics is not a book confined within a cover and bound between brazen clasps, whose contents it needs only patience to ransack; it is not a mine, whose treasures may take long to reduce into possession, but which fill only a limited number of veins and lodes; it is not a soil, whose fertility can be exhausted by the yield of successive harvests; it is not a continent or an ocean, whose area can be mapped out and its contour defined: it is limitless as that space which it finds too narrow for its aspirations; its possibilities are as infinite as the worlds which are forever crowding in and multiplying upon the astronomer's gaze; it is as incapable of being restricted within assigned boundaries or being reduced to definitions of permanent validity, as the consciousness of life, which seems to slumber in each monad, in every atom of matter, in each leaf and bud cell, and is forever ready to burst forth into new forms of vegetable and animal existence.[16] James Joseph Sylvester

What is mathematics? What is it for? What are mathematicians doing nowadays? Wasn't it all finished long ago? How many new numbers can you invent anyway? Is today's mathematics just a matter of huge calculations, with the mathematician as a kind of zookeeper, making sure the precious computers are fed and watered? If it's not, what is it other than the incomprehensible outpourings of superpowered brainboxes with their heads in the clouds and their feet dangling from the lofty balconies of their ivory towers? Mathematics is all of these, and none. Mostly, it's just different. It's not what you expect it to be, you turn your back for a moment and it's changed. It's certainly not just a fixed body of knowledge, its growth is not confined to inventing new numbers, and its hidden tendrils pervade every aspect of modern life.[16] Ian Stewart

2.5 See also

- Philosophy of mathematics

2.6 References

[1] Mura, Robert (Dec 1993). "Images of Mathematics Held by University Teachers of Mathematical Sciences". *Educational Studies in Mathematics* 25 (4): 375–385.

[2] Tobies, Renate and Helmut Neunzert (2012). *Iris Runge: A Life at the Crossroads of Mathematics, Science, and Industry.* Springer. pp. 9. ISBN 3-0348-0229-3. It is first necessary to ask what is meant by *mathematics* in general. Illustrious scholars have debated this matter until they were blue in the face, and yet no consensus has been reached about whether mathematics is a natural science, a branch of the humanities, or an art form.

[3] James Franklin, "Aristotelian Realism" in *Philosophy of Mathematics*", ed. A.D. Irvine, p. 104. Elsevier (2009).

[4] Arline Reilein Standley, *Auguste Comte*, p. 61. Twayne Publishers (1981).

[5] Florian Cajori *et al.*, *A History of Mathematics*, 5th ed., p. 285–6. American Mathematical Society (1991).

[6] Auguste Comte, *The Philosophy of Mathematics*, tr. W.M. Gillespie, pp. 17–25. Harper & Brothers, New York (1851).

[7] Bertrand Russell, *The Principles of Mathematics*, p. 5. University Press, Cambridge (1903)

[8] Foundations and fundamental concepts of mathematics By Howard Eves page 150

[9] Carl Boyer, Uta Merzbach, *A History of Mathematics*, p. 426. John Wiley and Sones (2011).

[10] Snapper, Ernst (September 1979), "The Three Crises in Mathematics: Logicism, Intuitionism, and Formalism", *Mathematics Magazine* 52 (4): 207–16, doi:10.2307/2689412, JSTOR 2689412.

[11] "Mathematics. *Encyclopaedia Britannica* from Encyclopaedia Britannica 2006 Ultimate reference Suite DVD.

[12] Russell, Bertrand (1901), "Recent Work on the Principles of Mathematics", *International Monthly* 4.

[13] 9.999...reasons that .999...=1, Vi Hart

[14] "Pi in the Sky", John Barrow

[15] What is mathematics?

[16] "From Here to Infinity", Ian Stewart

2.7   Further reading

- Courant, Richard; Robbins, Herbert (1996), *What is Mathematics?* (2nd ed.), Oxford University Press, ISBN 978-0-19-510519-3.

- Gowers, Timothy; Barrow-Green, June; Leader, Imre, eds. (2008), *The Princeton Companion to Mathematics*, Princeton University Press, ISBN 978-0-691-11880-2.

- Hersh, Reuben (1999), *What is Mathematics, Really?*, Oxford University Press, ISBN 978-0-19-513087-4.

- Paulos, John Allen (1991), *Beyond Numeracy*, Viking, ISBN 978-0-670-83654-3.

- Stewart, Ian (1996), *From Here to Infinity*, Oxford University Press, ISBN 0-19-283202-6.

Chapter 3

Mathematical beauty

Mathematical beauty describes the notion that some mathematicians may derive aesthetic pleasure from their work, and from mathematics in general. They express this pleasure by describing mathematics (or, at least, some aspect of mathematics) as *beautiful*. Mathematicians describe mathematics as an art form or, at a minimum, as a creative activity. Comparisons are often made with music and poetry.

Bertrand Russell expressed his sense of mathematical beauty in these words:

> Mathematics, rightly viewed, possesses not only truth, but supreme beauty — a beauty cold and austere, like that of sculpture, without appeal to any part of our weaker nature, without the gorgeous trappings of painting or music, yet sublimely pure, and capable of a stern perfection such as only the greatest art can show. The true spirit of delight, the exaltation, the sense of being more than Man, which is the touchstone of the highest excellence, is to be found in mathematics as surely as poetry.[1]

Paul Erdős expressed his views on the ineffability of mathematics when he said, "Why are numbers beautiful? It's like asking why is Beethoven's Ninth Symphony beautiful. If you don't see why, someone can't tell you. I *know* numbers are beautiful. If they aren't beautiful, nothing is".[2]

3.1   Beauty in method

Mathematicians describe an especially pleasing method of proof as *elegant*. Depending on context, this may mean:

- A proof that uses a minimum of additional assumptions or previous results.
- A proof that is unusually succinct.
- A proof that derives a result in a surprising way (e.g., from an apparently unrelated theorem or collection of theorems.)
- A proof that is based on new and original insights.
- A method of proof that can be easily generalized to solve a family of similar problems.

In the search for an elegant proof, mathematicians often look for different independent ways to prove a result—the first proof that is found may not be the best. The theorem for which the greatest number of different proofs have been discovered is possibly the Pythagorean theorem, with hundreds of proofs having been published.[3] Another theorem that has been proved in many different ways is the theorem of quadratic reciprocity—Carl Friedrich Gauss alone published eight different proofs of this theorem.

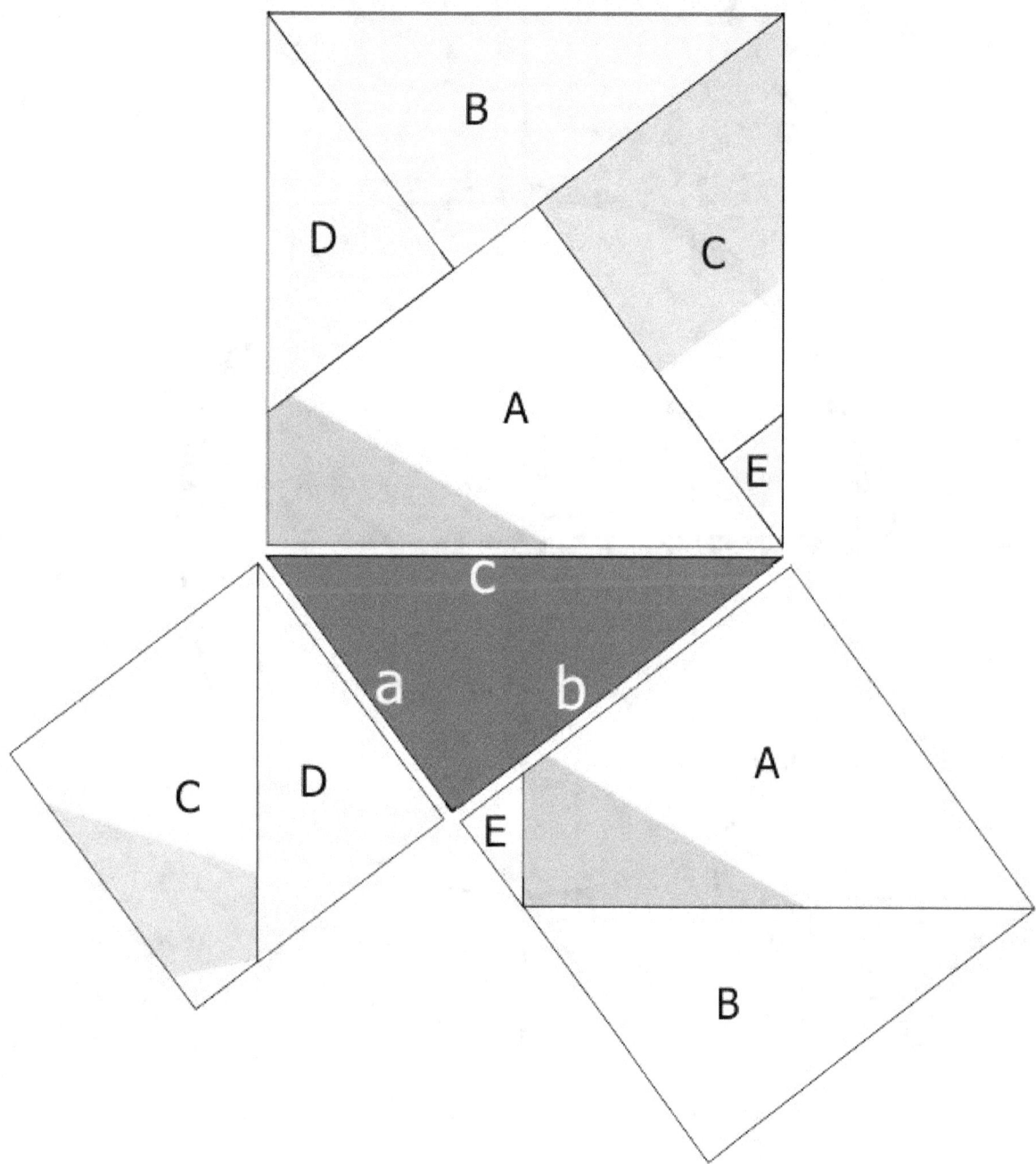

An example of "beauty in method"—a simple and elegant proof of the Pythagorean theorem.

Conversely, results that are logically correct but involve laborious calculations, over-elaborate methods, very conventional approaches, or that rely on a large number of particularly powerful axioms or previous results are not usually considered to be elegant, and may be called *ugly* or *clumsy*.

3.2 Beauty in results

Some mathematicians[4] see beauty in mathematical results that establish connections between two areas of mathematics that at first sight appear to be unrelated. These results are often described as *deep*.

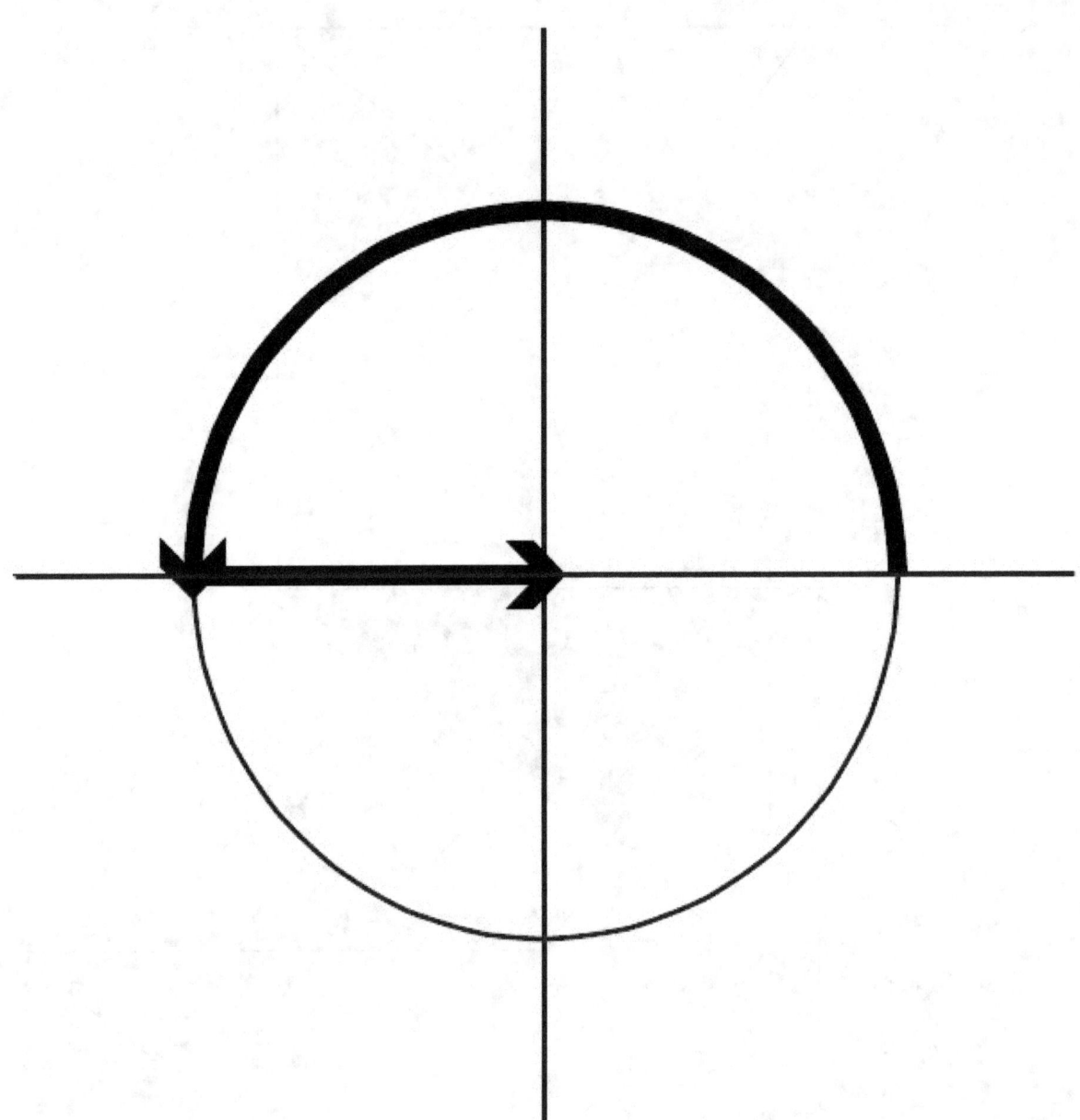

Starting at e^0 = 1, travelling at the velocity i relative to one's position for the length of time π, and adding 1, one arrives at 0. (The diagram is an Argand diagram)

While it is difficult to find universal agreement on whether a result is deep, some examples are often cited. One is Euler's identity:[5]

$$e^{i\pi} + 1 = 0.$$

This is a special case of Euler's formula, which the physicist Richard Feynman called "our jewel" and "the most remarkable formula in mathematics".[6] Modern examples include the modularity theorem, which establishes an important connection between elliptic curves and modular forms (work on which led to the awarding of the Wolf Prize to Andrew Wiles and Robert Langlands), and "monstrous moonshine", which connects the Monster group to modular functions via string theory for which Richard Borcherds was awarded the Fields Medal.

Other examples of deep results include unexpected insights into mathematical structures. For example, Gauss's Theorema Egregium is a deep theorem which relates a local phenomenon (curvature) to a global phenomenon (area) in a surprising way. In particular, the area of a triangle on a curved surface is proportional to the excess of the triangle and the proportionality is curvature. Another example is the fundamental theorem of calculus (and its vector versions including Green's theorem and Stokes' theorem).

The opposite of *deep* is *trivial*. A trivial theorem may be a result that can be derived in an obvious and straightforward way from other known results, or which applies only to a specific set of particular objects such as the empty set. Sometimes, however, a statement of a theorem can be original enough to be considered deep, even though its proof is fairly obvious.

In his *A Mathematician's Apology*, Hardy suggests that a beautiful proof or result possesses "inevitability", "unexpectedness", and "economy".[7]

Rota, however, disagrees with unexpectedness as a sufficient condition for beauty and proposes a counterexample:

> A great many theorems of mathematics, when first published, appear to be surprising; thus for example some twenty years ago [from 1977] the proof of the existence of non-equivalent differentiable structures on spheres of high dimension was thought to be surprising, but it did not occur to anyone to call such a fact beautiful, then or now.[8]

Perhaps ironically, Monastyrsky writes:

> It is very difficult to find an analogous invention in the past to Milnor's beautiful construction of the different differential structures on the seven-dimensional sphere....The original proof of Milnor was not very constructive, but later E. Briscorn showed that these differential structures can be described in an extremely explicit and beautiful form.[9]

This disagreement illustrates both the subjective nature of mathematical beauty and its connection with mathematical results: in this case, not only the existence of exotic spheres, but also a particular realization of them.

3.3   Beauty in experience

Interest in pure mathematics separate from empirical study has been part of the experience of various civilizations, including that of the Ancient Greeks, who "did mathematics for the beauty of it."[10] Mathematical beauty can also be experienced outside the confines of pure mathematics. For example, the aesthetic pleasure that mathematical physicists tend to experience in Einstein's theory of general relativity has been attributed (by Paul Dirac, among others) to its "great mathematical beauty."[11]

Some degree of delight in the manipulation of numbers and symbols is probably required to engage in any mathematics. Given the utility of mathematics in science and engineering, it is likely that any technological society will actively cultivate these aesthetics, certainly in its philosophy of science if nowhere else.

The most intense experience of mathematical beauty for most mathematicians comes from actively engaging in mathematics. It is very difficult to enjoy or appreciate mathematics in a purely passive way—in mathematics there is no real analogy of the role of the spectator, audience, or viewer.[12] Bertrand Russell referred to the *austere beauty* of mathematics.

3.4   Beauty and philosophy

Some mathematicians are of the opinion that the doing of mathematics is closer to discovery than invention, for example:

> There is no scientific discoverer, no poet, no painter, no musician, who will not tell you that he found ready made his discovery or poem or picture – that it came to him from outside, and that he did not consciously create it from within.
> —William Kingdon Clifford, from a lecture to the Royal Institution titled "Some of the conditions of mental development"

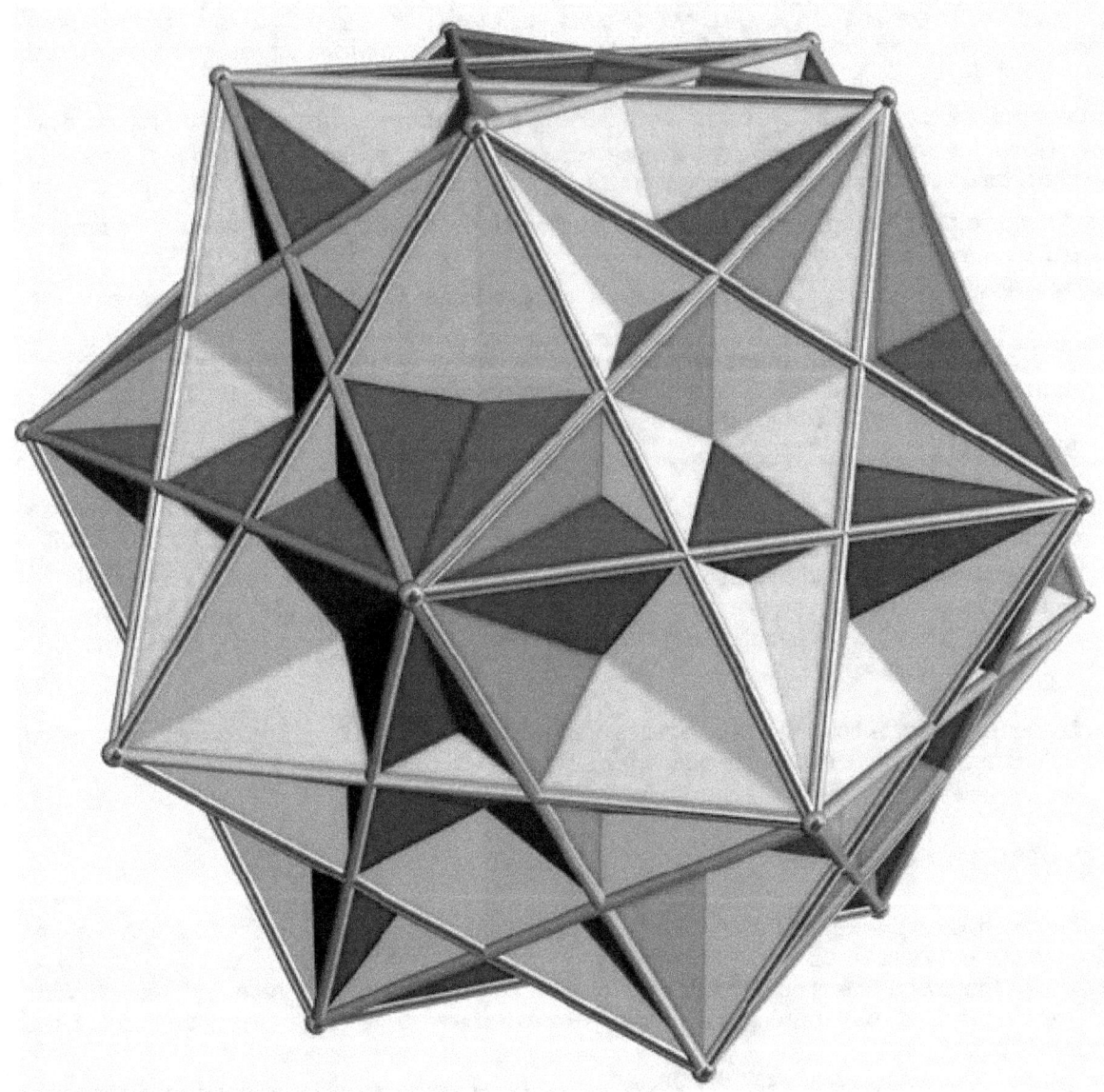

There is a certain "cold and austere" beauty in this compound of five cubes

These mathematicians believe that the detailed and precise results of mathematics may be reasonably taken to be true without any dependence on the universe in which we live. For example, they would argue that the theory of the natural numbers is fundamentally valid, in a way that does not require any specific context. Some mathematicians have extrapolated this viewpoint that mathematical beauty is truth further, in some cases becoming mysticism.

Pythagorean mathematicians believed in the literal reality of numbers. The discovery of the existence of irrational numbers was a shock to them, since they considered the existence of numbers not expressible as the ratio of two natural numbers to be a flaw in nature (the Pythagorean world view did not contemplate the limits of infinite sequences of ratios of natural numbers—the modern notion of a real number). From a modern perspective, their mystical approach to numbers may be viewed as numerology.

In Plato's philosophy there were two worlds, the physical one in which we live and another abstract world which contained unchanging truth, including mathematics. He believed that the physical world was a mere reflection of the more perfect abstract world.

Hungarian mathematician Paul Erdős[13] spoke of an imaginary book, in which God has written down all the most beautiful mathematical proofs. When Erdős wanted to express particular appreciation of a proof, he would exclaim "This one's from The Book!" This viewpoint expresses the idea that mathematics, as the intrinsically true foundation on which the laws of our universe are built, is a natural candidate for what has been personified as God by different religious believers.

Twentieth-century French philosopher Alain Badiou claims that ontology is mathematics. Badiou also believes in deep connections between mathematics, poetry and philosophy.

In some cases, natural philosophers and other scientists who have made extensive use of mathematics have made leaps of inference between beauty and physical truth in ways that turned out to be erroneous. For example, at one stage in his life, Johannes Kepler believed that the proportions of the orbits of the then-known planets in the Solar System have been arranged by God to correspond to a concentric arrangement of the five Platonic solids, each orbit lying on the circumsphere of one polyhedron and the insphere of another. As there are exactly five Platonic solids, Kepler's hypothesis could only accommodate six planetary orbits and was disproved by the subsequent discovery of Uranus.

3.5   Beauty and mathematical information theory

In the 1970s, Abraham Moles and Frieder Nake analyzed links between beauty, information processing, and information theory.[14][15] In the 1990s, Jürgen Schmidhuber formulated a mathematical theory of observer-dependent subjective beauty based on algorithmic information theory: the most beautiful objects among subjectively comparable objects have short algorithmic descriptions (i.e., Kolmogorov complexity) relative to what the observer already knows.[16][17][18] Schmidhuber explicitly distinguishes between beautiful and interesting. The latter corresponds to the first derivative of subjectively perceived beauty: the observer continually tries to improve the predictability and compressibility of the observations by discovering regularities such as repetitions and symmetries and fractal self-similarity. Whenever the observer's learning process (possibly a predictive artificial neural network) leads to improved data compression such that the observation sequence can be described by fewer bits than before, the temporary interestingness of the data corresponds to the compression progress, and is proportional to the observer's internal curiosity reward.[19][20]

3.6   Mathematics and the arts

Main articles: Mathematics and art and Mathematics and music

3.6.1   Music

Examples of the use of mathematics in music include the stochastic music of Iannis Xenakis, counterpoint of Johann Sebastian Bach, polyrhythmic structures (as in Igor Stravinsky's *The Rite of Spring*), the Metric modulation of Elliott Carter, permutation theory in serialism beginning with Arnold Schoenberg, and application of Shepard tones in Karlheinz Stockhausens *Hymnen*.

3.6.2   Visual arts

Examples of the use of mathematics in the visual arts include applications of chaos theory and fractal geometry to computer-generated art, symmetry studies of Leonardo da Vinci, projective geometries in development of the perspective theory of Renaissance art, grids in Op art, optical geometry in the camera obscura of Giambattista della Porta, and multiple perspective in analytic cubism and futurism.

The Dutch graphic designer M.C. Escher created mathematically inspired woodcuts, lithographs, and mezzotints. These feature impossible constructions, explorations of infinity, architecture, visual paradoxes and tessellations. British constructionist artist John Ernest created reliefs and paintings inspired by group theory.[21] A number of other British artists of the constructionist and systems schools also draw on mathematics models and structures as a source of inspiration, including Anthony Hill and Peter Lowe. Computer-generated art is based on mathematical algorithms.

3.7 See also

- Argument from beauty
- Cellular automaton
- Descriptive science
- Fluency heuristic
- Golden ratio
- Mathematics and architecture
- Normative science
- Philosophy of mathematics
- Processing fluency theory of aesthetic pleasure
- Pythagoreanism
- Theory of everything

3.8 Notes

[1] Russell, Bertrand (1919). "The Study of Mathematics". *Mysticism and Logic: And Other Essays.* Longman. p. 60. Retrieved 2008-08-22.

[2] Devlin, Keith (2000). "Do Mathematicians Have Different Brains?". *The Math Gene: How Mathematical Thinking Evolved And Why Numbers Are Like Gossip.* Basic Books. p. 140. ISBN 978-0-465-01619-8. Retrieved 2008-08-22.

[3] Elisha Scott Loomis published over 360 proofs in his book Pythagorean Proposition (ISBN 0-873-53036-5).

[4] Rota (1997), *The phenomenology of mathematical beauty,* p. 173

[5] Gallagher, James (13 February 2014). "Mathematics: Why the brain sees maths as beauty". *BBC News online.* Retrieved 13 February 2014.

[6] Feynman, Richard P. (1977). *The Feynman Lectures on Physics* I. Addison-Wesley. p. 22-10. ISBN 0-201-02010-6.

[7] Hardy, G.H. "18". Missing or empty |title= (help)

[8] Rota, *The phenomenology of mathematical beauty year = 1997,* p. 172

[9] Monastyrsky (2001), *Some Trends in Modern Mathematics and the Fields Medal*

[10] Lang, p. 3

[11] Chandrasekhar, p. 148

[12] Phillips, George (2005). "Preface". *Mathematics Is Not a Spectator Sport.* Springer Science+Business Media. ISBN 0-387-25528-1. Retrieved 2008-08-22. "...there is nothing in the world of mathematics that corresponds to an audience in a concert hall, where the passive listen to the active. Happily, mathematicians are all *doers,* not spectators.

[13] Schechter, Bruce (2000). *My brain is open: The mathematical journeys of Paul Erdős.* New York: Simon & Schuster. pp. 70–71. ISBN 0-684-85980-7.

[14] A. Moles: *Théorie de l'information et perception esthétique*, Paris, Denoël, 1973 (Information Theory and aesthetical perception)

[15] F Nake (1974). Ästhetik als Informationsverarbeitung. (Aesthetics as information processing). Grundlagen und Anwendungen der Informatik im Bereich ästhetischer Produktion und Kritik. Springer, 1974, ISBN 3-211-81216-4, ISBN 978-3-211-81216-7

[16] J. Schmidhuber. Low-complexity art. Leonardo, Journal of the International Society for the Arts, Sciences, and Technology, 30(2):97–103, 1997. http://www.jstor.org/pss/1576418

[17] J. Schmidhuber. Papers on the theory of beauty and low-complexity art since 1994: http://www.idsia.ch/~{}juergen/beauty.html

[18] J. Schmidhuber. Simple Algorithmic Principles of Discovery, Subjective Beauty, Selective Attention, Curiosity & Creativity. Proc. 10th Intl. Conf. on Discovery Science (DS 2007) p. 26-38, LNAI 4755, Springer, 2007. Also in Proc. 18th Intl. Conf. on Algorithmic Learning Theory (ALT 2007) p. 32, LNAI 4754, Springer, 2007. Joint invited lecture for DS 2007 and ALT 2007, Sendai, Japan, 2007. http://arxiv.org/abs/0709.0674

[19] .J. Schmidhuber. Curious model-building control systems. International Joint Conference on Neural Networks, Singapore, vol 2, 1458–1463. IEEE press, 1991

[20] Schmidhuber's theory of beauty and curiosity in a German TV show:http://www.br-online.de/bayerisches-fernsehen/faszination/schoenheit--aesthetik-wahrnehmung-ID1212005092828.xml

[21] John Ernest's use of mathematics and especially group theory in his art works is analysed in *John Ernest, A Mathematical Artist* by Paul Ernest in Philosophy of Mathematics Education Journal, No. 24 Dec. 2009 (Special Issue on Mathematics and Art): http://people.exeter.ac.uk/PErnest/pome24/index.htm

3.9   References

- Aigner, Martin, and Ziegler, Gunter M. (2003), *Proofs from THE BOOK*, 3rd edition, Springer-Verlag.

- Chandrasekhar, Subrahmanyan (1987), *Truth and Beauty: Aesthetics and Motivations in Science*, University of Chicago Press, Chicago, IL.

- Hadamard, Jacques (1949), *The Psychology of Invention in the Mathematical Field*, 1st edition, Princeton University Press, Princeton, NJ. 2nd edition, 1949. Reprinted, Dover Publications, New York, NY, 1954.

- Hardy, G.H. (1940), *A Mathematician's Apology*, 1st published, 1940. Reprinted, C.P. Snow (foreword), 1967. Reprinted, Cambridge University Press, Cambridge, UK, 1992.

- Hoffman, Paul (1992), *The Man Who Loved Only Numbers*, Hyperion.

- Huntley, H.E. (1970), *The Divine Proportion: A Study in Mathematical Beauty*, Dover Publications, New York, NY.

- Loomis, Elisha Scott (1968), *The Pythagorean Proposition*, The National Council of Teachers of Mathematics. Contains 365 proofs of the Pythagorean Theorem.

- Lang, Serge (1985). *The Beauty of Doing Mathematics: Three Public Dialogues*. New York: Springer-Verlag. ISBN 0-387-96149-6.

- Peitgen, H.-O., and Richter, P.H. (1986), *The Beauty of Fractals*, Springer-Verlag.

- Reber, R., Brun, M., & Mitterndorfer, K. (2008). The use of heuristics in intuitive mathematical judgment. *Psychonomic Bulletin & Review, 15*, 1174-1178.

- Strohmeier, John, and Westbrook, Peter (1999), *Divine Harmony, The Life and Teachings of Pythagoras*, Berkeley Hills Books, Berkeley, CA.

- Rota, Gian-Carlo (1997). "The phenomenology of mathematical beauty". *Synthese* 111(2): 171–182. doi:10.102234.

- Monastyrsky, Michael (2001). "Some Trends in Modern Mathematics and the Fields Medal" (PDF). *Can. Math. Soc. Notes* 33 (2 and 3).

3.10 External links

- Mathematics, Poetry and Beauty
- Is Mathematics Beautiful?
- The Beauty of Mathematics
- Justin Mullins
- Edna St. Vincent Millay (poet): *Euclid alone has looked on beauty bare*
- Terence Tao, *What is good mathematics?*
- Mathbeauty Blog
- The *Aesthetic Appeal* collection at the Internet Archive
- *A Mathematical Romance* Jim Holt December 5, 2013 issue of The New York Review of Books review of *Love and Math: The Heart of Hidden Reality* by Edward Frenkel

Chapter 4

Mathematical notation

For information on rendering mathematical formulae in Wikipedia, see Help:Formula.

Mathematical notation is a system of symbolic representations of mathematical objects and ideas. Mathematical notations are used in mathematics, the physical sciences, engineering, and economics. Mathematical notations include relatively simple symbolic representations, such as the numbers 0, 1 and 2, function symbols sin and +; conceptual symbols, such as lim, dy/dx, equations and variables; and complex diagrammatic notations such as Penrose graphical notation and Coxeter–Dynkin diagrams.

4.1 Definition

A mathematical notation is a writing system used for recording concepts in mathematics.

- The notation uses symbols or symbolic expressions which are intended to have a precise semantic meaning.

- In the history of mathematics, these symbols have denoted numbers, shapes, patterns, and change. The notation can also include symbols for parts of the conventional discourse between mathematicians, when viewing mathematics as a language.

The media used for writing are recounted below, but common materials currently include paper and pencil, board and chalk (or dry-erase marker), and electronic media. Systematic adherence to mathematical concepts is a fundamental concept of mathematical notation. (See also some related concepts: Logical argument, Mathematical logic, and Model theory.)

4.2 Expressions

A mathematical expression is a *sequence* of symbols which can be evaluated. For example, if the symbols represent numbers, the expressions are evaluated according to a conventional order of operations which provides for calculation, if possible, of any expressions within parentheses, followed by any exponents and roots, then multiplications and divisions and finally any additions or subtractions, all done from left to right. In a computer language, these rules are implemented by the compilers. For more on expression evaluation, see the computer science topics: eager evaluation, lazy evaluation, and evaluation operator.

4.3   Precise semantic meaning

Modern mathematics needs to be precise, because ambiguous notations do not allow formal proofs. Suppose that we have statements, denoted by some formal sequence of symbols, about some objects (for example, numbers, shapes, patterns). Until the statements can be shown to be valid, their meaning is not yet resolved. While reasoning, we might let the symbols refer to those denoted objects, perhaps in a model. The semantics of that object has a heuristic side and a deductive side. In either case, we might want to know the properties of that object, which we might then list in an intensional definition.

Those properties might then be expressed by some well-known and agreed-upon symbols from a table of mathematical symbols. This mathematical notation might include annotation such as

- "All x", "No x", "There is an x" (or its equivalent, "Some x"), "A set", "A function"
- "A mapping from the real numbers to the complex numbers"

In different contexts, the same symbol or notation can be used to represent different concepts. Therefore, to fully understand a piece of mathematical writing, it is important to first check the definitions that an author gives for the notations that are being used. This may be problematic if the author assumes the reader is already familiar with the notation in use.

4.4   History

Main article: History of mathematical notation

4.4.1   Counting

It is believed that a mathematical notation to represent counting was first developed at least 50,000 years ago[1] — early mathematical ideas such as finger counting[2] have also been represented by collections of rocks, sticks, bone, clay, stone, wood carvings, and knotted ropes. The tally stick is a timeless way of counting. Perhaps the oldest known mathematical texts are those of ancient Sumer. The Census Quipu of the Andes and the Ishango Bone from Africa both used the tally mark method of accounting for numerical concepts.

The development of zero as a number is one of the most important developments in early mathematics. It was used as a placeholder by the Babylonians and Greek Egyptians, and then as an integer by the Mayans, Indians and Arabs. (See The history of zero for more information.)

4.4.2   Geometry becomes analytic

The mathematical viewpoints in geometry did not lend themselves well to counting. The natural numbers, their relationship to fractions, and the identification of continuous quantities actually took millennia to take form, and even longer to allow for the development of notation. It was not until the invention of analytic geometry by René Descartes that geometry became more subject to a numerical notation.[3] Some symbolic shortcuts for mathematical concepts came to be used in the publication of geometric proofs. Moreover, the power and authority of geometry's theorem and proof structure greatly influenced non-geometric treatises, Isaac Newton's Principia Mathematica, for example.

4.4.3   Counting is mechanized

After the rise of Boolean algebra and the development of positional notation, it became possible to mechanize simple circuits for counting, first by mechanical means, such as gears and rods, using rotation and translation to represent changes of state, then by electrical means, using changes in voltage and current to represent the analogs of quantity. Today, computers use standard circuits to both store and change quantities, which represent not only numbers but pictures, sound, motion, and control.

4.4.4   Modern notation

The 18th and 19th centuries saw the creation and standardization of mathematical notation as used today. Euler was responsible for many of the notations in use today: the use of a, b, c for constants and x, y, z for unknowns, e for the base of the natural logarithm, sigma (Σ) for summation, i for the imaginary unit, and the functional notation $f(x)$. He also popularized the use of π for Archimedes constant (due to William Jones' proposal for the use of π in this way based on the earlier notation of William Oughtred). Many fields of mathematics bear the imprint of their creators for notation: the differential operator is due to Leibniz,[4] the cardinal infinities to Georg Cantor (in addition to the lemniscate (∞) of John Wallis), the congruence symbol ($\equiv$) to Gauss, and so forth.

4.4.5   Computerized notation

The rise of expression evaluators such as calculators and slide rules were only part of what was required to mathematicize civilization. Today, keyboard-based notations are used for the e-mail of mathematical expressions, the Internet shorthand notation. The wide use of programming languages, which teach their users the need for rigor in the statement of a mathematical expression (or else the compiler will not accept the formula) are all contributing toward a more mathematical viewpoint across all walks of life. Mathematically oriented markup languages such as TeX, LaTeX and, more recently, MathML are powerful enough that they qualify as mathematical notations in their own right.

For some people, computerized visualizations have been a boon to comprehending mathematics that mere symbolic notation could not provide. They can benefit from the wide availability of devices, which offer more graphical, visual, aural, and tactile feedback.

4.4.6   Ideographic notation

In the history of writing, ideographic symbols arose first, as more-or-less direct renderings of some concrete item. This has come full circle with the rise of computer visualization systems, which can be applied to abstract visualizations as well, such as for rendering some projections of a Calabi–Yau manifold.

Examples of abstract visualization which properly belong to the mathematical imagination can be found, for example in computer graphics. The need for such models abounds, for example, when the measures for the subject of study are actually random variables and not really ordinary mathematical functions.

4.5   Non-Latin-based mathematical notation

Modern Arabic mathematical notation is based mostly on the Arabic alphabet and is used widely in the Arab world, especially in pre-tertiary education. (Western notation uses Arabic numerals, but the Arabic notation also replaces Latin letters and related symbols with Arabic script.)

Some mathematical notations are mostly diagrammatic, and so are almost entirely script independent. Examples are Penrose graphical notation and Coxeter–Dynkin diagrams.

Braille-based mathematical notations used by blind people include Nemeth Braille and GS8 Braille.

4.6   See also

- Abuse of notation
- Begriffsschrift
- Bourbaki dangerous bend symbol
- History of mathematical notation

- ISO 31-11
- ISO/IEC 80000 –2
- Knuth's up-arrow notation
- Mathematical Alphanumeric Symbols
- Notation in probability
- Scientific notation
- Table of mathematical symbols
- Typographical conventions in mathematical formulae
- Vector notation
- Modern Arabic mathematical notation

4.7 Notes

[1] *An Introduction to the History of Mathematics* (6th Edition) by Howard Eves (1990)p.9

[2] Georges Ifrah notes that humans learned to count on their hands. Ifrah shows, for example, a picture of Boethius (who lived 480–524 or 525) reckoning on his fingers in Ifrah 2000, p. 48.

[3] Boyer, C. B. (1959), "Descartes and the geometrization of algebra", *The American Mathematical Monthly* 66: 390–393, JSTOR 2308751, MR 0105335, The great accomplishment of Descartes in mathematics invariably is described as the arithmetization of geometry.

[4] "Gottfried Wilhelm Leibnitz". Retrieved 5 October 2014.

4.8 References

- Florian Cajori, *A History of Mathematical Notations* (1929), 2 volumes. ISBN 0-486-67766-4

- Ifrah, Georges (2000), *The Universal History of Numbers: From prehistory to the invention of the computer*, John Wiley and Sons, p. 48, ISBN 0-471-39340-1. Translated from the French by David Bellos, E.F. Harding, Sophie Wood and Ian Monk. Ifrah supports his thesis by quoting idiomatic phrases from languages across the entire world.

4.9 External links

- Earliest Uses of Various Mathematical Symbols
- Mathematical ASCII Notation how to type math notation in any text editor.
- Mathematics as a Language at cut-the-knot
- Stephen Wolfram: Mathematical Notation: Past and Future. October 2000. Transcript of a keynote address presented at MathML and Math on the Web: MathML International Conference.

Chapter 5

Areas of mathematics

Mathematics has become a vastly diverse subject over history, and there is a corresponding need to categorize the different areas of mathematics. A number of different classification schemes have arisen, and though they share some similarities, there are differences due in part to the different purposes they serve. In addition, as mathematics evolves, these classification schemes must evolve as well to account for newly created areas or newly discovered links between different areas. Classification is made more difficult by some subjects, often the most active, which straddle the boundary between different areas.

A traditional division of mathematics is into pure mathematics, mathematics studied for its intrinsic interest, and applied mathematics, mathematics which can be directly applied to real world problems.[1] This division is not always clear and many subjects have been developed as pure mathematics to find unexpected applications later on. Broad divisions, such as discrete mathematics and computational mathematics, have emerged more recently.

5.1   Classification systems

- The Mathematics Subject Classification (MSC) is produced by the staff of the review databases Mathematical Reviews and Zentralblatt MATH. Many mathematics journals ask authors to label their papers with MSC subject codes. The MSC divides mathematics into over 60 areas, with further subdivisions within each area.

- In the Library of Congress Classification, mathematics is assigned the subclass QA within the class Q (Science). The LCC defines broad divisions, and individual subjects are assigned specific numerical values.

- The Dewey Decimal Classification assigns mathematics to division 510, with subdivisions for Algebra & number theory, Arithmetic, Topology, Analysis, Geometry, Numerical analysis, and Probabilities & applied mathematics.

- The Categories within Mathematics list is used by the Arxiv for categorising preprints. It is more modern than MSC and so includes things like quantum algebra.

- The IMU uses its programme structure for organizing the lectures at its four-yearly ICM. One of its top-level sections that MSC doesn't have is Lie theory.

- The ACM Computing Classification System includes a couple of mathematical categories F. Theory of Computation, and G. Mathematics of Computing.

- MathOverflow has a tag system.

- Mathematics book publishers such as Springer (subdisciplines), Cambridge (Browse Mathematics and statistics) and the AMS (subject area) use their own subject lists on their websites to enable customers to browse books or filter searches by subdiscipline, including topics such as mathematical biology and mathematical finance as top-level headings.

- Schools and other educational bodies have syllabuses.

- Research institutes and university mathematics departments often have sub-departments or study groups. e.g. SIAM has activity groups for its members.

- Wikipedia uses a Category:Mathematics system on its articles, and also has a list of mathematics lists.

5.2 Major divisions of mathematics

5.2.1 Foundations

Recreational mathematics From magic squares to the Mandelbrot set, numbers have been a source of amusement and delight for millions of people throughout the ages. Many important branches of "serious" mathematics have their roots in what was once a mere puzzle and/or game.

History and biography The history of mathematics is inextricably intertwined with the subject itself. This is perfectly natural: mathematics has an internal organic structure, deriving new theorems from those that have come before. As each new generation of mathematicians builds upon the achievements of our ancestors, the subject itself expands and grows new layers, like an onion.

Mathematical logic and foundations, including set theory Mathematicians have always worked with logic and symbols, but for centuries the underlying laws of logic were taken for granted, and never expressed symbolically. **Mathematical logic**, also known as symbolic logic, was developed when people finally realized that the tools of mathematics can be used to study the structure of logic itself. Areas of research in this field have expanded rapidly, and are usually subdivided into several distinct departments.

Model theory
Model theory studies mathematical structures in a general framework. Its main tool is first-order logic.

Set theory
A set can be thought of as a collection of distinct things united by some common feature. Set theory is subdivided into three main areas. Naive set theory is the original set theory developed by mathematicians at the end of the 19th century. Axiomatic set theory is a rigorous axiomatic theory developed in response to the discovery of serious flaws (such as Russell's paradox) in naive set theory. It treats sets as "whatever satisfies the axioms", and the notion of collections of things serves only as motivation for the axioms. Internal set theory is an axiomatic extension of set theory that supports a logically consistent identification of *illimited* (enormously large) and *infinitesimal* (unimaginably small) elements within the real numbers. See also List of set theory topics.

Proof theory and constructive mathematics
Proof theory grew out of David Hilbert's ambitious program to formalize all the proofs in mathematics. The most famous result in the field is encapsulated in Gödel's incompleteness theorems. A closely related and now quite popular concept is the idea of Turing machines. **Constructivism** is the outgrowth of Brouwer's unorthodox view of the nature of logic itself; constructively speaking, mathematicians cannot assert "Either a circle is round, or it is not" until they have actually exhibited a circle and measured its roundness.

5.2.2 Arithmetic

Arithmetic is the study of quantity.

5.2.3 Algebra

The study of structure begins with numbers, first the familiar natural numbers and integers and their arithmetical operations, which are recorded in elementary algebra. The deeper properties of these numbers are studied in number theory. The investigation of methods to solve equations leads to the field of abstract algebra, which, among other things, studies rings and fields, structures that generalize the properties possessed by everyday numbers. Long standing questions about compass and straightedge construction were finally settled by Galois theory. The physically important concept of vectors, generalized to vector spaces, is studied in linear algebra.

Order theory Any set of real numbers can be written out in ascending order. Order Theory extends this idea to sets in general. It includes notions like lattices and ordered algebraic structures. See also the order theory glossary and the list of order topics.

General algebraic systems Given a set, different ways of combining or relating members of that set can be defined. If these obey certain rules, then a particular algebraic structure is formed. Universal algebra is the more formal study of these structures and systems.

Number theory Number theory is traditionally concerned with the properties of integers. More recently, it has come to be concerned with wider classes of problems that have arisen naturally from the study of integers. It can be divided into elementary number theory (where the integers are studied without the aid of techniques from other mathematical fields); analytic number theory (where calculus and complex analysis are used as tools); algebraic number theory (which studies the algebraic numbers - the roots of polynomials with integer coefficients); geometric number theory; combinatorial number theory; transcendental number theory; and computational number theory. See also the list of number theory topics.

Field theory and polynomials Field theory studies the properties of fields. A field is a mathematical entity for which addition, subtraction, multiplication and division are well-defined. A polynomial is an expression in which constants and variables are combined using only addition, subtraction, and multiplication.

Commutative rings and algebras In ring theory, a branch of abstract algebra, a commutative ring is a ring in which the multiplication operation obeys the commutative law. This means that if a and b are any elements of the ring, then a×b=b×a. Commutative algebra is the field of study of commutative rings and their ideals, modules and algebras. It is foundational both for algebraic geometry and for algebraic number theory. The most prominent examples of commutative rings are rings of polynomials.

5.2.4 Analysis

Within the world of mathematics, analysis is the branch that focuses on change: rates of change, accumulated change, and multiple things changing relative to (or independently of) one another.

Modern analysis is a vast and rapidly expanding branch of mathematics that touches almost every other subdivision of the discipline, finding direct and indirect applications in topics as diverse as number theory, cryptography, and abstract algebra. It is also the language of science itself and is used across chemistry, biology, and physics, from astrophysics to X-ray crystallography.

5.2.5 Combinatorics

Combinatorics is the study of finite or discrete collections of objects that satisfy specified criteria. In particular, it is concerned with "counting" the objects in those collections (enumerative combinatorics) and with deciding whether certain "optimal" objects exist (extremal combinatorics). It includes graph theory, used to describe inter-connected objects (a graph in this sense is a network, or collection of connected points). See also the list of combinatorics topics, list of graph theory topics and glossary of graph theory. A *combinatorial flavour* is present in many parts of problem-solving.

5.2.6 Geometry and topology

Geometry deals with spatial relationships, using fundamental qualities or axioms. Such axioms can be used in conjunction with mathematical definitions for points, straight lines, curves, surfaces, and solids to draw logical conclusions. See also List of geometry topics

Convex geometry and discrete geometry Includes the study of objects such as polytopes and polyhedra. See also List of convexity topics

Discrete or combinatorial geometry The study of geometrical objects and properties that are discrete or combinatorial, either by their nature or by their representation. It includes the study of shapes such as the Platonic solids and the notion of tessellation.

Differential geometry The study of geometry using calculus, and is very closely related to differential topology. Covers such areas as Riemannian geometry, curvature and differential geometry of curves. See also the glossary of differential geometry and topology.

Algebraic geometry Given a polynomial of two real variables, then the points on a plane where that function is zero will form a curve. An algebraic curve extends this notion to polynomials over a field in a given number of variables. Algebraic geometry may be viewed as the study of these curves. See also the list of algebraic geometry topics and list of algebraic surfaces.

Topology Deals with the properties of a figure that do not change when the figure is continuously deformed. The main areas are point set topology (or general topology), algebraic topology, and the topology of manifolds, defined below.

General topology Also called *point set topology*. Properties of topological spaces. Includes such notions as open and closed sets, compact spaces, continuous functions, convergence, separation axioms, metric spaces, dimension theory. See also the glossary of general topology and the list of general topology topics.

Algebraic topology Properties of algebraic objects associated with a topological space and how these algebraic objects capture properties of such spaces. Contains areas like homology theory, cohomology theory, homotopy theory, and homological algebra, some of them examples of functors. Homotopy deals with homotopy groups (including the fundamental group) as well as simplicial complexes and CW complexes (also called *cell complexes*). See also the list of algebraic topology topics.

Manifolds A manifold can be thought of as an n-dimensional generalization of a surface in the usual 3-dimensional Euclidean space. The study of manifolds includes differential topology, which looks at the properties of differentiable functions defined over a manifold. See also complex manifolds.

5.2.7 Applied mathematics

Probability and statistics

See also glossary of probability and statistics

- Probability theory: The mathematical theory of random phenomena. Probability theory studies random variables and events, which are mathematical abstractions of non-deterministic events or measured quantities. See also Category:probability theory, and the list of probability topics.

  - Stochastic processes: An extension of probability theory that studies collections of random variables, such as time series or spatial processes. See also List of stochastic processes topics, and Category:Stochastic processes.

- Statistics: The science of making effective use of numerical data from experiments or from populations of individuals. Statistics includes not only the collection, analysis and interpretation of such data, but also the planning of the collection of data, in terms of the design of surveys and experiments. See also the list of statistical topics and Category:Statistics.

Computational sciences

Numerical analysis Many problems in mathematics cannot in general be solved exactly. Numerical analysis is the study of iterative methods and algorithms for approximately solving problems to a specified error bound. Includes numerical differentiation, numerical integration and numerical methods; c.f. scientific computing. See also List of numerical analysis topics

Computer algebra This area is also called **symbolic computation** or **algebraic computation**. It deals with exact computation, for example with integers of arbitrary size, polynomials or elements of finite fields. It includes also the computation with non numeric mathematical objects like polynomial ideals or series.

Physical sciences

Mechanics Addresses what happens when a real physical object is subjected to forces. This divides naturally into the study of rigid solids, deformable solids, and fluids, detailed below.

Particle mechanics In mathematics, a particle is a point-like, perfectly rigid, solid object. Particle mechanics deals with the results of subjecting particles to forces. It includes celestial mechanics—the study of the motion of celestial objects.

Mechanics of deformable solids Most real-world objects are not point-like nor perfectly rigid. More importantly, objects change shape when subjected to forces. This subject has a very strong overlap with continuum mechanics, which is concerned with continuous matter. It deals with such notions as stress, strain and elasticity. See also continuum mechanics.

Fluid mechanics Fluids in this sense includes not just liquids, but flowing gases, and even solids under certain situations. (For example, dry sand can behave like a fluid). It includes such notions as viscosity, turbulent flow and laminar flow (its opposite). See also fluid dynamics.

Other mathematical sciences

- Operations research (OR), also known as operational research, provides optimal or near-optimal solutions to complex problems. OR uses mathematical modeling, statistical analysis, and mathematical optimization.

- Mathematical programming (or mathematical optimization) minimizes (or maximizes) a real-valued function over a domain that is often specified by constraints on the variables. Mathematical programming studies these problems and develops iterative methods and algorithms for their solution.

5.3 See also

- Glossary of areas of mathematics

- Outline of mathematics

5.4   Notes

[1] For example the Encyclopædia Britannica Eleventh Edition groups its mathematics articles as Pure, Applied, and Biographies. See .

5.5   External links

- The Divisions of Mathematics (Forbidden)

Chapter 6

Philosophy of mathematics

The **philosophy of mathematics** is the branch of philosophy that studies the philosophical assumptions, foundations, and implications of mathematics. The aim of the philosophy of mathematics is to provide an account of the nature and methodology of mathematics and to understand the place of mathematics in people's lives. The logical and structural nature of mathematics itself makes this study both broad and unique among its philosophical counterparts.

The terms *philosophy of mathematics* and *mathematical philosophy* are frequently used as synonyms.[1] The latter, however, may be used to refer to several other areas of study. One refers to a project of formalizing a philosophical subject matter, say, aesthetics, ethics, logic, metaphysics, or theology, in a purportedly more exact and rigorous form, as for example the labors of scholastic theologians, or the systematic aims of Leibniz and Spinoza. Another refers to the working philosophy of an individual practitioner or a like-minded community of practicing mathematicians. Additionally, some understand the term "mathematical philosophy" to be an allusion to the approach to the foundations of mathematics taken by Bertrand Russell in his books *The Principles of Mathematics* and *Introduction to Mathematical Philosophy*.

6.1 Recurrent themes

Recurrent themes include:

- What is the role of Mankind in developing mathematics?

- What are the sources of mathematical subject matter?

- What is the ontological status of mathematical entities?

- What does it mean to refer to a mathematical object?

- What is the character of a mathematical proposition?

- What is the relation between logic and mathematics?

- What is the role of hermeneutics in mathematics?

- What kinds of inquiry play a role in mathematics?

- What are the objectives of mathematical inquiry?

- What gives mathematics its hold on experience?

- What are the human traits behind mathematics?

- What is mathematical beauty?

- What is the source and nature of mathematical truth?

- What is the relationship between the abstract world of mathematics and the material universe?

6.2 History

The origin of mathematics is subject to argument. Whether the birth of mathematics was a random happening or induced by necessity duly contingent upon other subjects, say for example physics, is still a matter of prolific debates.

Many thinkers have contributed their ideas concerning the nature of mathematics. Today, some philosophers of mathematics aim to give accounts of this form of inquiry and its products as they stand, while others emphasize a role for themselves that goes beyond simple interpretation to critical analysis. There are traditions of mathematical philosophy in both Western philosophy and Eastern philosophy. Western philosophies of mathematics go as far back as Plato, who studied the ontological status of mathematical objects, and Aristotle, who studied logic and issues related to infinity (actual versus potential).

Greek philosophy on mathematics was strongly influenced by their study of geometry. For example, at one time, the Greeks held the opinion that 1 (one) was not a number, but rather a unit of arbitrary length. A number was defined as a multitude. Therefore 3, for example, represented a certain multitude of units, and was thus not "truly" a number. At another point, a similar argument was made that 2 was not a number but a fundamental notion of a pair. These views come from the heavily geometric straight-edge-and-compass viewpoint of the Greeks: just as lines drawn in a geometric problem are measured in proportion to the first arbitrarily drawn line, so too are the numbers on a number line measured in proportion to the arbitrary first "number" or "one".

These earlier Greek ideas of numbers were later upended by the discovery of the irrationality of the square root of two. Hippasus, a disciple of Pythagoras, showed that the diagonal of a unit square was incommensurable with its (unit-length) edge: in other words he proved there was no existing (rational) number that accurately depicts the proportion of the diagonal of the unit square to its edge. This caused a significant re-evaluation of Greek philosophy of mathematics. According to legend, fellow Pythagoreans were so traumatized by this discovery that they murdered Hippasus to stop him from spreading his heretical idea. Simon Stevin was one of the first in Europe to challenge Greek ideas in the 16th century. Beginning with Leibniz, the focus shifted strongly to the relationship between mathematics and logic. This perspective dominated the philosophy of mathematics through the time of Frege and of Russell, but was brought into question by developments in the late 19th and early 20th centuries.

6.2.1 20th century

A perennial issue in the philosophy of mathematics concerns the relationship between logic and mathematics at their joint foundations. While 20th century philosophers continued to ask the questions mentioned at the outset of this article, the philosophy of mathematics in the 20th century was characterized by a predominant interest in formal logic, set theory, and foundational issues.

It is a profound puzzle that on the one hand mathematical truths seem to have a compelling inevitability, but on the other hand the source of their "truthfulness" remains elusive. Investigations into this issue are known as the foundations of mathematics program.

At the start of the 20th century, philosophers of mathematics were already beginning to divide into various schools of thought about all these questions, broadly distinguished by their pictures of mathematical epistemology and ontology. Three schools, formalism, intuitionism, and logicism, emerged at this time, partly in response to the increasingly widespread worry that mathematics as it stood, and analysis in particular, did not live up to the standards of certainty and rigor that had been taken for granted. Each school addressed the issues that came to the fore at that time, either attempting to resolve them or claiming that mathematics is not entitled to its status as our most trusted knowledge.

Surprising and counter-intuitive developments in formal logic and set theory early in the 20th century led to new questions concerning what was traditionally called the *foundations of mathematics*. As the century unfolded, the initial focus of concern expanded to an open exploration of the fundamental axioms of mathematics, the axiomatic approach having been taken for granted since the time of Euclid around 300 BCE as the natural basis for mathematics. Notions of axiom, proposition and proof, as well as the notion of a proposition being true of a mathematical object (see Assignment (mathematical logic)), were formalized, allowing them to be treated mathematically. The Zermelo–Fraenkel axioms for set theory were formulated which provided a conceptual framework in which much mathematical discourse would be interpreted. In mathematics, as in physics, new and unexpected ideas had arisen and significant changes were coming. With

Gödel numbering, propositions could be interpreted as referring to themselves or other propositions, enabling inquiry into the consistency of mathematical theories. This reflective critique in which the theory under review "becomes itself the object of a mathematical study" led Hilbert to call such study *metamathematics* or *proof theory*.[2]

At the middle of the century, a new mathematical theory was created by Samuel Eilenberg and Saunders Mac Lane, known as category theory, and it became a new contender for the natural language of mathematical thinking.[3] As the 20th century progressed, however, philosophical opinions diverged as to just how well-founded were the questions about foundations that were raised at the century's beginning. Hilary Putnam summed up one common view of the situation in the last third of the century by saying:

> When philosophy discovers something wrong with science, sometimes science has to be changed—Russell's paradox comes to mind, as does Berkeley's attack on the actual infinitesimal—but more often it is philosophy that has to be changed. I do not think that the difficulties that philosophy finds with classical mathematics today are genuine difficulties; and I think that the philosophical interpretations of mathematics that we are being offered on every hand are wrong, and that "philosophical interpretation" is just what mathematics doesn't need.[4]:169–170

Philosophy of mathematics today proceeds along several different lines of inquiry, by philosophers of mathematics, logicians, and mathematicians, and there are many schools of thought on the subject. The schools are addressed separately in the next section, and their assumptions explained.

6.3   Major themes

6.3.1   Mathematical realism

Mathematical realism, like realism in general, holds that mathematical entities exist independently of the human mind. Thus humans do not invent mathematics, but rather discover it, and any other intelligent beings in the universe would presumably do the same. In this point of view, there is really one sort of mathematics that can be discovered; triangles, for example, are real entities, not the creations of the human mind.

Many working mathematicians have been mathematical realists; they see themselves as discoverers of naturally occurring objects. Examples include Paul Erdős and Kurt Gödel. Gödel believed in an objective mathematical reality that could be perceived in a manner analogous to sense perception. Certain principles (e.g., for any two objects, there is a collection of objects consisting of precisely those two objects) could be directly seen to be true, but the continuum hypothesis conjecture might prove undecidable just on the basis of such principles. Gödel suggested that quasi-empirical methodology could be used to provide sufficient evidence to be able to reasonably assume such a conjecture.

Within realism, there are distinctions depending on what sort of existence one takes mathematical entities to have, and how we know about them. Major forms of mathematical realism include Platonism and empiricism.

6.3.2   Mathematical anti-realism

Mathematical anti-realism generally holds that mathematical statements have truth-values, but that they do not do so by corresponding to a special realm of immaterial or non-empirical entities. Major forms of mathematical anti-realism include Formalism and Fictionalism.

6.4   Contemporary schools of thought

6.4.1   Platonism

Mathematical Platonism is the form of realism that suggests that mathematical entities are abstract, have no spatiotemporal or causal properties, and are eternal and unchanging. This is often claimed to be the view most people have of numbers.

The term *Platonism* is used because such a view is seen to parallel Plato's Theory of Forms and a "World of Ideas" (Greek: *eidos* (εἶδος)) described in Plato's allegory of the cave: the everyday world can only imperfectly approximate an unchanging, ultimate reality. Both *Plato's cave* and *Platonism* have meaningful, not just superficial connections, because Plato's ideas were preceded and probably influenced by the hugely popular *Pythagoreans* of ancient Greece, who believed that the world was, quite literally, generated by numbers.

A major question considered in mathematical platonism is this: precisely where and how do the mathematical entities exist, and how do we know about them? Is there a world, completely separate from our physical one, that is occupied by the mathematical entities? How can we gain access to this separate world and discover truths about the entities? One answer might be the Ultimate Ensemble, which is a theory that postulates all structures that exist mathematically also exist physically in their own universe.

Plato spoke of mathematics by:

> How do you mean?
> I mean, as I was saying, that arithmetic has a very great and elevating effect, compelling the soul to reason about abstract number, and rebelling against the introduction of visible or tangible objects into the argument. You know how steadily the masters of the art repel and ridicule any one who attempts to divide absolute unity when he is calculating, and if you divide, they multiply, taking care that one shall continue one and not become lost in fractions.
> That is very true.
> Now, suppose a person were to say to them: O my friends, what are these wonderful numbers about which you are reasoning, in which, as you say, there is a unity such as you demand, and each unit is equal, invariable, indivisible, --what would they answer?
> —Plato, Chapter 7. "The Republic" (Jowett translation).

In context, chapter 8, of H.D.P. Lee's translation, reports the education of a philosopher contains five mathematical disciplines:

1. mathematics;

2. arithmetic, written in unit fraction "parts" using theoretical unities and abstract numbers;

3. plane geometry and solid geometry also considered the line to be segmented into rational and irrational unit "parts";

4. astronomy

5. harmonics

Translators of the works of Plato rebelled against practical versions of his culture's practical mathematics. However, Plato himself and Greeks had copied 1,500 older Egyptian fraction abstract unities, one being a hekat unity scaled to (64/64) in the Akhmim Wooden Tablet, thereby not getting lost in fractions.

Gödel's Platonism postulates a special kind of mathematical intuition that lets us perceive mathematical objects directly. (This view bears resemblances to many things Husserl said about mathematics, and supports Kant's idea that mathematics is synthetic *a priori*.) Davis and Hersh have suggested in their book *The Mathematical Experience* that most mathematicians act as though they are Platonists, even though, if pressed to defend the position carefully, they may retreat to formalism (see below).

Some mathematicians hold opinions that amount to more nuanced versions of Platonism.

Full-blooded Platonism is a modern variation of Platonism, which is in reaction to the fact that different sets of mathematical entities can be proven to exist depending on the axioms and inference rules employed (for instance, the law of the excluded middle, and the axiom of choice). It holds that all mathematical entities exist, however they may be provable, even if they cannot all be derived from a single consistent set of axioms.

6.4.2 Empiricism

Empiricism is a form of realism that denies that mathematics can be known *a priori* at all. It says that we discover mathematical facts by empirical research, just like facts in any of the other sciences. It is not one of the classical three positions advocated in the early 20th century, but primarily arose in the middle of the century. However, an important early proponent of a view like this was John Stuart Mill. Mill's view was widely criticized, because, according to critics, it makes statements like "2 + 2 = 4" come out as uncertain, contingent truths, which we can only learn by observing instances of two pairs coming together and forming a quartet.

Contemporary mathematical empiricism, formulated by Quine and Putnam, is primarily supported by the indispensability argument: mathematics is indispensable to all empirical sciences, and if we want to believe in the reality of the phenomena described by the sciences, we ought also believe in the reality of those entities required for this description. That is, since physics needs to talk about electrons to say why light bulbs behave as they do, then electrons must exist. Since physics needs to talk about numbers in offering any of its explanations, then numbers must exist. In keeping with Quine and Putnam's overall philosophies, this is a naturalistic argument. It argues for the existence of mathematical entities as the best explanation for experience, thus stripping mathematics of being distinct from the other sciences.

Putnam strongly rejected the term "Platonist" as implying an over-specific ontology that was not necessary to mathematical practice in any real sense. He advocated a form of "pure realism" that rejected mystical notions of truth and accepted much quasi-empiricism in mathematics. Putnam was involved in coining the term "pure realism" (see below).

The most important criticism of empirical views of mathematics is approximately the same as that raised against Mill. If mathematics is just as empirical as the other sciences, then this suggests that its results are just as fallible as theirs, and just as contingent. In Mill's case the empirical justification comes directly, while in Quine's case it comes indirectly, through the coherence of our scientific theory as a whole, i.e. consilience after E.O. Wilson. Quine suggests that mathematics seems completely certain because the role it plays in our web of belief is incredibly central, and that it would be extremely difficult for us to revise it, though not impossible.

For a philosophy of mathematics that attempts to overcome some of the shortcomings of Quine and Gödel's approaches by taking aspects of each see Penelope Maddy's *Realism in Mathematics*. Another example of a realist theory is the embodied mind theory (below). For a modern revision of mathematical empiricism see New Empiricism (below).

For experimental evidence suggesting that human infants can do elementary arithmetic, see Brian Butterworth.

6.4.3 Mathematical monism

Max Tegmark's mathematical universe hypothesis goes further than full-blooded Platonism in asserting that not only do all mathematical objects exist, but nothing else does. Tegmark's sole postulate is: *All structures that exist mathematically also exist physically*. That is, in the sense that "in those [worlds] complex enough to contain self-aware substructures [they] will subjectively perceive themselves as existing in a physically 'real' world".[5][6]

6.4.4 Logicism

Logicism is the thesis that mathematics is reducible to logic, and hence nothing but a part of logic.[7]:41 Logicists hold that mathematics can be known *a priori*, but suggest that our knowledge of mathematics is just part of our knowledge of logic in general, and is thus analytic, not requiring any special faculty of mathematical intuition. In this view, logic is the proper foundation of mathematics, and all mathematical statements are necessary logical truths.

Rudolf Carnap (1931) presents the logicist thesis in two parts:[7]

1. The *concepts* of mathematics can be derived from logical concepts through explicit definitions.

2. The *theorems* of mathematics can be derived from logical axioms through purely logical deduction.

Gottlob Frege was the founder of logicism. In his seminal *Die Grundgesetze der Arithmetik* (*Basic Laws of Arithmetic*) he built up arithmetic from a system of logic with a general principle of comprehension, which he called "Basic Law V" (for concepts F and G, the extension of F equals the extension of G if and only if for all objects a, Fa if and only if Ga), a principle that he took to be acceptable as part of logic.

Frege's construction was flawed. Russell discovered that Basic Law V is inconsistent (this is Russell's paradox). Frege abandoned his logicist program soon after this, but it was continued by Russell and Whitehead. They attributed the paradox to "vicious circularity" and built up what they called ramified type theory to deal with it. In this system, they were eventually able to build up much of modern mathematics but in an altered, and excessively complex form (for example, there were different natural numbers in each type, and there were infinitely many types). They also had to make several compromises in order to develop so much of mathematics, such as an "axiom of reducibility". Even Russell said that this axiom did not really belong to logic.

Modern logicists (like Bob Hale, Crispin Wright, and perhaps others) have returned to a program closer to Frege's. They have abandoned Basic Law V in favor of abstraction principles such as Hume's principle (the number of objects falling under the concept F equals the number of objects falling under the concept G if and only if the extension of F and the extension of G can be put into one-to-one correspondence). Frege required Basic Law V to be able to give an explicit definition of the numbers, but all the properties of numbers can be derived from Hume's principle. This would not have been enough for Frege because (to paraphrase him) it does not exclude the possibility that the number 3 is in fact Julius Caesar. In addition, many of the weakened principles that they have had to adopt to replace Basic Law V no longer seem so obviously analytic, and thus purely logical.

6.4.5   Formalism

Main article: Formalism (mathematics)

Formalism holds that mathematical statements may be thought of as statements about the consequences of certain string manipulation rules. For example, in the "game" of Euclidean geometry (which is seen as consisting of some strings called "axioms", and some "rules of inference" to generate new strings from given ones), one can prove that the Pythagorean theorem holds (that is, you can generate the string corresponding to the Pythagorean theorem). According to formalism, mathematical truths are not about numbers and sets and triangles and the like—in fact, they aren't "about" anything at all.

Another version of formalism is often known as deductivism. In deductivism, the Pythagorean theorem is not an absolute truth, but a relative one: *if* you assign meaning to the strings in such a way that the rules of the game become true (i.e., true statements are assigned to the axioms and the rules of inference are truth-preserving), *then* you have to accept the theorem, or, rather, the interpretation you have given it must be a true statement. The same is held to be true for all other mathematical statements. Thus, formalism need not mean that mathematics is nothing more than a meaningless symbolic game. It is usually hoped that there exists some interpretation in which the rules of the game hold. (Compare this position to structuralism.) But it does allow the working mathematician to continue in his or her work and leave such problems to the philosopher or scientist. Many formalists would say that in practice, the axiom systems to be studied will be suggested by the demands of science or other areas of mathematics.

A major early proponent of formalism was David Hilbert, whose program was intended to be a complete and consistent axiomatization of all of mathematics. Hilbert aimed to show the consistency of mathematical systems from the assumption that the "finitary arithmetic" (a subsystem of the usual arithmetic of the positive integers, chosen to be philosophically uncontroversial) was consistent. Hilbert's goals of creating a system of mathematics that is both complete and consistent were dealt a fatal blow by the second of Gödel's incompleteness theorems, which states that sufficiently expressive consistent axiom systems can never prove their own consistency. Since any such axiom system would contain the finitary arithmetic as a subsystem, Gödel's theorem implied that it would be impossible to prove the system's consistency relative to that (since it would then prove its own consistency, which Gödel had shown was impossible). Thus, in order to show that any axiomatic system of mathematics is in fact consistent, one needs to first assume the consistency of a system of mathematics that is in a sense stronger than the system to be proven consistent.

Hilbert was initially a deductivist, but, as may be clear from above, he considered certain metamathematical methods to yield intrinsically meaningful results and was a realist with respect to the finitary arithmetic. Later, he held the opinion that there was no other meaningful mathematics whatsoever, regardless of interpretation.

Other formalists, such as Rudolf Carnap, Alfred Tarski, and Haskell Curry, considered mathematics to be the investigation

David Hilbert

of formal axiom systems. Mathematical logicians study formal systems but are just as often realists as they are formalists.

Formalists are relatively tolerant and inviting to new approaches to logic, non-standard number systems, new set theories etc. The more games we study, the better. However, in all three of these examples, motivation is drawn from existing mathematical or philosophical concerns. The "games" are usually not arbitrary.

The main critique of formalism is that the actual mathematical ideas that occupy mathematicians are far removed from the string manipulation games mentioned above. Formalism is thus silent on the question of which axiom systems ought to be studied, as none is more meaningful than another from a formalistic point of view.

Recently, some formalist mathematicians have proposed that all of our *formal* mathematical knowledge should be systematically encoded in computer-readable formats, so as to facilitate automated proof checking of mathematical proofs and the use of interactive theorem proving in the development of mathematical theories and computer software. Because of their close connection with computer science, this idea is also advocated by mathematical intuitionists and constructivists in the "computability" tradition (see below). See QED project for a general overview.

6.4.6 Conventionalism

The French mathematician Henri Poincaré was among the first to articulate a conventionalist view. Poincaré's use of non-Euclidean geometries in his work on differential equations convinced him that Euclidean geometry should not be regarded as *a priori* truth. He held that axioms in geometry should be chosen for the results they produce, not for their apparent coherence with human intuitions about the physical world.

6.4.7 Psychologism

Psychologism in the philosophy of mathematics is the position that mathematical concepts and/or truths are grounded in, derived from or explained by psychological facts (or laws).

John Stuart Mill seems to have been an advocate of a type of logical psychologism, as were many 19th-century German logicians such as Sigwart and Erdmann as well as a number of psychologists, past and present: for example, Gustave Le Bon. Psychologism was famously criticized by Frege in his *The Foundations of Arithmetic*, and many of his works and essays, including his review of Husserl's *Philosophy of Arithmetic*. Edmund Husserl, in the first volume of his *Logical Investigations*, called "The Prolegomena of Pure Logic", criticized psychologism thoroughly and sought to distance himself from it. The "Prolegomena" is considered a more concise, fair, and thorough refutation of psychologism than the criticisms made by Frege, and also it is considered today by many as being a memorable refutation for its decisive blow to psychologism. Psychologism was also criticized by Charles Sanders Peirce and Maurice Merleau-Ponty.

6.4.8 Intuitionism

Main article: Mathematical intuitionism

In mathematics, intuitionism is a program of methodological reform whose motto is that "there are no non-experienced mathematical truths" (L.E.J. Brouwer). From this springboard, intuitionists seek to reconstruct what they consider to be the corrigible portion of mathematics in accordance with Kantian concepts of being, becoming, intuition, and knowledge. Brouwer, the founder of the movement, held that mathematical objects arise from the *a priori* forms of the volitions that inform the perception of empirical objects.[8]

A major force behind intuitionism was L.E.J. Brouwer, who rejected the usefulness of formalized logic of any sort for mathematics. His student Arend Heyting postulated an intuitionistic logic, different from the classical Aristotelian logic; this logic does not contain the law of the excluded middle and therefore frowns upon proofs by contradiction. The axiom of choice is also rejected in most intuitionistic set theories, though in some versions it is accepted. Important work was later done by Errett Bishop, who managed to prove versions of the most important theorems in real analysis within this framework.

In intuitionism, the term "explicit construction" is not cleanly defined, and that has led to criticisms. Attempts have been made to use the concepts of Turing machine or computable function to fill this gap, leading to the claim that only questions regarding the behavior of finite algorithms are meaningful and should be investigated in mathematics. This has led to the study of the computable numbers, first introduced by Alan Turing. Not surprisingly, then, this approach to mathematics is sometimes associated with theoretical computer science.

Constructivism

Main article: Mathematical constructivism

Like intuitionism, constructivism involves the regulative principle that only mathematical entities which can be explicitly constructed in a certain sense should be admitted to mathematical discourse. In this view, mathematics is an exercise of the human intuition, not a game played with meaningless symbols. Instead, it is about entities that we can create directly through mental activity. In addition, some adherents of these schools reject non-constructive proofs, such as a proof by contradiction.

Finitism

Finitism is an extreme form of constructivism, according to which a mathematical object does not exist unless it can be constructed from natural numbers in a finite number of steps. In her book *Philosophy of Set Theory*, Mary Tiles characterized those who allow countably infinite objects as classical finitists, and those who deny even countably infinite objects as strict finitists.

The most famous proponent of finitism was Leopold Kronecker,[9] who said:

> God created the natural numbers, all else is the work of man.

Ultrafinitism is an even more extreme version of finitism, which rejects not only infinities but finite quantities that cannot feasibly be constructed with available resources.

6.4.9   Structuralism

Main article: Mathematical structuralism

Structuralism is a position holding that mathematical theories describe structures, and that mathematical objects are exhaustively defined by their *places* in such structures, consequently having no intrinsic properties. For instance, it would maintain that all that needs to be known about the number 1 is that it is the first whole number after 0. Likewise all the other whole numbers are defined by their places in a structure, the number line. Other examples of mathematical objects might include lines and planes in geometry, or elements and operations in abstract algebra.

Structuralism is an epistemologically realistic view in that it holds that mathematical statements have an objective truth value. However, its central claim only relates to what *kind* of entity a mathematical object is, not to what kind of *existence* mathematical objects or structures have (not, in other words, to their ontology). The kind of existence mathematical objects have would clearly be dependent on that of the structures in which they are embedded; different sub-varieties of structuralism make different ontological claims in this regard.[10]

The *Ante Rem*, or fully realist, variation of structuralism has a similar ontology to Platonism in that structures are held to have a real but abstract and immaterial existence. As such, it faces the usual problems of explaining the interaction between such abstract structures and flesh-and-blood mathematicians.

In Re, or moderately realistic, structuralism is the equivalent of Aristotelian realism. Structures are held to exist inasmuch as some concrete system exemplifies them. This incurs the usual issues that some perfectly legitimate structures might accidentally happen not to exist, and that a finite physical world might not be "big" enough to accommodate some otherwise legitimate structures.

The *Post Res* or eliminative variant of structuralism is anti-realist about structures in a way that parallels nominalism. According to this view mathematical *systems* exist, and have structural features in common. If something is true of a structure, it will be true of all systems exemplifying the structure. However, it is merely convenient to talk of structures being "held in common" between systems: they in fact have no independent existence.

6.4.10 Embodied mind theories

Embodied mind theories hold that mathematical thought is a natural outgrowth of the human cognitive apparatus which finds itself in our physical universe. For example, the abstract concept of number springs from the experience of counting discrete objects. It is held that mathematics is not universal and does not exist in any real sense, other than in human brains. Humans construct, but do not discover, mathematics.

With this view, the physical universe can thus be seen as the ultimate foundation of mathematics: it guided the evolution of the brain and later determined which questions this brain would find worthy of investigation. However, the human mind has no special claim on reality or approaches to it built out of math. If such constructs as Euler's identity are true then they are true as a map of the human mind and cognition.

Embodied mind theorists thus explain the effectiveness of mathematics—mathematics was constructed by the brain in order to be effective in this universe.

The most accessible, famous, and infamous treatment of this perspective is *Where Mathematics Comes From*, by George Lakoff and Rafael E. Núñez. In addition, mathematician Keith Devlin has investigated similar concepts with his book *The Math Instinct*, as has neuroscientist Stanislas Dehaene with his book *The Number Sense*. For more on the philosophical ideas that inspired this perspective, see cognitive science of mathematics.

New empiricism

A more recent empiricism returns to the principle of the English empiricists of the 18th and 19th centuries, in particular John Stuart Mill, who asserted that all knowledge comes to us from observation through the senses. This applies not only to matters of fact, but also to "relations of ideas", as Hume called them: the structures of logic which interpret, organize and abstract observations.

To this principle it adds a materialist connection: all the processes of logic which interpret, organize and abstract observations, are physical phenomena which take place in real time and physical space: namely, in the brains of human beings. Abstract objects, such as mathematical objects, are ideas, which in turn exist as electrical and chemical states of the billions of neurons in the human brain.

This second concept is reminiscent of the social constructivist approach, which holds that mathematics is produced by humans rather than being "discovered" from abstract, *a priori* truths. However, it differs sharply from the constructivist implication that humans arbitrarily construct mathematical principles that have no inherent truth but which instead are created on a conveniency basis. On the contrary, new empiricism shows how mathematics, although constructed by humans, follows rules and principles that will be agreed on by all who participate in the process, with the result that everyone practicing mathematics comes up with the same answer—except in those areas where there is philosophical disagreement on the meaning of fundamental concepts. This is because the new empiricism perceives this agreement as being a physical phenomenon, one which is observed by other humans in the same way that other physical phenomena, like the motions of inanimate bodies, or the chemical interaction of various elements, are observed.

Combining the materialist principle with Millisian epistemology evades the principal difficulty with classical empiricism—that all knowledge comes from the senses. That difficulty lies in the observation that mathematical truths based on logical deduction appear to be more certainly true than knowledge of the physical world itself. (The physical world in this case is taken to mean the portion of it lying outside the human brain.)

Kant argued that the structures of logic which organize, interpret and abstract observations were built into the human mind and were true and valid *a priori*. Mill, on the contrary, said that we believe them to be true because we have enough individual instances of their truth to generalize: in his words, "From instances we have observed, we feel warranted in concluding that what we found true in those instances holds in all similar ones, past, present and future, however numerous they may be".[11] Although the psychological or epistemological specifics given by Mill through which we build our logical apparatus may not be completely warranted, his explanation still nonetheless manages to demonstrate that there is no way around Kant's *a priori* logic. To recant Mill's original idea in an empiricist twist: "*Indeed, the very principles of logical deduction are true because we observe that using them leads to true conclusions*", which is itself an *a priori* presupposition.

If all this is true, then where do the world senses come in? The early empiricists all stumbled over this point. Hume asserted that all knowledge comes from the senses, and then gave away the ballgame by excepting abstract propositions, which he called "relations of ideas". These, he said, were absolutely true (although the mathematicians who thought them up, being human, might get them wrong). Mill, on the other hand, tried to deny that abstract ideas exist outside the physical world: all numbers, he said, "must be numbers of something: there are no such things as numbers in the abstract". When we count to eight or add five and three we are really counting spoons or bumblebees. "All things possess quantity", he said, so that propositions concerning numbers are propositions concerning "all things whatever". But then in almost a contradiction of himself he went on to acknowledge that numerical and algebraic expressions are not necessarily attached to real world objects: they "do not excite in our minds ideas of any things in particular". Mill's low reputation as a philosopher of logic, and the low estate of empiricism in the century and a half following him, derives from this failed attempt to link abstract thoughts to the physical world, when it may be more plausibly arguable that abstraction consists precisely of separating the thought from its physical foundations.

The conundrum created by our certainty that abstract deductive propositions, if valid (i.e. if we can "prove" them), are true, exclusive of observation and testing in the physical world, gives rise to a further reflection ... What if thoughts themselves, and the minds that create them, are physical objects, existing only in the physical world?

This would reconcile the contradiction between our belief in the certainty of abstract deductions and the empiricist principle that knowledge comes from observation of individual instances. We know that Euler's equation is true because every time a human mind derives the equation, it gets the same result, unless it has made a mistake, which can be acknowledged and corrected. We observe this phenomenon, and we extrapolate to the general proposition that it is always true.

This applies not only to physical principles, like the law of gravity, but to abstract phenomena that we observe only in human brains: in ours and in those of others.

Aristotelian realism

Main article: Aristotle's theory of universals

Similar to empiricism in emphasizing the relation of mathematics to the real world, Aristotelian realism holds that mathematics studies properties such as symmetry, continuity and order that can be literally realized in the physical world (or in any other world there might be). It contrasts with Platonism in holding that the objects of mathematics, such as numbers, do not exist in an "abstract" world but can be physically realized. For example, the number 4 is realized in the relation between a heap of parrots and the universal "being a parrot" that divides the heap into so many parrots.[12] Aristotelian realism is defended by James Franklin and the Sydney School in the philosophy of mathematics and is close to the view of Penelope Maddy that when an egg carton is opened, a set of three eggs is perceived (that is, a mathematical entity realized in the physical world).[13] A problem for Aristotelian realism is what account to give of higher infinities, which may not be realizable in the physical world.

6.4.11   Fictionalism

Fictionalism in mathematics was brought to fame in 1980 when Hartry Field published *Science Without Numbers*, which rejected and in fact reversed Quine's indispensability argument. Where Quine suggested that mathematics was indispensable for our best scientific theories, and therefore should be accepted as a body of truths talking about independently existing entities, Field suggested that mathematics was dispensable, and therefore should be considered as a body of falsehoods not talking about anything real. He did this by giving a complete axiomatization of Newtonian mechanics that didn't reference numbers or functions at all. He started with the "betweenness" of Hilbert's axioms to characterize space without coordinatizing it, and then added extra relations between points to do the work formerly done by vector fields. Hilbert's geometry is mathematical, because it talks about abstract points, but in Field's theory, these points are the concrete points of physical space, so no special mathematical objects at all are needed.

Having shown how to do science without using numbers, Field proceeded to rehabilitate mathematics as a kind of useful fiction. He showed that mathematical physics is a conservative extension of his non-mathematical physics (that is, every physical fact provable in mathematical physics is already provable from Field's system), so that mathematics is a reliable process whose physical applications are all true, even though its own statements are false. Thus, when doing mathematics, we can see ourselves as telling a sort of story, talking as if numbers existed. For Field, a statement like "2 + 2 = 4" is just as fictitious as "Sherlock Holmes lived at 221B Baker Street"—but both are true according to the relevant fictions.

By this account, there are no metaphysical or epistemological problems special to mathematics. The only worries left are the general worries about non-mathematical physics, and about fiction in general. Field's approach has been very influential, but is widely rejected. This is in part because of the requirement of strong fragments of second-order logic to carry out his reduction, and because the statement of conservativity seems to require quantification over abstract models or deductions.

6.4.12   Social constructivism or social realism

Social constructivism or *social realism* theories see mathematics primarily as a social construct, as a product of culture, subject to correction and change. Like the other sciences, mathematics is viewed as an empirical endeavor whose results are constantly evaluated and may be discarded. However, while on an empiricist view the evaluation is some sort of comparison with "reality", social constructivists emphasize that the direction of mathematical research is dictated by the fashions of the social group performing it or by the needs of the society financing it. However, although such external forces may change the direction of some mathematical research, there are strong internal constraints—the mathematical traditions, methods, problems, meanings and values into which mathematicians are enculturated—that work to conserve the historically defined discipline.

This runs counter to the traditional beliefs of working mathematicians, that mathematics is somehow pure or objective. But social constructivists argue that mathematics is in fact grounded by much uncertainty: as mathematical practice evolves, the status of previous mathematics is cast into doubt, and is corrected to the degree it is required or desired by the current mathematical community. This can be seen in the development of analysis from reexamination of the calculus of Leibniz and Newton. They argue further that finished mathematics is often accorded too much status, and folk mathematics not enough, due to an overemphasis on axiomatic proof and peer review as practices. However, this might be seen as merely saying that rigorously proven results are overemphasized, and then "look how chaotic and uncertain the rest of it all is!"

The social nature of mathematics is highlighted in its subcultures. Major discoveries can be made in one branch of mathematics and be relevant to another, yet the relationship goes undiscovered for lack of social contact between mathematicians. Social constructivists argue each speciality forms its own epistemic community and often has great difficulty communicating, or motivating the investigation of unifying conjectures that might relate different areas of mathematics. Social constructivists see the process of "doing mathematics" as actually creating the meaning, while social realists see a deficiency either of human capacity to abstractify, or of human's cognitive bias, or of mathematicians' collective intelligence as preventing the comprehension of a real universe of mathematical objects. Social constructivists sometimes reject the search for foundations of mathematics as bound to fail, as pointless or even meaningless.

Contributions to this school have been made by Imre Lakatos and Thomas Tymoczko, although it is not clear that either would endorse the title. More recently Paul Ernest has explicitly formulated a social constructivist philosophy of mathematics.[14] Some consider the work of Paul Erdős as a whole to have advanced this view (although he personally rejected it) because of his uniquely broad collaborations, which prompted others to see and study "mathematics as a social activity", e.g., via the Erdős number. Reuben Hersh has also promoted the social view of mathematics, calling it a "humanistic" approach,[15] similar to but not quite the same as that associated with Alvin White;[16] one of Hersh's co-authors, Philip J. Davis, has expressed sympathy for the social view as well.

A criticism of this approach is that it is trivial, based on the trivial observation that mathematics is a human activity. To observe that rigorous proof comes only after unrigorous conjecture, experimentation and speculation is true, but it is trivial and no-one would deny this. So it's a bit of a stretch to characterize a philosophy of mathematics in this way, on something trivially true. The calculus of Leibniz and Newton was reexamined by mathematicians such as Weierstrass in order to rigorously prove the theorems thereof. There is nothing special or interesting about this, as it fits in with the more general trend of unrigorous ideas which are later made rigorous. There needs to be a clear distinction between the objects of study of mathematics and the study of the objects of study of mathematics. The former doesn't seem to change a great deal; the latter is forever in flux. The latter is what the social theory is about, and the former is what Platonism *et al.* are about.

However, this criticism is rejected by supporters of the social constructivist perspective because it misses the point that the very objects of mathematics are social constructs. These objects, it asserts, are primarily semiotic objects existing in the sphere of human culture, sustained by social practices (after Wittgenstein) that utilize physically embodied signs and give rise to intrapersonal (mental) constructs. Social constructivists view the reification of the sphere of human culture into a Platonic realm, or some other heaven-like domain of existence beyond the physical world, a long-standing category error.

6.4.13   Beyond the traditional schools

Rather than focus on narrow debates about the true nature of mathematical truth, or even on practices unique to mathematicians such as the proof, a growing movement from the 1960s to the 1990s began to question the idea of seeking foundations or finding any one right answer to why mathematics works. The starting point for this was Eugene Wigner's famous 1960 paper *The Unreasonable Effectiveness of Mathematics in the Natural Sciences*, in which he argued that the happy coincidence of mathematics and physics being so well matched seemed to be unreasonable and hard to explain.

The embodied-mind or cognitive school and the social school were responses to this challenge, but the debates raised were difficult to confine to those.

Quasi-empiricism

One parallel concern that does not actually challenge the schools directly but instead questions their focus is the notion of quasi-empiricism in mathematics. This grew from the increasingly popular assertion in the late 20th century that no one foundation of mathematics could be ever proven to exist. It is also sometimes called "postmodernism in mathematics" although that term is considered overloaded by some and insulting by others. Quasi-empiricism argues that in doing their research, mathematicians test hypotheses as well as prove theorems. A mathematical argument can transmit falsity from the conclusion to the premises just as well as it can transmit truth from the premises to the conclusion. Quasi-empiricism was developed by Imre Lakatos, inspired by the philosophy of science of Karl Popper.

Lakatos' philosophy of mathematics is sometimes regarded as a kind of social constructivism, but this was not his intention.

Such methods have always been part of folk mathematics by which great feats of calculation and measurement are sometimes achieved. Indeed, such methods may be the only notion of proof a culture has.

Hilary Putnam has argued that any theory of mathematical realism would include quasi-empirical methods. He proposed that an alien species doing mathematics might well rely on quasi-empirical methods primarily, being willing often to forgo rigorous and axiomatic proofs, and still be doing mathematics—at perhaps a somewhat greater risk of failure of their calculations. He gave a detailed argument for this in *New Directions*.[17]

Popper's "two senses" theory

Realist and constructivist theories are normally taken to be contraries. However, Karl Popper[18] argued that a number statement such as "2 apples + 2 apples = 4 apples" can be taken in two senses. In one sense it is irrefutable and logically true. In the second sense it is factually true and falsifiable. Another way of putting this is to say that a single number statement can express two propositions: one of which can be explained on constructivist lines; the other on realist lines.[19]

Language

Main article: Philosophy of language

Innovations in the philosophy of language during the 20th century renewed interest in whether mathematics is, as is often said, the *language* of science. Although some mathematicians and philosophers would accept the statement "mathematics is a language", linguists believe that the implications of such a statement must be considered. For example, the tools of linguistics are not generally applied to the symbol systems of mathematics, that is, mathematics is studied in a markedly different way than other languages. If mathematics is a language, it is a different type of language than natural languages. Indeed, because of the need for clarity and specificity, the language of mathematics is far more constrained than natural languages studied by linguists. However, the methods developed by Frege and Tarski for the study of mathematical language have been extended greatly by Tarski's student Richard Montague and other linguists working in formal semantics to show that the distinction between mathematical language and natural language may not be as great as it seems.

6.5 Arguments

6.5.1 Indispensability argument for realism

This argument, associated with Willard Quine and Hilary Putnam, is considered by Stephen Yablo to be one of the most challenging arguments in favor of the acceptance of the existence of abstract mathematical entities, such as numbers and sets.[20] The form of the argument is as follows.

1. One must have ontological commitments to *all* entities that are indispensable to the best scientific theories, and to those entities *only* (commonly referred to as "all and only").

2. Mathematical entities are indispensable to the best scientific theories. Therefore,

3. One must have ontological commitments to mathematical entities.[21]

The justification for the first premise is the most controversial. Both Putnam and Quine invoke naturalism to justify the exclusion of all non-scientific entities, and hence to defend the "only" part of "all and only". The assertion that "all" entities postulated in scientific theories, including numbers, should be accepted as real is justified by confirmation holism. Since theories are not confirmed in a piecemeal fashion, but as a whole, there is no justification for excluding any of the entities referred to in well-confirmed theories. This puts the nominalist who wishes to exclude the existence of sets and non-Euclidean geometry, but to include the existence of quarks and other undetectable entities of physics, for example, in a difficult position.[21]

6.5.2 Epistemic argument against realism

The anti-realist "epistemic argument" against Platonism has been made by Paul Benacerraf and Hartry Field. Platonism posits that mathematical objects are *abstract* entities. By general agreement, abstract entities cannot interact causally with concrete, physical entities. ("the truth-values of our mathematical assertions depend on facts involving Platonic entities that reside in a realm outside of space-time"[22]) Whilst our knowledge of concrete, physical objects is based on our ability to perceive them, and therefore to causally interact with them, there is no parallel account of how mathematicians come to have knowledge of abstract objects.[23][24][25] ("An account of mathematical truth ... must be consistent with the possibility of mathematical knowledge."[26]) Another way of making the point is that if the Platonic world were to disappear, it would make no difference to the ability of mathematicians to generate proofs, etc., which is already fully accountable in terms of physical processes in their brains.

Field developed his views into fictionalism. Benacerraf also developed the philosophy of mathematical structuralism, according to which there are no mathematical objects. Nonetheless, some versions of structuralism are compatible with some versions of realism.

The argument hinges on the idea that a satisfactory naturalistic account of thought processes in terms of brain processes can be given for mathematical reasoning along with everything else. One line of defense is to maintain that this is false, so that mathematical reasoning uses some special intuition that involves contact with the Platonic realm. A modern form of this argument is given by Sir Roger Penrose.[27]

Another line of defense is to maintain that abstract objects are relevant to mathematical reasoning in a way that is non-causal, and not analogous to perception. This argument is developed by Jerrold Katz in his book *Realistic Rationalism*.

A more radical defense is denial of physical reality, i.e. the mathematical universe hypothesis. In that case, a mathematician's knowledge of mathematics is one mathematical object making contact with another.

6.6   Aesthetics

Many practicing mathematicians have been drawn to their subject because of a sense of beauty they perceive in it. One sometimes hears the sentiment that mathematicians would like to leave philosophy to the philosophers and get back to mathematics—where, presumably, the beauty lies.

In his work on the divine proportion, H.E. Huntley relates the feeling of reading and understanding someone else's proof of a theorem of mathematics to that of a viewer of a masterpiece of art—the reader of a proof has a similar sense of exhilaration at understanding as the original author of the proof, much as, he argues, the viewer of a masterpiece has a sense of exhilaration similar to the original painter or sculptor. Indeed, one can study mathematical and scientific writings as literature.

Philip J. Davis and Reuben Hersh have commented that the sense of mathematical beauty is universal amongst practicing mathematicians. By way of example, they provide two proofs of the irrationality of the $\sqrt{2}$. The first is the traditional proof by contradiction, ascribed to Euclid; the second is a more direct proof involving the fundamental theorem of arithmetic that, they argue, gets to the heart of the issue. Davis and Hersh argue that mathematicians find the second proof more aesthetically appealing because it gets closer to the nature of the problem.

Paul Erdős was well known for his notion of a hypothetical "Book" containing the most elegant or beautiful mathematical proofs. There is not universal agreement that a result has one "most elegant" proof; Gregory Chaitin has argued against this idea.

Philosophers have sometimes criticized mathematicians' sense of beauty or elegance as being, at best, vaguely stated. By the same token, however, philosophers of mathematics have sought to characterize what makes one proof more desirable than another when both are logically sound.

Another aspect of aesthetics concerning mathematics is mathematicians' views towards the possible uses of mathematics for purposes deemed unethical or inappropriate. The best-known exposition of this view occurs in G.H. Hardy's book *A Mathematician's Apology*, in which Hardy argues that pure mathematics is superior in beauty to applied mathematics precisely because it cannot be used for war and similar ends. Some later mathematicians have characterized Hardy's views as mildly dated, with the applicability of number theory to modern-day cryptography.

6.7   See also

6.7.1   Related works

6.7.2   Historical topics

- History and philosophy of science

- History of mathematics

- History of philosophy

6.8   Notes

[1] Maziars, Edward A. (1969). "Problems in the Philosophy of Mathematics (Book Review)". *Philosophy of Science* 36 (3): 325.. For example, when Edward Maziars proposes in a 1969 book review "*to distinguish philosophical mathematics (which is primarily a specialised task for a mathematician) from mathematical philosophy (which ordinarily may be the philosopher's metier)*", he uses the term *mathematical philosophy* as being synonymous with *philosophy of mathematics*.

[2] Kleene, Stephen (1971). *Introduction to Metamathematics*. Amsterdam, Netherlands: North-Holland Publishing Company. p. 5.

[3] Mac Lane, Saunders (1998), *Categories for the Working Mathematician*, 2nd edition, Springer-Verlag, New York, NY.

[4] • Putnam, Hilary (1967), "Mathematics Without Foundations", *Journal of Philosophy* 64/1, 5-22. Reprinted, pp. 168–184 in W.D. Hart (ed., 1996).

[5] Tegmark, Max (February 2008). "The Mathematical Universe". *Foundations of Physics* 38 (2): 101–150. arXiv:0704.0646. Bibcode:2008FoPh...38..101T. doi:10.1007/s10701-007-9186-9.

[6] Tegmark (1998), p. 1.

[7] Carnap, Rudolf (1931), "Die logizistische Grundlegung der Mathematik", *Erkenntnis* 2, 91-121. Republished, "The Logicist Foundations of Mathematics", E. Putnam and G.J. Massey (trans.), in Benacerraf and Putnam (1964). Reprinted, pp. 41–52 in Benacerraf and Putnam (1983).

[8] Audi, Robert (1999), *The Cambridge Dictionary of Philosophy*, Cambridge University Press, Cambridge, UK, 1995. 2nd edition. Page 542.

[9] From an 1886 lecture at the 'Berliner Naturforscher-Versammlung', according to H. M. Weber's memorial article, as quoted and translated in Gonzalez Cabillon, Julio (2000-02-03). "FOM: What were Kronecker's f.o.m.?". Retrieved 2008-07-19. Gonzalez gives as the sources for the memorial article, the following: 'Weber, H: "Leopold Kronecker", _Jahresberichte der Deutschen Mathematiker Vereinigung_, vol ii (1893) pp 5-31. Cf page 19. See also _Mathematische Annalen_ vol xliii (1893) pp 1-25'.

[10] Brown, James (2008). *Philosophy of Mathematics*. New York: Routledge. ISBN 978-0-415-96047-2.

[11] A System of Logic Ratiocinative and Inductive, The Collected Works of John Stuart Mill published by the University of Toronto Press in 1973. Book II, Chapter vi, Section 2 (Toronto edition 1975, Vol.7, p. 254)

[12] Franklin, James (2014), "An Aristotelian Realist Philosophy of Mathematics", Palgrave Macmillan, Basingstoke; Franklin, James (2011), "Aristotelianism in the philosophy of mathematics," *Studia Neoaristotelica* 8, 3-15.

[13] Maddy, Penelope (1990), *Realism in Mathematics*, Oxford University Press, Oxford, UK.

[14] Ernest, Paul. "Is Mathematics Discovered or Invented?". University of Exeter. Retrieved 2008-12-26.

[15] Hersh, Reuben (February 10, 1997). *What Kind of a Thing is a Number?*. Interview with John Brockman. Edge Foundation. Retrieved 2008-12-26.

[16] "Humanism and Mathematics Education". *Math Forum*. Humanistic Mathematics Network Journal. Retrieved 2008-12-26.

[17] Tymoczko, Thomas (1998), *New Directions in the Philosophy of Mathematics*. ISBN 978-0691034980.

[18] Popper, Karl Raimund (1946) Aristotelian Society Supplementary Volume XX.

[19] Gregory, Frank Hutson (1996) Arithmetic and Reality: A Development of Popper's Ideas. City University of Hong Kong. Republished in Philosophy of Mathematics Education Journal No. 26 (December 2011)

[20] Yablo, S. (November 8, 1998). "A Paradox of Existence".

[21] Putnam, H. *Mathematics, Matter and Method. Philosophical Papers, vol. 1.* Cambridge: Cambridge University Press, 1975. 2nd. ed., 1985.

[22] Field, Hartry, 1989, Realism, Mathematics, and Modality, Oxford: Blackwell, p. 68

[23] "Since abstract objects are outside the nexus of causes and effects, and thus perceptually inaccessible, they cannot be known through their effects on us" Katz, J. *Realistic Rationalism*, p15

[24] ,Philosophy Now: *Mathematical_Knowledge_A_Dilemma Mathematical Knowledge: A dilemma*

[25] Standard Encyclopaedia of Philosophy

[26] Benacceraf, 1973, p409

[27] Review of The Emperor's New Mind

6.9   Further reading

- Aristotle, "Prior Analytics", Hugh Tredennick (trans.), pp. 181–531 in *Aristotle, Volume 1*, Loeb Classical Library, William Heinemann, London, UK, 1938.

- Benacerraf, Paul, and Putnam, Hilary (eds., 1983), *Philosophy of Mathematics, Selected Readings*, 1st edition, Prentice-Hall, Englewood Cliffs, NJ, 1964. 2nd edition, Cambridge University Press, Cambridge, UK, 1983.

- Berkeley, George (1734), *The Analyst; or, a Discourse Addressed to an Infidel Mathematician. Wherein It is examined whether the Object, Principles, and Inferences of the modern Analysis are more distinctly conceived, or more evidently deduced, than Religious Mysteries and Points of Faith*, London & Dublin. Online text, David R. Wilkins (ed.), Eprint.

- Bourbaki, N. (1994), *Elements of the History of Mathematics*, John Meldrum (trans.), Springer-Verlag, Berlin, Germany.

- Chandrasekhar, Subrahmanyan (1987), *Truth and Beauty. Aesthetics and Motivations in Science*, University of Chicago Press, Chicago, IL.

- Colyvan, Mark (2004), "Indispensability Arguments in the Philosophy of Mathematics", *Stanford Encyclopedia of Philosophy*, Edward N. Zalta (ed.), Eprint.

- Davis, Philip J. and Hersh, Reuben (1981), *The Mathematical Experience*, Mariner Books, New York, NY.

- Devlin, Keith (2005), *The Math Instinct: Why You're a Mathematical Genius (Along with Lobsters, Birds, Cats, and Dogs)*, Thunder's Mouth Press, New York, NY.

- Dummett, Michael (1991 a), *Frege, Philosophy of Mathematics*, Harvard University Press, Cambridge, MA.

- Dummett, Michael (1991 b), *Frege and Other Philosophers*, Oxford University Press, Oxford, UK.

- Dummett, Michael (1993), *Origins of Analytical Philosophy*, Harvard University Press, Cambridge, MA.

- Ernest, Paul (1998), *Social Constructivism as a Philosophy of Mathematics*, State University of New York Press, Albany, NY.

- George, Alexandre (ed., 1994), *Mathematics and Mind*, Oxford University Press, Oxford, UK.

- Hadamard, Jacques (1949), *The Psychology of Invention in the Mathematical Field*, 1st edition, Princeton University Press, Princeton, NJ. 2nd edition, 1949. Reprinted, Dover Publications, New York, NY, 1954.

- Hardy, G.H. (1940), *A Mathematician's Apology*, 1st published, 1940. Reprinted, C.P. Snow (foreword), 1967. Reprinted, Cambridge University Press, Cambridge, UK, 1992.

- Hart, W.D. (ed., 1996), *The Philosophy of Mathematics*, Oxford University Press, Oxford, UK.

- Hendricks, Vincent F. and Hannes Leitgeb (eds.). *Philosophy of Mathematics: 5 Questions*, New York: Automatic Press / VIP, 2006.

- Huntley, H.E. (1970), *The Divine Proportion: A Study in Mathematical Beauty*, Dover Publications, New York, NY.

- Irvine, A., ed (2009), *The Philosophy of Mathematics*, in *Handbook of the Philosophy of Science* series, North-Holland Elsevier, Amsterdam.

- Klein, Jacob (1968), *Greek Mathematical Thought and the Origin of Algebra*, Eva Brann (trans.), MIT Press, Cambridge, MA, 1968. Reprinted, Dover Publications, Mineola, NY, 1992.

- Kline, Morris (1959), *Mathematics and the Physical World*, Thomas Y. Crowell Company, New York, NY, 1959. Reprinted, Dover Publications, Mineola, NY, 1981.

- Kline, Morris (1972), *Mathematical Thought from Ancient to Modern Times*, Oxford University Press, New York, NY.

- König, Julius (Gyula) (1905), "Über die Grundlagen der Mengenlehre und das Kontinuumproblem", *Mathematische Annalen* 61, 156-160. Reprinted, "On the Foundations of Set Theory and the Continuum Problem", Stefan Bauer-Mengelberg (trans.), pp. 145–149 in Jean van Heijenoort (ed., 1967).

- Körner, Stephan, *The Philosophy of Mathematics, An Introduction*. Harper Books, 1960.

- Lakoff, George, and Núñez, Rafael E. (2000), *Where Mathematics Comes From: How the Embodied Mind Brings Mathematics into Being*, Basic Books, New York, NY.

- Lakatos, Imre 1976 *Proofs and Refutations: The Logic of Mathematical Discovery* (Eds) J. Worrall & E. Zahar Cambridge University Press

- Lakatos, Imre 1978 *Mathematics, Science and Epistemology: Philosophical Papers* Volume 2 (Eds) J.Worrall & G.Currie Cambridge University Press

- Lakatos, Imre 1968 *Problems in the Philosophy of Mathematics* North Holland

- Leibniz, G.W., *Logical Papers* (1666–1690), G.H.R. Parkinson (ed., trans.), Oxford University Press, London, UK, 1966.

- Maddy, Penelope (1997), *Naturalism in Mathematics*, Oxford University Press, Oxford, UK.

- Maziarz, Edward A., and Greenwood, Thomas (1995), *Greek Mathematical Philosophy*, Barnes and Noble Books.

- Mount, Matthew, *Classical Greek Mathematical Philosophy*, .

- Parsons, Charles (2014). *Philosophy of Mathematics in the Twentieth Century: Selected Essays*. Cambridge, MA: Harvard University Press. ISBN 978-0-674-72806-6.

- Peirce, Benjamin (1870), "Linear Associative Algebra", § 1. See *American Journal of Mathematics* 4 (1881).

- Peirce, C.S., *Collected Papers of Charles Sanders Peirce*, vols. 1-6, Charles Hartshorne and Paul Weiss (eds.), vols. 7-8, Arthur W. Burks (ed.), Harvard University Press, Cambridge, MA, 1931 – 1935, 1958. Cited as CP (volume).(paragraph).

- Peirce, C.S., various pieces on mathematics and logic, many readable online through links at the Charles Sanders Peirce bibliography, especially under Books authored or edited by Peirce, published in his lifetime and the two sections following it.

- Plato, "The Republic, Volume 1", Paul Shorey (trans.), pp. 1–535 in *Plato, Volume 5*, Loeb Classical Library, William Heinemann, London, UK, 1930.

- Plato, "The Republic, Volume 2", Paul Shorey (trans.), pp. 1–521 in *Plato, Volume 6*, Loeb Classical Library, William Heinemann, London, UK, 1935.

- Resnik, Michael D. *Frege and the Philosophy of Mathematics*, Cornell University, 1980.

- Resnik, Michael (1997), *Mathematics as a Science of Patterns*, Clarendon Press, Oxford, UK, ISBN 978-0-19-825014-2

- Robinson, Gilbert de B. (1959), *The Foundations of Geometry*, University of Toronto Press, Toronto, Canada, 1940, 1946, 1952, 4th edition 1959.

- Raymond, Eric S. (1993), "The Utility of Mathematics", Eprint.

- Smullyan, Raymond M. (1993), *Recursion Theory for Metamathematics*, Oxford University Press, Oxford, UK.

- Russell, Bertrand (1919), *Introduction to Mathematical Philosophy*, George Allen and Unwin, London, UK. Reprinted, John G. Slater (intro.), Routledge, London, UK, 1993.

- Shapiro, Stewart (2000), *Thinking About Mathematics: The Philosophy of Mathematics*, Oxford University Press, Oxford, UK

- Strohmeier, John, and Westbrook, Peter (1999), *Divine Harmony, The Life and Teachings of Pythagoras*, Berkeley Hills Books, Berkeley, CA.

- Styazhkin, N.I. (1969), *History of Mathematical Logic from Leibniz to Peano*, MIT Press, Cambridge, MA.

- Tait, William W. (1986), "Truth and Proof: The Platonism of Mathematics", *Synthese* 69 (1986), 341-370. Reprinted, pp. 142–167 in W.D. Hart (ed., 1996).

- Tarski, A. (1983), *Logic, Semantics, Metamathematics: Papers from 1923 to 1938*, J.H. Woodger (trans.), Oxford University Press, Oxford, UK, 1956. 2nd edition, John Corcoran (ed.), Hackett Publishing, Indianapolis, IN, 1983.

- Ulam, S.M. (1990), *Analogies Between Analogies: The Mathematical Reports of S.M. Ulam and His Los Alamos Collaborators*, A.R. Bednarek and Françoise Ulam (eds.), University of California Press, Berkeley, CA.

- van Heijenoort, Jean (ed. 1967), *From Frege To Gödel: A Source Book in Mathematical Logic, 1879-1931*, Harvard University Press, Cambridge, MA.

- Wigner, Eugene (1960), "The Unreasonable Effectiveness of Mathematics in the Natural Sciences", *Communications on Pure and Applied Mathematics* 13(1): 1-14. Eprint

- Wilder, Raymond L. *Mathematics as a Cultural System*, Pergamon, 1980.

6.10 External links

- Philosophy of mathematics at PhilPapers

- Philosophy of mathematics at the Indiana Philosophy Ontology Project

- Philosophy of Mathematics entry by Leon Horsten in the *Stanford Encyclopedia of Philosophy*

- Philosophy of mathematics entry in the *Internet Encyclopedia of Philosophy*

- The London Philosophy Study Guide offers many suggestions on what to read, depending on the student's familiarity with the subject:

    - Philosophy of Mathematics

    - Mathematical Logic

    - Set Theory & Further Logic

- R.B. Jones' philosophy of mathematics page

- Philosophy of mathematics at DMOZ

- The Philosophy of Real Mathematics Blog

- Kaina Stoicheia by C.S. Peirce.

6.10.1 Journals

- Philosophia Mathematica journal

- The Philosophy of Mathematics Education Journal homepage

Chapter 7

Glossary of areas of mathematics

This is a glossary of terms that are or have been considered areas of study in mathematics.

7.1 A

- **Absolute differential calculus** — the original name for tensor calculus developed around 1890.

- **Absolute geometry** — an extension of ordered geometry that is sometimes referred to as *neutral geometry* because its axiom system is neutral to the parallel postulate.

- **Abstract algebra** — the study of algebraic structures and their properties. Originally it was known as *modern algebra*.

- **Abstract analytic number theory** — a branch mathematics that take ideas from classical analytic number theory and applies them to various other areas of mathematics.

- **Abstract differential geometry** — a form of differential geometry without the notion of smoothness from calculus. Instead it is built using sheaf theory and sheaf cohomology.

- **Abstract harmonic analysis** — a modern branch of harmonic analysis that extends upon the generalized Fourier transforms that can be defined on locally compact groups.

- **Abstract homotopy theory**

- **Additive combinatorics** — the part of arithmetic combinatorics devoted to the operations of addition and subtraction.

- **Additive number theory** — a part of number theory that studies subsets of integers and their behaviour under addition.

- **Affine geometry** — a branch of geometry that is centered on the study of geometric properties that remain unchanged by affine transformations. It can be described as a generalization of Euclidean geometry.

- **Affine geometry of curves** — the study of curves in affine space.

- **Affine differential geometry** — a type of differential geometry dedicated to differential invariants under volume-preserving affine transformations.

- **Ahlfors theory** — a part of complex analysis being the geometric counterpart of Nevanlinna theory. It was invented by Lars Ahlfors

- **Algebra** — a major part of pure mathematics centered around operations and relations. Beginning with elementary algebra, it introduces the concept of variables and how these can be manipulated towards problem solving; known as equation solving. Generalizations of operations and relations defined on sets have led to the idea of an algebraic structure which are studied in abstract algebra. Other branches of algebra include universal algebra, linear algebra and multilinear algebra.

- **Algebraic analysis** — motivated by systems of linear partial differential equations, it is a branch of algebraic geometry and algebraic topology that uses methods from sheaf theory and complex analysis, to study the properties and generalizations of functions. It was started by Mikio Sato.

- **Algebraic combinatorics** — an area that employs methods of abstract algebra to problems of combinatorics. It also refers to the application of methods from combinatorics to problems in abstract algebra.

- **Algebraic computation** — see *symbolic computation*.

- **Algebraic geometry** — a branch that combines techniques from abstract algebra with the language and problems of geometry. Fundamentally, it studies algebraic varieties.

- **Algebraic graph theory** — a branch of graph theory in which methods are taken from algebra and employed to problems about graphs. The methods are commonly taken from group theory and linear algebra.

- **Algebraic K-theory** — an important part of homological algebra concerned with defining and applying a certain sequence of functors from rings to abelian groups.

- **Algebraic number theory** — a part of algebraic geometry devoted to the study of the points of the algebraic varieties whose coordinates belong to an algebraic number field. It is a major branch of number theory and is also said to study algebraic structures related to algebraic integers.

- **Algebraic statistics** — the use of algebra to advance statistics, although the term is sometimes restricted to label the use of algebraic geometry and commutative algebra in statistics.

- **Algebraic topology** — a branch that uses tools from abstract algebra for topology to study topological spaces.

- **Algorithmic number theory** — also known as *computational number theory*, it is the study of algorithms for performing number theoretic computations.

- **Anabelian geometry** — an area of study based on the theory proposed by Alexander Grothendieck in the 1980s that describes the way a geometric object of an algebraic variety (such as an algebraic fundamental group) can be mapped into another object, without it being an abelian group.

- **Analysis** — a rigorous branch of pure mathematics that had its beginnings in the formulation of infinitesimal calculus. (Then it was known as *infinitesimal analysis.*) The classical forms of analysis are real analysis and its extension complex analysis, whilst more modern forms are those such as functional analysis.

- **Analytic combinatorics** — part of enumerative combinatorics where methods of complex analysis are applied to generating functions.

- **Analytic geometry** — usually this refer to the study of geometry using a coordinate system (also known as *Cartesian geometry*). Alternatively it can refer to the geometry of analytic varieties. In this respect it is essentially equivalent to real and complex algebraic geometry.

- **Analytic number theory** — part of number theory using methods of analysis (as opposed to algebraic number theory)

- **Applied mathematics** — a combination of various parts of mathematics that concern a variety of mathematical methods that can be applied to practical and theoretical problems. Typically the methods used are for science, engineering, finance, economics and logistics.

- **Approximation theory** — part of analysis that studies how well functions can be approximated by simpler ones (such as polynomials or trigonometric polynomials)

- **Arakelov geometry** — also known as *Arakelov theory*

- **Arakelov theory** — an approach to Diophantine geometry used to study Diophantine equations in higher dimensions (using techniques from algebraic geometry). It is named after Suren Arakelov.

- **Arithmetic** — to most people this refers to the branch known as elementary arithmetic dedicated to the usage of addition, subtraction, multiplication and division. However arithmetic also includes higher arithmetic referring to advanced results from number theory.

- **Arithmetic algebraic geometry** — see *arithmetic geometry*

- **Arithmetic combinatorics** — the study of the estimates from combinatorics that are associated with arithmetic operations such as addition, subtraction, multiplication and division.

- **Arithmetic dynamics**

- **Arithmetic geometry** — the study of schemes of finite type over the spectrum of the ring of integers

- **Arithmetic topology** — a combination of algebraic number theory and topology studying analogies between prime ideals and knots

- **Arithmetical algebraic geometry** — an alternative name for *arithmetic algebraic geometry*

- **Asymptotic combinatorics**

- **Asymptotic geometric analysis**

- **Asymptotic theory** — the study of asymptotic expansions

- **Auslander–Reiten theory** — the study of the representation theory of Artinian rings

- **Axiomatic geometry** — see *synthetic geometry*.

- **Axiomatic homology theory**

- **Axiomatic set theory** — the study of systems of axioms in a context relevant to set theory and mathematical logic.

7.2   B

- **Bifurcation theory** — the study of changes in the qualitative or topological structure of a given family. It is a part of dynamical systems theory

- **Birational geometry** — a part of algebraic geometry that deals with the geometry (of an algebraic variety) that is dependent only on its function field.

- **Bolyai-Lobachevskian geometry** — see *hyperbolic geometry*.

7.3   C

- **C*-algebra theory**

- **Cartesian geometry** — see *analytic geometry*

- **Calculus** — a branch usually associated with limits, functions, derivatives, integrals and infinite series. It forms the basis of classical analysis, and historically was called the *calculus of infinitesimals* or *infinitesimal calculus*. Now it can refer to a system of calculation guided by symbolic manipulation.

- **Calculus of infinitesimals** — also known as *infinitesimal calculus*. It is a branch of calculus built upon the concepts of infinitesimals.

- **Calculus of moving surfaces** — an extension of the theory of tensor calculus to include deforming manifolds.

- **Calculus of variations** — the field dedicated to maximizing or minimizing functionals. It used to be called *functional calculus*.

- **Catastrophe theory** — a branch of bifurcation theory from dynamical systems theory, and also a special case of the more general singularity theory from geometry. It analyses the germs of the catastrophe geometries.

- **Categorical logic** — a branch of category theory adjacent to the mathematical logic. It is based on type theory for intuitionistic logics.

- **Category theory** — the study of the properties of particular mathematical concepts by formalising them as collections of objects and arrows.

- **Chaos theory** — the study of the behaviour of dynamical systems that are highly sensitive to their initial conditions.

- **Character theory** — a branch of group theory that studies the characters of group representations or modular representations.

- **Class field theory** — a branch of algebraic number theory that studies abelian extensions of number fields.

- **Classical differential geometry** — also known as Euclidean differential geometry. see *Euclidean differential geometry*.

- **Classical algebraic topology**

- **Classical analysis** — usually refers to the more traditional topics of analysis such as real analysis and complex analysis. It includes any work that does not use techniques from functional analysis and is sometimes called *hard analysis*. However it may also refer to mathematical analysis done according to the principles of classical mathematics.

- **Classical analytic number theory**

- **Classical differential calculus**

- **Classical Diophantine geometry**

- **Classical Euclidean geometry** — see *Euclidean geometry*

- **Classical geometry** — may refer to solid geometry or classical Euclidean geometry. See *geometry*

- **Classical invariant theory** — the form of invariant theory that deals with describing polynomial functions that are invariant under transformations from a given linear group.

- **Classical mathematics** — the standard approach to mathematics based on classical logic and ZFC set theory.

- **Classical projective geometry**

- **Classical tensor calculus**

- **Clifford analysis** — the study of Dirac operators and Dirac type operators from geometry and analysis using clifford algebras.

- **Clifford theory** is a branch of representation theory spawned from Cliffords theorem.

- **Cobordism theory**

- **Cohomology theory**

- **Combinatorial analysis**

- **Combinatorial commutative algebra** — a discipline viewed as the intersection between commutative algebra and combinatorics. It frequently employs methods from one to address problems arising in the other. Polyhedral geometry also plays a significant role.

- **Combinatorial design theory** — a part of combinatorial mathematics that deals with the existence and construction of systems of fintie sets whose intersections have certain properties.

- **Combinatorial game theory**

- **Combinatorial geometry** — see *discrete geometry*

- **Combinatorial group theory** — the theory of free groups and the presentation of a group. It is closely related to geometric group theory and is applied in geometric topology.

- **Combinatorial mathematics**

- **Combinatorial number theory**

- **Combinatorial set theory** — also known as Infinitary combinatorics. see *infinitary combinatorics*

- **Combinatorial theory**

- **Combinatorial topology** — an old name for algebraic topology, when topological invariants of spaces were regarded as derived from combinatorial decompositions.

- **Combinatorics** — a branch of discrete mathematics concerned with countable structures. Branches of it include enumerative combinatorics, combinatorial design theory, matroid theory, extremal combinatorics and algebraic combinatorics, as well as many more.

- **Commutative algebra** — a branch of abstract algebra studying commutative rings.

- **Complex algebra**

- **Complex algebraic geometry** — the main stream of algebraic geometry devoted to the study of the complex points of algebraic varieties.

- **Complex analysis** — a part of analysis that deals with functions of a complex variable.

- **Complex analytic dynamics** — a subdivision of complex dynamics being the study of the dynamic systems defined by analytic functions.

- **Complex analytic geometry** — the application of complex numbers to plane geometry.

- **Complex diff erential geometry** — a branch of differential geometry that studies complex manifolds.

- **Complex dynamics** — the study of dynamical systems defined by iterated functions on complex number spaces.

- **Complex geometry** — the study of complex manifolds and functions of complex variables. It includes complex algebraic geometry and complex analytic geometry.

- **Complexity theory** — the study of complex systems with the inclusion of the theory of complex systems.

- **Computable analysis** — the study of which parts of real analysis and functional analysis can be carried out in a computable manner. It is closely related to constructive analysis.

- **Computable model theory** — a branch of model theory dealing with the relevant questions computability.

- **Computability theory** — a branch of mathematical logic originating in the 1930s with the study of computable functions and Turing degrees, but now includes the study of generalized computability and definability. It overlaps with proof theory and effective descriptive set theory.

- **Computational algebraic geometry**

- **Computational complexity theory** — a branch of mathematics and theoretical computer science that focuses on classifying computational problems according to their inherent difficulty, and relating those classes to each other.

- **Computational geometry**

- **Computational group theory** — the study of groups by means of computers.

- **Computational mathematics** — the mathematical research in areas of science where computing plays an essential role.

- **Computational number theory** — also known as *algorithmic number theory*, it is the study of algorithms for performing number theoretic computations.

- **Computational real algebraic geometry**

- **Computational synthetic geometry**

- **Computational topology**

- **Computer algebra** — see *symbolic computation*

- **Conformal geometry** — the study of conformal transformations on a space.

- **Constructive analysis** — mathematical analysis done according to the principles of constructive mathematics. This differs from *classical analysis*.

- **Constructive function theory** — a branch of analysis that is closely related to approximation theory, studying the connection between the smoothness of a function and its degree of approximation

- **Constructive mathematics** — mathematics which tends to use intuitionistic logic. Essentially that is classical logic but without the assumption that the law of the excluded middle is an axiom.

- **Constructive quantum field theory** — a branch of mathematical physics that is devoted to showing that quantum theory is mathematically compatible with special relativity.

- **Constructive set theory**

- **Contact geometry** — a branch of differential geometry and topology, closely related to and considered the odd-dimensional counterpart of symplectic geometry. It is the study of a geometric structure called a contact structure on a differentiable manifold.

- **Convex analysis** — the study of properties of convex functions and convex sets.

- **Convex geometry** — part of geometry devoted to the study of convex sets.

- **Coordinate geometry** — see *analytic geometry*

- **CR geometry** — a branch of differential geometry, being the study of CR manifolds.

7.4   D

- **Derived noncommutative algebraic geometry**

- **Descriptive set theory** — a part of mathematical logic, more specifically a part of set theory dedicated to the study of Polish spaces.

- **Differential algebraic geometry** — the adaption of methods and concepts from algebraic geometry to systems of algebraic differential equations.

- **Differential calculus** — a subfield of calculus concerned with derivatives or the rates that quantities change. It is one of two traditional divisions of calculus, the other being integral calculus.

- **Differential Galois theory** — the study of the Galois groups of differential fields.

- **Differential geometry** — a form of geometry that uses techniques from integral and differential calculus as well as linear and multilinear algebra to study problems in geometry. Classically, these were problems of Euclidean geometry, although now it has been expanded. It is generally concerned with geometric structures on differentiable manifolds. It is closely related to differential topology.

- **Differential geometry of curves** — the study of smooth curves in Euclidean space by using techniques from differential geometry

- **Differential geometry of surfaces** — the study of smooth surfaces with various additional structures using the techniques of differential geometry.

- **Differential topology** — a branch of topology that deals with differentiable functions on differentiable manifolds.

- **Diffiety theory**

- **Diophantine geometry** — in general the study of algebraic varieties over fields that are finitely generated over their prime fields.

- **Discrepancy theory**

- **Discrete computational geometry**

- **Discrete differential geometry**

- **Discrete dynamics**

- **Discrete exterior calculus**

- **Discrete geometry**

- **Discrete mathematics**

- **Discrete Morse theory** — a combinatorial adaption of Morse theory.

- **Distance geometry**

- **Domain theory**

- **Donaldson theory** — the study of smooth 4-manifolds using gauge theory.

- **Dynamical systems theory**

7.5 E

- **Econometrics** — the application of mathematical and statistical methods to economic data.

- **Effective descriptive set theory** — a branch of descriptive set theory dealing with set of real numbers that have lightface definitions. It uses aspects of computability theory.

- **Elementary algebra** — a fundamental form of algebra extending on elementary arithmetic to include the concept of variables.

- **Elementary arithmetic** — the simplified portion of arithmetic considered necessary for primary education. It includes the usage addition, subtraction, multiplication and division of the natural numbers. It also includes the concept of fractions and negative numbers.

- **Elementary mathematics** — parts of mathematics frequently taught at the primary and secondary school levels. This includes elementary arithmetic, geometry, probability and statistics, elementary algebra and trigonometry. (calculus is not usually considered a part)

- **Elementary group theory** — the study of the basics of group theory

- **Elimination theory** — the classical name for algorithmic approaches to eliminating between polynomials of several variables. It is a part of commutative algebra and algebraic geometry.

- **Elliptic geometry** — a type of non-Euclidean geometry (it violates Euclid's parallel postulate) and is based on spherical geometry. It is constructed in elliptic space.

- **Enumerative combinatorics** — an area of combinatorics that deals with the number of ways that certain patterns can be formed.

- **Enumerative geometry** — a branch of algebraic geometry concerned with counting the number of solutions to geometric questions. This is usually done by means of intersection theory.

- **Equivariant noncommutative algebraic geometry**

- **Ergodic Ramsey theory** — a branch where problems are motivated by additive combinatorics and solved using ergodic theory.

- **Ergodic theory** — the study of dynamical systems with an invariant measure, and related problems.

- **Euclidean geometry**

- **Euclidean differential geometry** — also known as *classical differential geometry*. See *differential geometry*.

- **Euler calculus**

- **Experimental mathematics**

- **Extraordinary cohomology theory**

- **Extremal combinatorics** — a branch of combinatorics, it is the study of the possible sizes of a collection of finite objects given certain restrictions.

- **Extremal graph theory**

7.6 F

- **Field theory** — branch of abstract algebra studying fields.

- **Finite geometry**

- **Finite model theory**

- **Finsler geometry** — a branch of differential geometry whose main object of study is the Finsler manifold (a generalisation of a Riemannian manifold).

- **First order arithmetic**

- **Fourier analysis**

- **Fractional calculus** — a branch of analysis that studies the possibility of taking real or complex powers of the differentiation operator.

- **Fractional dynamics** — investigates the behaviour of objects and systems that are described by differentiation and integration of fractional orders using methods of fractional calculus.

- **Fredholm theory** — part of spectral theory studying integral equations.

- **Function theory** — part of analysis devoted to properties of functions, especially functions of a complex variable (see *complex analysis*).

- **Functional analysis**

- **Functional calculus** — historically the term was used synonymously with calculus of variations, but now refers to a branch of functional analysis connected with spectral theory

- **Fuzzy arithmetic**

- **Fuzzy geometry**

- **Fuzzy Galois theory**

- **Fuzzy mathematics** — a branch of mathematics based on fuzzy set theory and fuzzy logic.

- **Fuzzy measure theory**

- **Fuzzy qualitative trigonometry**

- **Fuzzy set theory** — a form of set theory that studies fuzzy sets, that is sets that have degrees of membership.

- **Fuzzy topology**

7.7   G

- **Galois cohomology** — an application of homological algebra, it is the study of group cohomology of Galois modules.

- **Galois theory** — named after Évariste Galois, it is a branch of abstract algebra providing a connection between field theory and group theory.

- **Galois geometry** — a branch of finite geometry concerned with algebraic and analytic geometry over a Galois field.

- **Game theory**

- **Gauge theory**

- **General topology** — also known as *point-set topology*, it is a branch of topology studying the properties of topological spaces and structures defined on them. It differs from other branches of topology as the topological spaces do not have to be similar to manifolds.

- **Generalized trigonometry** — developments of trigonometric methods from the application to real numbers of Euclidean geometry to any geometry or space. This includes spherical trigonometry, hyperbolic trigonometry, gyrotrigonometry, rational trigonometry, universal hyperbolic trigonometry, fuzzy qualitative trigonometry, operator trigonometry and lattice trigonometry.

- **Geometric algebra** — an alternative approach to classical, computational and relativistic geometry. It shows a natural correspondence between geometric entities and elements of algebra.

- **Geometric analysis** — a discipline that uses methods from differential geometry to study partial differential equations as well as the applications to geometry.

- **Geometric calculus**

- **Geometric combinatorics**

- **Geometric function theory** — the study of geometric properties of analytic functions.

- **Geometric homology theory**

- **Geometric invariant theory**

- Geometric graph theory

- Geometric group theory

- Geometric measure theory

- Geometric topology — a branch of topology studying manifolds and mappings between them; in particular the embedding of one manifold into another.

- Geometry— a branch of mathematics concerned with shape and the properties of space. Classically it arose as what is now known as solid geometry; this was concerning practical knowledge of length, area and volume. It was then put into an axiomatic form by Euclid, giving rise to what is now known as classical Euclidean geometry. The use of coordinates by René Descartes gave rise to Cartesian geometry enabling a more analytical approach to geometric entities. Since then many other branches have appeared including projective geometry, differential geometry, non-Euclidean geometry, Fractal geometry and algebraic geometry. Geometry also gave rise to the modern discipline of topology.

- Geometry of numbers — initiated by Hermann Minkowski, it is a branch of number theory studying convex bodies and integer vectors.

- Global analysis — the study of differential equations on manifolds and the relationship between differential equations and topology.

- Global arithmetic dynamics

- Graph theory — a branch of discrete mathematics devoted to the study of graphs. It has many applications in physical, biological and social systems.

- Group-character theory — the part of character theory dedicated to the study of characters of group representations.

- Group representation theory

- Group theory

- Gyrotrigonometry — a form of trigonometry used in gyrovector space for hyperbolic geometry. (An analogy of the vector space in Euclidean geometry.)

7.8   H

- Hard analysis — see *classical analysis*

- Harmonic analysis — part of analysis concerned with the representations of functions in terms of waves. It generalizes the notions of Fourier series and Fourier transforms from the Fourier analysis.

- High-dimensional topology

- Higher arithmetic

- Higher category theory

- Higher-dimensional algebra

- Hodge theory

- Holomorphic functional calculus — a branch of functional calculus starting with holomorphic functions.

- Homological algebra — the study of homology in general algebraic settings.

- Homology theory

- **Homotopy theory**

- **Hyperbolic geometry** — also known as *Lobachevskian geometry* or *Bolyai-Lobachevskian geometry*. It is a non-Euclidean geometry looking at hyperbolic space.

- **hyperbolic trigonometry** — the study of hyperbolic triangles in hyperbolic geometry, or hyperbolic functions in Euclidean geometry. Other forms include gyrotrigonometry and universal hyperbolic trigonometry.

- **Hypercomplex analysis**

- **Hyperfunction theory**

7.9   I

- **Ideal theory** — once the precursor name for what is now known as commutative algebra; it is the theory of ideals in commutative rings.

- **Idempotent analysis**

- **Incidence geometry** — the study of relations of incidence between various geometric objects, like curves and lines.

- **Inconsistent mathematics** — see *paraconsistent mathematics*.

- **Infinitary combinatorics** — an expansion of ideas in combinatorics to account for infinite sets.

- **Infinitesimal analysis** — once a synonym for *infinitesimal calculus*

- **Infinitesimal calculus** — see *calculus of infinitesimals*

- **Information geometry**

- **Integral calculus**

- **Integral geometry**

- **Intersection theory** — a branch of algebraic geometry and algebraic topology

- **Intuitionistic type theory**

- **Invariant theory** — studies how group actions on algebraic varieties affect functions.

- **Inversive geometry** — the study of invariants preserved by a type of transformation known as inversion

- **Inversive plane geometry** — inversive geometry that is limited to two dimensions

- **Inversive ring geometry**

- **Itō calculus**

- **Iwasawa theory**

7.10   J

7.11   K

- **K-theory** — originated as the study of a ring generated by vector bundles over a topological space or scheme. In algebraic topology it is an extraordinary cohomology theory known as topological K-theory. In algebra and algebraic geometry it is referred to as algebraic K-theory. In physics, K-theory has appeared in type II string theory. (In particular twisted K-theory.)

- **K-homology**

- **Kähler geometry** — a branch of differential geometry, more specifically a union of Riemannian geometry, complex differential geometry and symplectic geometry. It is the study of Kähler manifolds. (named after Erich Kähler)

- **KK-theory**

- **Klein geometry**

- **Knot theory** — part of topology dealing with knots

- **Kummer theory**

7.12   L

- **L-theory**

- **Large deviations theory** — part of probability theory studying events of small probability (tail events).

- **Large sample theory** — also known as *asymptotic theory*

- **Lattice theory** — the study of lattices, being important in order theory and universal algebra

- **Lattice trigonometry**

- **Lie algebra theory**

- **Lie group theory**

- **Lie sphere geometry**

- **Lie theory**

- **Line geometry**

- **Linear algebra** – a branch of algebra studying linear spaces and linear maps. It has applications in fields such as abstract algebra and functional analysis; it can be represented in analytic geometry and it is generalized in operator theory and in module theory. Sometimes matrix theory is considered a branch, although linear algebra is restricted to only finite dimensions. Extensions of the methods used belong to multilinear algebra.

- **Linear functional analysis**

- **Local algebra** — a term sometimes applied to the theory of local rings.

- **Local arithmetic dynamics** — also known as *p-adic dynamics* or *nonarchimedean dynamics*.

- **Local class field theory**

- **Low-dimensional topology**

7.13   M

- **Malliavin calculus**

- **Mathematical logic**

- **Mathematical physics** — a part of mathematics that develops mathematical methods motivated by problems in physics.

- **Mathematical sciences** — refers to academic disciplines that are mathematical in nature, but are not considered proper subfields of mathematics. Examples include statistics, cryptography, game theory and actuarial science.

- **Matrix algebra**

- **Matrix calculus**

- **Matrix theory**

- **Matroid theory**

- **Measure theory**

- **Metric geometry**

- **Microlocal analysis**

- **Model theory**

- **Modern algebra** — see *abstract algebra*

- **Modern algebraic geometry** — the form of algebraic geometry given by Alexander Grothendieck and Jean-Pierre Serre drawing on sheaf theory.

- **Modern invariant theory** — the form of invariant theory that analyses the decomposition of representations into irreducibles.

- **Modular representation theory**

- **Module theory**

- **Molecular geometry**

- **Morse theory** — a part of differential topology, it analyzes the topological space of a manifold by studying differentiable functions on that manifold.

- **Motivic cohomology**

- **Multilinear algebra** — an extension of linear algebra building upon concepts of p-vectors and multivectors with Grassmann algebra.

- **Multivariable calculus**

- **Multiplicative number theory** — a subfield of analytic number theory that deals with prime numbers, factorization and divisors.

- **Multiple-scale analysis**

7.14   N

- Neutral geometry — see *absolute geometry*

- Nevanlinna theory — part of complex analysis studying the value distribution of meromorphic functions. It is named after Rolf Nevanlinna

- Nielsen theory — an area of mathematical research with its origins in fixed point topology, developed by Jakob Nielsen

- Non-abelian class field theory

- Non-classical analysis

- Non-Euclidean geometry

- Non-standard analysis

- Non-standard calculus

- Nonarchimedean dynamics — also known as *p-adic analysis* or *local arithmetic dynamics*

- Noncommutative algebraic geometry — a direction in noncommutative geometry studying the geometric properties of formal duals of non-commutative algebraic objects.

- Noncommutative geometry

- Noncommutative harmonic analysis — see *representation theory*

- Noncommutative topology

- Nonlinear analysis

- Nonlinear functional analysis

- Number theory — a branch of pure mathematics primarily devoted to the study of the integers. Originally it was known as *arithmetic* or *higher arithmetic*.

- Numerical analysis

- Numerical geometry

- Numerical linear algebra

7.15   O

- Operad theory — a type of abstract algebra concerned with prototypical algebras.

- Operator geometry

- Operator K-theory

- Operator theory — part of functional analysis studying operators.

- Operator trigonometry

- Optimal control theory — a generalization of the calculus of variations.

- Orbifold theory

- Order theory — a branch that investigates the intuitive notion of order using binary relations.

- Ordered geometry — a form of geometry omitting the notion of measurement but featuring the concept of intermediacy. It is a fundamental geometry forming a common framework for affine geometry, Euclidean geometry, absolute geometry and hyperbolic geometry.

- **Oriented elliptic geometry**

- **Oriented spherical geometry**

7.16   P

- p-adic analysis — a branch of number theory that deals with the analysis of functions of p-adic numbers.

- p-adic dynamics — an application of p-adic analysis looking at p-adic differential equations.

- **p-adic Hodge theory**

- **Parabolic geometry**

- Paraconsistent mathematics — sometimes called *inconsistent mathematics*, it is an attempt to develop the classical infrastructure of mathematics based on a foundation of paraconsistent logic instead of classical logic.

- **Partition theory**

- **Perturbation theory**

- **Picard–Vessiot theory**

- **Plane geometry**

- Point-set topology — see *general topology*

- **Pointless topology**

- **Poisson geometry**

- Polyhedral combinatorics — a branch within combinatorics and discrete geometry that studies the problems of describing convex polytopes.

- **Polyhedral geometry**

- **Possibility theory**

- **Potential theory**

- **Precalculus**

- **Predicative mathematics**

- **Probability theory**

- **Probabilistic combinatorics**

- **Probabilistic graph theory**

- **Probabilistic number theory**

- Projective geometry — a form of geometry that studies geometric properties that are invariant under a projective transformation.

- **Projective differential geometry**

- **Proof theory**

- Pseudo-Riemannian geometry — generalizes Riemannian geometry to the study of pseudo-Riemannian manifolds.

- Pure mathematics — the part of mathematics that studies entirely abstract concepts.

7.17   Q

- **Quantum calculus** — a form of calculus without the notion of limits. There are 2 forms known as q-calculus and h-calculus

- **Quantum geometry** — the generalization of concepts of geometry used to describe the physical phenomena of quantum physics

- **Quaternionic analysis**

7.18   R

- **Ramsey theory** — the study of the conditions in which order must appear. It is named after Frank P. Ramsey.

- **Rational geometry**

- **Rational trigonometry** — a reformulation of trigonometry in terms of spread and quadrance instead of angle and length.

- **Real algebra** — the study of the part of algebra relevant to real algebraic geometry.

- **Real algebraic geometry** — the part of algebraic geometry that studies real points of the algebraic varieties.

- **Real analysis** — a branch of mathematical analysis; in particular *hard analysis*, that is the study of real numbers and functions of Real values. It provides a rigorous formulation of the calculus of real numbers in terms of continuity and smoothness, whilst the theory is extended to the complex numbers in complex analysis.

- **Real analytic geometry**

- **Real K-theory**

- **Recreational mathematics** — the area dedicated to mathematical puzzles and mathematical games.

- **Recursion theory** — see *computability theory*

- **Representation theory** — a subfield of abstract algebra; it studies algebraic structures by representing their elements as linear transformations of vector spaces. It also studies modules over these algebraic structures, providing a way of reducing problems in abstract algebra to problems in linear algebra.

- **Representation theory of algebraic groups**

- **Representation theory of algebras**

- **Representation theory of diff eomorphism groups**

- **Representation theory of finite groups**

- **Representation theory of groups**

- **Representation theory of Hopf algebras**

- **Representation theory of Lie algebras**

- **Representation theory of Lie groups**

- **Representation theory of the Galilean group**

- **Representation theory of the Lorentz group**

- **Representation theory of the Poincaré group**

- **Representation theory of the symmetric group**

- **Ribbon theory** — a branch of topology studying ribbons.

- **Riemannian geometry** — a branch of differential geometry that is more specifically, the study of Riemannian manifolds. It is named after Bernhard Riemann and it features many generalizations of concepts from Euclidean geometry, analysis and calculus.

- **Rough set theory** — the a form of set theory based on rough sets.

7.19   S

- **Scheme theory** — the study of schemes introduced by Alexander Grothendieck. It allows the use of sheaf theory to study algebraic varieties and is considered the central part of *modern algebraic geometry*.

- **Secondary calculus**

- **Semialgebraic geometry** — a part of algebraic geometry; more specifically a branch of real algebraic geometry that studies semialgebraic sets.

- **Set-theoretic topology**

- **Set theory**

- **Sheaf theory**

- **Sheaf cohomology**

- **Sieve theory**

- **Single operator theory** — deals with the properties and classifications of single operators.

- **Singularity theory** — a branch, notably of geometry; that studies the failure of manifold structure.

- **Smooth infinitesimal analysis** — a rigorous reformation of infinitesimal calculus employing methods of category theory. As a theory, it is a subset of synthetic differential geometry.

- **Solid geometry**

- **Spatial geometry**

- **Spectral geometry** — a field that concerns the relationships between geometric structures of manifolds and spectra of canonically defined differential operators.

- **Spectral graph theory** — the study of properties of a graph using methods from matrix theory.

- **Spectral theory** — part of *operator theory* extending the concepts of eigenvalues and eigenvectors from linear algebra and matrix theory.

- **Spectral theory of ordinary differential equations** — part of spectral theory concerned with the spectrum and eigenfunction expansion associated with linear ordinary differential equations.

- **Spectrum continuation analysis** — generalizes the concept of a Fourier series to non-periodic functions.

- **Spherical geometry** — a branch of non-Euclidean geometry, studying the 2-dimensional surface of a sphere.

- **Spherical trigonometry** — a branch of spherical geometry that studies polygons on the surface of a sphere. Usually the polygons are triangles.

- **Statistics** — although the term may refer to the more general study of statistics, the term is used in mathematics to refer to the mathematical study of statistics and related fields. This includes probability theory.

- Stochastic calculus

- Stochastic calculus of variations

- Stochastic geometry — the study of random patterns of points

- Stratified Morse theory

- Super category theory

- Super linear algebra

- Surgery theory — a part of geometric topology referring to methods used to produce one manifold from another (in a controlled way.)

- Symbolic computation — also known as *algebraic computation* and *computer algebra*. It refers to the techniques used to manipulate mathematical expressions and equations in symbolic form as opposed to manipulating them by the numerical quantities represented by them.

- Symbolic dynamics

- Symmetric function theory

- Symplectic geometry — a branch of differential geometry and topology whose main object of study is the symplectic manifold.

- Symplectic topology

- Synthetic differential geometry — a reformulation of differential geometry in the language of topos theory and in the context of an intuitionistic logic.

- Synthetic geometry — also known as *axiomatic geometry*, it is a branch of geometry that uses axioms and logical arguments to draw conclusions as opposed to analytic and algebraic methods.

- Systolic geometry — a branch of differential geometry studying systolic invariants of manifolds and polyhedra.

- Systolic hyperbolic geometry — the study of systoles in hyperbolic geometry.

7.20   T

- Tensor analysis — the study of tensors, which play a role in subjects such as differential geometry, mathematical physics, algebraic topology, multilinear algebra, homological algebra and representation theory.

- Tensor calculus — an older term for *tensor analysis*.

- Tensor theory — an alternative name for *tensor analysis*.

- Theoretical physics — a branch primarily of the science physics that uses mathematical models and abstraction of physics to rationalize and predict phenomena.

- Time-scale calculus

- Topology

- Topological combinatorics — the application of methods from algebraic topology to solve problems in combinatorics.

- Topological degree theory

- Topological fixed point theory

- Topological graph theory

- Topological K-theory

- Topos theory

- Toric geometry

- Transcendence theory — a branch of number theory that revolves around the transcendental numbers.

- Transfinite order theory

- Transformation geometry

- Trigonometry — the study of triangles and the relationships between the length of their sides, and the angles between them. It is essential to many parts of applied mathematics.

- Tropical analysis — see *idempotent analysis*

- Tropical geometry

- Twisted K-theory — a variation on K-theory, spanning abstract algebra, algebraic topology and operator theory.

- Type theory

7.21 U

- Umbral calculus — the study of Sheffer sequences

- Uncertainty theory— a new branch of mathematics based on normality, monotonicity, self-duality, countable subadditivity, and product measure axioms.

- Unitary representation theory

- Universal algebra — a field studying the formalization of algebraic structures itself.

- Universal hyperbolic trigonometry — an approach to hyperbolic trigonometry based on rational geometry.

7.22 V

- Valuation theory

- Variational analysis

- Vector algebra — a part of linear algebra concerned with the operations of vector addition and scalar multiplication, although it may also refer to vector operations of vector calculus, including the dot and cross product. In this case it can be contrasted with geometric algebra which generalizes into higher dimensions.

- Vector analysis — also known as vector calculus, see *vector calculus*.

- Vector calculus — a branch of multivariable calculus concerned with differentiation and integration of vector fields. Primarily it is concerned with 3-dimensional Euclidean space.

7.23 W

- Wavelet

- Windowed Fourier transform

- Window function

Glossary of engineering

Chapter 8

Arithmetic

For the song by Brooke Fraser, see Arithmetic (song).

Arithmetic or arithmetics (from the Greek ἀριθμός, *arithmos*, "number") is the oldest[1] and most elementary branch of mathematics. It consists of the study of numbers, especially the properties of the traditional operations between them—addition, subtraction, multiplication and division. Arithmetic is an elementary part of number theory, and number theory is considered to be one of the top-level divisions of modern mathematics, along with algebra, geometry, and analysis. The terms *arithmetic* and *higher arithmetic* were used until the beginning of the 20th century as synonyms for *number theory* and are sometimes still used to refer to a wider part of number theory.[2]

8.1   History

The prehistory of arithmetic is limited to a small number of artifacts which may indicate the conception of addition and subtraction, the best-known being the Ishango bone from central Africa, dating from somewhere between 20,000 and 18,000 BC, although its interpretation is disputed.[3]

The earliest written records indicate the Egyptians and Babylonians used all the elementary arithmetic operations as early as 2000 BC. These artifacts do not always reveal the specific process used for solving problems, but the characteristics of the particular numeral system strongly influence the complexity of the methods. The hieroglyphic system for Egyptian numerals, like the later Roman numerals, descended from tally marks used for counting. In both cases, this origin resulted in values that used a decimal base but did not include positional notation. Complex calculations with Roman numerals required the assistance of a counting board or the Roman abacus to obtain the results.

Early number systems that included positional notation were not decimal, including the sexagesimal (base 60) system for Babylonian numerals and the vigesimal (base 20) system that defined Maya numerals. Because of this place-value concept, the ability to reuse the same digits for different values contributed to simpler and more efficient methods of calculation.

The continuous historical development of modern arithmetic starts with the Hellenistic civilization of ancient Greece, although it originated much later than the Babylonian and Egyptian examples. Prior to the works of Euclid around 300 BC, Greek studies in mathematics overlapped with philosophical and mystical beliefs. For example, Nicomachus summarized the viewpoint of the earlier Pythagorean approach to numbers, and their relationships to each other, in his *Introduction to Arithmetic*.

Greek numerals were used by Archimedes, Diophantus and others in a positional notation not very different from ours. Because the ancient Greeks lacked a symbol for zero (until the Hellenistic period), they used three separate sets of symbols. One set for the unit's place, one for the ten's place, and one for the hundred's. Then for the thousand's place they would reuse the symbols for the unit's place, and so on. Their addition algorithm was identical to ours, and their multiplication algorithm was only very slightly different. Their long division algorithm was the same, and the square root algorithm that was once taught in school was known to Archimedes, who may have invented it. He preferred it to Hero's method of successive approximation because, once computed, a digit doesn't change, and the square roots of perfect squares, such as 7485696, terminate immediately as 2736. For numbers with a fractional part, such as 546.934, they used

83

negative powers of 60 instead of negative powers of 10 for the fractional part 0.934.[4] The ancient Chinese used a similar positional notation. Because they also lacked a symbol for zero, they had one set of symbols for the unit's place, and a second set for the ten's place. For the hundred's place they then reused the symbols for the unit's place, and so on. Their symbols were based on the ancient counting rods. It is a complicated question to determine exactly when the Chinese started calculating with positional representation, but it was definitely before 400 BC.[5] The Bishop of Syria, Severus Sebokht (650 AD), "Indians possess a method of calculation that no word can praise enough. Their rational system of mathematics, or of their method of calculation. I mean the system using nine symbols."[6]

Leonardo of Pisa (Fibonacci) in 1200 AD wrote in *Liber Abaci* "The method of the Indians (Modus Indoram) surpasses any known method to compute. It's a marvelous method. They do their computations using nine figures and symbol zero".[7]

The gradual development of Hindu–Arabic numerals independently devised the place-value concept and positional notation, which combined the simpler methods for computations with a decimal base and the use of a digit representing 0. This allowed the system to consistently represent both large and small integers. This approach eventually replaced all other systems. In the early 6th century AD, the Indian mathematician Aryabhata incorporated an existing version of this system in his work, and experimented with different notations. In the 7th century, Brahmagupta established the use of 0 as a separate number and determined the results for multiplication, division, addition and subtraction of zero and all other numbers, except for the result of division by 0. His contemporary, the Syriac bishop Severus Sebokht described the excellence of this system as "... valuable methods of calculation which surpass description". The Arabs also learned this new method and called it *hesab*.

Although the Codex Vigilanus described an early form of Arabic numerals (omitting 0) by 976 AD, Fibonacci was primarily responsible for spreading their use throughout Europe after the publication of his book *Liber Abaci* in 1202. He considered the significance of this "new" representation of numbers, which he styled the "Method of the Indians" (Latin *Modus Indorum*), so fundamental that all related mathematical foundations, including the results of Pythagoras and the algorism describing the methods for performing actual calculations, were "almost a mistake" in comparison.

In the Middle Ages, arithmetic was one of the seven liberal arts taught in universities.

The flourishing of algebra in the medieval Islamic world and in Renaissance Europe was an outgrowth of the enormous simplification of computation through decimal notation.

Various types of tools exist to assist in numeric calculations. Examples include slide rules (for multiplication, division, and trigonometry) and nomographs in addition to the electrical calculator.

8.2   Arithmetic operations

See also: Algebraic operation

The basic arithmetic operations are addition, subtraction, multiplication and division, although this subject also includes more advanced operations, such as manipulations of percentages, square roots, exponentiation, and logarithmic functions. Arithmetic is performed according to an order of operations. Any set of objects upon which all four arithmetic operations (except division by 0) can be performed, and where these four operations obey the usual laws, is called a field.[8]

8.2.1   Addition (+)

Main article: Addition

Addition is the basic operation of arithmetic. In its simplest form, addition combines two numbers, the *addends* or *terms*, into a single number, the *sum* of the numbers (Such as 2 + 2 = 4 or 3 + 5 = 8).

Adding more than two numbers can be viewed as repeated addition; this procedure is known as summation and includes ways to add infinitely many numbers in an infinite series; repeated addition of the number 1 is the most basic form of counting.

Addition is commutative and associative so the order the terms are added in does not matter. The identity element of addition (the additive identity) is 0, that is, adding 0 to any number yields that same number. Also, the inverse element of addition (the additive inverse) is the opposite of any number, that is, adding the opposite of any number to the number itself yields the additive identity, 0. For example, the opposite of 7 is −7, so 7 + (−7) = 0.

Addition can be given geometrically as in the following example:

> If we have two sticks of lengths 2 and 5, then if we place the sticks one after the other, the length of the stick thus formed is 2 + 5 = 7.

8.2.2   Subtraction (−)

Main article: Subtraction
See also: Method of complements

Subtraction is the inverse of addition. Subtraction finds the *difference* between two numbers, the *minuend* minus the *subtrahend*. If the minuend is larger than the subtrahend, the difference is positive; if the minuend is smaller than the subtrahend, the difference is negative; if they are equal, the difference is 0.

Subtraction is neither commutative nor associative. For that reason, it is often helpful to look at subtraction as addition of the minuend and the opposite of the subtrahend, that is $a - b = a + (-b)$. When written as a sum, all the properties of addition hold.

There are several methods for calculating results, some of which are particularly advantageous to machine calculation. For example, digital computers employ the method of two's complement. Of great importance is the counting up method by which change is made. Suppose an amount P is given to pay the required amount Q, with P greater than Q. Rather than performing the subtraction $P - Q$ and counting out that amount in change, money is counted out starting at Q and continuing until reaching P. Although the amount counted out must equal the result of the subtraction $P - Q$, the subtraction was never really done and the value of $P - Q$ might still be unknown to the change-maker.

8.2.3   Multiplication (× or · or *)

Main article: Multiplication

Multiplication is the second basic operation of arithmetic. Multiplication also combines two numbers into a single number, the *product*. The two original numbers are called the *multiplier* and the *multiplicand*, sometimes both simply called *factors*.

Multiplication may be viewed as a scaling operation. If the numbers are imagined as lying in a line, multiplication by a number, say x, greater than 1 is the same as stretching everything away from 0 uniformly, in such a way that the number 1 itself is stretched to where x was. Similarly, multiplying by a number less than 1 can be imagined as squeezing towards 0. (Again, in such a way that 1 goes to the multiplicand.)

Multiplication is commutative and associative; further it is distributive over addition and subtraction. The multiplicative identity is 1, that is, multiplying any number by 1 yields that same number. Also, the multiplicative inverse is the reciprocal of any number (except 0; 0 is the only number without a multiplicative inverse), that is, multiplying the reciprocal of any number by the number itself yields the multiplicative identity.

The product of a and b is written as $a \times b$ or $a \cdot b$. When a or b are expressions not written simply with digits, it is also written by simple juxtaposition: ab. In computer programming languages and software packages in which one can only use characters normally found on a keyboard, it is often written with an asterisk: $a * b$.

8.2.4   Division (÷ or /)

Main article: Division (mathematics)

Division is essentially the inverse of multiplication. Division finds the *quotient* of two numbers, the *dividend* divided by the *divisor*. Any dividend divided by 0 is undefined. For distinct positive numbers, if the dividend is larger than the divisor, the quotient is greater than 1, otherwise it is less than 1 (a similar rule applies for negative numbers). The quotient multiplied by the divisor always yields the dividend.

Division is neither commutative nor associative. As it is helpful to look at subtraction as addition, it is helpful to look at division as multiplication of the dividend times the reciprocal of the divisor, that is $a \div b = a \times \frac{1}{b}$. When written as a product, it obeys all the properties of multiplication.

8.3   Decimal arithmetic

Decimal representation refers exclusively, in common use, to the written numeral system employing arabic numerals as the digits for a radix 10 ("decimal") positional notation; however, any numeral system based on powers of 10, e.g., Greek, Cyrillic, Roman, or Chinese numerals may conceptually be described as "decimal notation" or "decimal representation".

Modern methods for four fundamental operations (addition, subtraction, multiplication and division) were first devised by Brahmagupta of India. This was known during medieval Europe as "Modus Indoram" or Method of the Indians. Positional notation (also known as "place-value notation") refers to the representation or encoding of numbers using the same symbol for the different orders of magnitude (e.g., the "ones place", "tens place", "hundreds place") and, with a radix point, using those same symbols to represent fractions (e.g., the "tenths place", "hundredths place"). For example, 507.36 denotes 5 hundreds (10^2), plus 0 tens (10^1), plus 7 units (10^0), plus 3 tenths (10^{-1}) plus 6 hundredths (10^{-2}).

The concept of 0 as a number comparable to the other basic digits is essential to this notation, as is the concept of 0's use as a placeholder, and as is the definition of multiplication and addition with 0. The use of 0 as a placeholder and, therefore, the use of a positional notation is first attested to in the Jain text from India entitled the *Lokavibhâga*, dated 458 AD and it was only in the early 13th century that these concepts, transmitted via the scholarship of the Arabic world, were introduced into Europe by Fibonacci[9] using the Hindu–Arabic numeral system.

Algorism comprises all of the rules for performing arithmetic computations using this type of written numeral. For example, addition produces the sum of two arbitrary numbers. The result is calculated by the repeated addition of single digits from each number that occupies the same position, proceeding from right to left. An addition table with ten rows and ten columns displays all possible values for each sum. If an individual sum exceeds the value 9, the result is represented with two digits. The rightmost digit is the value for the current position, and the result for the subsequent addition of the digits to the left increases by the value of the second (leftmost) digit, which is always one. This adjustment is termed a *carry* of the value 1.

The process for multiplying two arbitrary numbers is similar to the process for addition. A multiplication table with ten rows and ten columns lists the results for each pair of digits. If an individual product of a pair of digits exceeds 9, the *carry* adjustment increases the result of any subsequent multiplication from digits to the left by a value equal to the second (leftmost) digit, which is any value from 1 to 8 ($9 \times 9 = 81$). Additional steps define the final result.

Similar techniques exist for subtraction and division.

The creation of a correct process for multiplication relies on the relationship between values of adjacent digits. The value for any single digit in a numeral depends on its position. Also, each position to the left represents a value ten times larger than the position to the right. In mathematical terms, the exponent for the radix (base) of 10 increases by 1 (to the left) or decreases by 1 (to the right). Therefore, the value for any arbitrary digit is multiplied by a value of the form 10^n with integer n. The list of values corresponding to all possible positions for a single digit is written as $\{..., 10^2, 10, 1, 10^{-1}, 10^{-2}, ...\}$.

Repeated multiplication of any value in this list by 10 produces another value in the list. In mathematical terminology, this characteristic is defined as closure, and the previous list is described as **closed under multiplication**. It is the basis for correctly finding the results of multiplication using the previous technique. This outcome is one example of the uses of number theory.

8.4 Compound unit arithmetic

Compound[10] unit arithmetic is the application of arithmetic operations to mixed radix quantities such as feet and inches, gallons and pints, pounds shillings and pence, and so on. Prior to the use of decimal-based systems of money and units of measure, the use of compound unit arithmetic formed a significant part of commerce and industry.

8.4.1 Basic arithmetic operations

The techniques used for compound unit arithmetic were developed over many centuries and are well-documented in many textbooks in many different languages.[11][12][13][14] In addition to the basic arithmetic functions encountered in decimal arithmetic, compound unit arithmetic employs three more functions:

- **Reduction** where a compound quantity is reduced to a single quantity, for example conversion of a distance expressed in yards, feet and inches to one expressed in inches.[15]

- **Expansion**, the inverse function to reduction, is the conversion of a quantity that is expressed as a single unit of measure to a compound unit, such as expanding 24 oz to 1 lb, 8 oz.

- **Normalization** is the conversion of a set of compound units to a standard form – for example rewriting "1 ft 13 in" as "2 ft 1 in".

Knowledge of the relationship between the various units of measure, their multiples and their submultiples forms an essential part of compound unit arithmetic.

8.4.2 Principles of compound unit arithmetic

There are two basic approaches to compound unit arithmetic:

- **Reduction–expansion method** where all the compound unit variables are reduced to single unit variables, the calculation performed and the result expanded back to compound units. This approach is suited for automated calculations. A typical example is the handling of time by Microsoft Excel where all time intervals are processed internally as days and decimal fractions of a day.

- **On-going normalization method** in which each unit is treated separately and the problem is continuously normalized as the solution develops. This approach, which is widely described in classical texts, is best suited for manual calculations. An example of the ongoing normalization method as applied to addition is shown below.

8.4.3 Operations in practice

During the 19th and 20th centuries various aids were developed to aid the manipulation of compound units, particularly in commercial applications. The most common aids were mechanical tills which were adapted in countries such as the United Kingdom to accommodate pounds, shillings, pennies and farthings and "Ready Reckoners" – books aimed at traders that catalogued the results of various routine calculations such as the percentages or multiples of various sums of money. One typical booklet[16] that ran to 150 pages tabulated multiples "from one to ten thousand at the various prices from one farthing to one pound".

The cumbersome nature of compound unit arithmetic has been recognized for many years – in 1586, the Flemish mathematician Simon Stevin published a small pamphlet called *De Thiende* ("the tenth")[17] in which he declared that the universal introduction of decimal coinage, measures, and weights to be merely a question of time while in the modern era, many conversion programs, such as that embedded in the calculator supplied as a standard part of the Microsoft Windows 7 operating system display compound units in a reduced decimal format rather than using an expanded format (i.e. "2.5 ft" is displayed rather than "2 ft 6 in").

8.5   Number theory

Main article: Number theory

Until the 19th century, *number theory* was a synonym of "arithmetic". The addressed problems were directly related to the basic operations and concerned primality, divisibility, and the solution of equations in integers, such as Fermat's last theorem. It appeared that most of these problems, although very elementary to state, are very difficult and may not be solved without very deep mathematics involving concepts and methods from many other branches of mathematics. This led to new branches of number theory such as analytic number theory, algebraic number theory, Diophantine geometry and arithmetic algebraic geometry. Wiles' proof of Fermat's Last Theorem is a typical example of the necessity of sophistical methods, which go far beyond the classical methods of arithmetic, for solving problems that can be stated in elementary arithmetic.

8.6   Arithmetic in education

Primary education in mathematics often places a strong focus on algorithms for the arithmetic of natural numbers, integers, fractions, and decimals (using the decimal place-value system). This study is sometimes known as algorism.

The difficulty and unmotivated appearance of these algorithms has long led educators to question this curriculum, advocating the early teaching of more central and intuitive mathematical ideas. One notable movement in this direction was the New Math of the 1960s and 1970s, which attempted to teach arithmetic in the spirit of axiomatic development from set theory, an echo of the prevailing trend in higher mathematics.[18]

Also, arithmetic was used by Islamic Scholars in order to teach application of the rulings related to Zakat and Irth. This was done in a book entitled *The Best of Arithmetic* by Abd-al-Fattah-al-Dumyati.[19]

The book begins with the foundations of mathematics and proceeds to its application in the later chapters.

8.7   See also

- Lists of mathematics topics
- Mathematics
- Outline of arithmetic

8.7.1   Related topics

- Addition of natural numbers
- Additive inverse
- Arithmetic coding
- Arithmetic mean
- Arithmetic progression
- Arithmetic properties
- Associativity
- Commutativity
- Distributivity

- Elementary arithmetic
- Finite field arithmetic
- Integer
- List of important publications in mathematics
- Mental calculation
- Number line

8.8 Notes

[1] "Mathematics". Science Clarified. Retrieved 23 October 2012.

[2] Davenport, Harold, *The Higher Arithmetic: An Introduction to the Theory of Numbers* (7th ed.), Cambridge University Press, Cambridge, UK, 1999, ISBN 0-521-63446-6

[3] Rudman, Peter Strom (2007). *How Mathematics Happened: The First 50,000 Years.* Prometheus Books. p. 64. ISBN 978-1-59102-477-4.

[4] *The Works of Archimedes,* Chapter IV, *Arithmetic in Archimedes,* edited by T.L. Heath, Dover Publications Inc, New York, 2002.

[5] Joseph Needham, *Science and Civilization in China,* Vol. 3, page 9, Cambridge University Press, 1959.

[6] Reference: Revue de l'Orient Chretien by François Nau pp.327-338. (1929)

[7] Reference: Sigler, L., "Fibonacci's Liber Abaci", Springer, 2003.

[8] Tapson, Frank (1996). *The Oxford Mathematics Study Dictionary.* Oxford University Press. ISBN 0 19 914551 2.

[9] Leonardo Pisano - page 3: "Contributions to number theory". *Encyclopædia Britannica* Online, 2006. Retrieved 18 September 2006.

[10] Walkingame, Francis (1860). "The Tutor's Companion; or, Complete Practical Arithmetic" (PDF). Webb, Millington & Co. pp. 24–39.

[11] Palaiseau, JFG (October 1816). *Métrologie universelle, ancienne et moderne: ou rapport des poids et mesures des empires, royaumes, duchés et principautés des quatre parties du monde* [*Universal, ancient and modern metrology: or report of weights and measurements of empires, kingdoms, duchies and principalities of all parts of the world*] (in French). Bordeaux. Retrieved October 30, 2011.

[12] Jacob de Gelder (1824). *Allereerste Gronden der Cijferkunst* [*Introduction to Numeracy*] (in Dutch). 's Gravenhage and Amsterdam: de Gebroeders van Cleef. pp. 163–176. Retrieved March 2, 2011.

[13] Malaisé, Ferdinand (1842). *Theoretisch-Praktischer Unterricht im Rechnen für die niederen Classen der Regimentsschulen der Königl. Bayer. Infantrie und Cavalerie* [*Theoretical and practical instruction in arithmetic for the lower classes of the Royal Bavarian Infantry and Cavalry School*] (in German). Munich. Retrieved 20 March 2012.

[14] *Encyclopædia Britannica*, Vol I, Edinburgh, 1772, Arithmetick

[15] Walkingame, Francis (1860). "The Tutor's Companion; or, Complete Practical Arithmetic" (PDF). Webb, Millington & Co. pp. 43–50.

[16] Thomson, J (1824). *The Ready Reckoner in miniature containing accurate table from one to the thousand at the various prices from one farthing to one pound.* Montreal. Retrieved 25 March 2012.

[17] O'Connor, John J.; Robertson, Edmund F. (January 2004), "Arithmetic", *MacTutor History of Mathematics archive,* University of St Andrews.

[18] Mathematically Correct: Glossary of Terms

[19] Abd-al-Fattah Bin Abd-al-Rahman al-Banna al-Dumyati (1887). "The Best of Arithmetic". *World Digital Library* (in Arabic). Retrieved 30 June 2013.

8.9   References

- Cunnington, Susan, *The Story of Arithmetic: A Short History of Its Origin and Development*, Swan Sonnenschein, London, 1904

- Dickson, Leonard Eugene, *History of the Theory of Numbers* (3 volumes), reprints: Carnegie Institute of Washington, Washington, 1932; Chelsea, New York, 1952, 1966

- Euler, Leonhard, *Elements of Algebra*, Tarquin Press, 2007

- Fine, Henry Burchard (1858–1928), *The Number System of Algebra Treated Theoretically and Historically*, Leach, Shewell & Sanborn, Boston, 1891

- Karpinski, Louis Charles (1878–1956), *The History of Arithmetic*, Rand McNally, Chicago, 1925; reprint: Russell & Russell, New York, 1965

- Ore, Øystein, *Number Theory and Its History*, McGraw–Hill, New York, 1948

- Weil, André, *Number Theory: An Approach through History*, Birkhauser, Boston, 1984; reviewed: Mathematical Reviews 85c:01004

8.10   External links

- MathWorld article about arithmetic

- The New Student's Reference Work/Arithmetic (historical)

- The Great Calculation According to the Indians, of Maximus Planudes – an early Western work on arithmetic at Convergence

- ⊕ P. H. Vander Weyde (1879). "Arithmetic". *The American Cyclopaedia*.

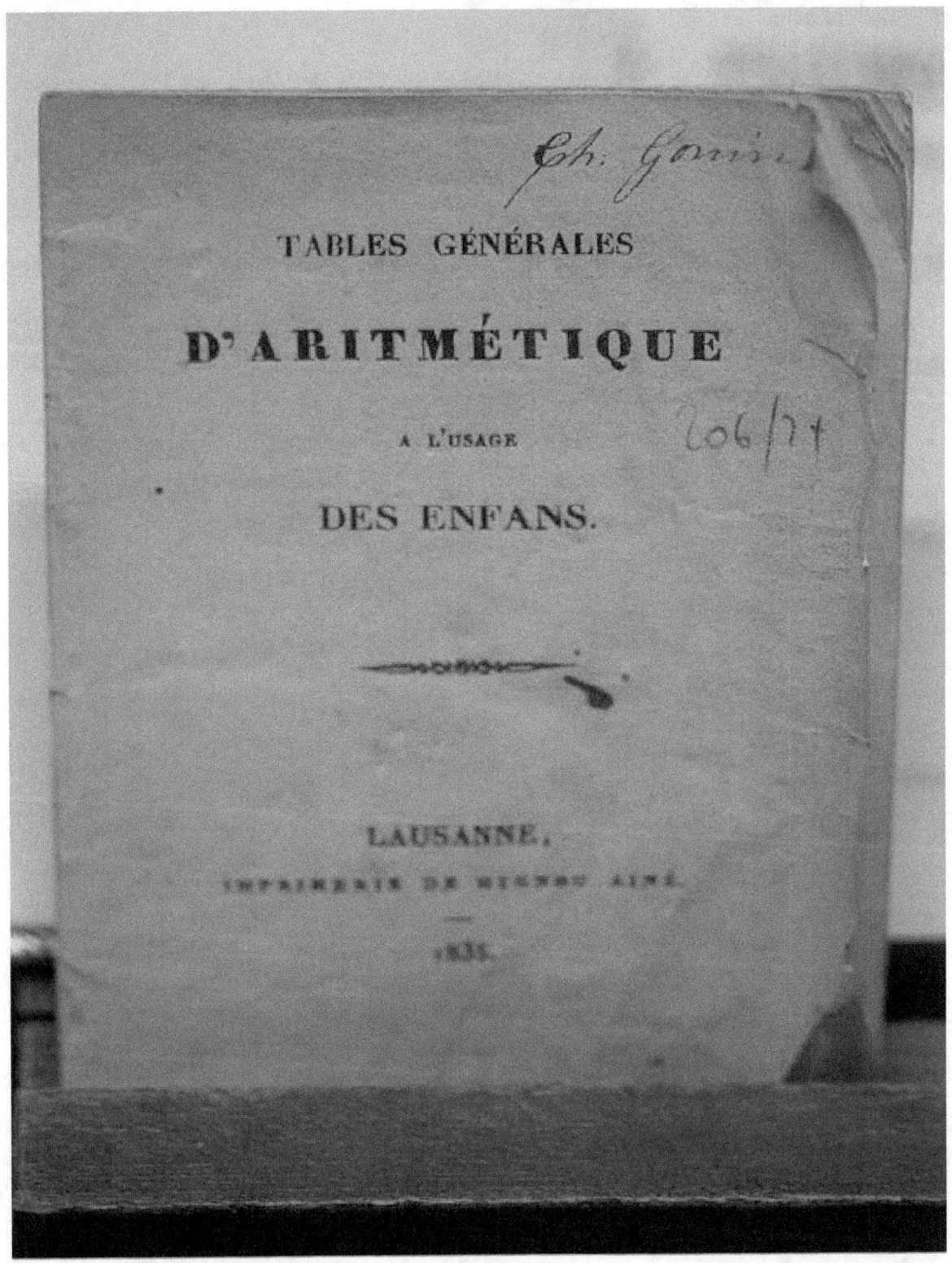

Arithmetic tables for children, Lausanne, 1835

Leibniz's Stepped Reckoner was the first calculator that could perform all four arithmetic operations.

A scale calibrated in imperial units with an associated cost display.

Chapter 9

Order theory

For a topical guide to this subject, see Outline of order theory.

Order theory is a branch of mathematics which investigates our intuitive notion of order using binary relations. It provides a formal framework for describing statements such as "this is less than that" or "this precedes that". This article introduces the field and provides basic definitions. A list of order-theoretic terms can be found in the order theory glossary.

9.1   Background and motivation

Orders are everywhere in mathematics and related fields like computer science. The first order often discussed in primary school is the standard order on the natural numbers e.g. "2 is less than 3", "10 is greater than 5", or "Does Tom have fewer cookies than Sally?". This intuitive concept can be extended to orders on other sets of numbers, such as the integers and the reals. The idea of being greater than or less than another number is one of the basic intuitions of number systems (compare with numeral systems) in general (although one usually is also interested in the actual difference of two numbers, which is not given by the order). Another familiar example of an ordering is the lexicographic order of words in a dictionary.

The above types of orders have a special property: each element can be *compared* to any other element, i.e. it is greater, smaller, or equal. However, this is not always a desired requirement. For example, consider the subset ordering of sets. If a set A contains all the elements of a set B, then B is said to be smaller than or equal to A. Yet there are some sets that cannot be related in this fashion. Whenever both contain some elements that are not in the other, the two sets are not related by subset-inclusion. Hence, subset-inclusion is only a *partial order*, as opposed to the *total orders* given before.

Order theory captures the intuition of orders that arises from such examples in a general setting. This is achieved by specifying properties that a relation ≤ must have to be a mathematical order. This more abstract approach makes much sense, because one can derive numerous theorems in the general setting, without focusing on the details of any particular order. These insights can then be readily transferred to many less abstract applications.

Driven by the wide practical usage of orders, numerous special kinds of ordered sets have been defined, some of which have grown into mathematical fields of their own. In addition, order theory does not restrict itself to the various classes of ordering relations, but also considers appropriate functions between them. A simple example of an order theoretic property for functions comes from analysis where monotone functions are frequently found.

9.2   Basic definitions

This section introduces ordered sets by building upon the concepts of set theory, arithmetic, and binary relations.

9.2.1   Partially ordered sets

Orders are special binary relations. Suppose that P is a set and that $\leq$ is a relation on P. Then $\leq$ is a **partial order** if it is reflexive, antisymmetric, and transitive, i.e., for all a, b and c in P, we have that:

$a \leq a$ (reflexivity)

if $a \leq b$ and $b \leq a$ then $a = b$ (antisymmetry)

if $a \leq b$ and $b \leq c$ then $a \leq c$ (transitivity).

A set with a partial order on it is called a **partially ordered set, poset,** or just an **ordered set** if the intended meaning is clear. By checking these properties, one immediately sees that the well-known orders on natural numbers, integers, rational numbers and reals are all orders in the above sense. However, they have the additional property of being **total**, i.e., for all a and b in P, we have that:

$a \leq b$ or $b \leq a$ (totality).

These orders can also be called **linear orders** or **chains**. While many classical orders are linear, the subset order on sets provides an example where this is not the case. Another example is given by the divisibility relation "|". For two natural numbers n and m, we write $n|m$ if n divides m without remainder. One easily sees that this yields a partial order. The identity relation = on any set is also a partial order in which every two distinct elements are incomparable. It is also the only relation that is both a partial order and an equivalence relation. Many advanced properties of posets are interesting mainly for non-linear orders.

9.2.2   Visualizing a poset

Hasse diagrams can visually represent the elements and relations of a partial ordering. These are graph drawings where the vertices are the elements of the poset and the ordering relation is indicated by both the edges and the relative positioning of the vertices. Orders are drawn bottom-up: if an element x is smaller than (precedes) y then there exists a path from x to y that is directed upwards. It is often necessary for the edges connecting elements to cross each other, but elements must never be located within an edge. An instructive exercise is to draw the Hasse diagram for the set of natural numbers that are smaller than or equal to 13, ordered by | (the *divides* relation).

Even some infinite sets can be diagrammed by superimposing an ellipsis (...) on a finite sub-order. This works well for the natural numbers, but it fails for the reals, where there is no immediate successor above 0; however, quite often one can obtain an intuition related to diagrams of a similar kind.

9.2.3   Special elements within an order

In a partially ordered set there may be some elements that play a special role. The most basic example is given by the **least element** of a poset. For example, 1 is the least element of the positive integers and the empty set is the least set under the subset order. Formally, an element m is a least element if:

$m \leq a$, for all elements a of the order.

The notation 0 is frequently found for the least element, even when no numbers are concerned. However, in orders on sets of numbers, this notation might be inappropriate or ambiguous, since the number 0 is not always least. An example is given by the above divisibility order |, where 1 is the least element since it divides all other numbers. In contrast, 0 is the number that is divided by all other numbers. Hence it is the **greatest element** of the order. Other frequent terms for the least and greatest elements is **bottom** and **top** or **zero** and **unit**.

Least and greatest elements may fail to exist, as the example of the real numbers shows. But if they exist, they are always unique. In contrast, consider the divisibility relation | on the set {2,3,4,5,6}. Although this set has neither top nor bottom, the elements 2, 3, and 5 have no elements below them, while 4, 5, and 6 have none above. Such elements are called **minimal and maximal**, respectively. Formally, an element m is minimal if:

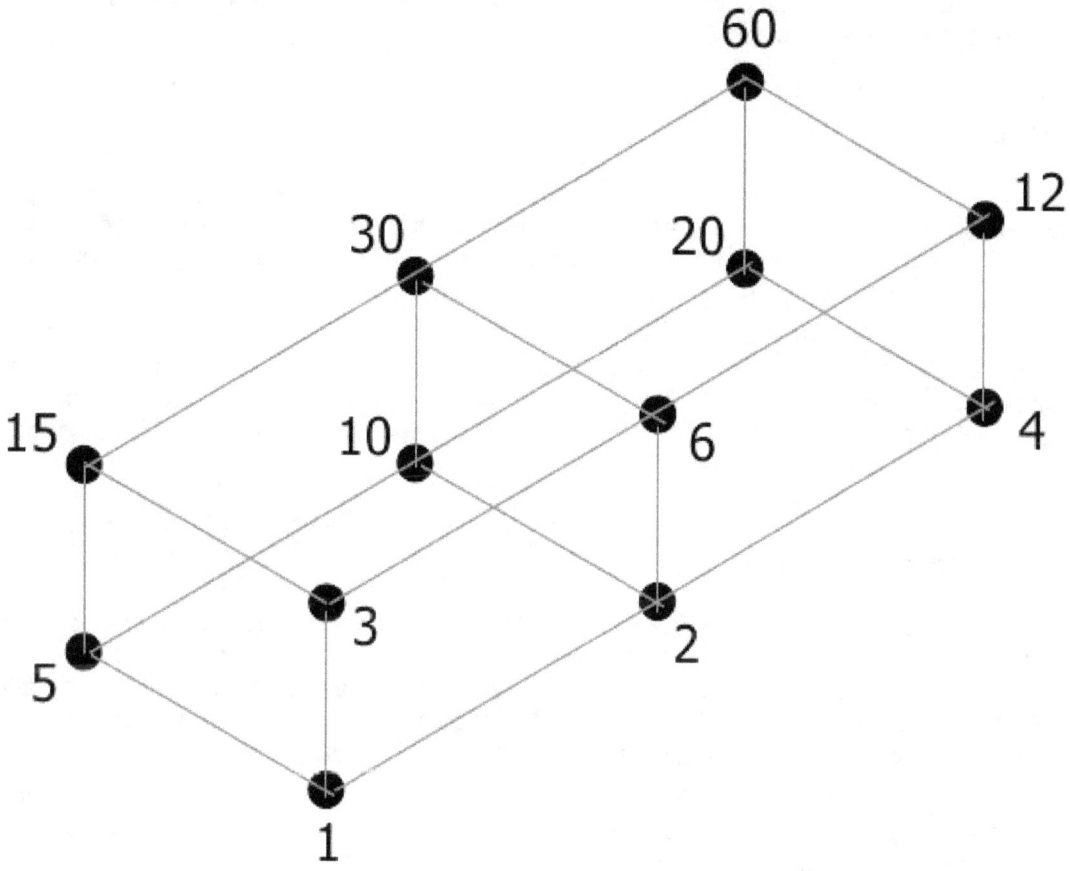

Hasse diagram of the set of all divisors of 60, partially ordered by divisibility

$a \le m$ implies $a = m$, for all elements a of the order.

Exchanging $\le$ with $\ge$ yields the definition of maximality. As the example shows, there can be many maximal elements and some elements may be both maximal and minimal (e.g. 5 above). However, if there is a least element, then it is the only minimal element of the order. Again, in infinite posets maximal elements do not always exist - the set of all *finite* subsets of a given infinite set, ordered by subset inclusion, provides one of many counterexamples. An important tool to ensure the existence of maximal elements under certain conditions is **Zorn's Lemma**.

Subsets of partially ordered sets inherit the order. We already applied this by considering the subset {2,3,4,5,6} of the natural numbers with the induced divisibility ordering. Now there are also elements of a poset that are special with respect to some subset of the order. This leads to the definition of **upper bounds**. Given a subset S of some poset P, an upper bound of S is an element b of P that is above all elements of S. Formally, this means that

$s \le b$, for all s in S.

Lower bounds again are defined by inverting the order. For example, -5 is a lower bound of the natural numbers as a subset of the integers. Given a set of sets, an upper bound for these sets under the subset ordering is given by their union. In fact, this upper bound is quite special: it is the smallest set that contains all of the sets. Hence, we have found the **least upper bound** of a set of sets. This concept is also called **supremum** or **join**, and for a set S one writes $\sup(S)$ or $\vee S$ for its least upper bound. Conversely, the **greatest lower bound** is known as **infimum** or **meet** and denoted $\inf(S)$ or $\wedge S$. These concepts play an important role in many applications of order theory. For two elements x and y, one also writes $x \vee y$ and $x \wedge y$ for $\sup(\{x,y\})$ and $\inf(\{x,y\})$, respectively.

For another example, consider again the relation | on natural numbers. The least upper bound of two numbers is the smallest number that is divided by both of them, i.e. the least common multiple of the numbers. Greatest lower bounds in turn are given by the greatest common divisor.

9.2.4 Duality

In the previous definitions, we often noted that a concept can be defined by just inverting the ordering in a former definition. This is the case for "least" and "greatest", for "minimal" and "maximal", for "upper bound" and "lower bound", and so on. This is a general situation in order theory: A given order can be inverted by just exchanging its direction, pictorially flipping the Hasse diagram top-down. This yields the so-called **dual, inverse,** or **opposite order.**

Every order theoretic definition has its dual: it is the notion one obtains by applying the definition to the inverse order. Since all concepts are symmetric, this operation preserves the theorems of partial orders. For a given mathematical result, one can just invert the order and replace all definitions by their duals and one obtains another valid theorem. This is important and useful, since one obtains two theorems for the price of one. Some more details and examples can be found in the article on duality in order theory.

9.2.5 Constructing new orders

There are many ways to construct orders out of given orders. The dual order is one example. Another important construction is the cartesian product of two partially ordered sets, taken together with the product order on pairs of elements. The ordering is defined by $(a, x) \leq (b, y)$ if (and only if) $a \leq b$ and $x \leq y$. (Notice carefully that there are three distinct meanings for the relation symbol $\leq$ in this definition.) The disjoint union of two posets is another typical example of order construction, where the order is just the (disjoint) union of the original orders.

Every partial order $\leq$ gives rise to a so-called strict order $<$, by defining $a < b$ if $a \leq b$ and not $b \leq a$. This transformation can be inverted by setting $a \leq b$ if $a < b$ or $a = b$. The two concepts are equivalent although in some circumstances one can be more convenient to work with than the other.

9.3 Functions between orders

It is reasonable to consider functions between partially ordered sets having certain additional properties that are related to the ordering relations of the two sets. The most fundamental condition that occurs in this context is monotonicity. A function f from a poset P to a poset Q is **monotone,** or **order-preserving,** if $a \leq b$ in P implies $f(a) \leq f(b)$ in Q (Noting that, strictly, the two relations here are different since they apply to different sets.). The converse of this implication leads to functions that are **order-reflecting,** i.e. functions f as above for which $f(a) \leq f(b)$ implies $a \leq b$. On the other hand, a function may also be **order-reversing** or **antitone,** if $a \leq b$ implies $f(b) \leq f(a)$.

An **order-embedding** is a function f between orders that is both order-preserving and order-reflecting. Examples for these definitions are found easily. For instance, the function that maps a natural number to its successor is clearly monotone with respect to the natural order. Any function from a discrete order, i.e. from a set ordered by the identity order "=", is also monotone. Mapping each natural number to the corresponding real number gives an example for an order embedding. The set complement on a powerset is an example of an antitone function.

An important question is when two orders are "essentially equal", i.e. when they are the same up to renaming of elements. **Order isomorphisms** are functions that define such a renaming. An order-isomorphism is a monotone bijective function that has a monotone inverse. This is equivalent to being a surjective order-embedding. Hence, the image $f(P)$ of an order-embedding is always isomorphic to P, which justifies the term "embedding".

A more elaborate type of functions is given by so-called **Galois connections.** Monotone Galois connections can be viewed as a generalization of order-isomorphisms, since they constitute of a pair of two functions in converse directions, which are "not quite" inverse to each other, but that still have close relationships.

Another special type of self-maps on a poset are **closure operators,** which are not only monotonic, but also idempotent, i.e. $f(x) = f(f(x))$, and extensive (or *inflationary*), i.e. $x \le f(x)$. These have many applications in all kinds of "closures" that appear in mathematics.

Besides being compatible with the mere order relations, functions between posets may also behave well with respect to special elements and constructions. For example, when talking about posets with least element, it may seem reasonable to consider only monotonic functions that preserve this element, i.e. which map least elements to least elements. If binary infima $\wedge$ exist, then a reasonable property might be to require that $f(x \wedge y) = f(x) \wedge f(y)$, for all x and y. All of these properties, and indeed many more, may be compiled under the label of limit-preserving functions.

Finally, one can invert the view, switching from *functions of orders* to *orders of functions*. Indeed, the functions between two posets P and Q can be ordered via the pointwise order. For two functions f and g, we have $f \le g$ if $f(x) \le g(x)$ for all elements x of P. This occurs for example in domain theory, where function spaces play an important role.

9.4 Special types of orders

Many of the structures that are studied in order theory employ order relations with further properties. In fact, even some relations that are not partial orders are of special interest. Mainly the concept of a preorder has to be mentioned. A preorder is a relation that is reflexive and transitive, but not necessarily antisymmetric. Each preorder induces an equivalence relation between elements, where a is equivalent to b, if $a \le b$ and $b \le a$. Preorders can be turned into orders by identifying all elements that are equivalent with respect to this relation.

Several types of orders can be defined from numerical data on the items of the order: a total order results from attaching distinct real numbers to each item and using the numerical comparisons to order the items; instead, if distinct items are allowed to have equal numerical scores, one obtains a strict weak ordering. Requiring two scores to be separated by a fixed threshold before they may be compared leads to the concept of a semiorder, while allowing the threshold to vary on a per-item basis produces an interval order.

An additional simple but useful property leads to so-called **well-orders**, for which all non-empty subsets have a minimal element. Generalizing well-orders from linear to partial orders, a set is **well partially ordered** if all its non-empty subsets have a finite number of minimal elements.

Many other types of orders arise when the existence of infima and suprema of certain sets is guaranteed. Focusing on this aspect, usually referred to as completeness of orders, one obtains:

- Bounded posets, i.e. posets with a least and greatest element (which are just the supremum and infimum of the empty subset),

- Lattices, in which every non-empty finite set has a supremum and infimum,

- Complete lattices, where every set has a supremum and infimum, and

- Directed complete partial orders (dcpos), that guarantee the existence of suprema of all directed subsets and that are studied in domain theory.

- Partial orders with complements, or *poc sets*[1] are posets S having a unique bottom element $0 \in S$, along with an order-reversing involution, such that $a \le \overline{a} \Rightarrow a = 0$.

However, one can go even further: if all finite non-empty infima exist, then $\wedge$ can be viewed as a total binary operation in the sense of universal algebra. Hence, in a lattice, two operations $\wedge$ and $\vee$ are available, and one can define new properties by giving identities, such as

$$x \wedge (y \vee z) = (x \wedge y) \vee (x \wedge z), \text{ for all } x, y, \text{ and } z.$$

This condition is called **distributivity** and gives rise to distributive lattices. There are some other important distributivity laws which are discussed in the article on distributivity in order theory. Some additional order structures that are often specified via algebraic operations and defining identities are

- Heyting algebras and

- Boolean algebras,

which both introduce a new operation ~ called **negation**. Both structures play a role in mathematical logic and especially Boolean algebras have major applications in computer science. Finally, various structures in mathematics combine orders with even more algebraic operations, as in the case of quantales, that allow for the definition of an addition operation.

Many other important properties of posets exist. For example, a poset is **locally finite** if every closed interval $[a, b]$ in it is finite. Locally finite posets give rise to incidence algebras which in turn can be used to define the Euler characteristic of finite bounded posets.

9.5   Subsets of ordered sets

In an ordered set, one can define many types of special subsets based on the given order. A simple example are **upper sets**; i.e. sets that contain all elements that are above them in the order. Formally, the **upper closure** of a set S in a poset P is given by the set $\{x$ in $P \mid$ there is some y in S with $y \le x\}$. A set that is equal to its upper closure is called an upper set. **Lower sets** are defined dually.

More complicated lower subsets are ideals, which have the additional property that each two of their elements have an upper bound within the ideal. Their duals are given by filters. A related concept is that of a directed subset, which like an ideal contains upper bounds of finite subsets, but does not have to be a lower set. Furthermore it is often generalized to preordered sets.

A subset which is - as a sub-poset - linearly ordered, is called a chain. The opposite notion, the antichain, is a subset that contains no two comparable elements; i.e. that is a discrete order.

9.6   Related mathematical areas

Although most mathematical areas *use* orders in one or the other way, there are also a few theories that have relationships which go far beyond mere application. Together with their major points of contact with order theory, some of these are to be presented below.

9.6.1   Universal algebra

As already mentioned, the methods and formalisms of universal algebra are an important tool for many order theoretic considerations. Beside formalizing orders in terms of algebraic structures that satisfy certain identities, one can also establish other connections to algebra. An example is given by the correspondence between Boolean algebras and Boolean rings. Other issues are concerned with the existence of free constructions, such as *free lattices* based on a given set of generators. Furthermore, closure operators are important in the study of universal algebra.

9.6.2   Topology

In topology orders play a very prominent role. In fact, the set of open sets provides a classical example of a complete lattice, more precisely a complete Heyting algebra (or "frame" or "locale"). Filters and nets are notions closely related to order theory and the closure operator of sets can be used to define topology. Beyond these relations, topology can be looked at solely in terms of the open set lattices, which leads to the study of pointless topology. Furthermore, a natural preorder of elements of the underlying set of a topology is given by the so-called specialization order, that is actually a partial order if the topology is T₀.

Conversely, in order theory, one often makes use of topological results. There are various ways to define subsets of an order which can be considered as open sets of a topology. Especially, it is interesting to consider topologies on a poset

$(X, \leq)$ that in turn induce $\leq$ as their specialization order. The *finest* such topology is the Alexandrov topology, given by taking all upper sets as opens. Conversely, the *coarsest* topology that induces the specialization order is the upper topology, having the complements of principal ideals (i.e. sets of the form $\{y \text{ in } X \mid y \leq x\}$ for some x) as a subbase. Additionally, a topology with specialization order $\leq$ may be order consistent, meaning that their open sets are "inaccessible by directed suprema" (with respect to $\leq$). The finest order consistent topology is the Scott topology, which is coarser than the Alexandrov topology. A third important topology in this spirit is the Lawson topology. There are close connections between these topologies and the concepts of order theory. For example, a function preserves directed suprema iff it is continuous with respect to the Scott topology (for this reason this order theoretic property is also called Scott-continuity).

9.6.3   Category theory

The visualization of orders with Hasse diagrams has a straightforward generalization: instead of displaying lesser elements *below* greater ones, the direction of the order can also be depicted by giving directions to the edges of a graph. In this way, each order is seen to be equivalent to a directed acyclic graph, where the nodes are the elements of the poset and there is a directed path from a to b if and only if $a \leq b$. Dropping the requirement of being acyclic, one can also obtain all preorders.

When equipped with all transitive edges, these graphs in turn are just special categories, where elements are objects and each set of morphisms between two elements is at most singleton. Functions between orders become functors between categories. Interestingly, many ideas of order theory are just concepts of category theory in small. For example, an infimum is just a categorical product. More generally, one can capture infima and suprema under the abstract notion of a categorical limit (or *colimit*, respectively). Another place where categorical ideas occur is the concept of a (monotone) Galois connection, which is just the same as a pair of adjoint functors.

But category theory also has its impact on order theory on a larger scale. Classes of posets with appropriate functions as discussed above form interesting categories. Often one can also state constructions of orders, like the product order, in terms of categories. Further insights result when categories of orders are found categorically equivalent to other categories, for example of topological spaces. This line of research leads to various *representation theorems*, often collected under the label of Stone duality.

9.7   History

As explained before, orders are ubiquitous in mathematics. However, earliest explicit mentionings of partial orders are probably to be found not before the 19th century. In this context the works of George Boole are of great importance. Moreover, works of Charles Sanders Peirce, Richard Dedekind, and Ernst Schröder also consider concepts of order theory. Certainly, there are others to be named in this context and surely there exists more detailed material on the history of order theory.

The term *poset* as an abbreviation for partially ordered set was coined by Garrett Birkhoff in the second edition of his influential book *Lattice Theory*.[2][3]

9.8   See also

- Cyclic order

- Hierarchy

- Incidence algebra

- Important publications in order theory

- Causal Sets

9.9 Notes

[1] Roller, Martin A. (1998), *Poc sets, median algebras and group actions. An extended study of Dunwoody's construction and Sageev's theorem* (PDF), Southampton Preprint Archive

[2] Birkhoff 1948, p.1

[3] Earliest Known Uses of Some of the Words of Mathematics

9.10 References

- Birkhoff, Garrett (1940). *Lattice Theory* 25 (3rd Revised ed.). American Mathematical Society. ISBN 978-0-8218-1025-5.

- Burris, S. N.; Sankappanavar, H. P. (1981). *A Course in Universal Algebra*. Springer. ISBN 978-0-387-90578-5.

- Davey, B. A.; Priestley, H. A. (2002). *Introduction to Lattices and Order* (2nd ed.). Cambridge University Press. ISBN 0-521-78451-4.

- Gierz, G.; Hofmann, K. H.; Keimel, K.; Mislove, M.; Scott, D. S. (2003). *Continuous Lattices and Domains*. Encyclopedia of Mathematics and its Applications 93. Cambridge University Press. ISBN 978-0-521-80338-0.

9.11 External links

- Orders at ProvenMath partial order, linear order, well order, initial segment; formal definitions and proofs within the axioms of set theory.

- Nagel, Felix (2013). Set Theory and Topology. An Introduction to the Foundations of Analysis

Chapter 10

Algebraic structure

In mathematics, and more specifically in abstract algebra, the term **algebraic structure** generally refers to a set (called **carrier set** or **underlying set**) with one or more finitary operations defined on it.[1]

Examples of algebraic structures include groups, rings, fields, and lattices. More complex structures can be defined by introducing multiple operations, different underlying sets, or by altering the defining axioms. Examples of more complex algebraic structures include vector spaces, modules, and algebras.

The properties of specific algebraic structures are studied in abstract algebra. The general theory of algebraic structures has been formalized in universal algebra. Category theory is used to study the relationships between two or more classes of algebraic structures, often of different kinds. For example, Galois theory studies the connection between certain fields and groups, algebraic structures of two different kinds.

In a slight abuse of notation, the word "structure" can also refer only to the operations on a structure, and not the underlying set itself. For example, a phrase "we have defined a ring *structure* (a *structure* of ring) on the set A " means that we have defined ring operations on the set A. For another example, the group $(Z, +)$ can be seen as a set Z that is equipped with an *algebraic structure*, namely the operation $+$.

10.1 Introduction

Addition and multiplication on numbers are the prototypical example of an operation that combines two elements of a set to produce a third. These operations obey several algebraic laws. For example $a + (b + c) = (a + b) + c$ and $a(bc) = (ab)c$, both examples of the *associative law*. Also $a + b = b + a$, and $ab = ba$, the *commutative law*. Many systems studied by mathematicians have operations that obey some, but not necessarily all, of the laws of ordinary arithmetic. For example, rotations of objects in three-dimensional space can be combined by performing the first rotation and then applying the second rotation to the object in its new orientation. This operation on rotations obeys the associative law, but can fail the commutative law.

Mathematicians give names to sets with one or more operations that obey a particular collection of laws, and study them in the abstract as algebraic structures. When a new problem can be shown to follow the laws of one of these algebraic structures, all the work that has been done on that category in the past can be applied to the new problem.

In full generality, algebraic structures may involve an arbitrary number of sets and operations that can combine more than two elements (higher arity), but this article focuses on binary operations on one or two sets. The examples here are by no means a complete list, but they are meant to be a representative list and include the most common structures. Longer lists of algebraic structures may be found in the external links and within *Category:Algebraic structures*. Structures are listed in approximate order of increasing complexity.

10.2   Examples

10.2.1   One set with operations

Simple structures: No binary operation:

- Set: a degenerate algebraic structure having no operations.
- Pointed set: *S* has one or more distinguished elements, often 0, 1, or both.
- Unary system: *S* and a single unary operation over *S*.
- Pointed unary system: a unary system with *S* a pointed set.

Group-like structures: One binary operation. The binary operation can be indicated by any symbol, or with no symbol (juxtaposition) as is done for ordinary multiplication of real numbers.

- Magma or groupoid: *S* and a single binary operation over *S*.
- Semigroup: an associative magma.
- Monoid: a semigroup with identity.
- Group: a monoid with a unary operation (inverse), giving rise to inverse elements.
- Abelian group: a group whose binary operation is commutative.
- Semilattice: a semigroup whose operation is idempotent and commutative. The binary operation can be called either meet or join.
- Quasigroup: a magma obeying the latin square property. A quasigroup may also be represented using three binary operations.[2]
- Loop: a quasigroup with identity.

Ring-like structures or Ringoids: Two binary operations, often called addition and multiplication, with multiplication distributing over addition.

- Semiring: a ringoid such that *S* is a monoid under each operation. Addition is typically assumed to be commutative and associative, and the monoid product is assumed to distribute over the addition on both sides, and the additive identity satisfies $0\,x = 0$ for all x.
- Near-ring: a semiring whose additive monoid is a (not necessarily Abelian) group.
- Ring: a semiring whose additive monoid is an Abelian group.
- Lie ring: a ringoid whose additive monoid is an abelian group, but whose multiplicative operation satisfies the Jacobi identity rather than associativity.
- Boolean ring: a commutative ring with idempotent multiplication operation.
- Field: a commutative ring which contains a multiplicative inverse for every nonzero element
- Kleene algebras: a semiring with idempotent addition and a unary operation, the Kleene star, satisfying additional properties.
- *-algebra: a ring with an additional unary operation (*) satisfying additional properties.

Lattice structures: Two or more binary operations, including operations called meet and join, connected by the absorption law.[3]

- Complete lattice: a lattice in which arbitrary meet and joins exist.

- Bounded lattice: a lattice with a greatest element and least element.

- Complemented lattice: a bounded lattice with a unary operation, complementation, denoted by postfix ′. The join of an element with its complement is the greatest element, and the meet of the two elements is the least element.

- Modular lattice: a lattice whose elements satisfy the additional *modular identity*.

- Distributive lattice: a lattice in which each of meet and join distributes over the other. Distributive lattices are modular, but the converse does not hold.

- Boolean algebra: a complemented distributive lattice. Either of meet or join can be defined in terms of the other and complementation. This can be shown to be equivalent with the ring-like structure of the same name above.

- Heyting algebra: a bounded distributive lattice with an added binary operation, relative pseudo-complement, denoted by infix →, and governed by the axioms $x \rightarrow x = 1$, $x(x \rightarrow y) = xy$, $y(x \rightarrow y) = y$, $x \rightarrow (yz) = (x \rightarrow y)(x \rightarrow z)$.

Arithmetics: Two binary operations, addition and multiplication. S is an infinite set. Arithmetics are pointed unary systems, whose unary operation is injective successor, and with distinguished element 0.

- Robinson arithmetic. Addition and multiplication are recursively defined by means of successor. 0 is the identity element for addition, and annihilates multiplication. Robinson arithmetic is listed here even though it is a variety, because of its closeness to Peano arithmetic.

- Peano arithmetic. Robinson arithmetic with an axiom schema of induction. Most ring and field axioms bearing on the properties of addition and multiplication are theorems of Peano arithmetic or of proper extensions thereof.

10.2.2   Two sets with operations

Module-like structures: composite systems involving two sets and employing at least two binary operations.

- Group with operators: a group G with a set Ω and a binary operation $\Omega \times G \rightarrow G$ satisfying certain axioms.

- Module: an Abelian group M and a ring R acting as operators on M. The members of R are sometimes called scalars, and the binary operation of *scalar multiplication* is a function $R \times M \rightarrow M$, which satisfies several axioms. Counting the ring operations these systems have at least three operations.

- Vector space: a module where the ring R is a division ring or field.

- Graded vector space: a vector space with a direct sum decomposition breaking the space into "grades".

- Quadratic space: a vector space V over a field F with a function from V into F satisfying certain properties. Every quadratic space is also an inner product space (see below).

Algebra-like structures: composite system defined over two sets, a ring R and a R module M equipped with an operation called multiplication. This can be viewed as a system with five binary operations: two operations on R, two on M and one involving both R and M.

- Algebra over a ring (also *R-algebra*): a module over a commutative ring R, which also carries a multiplication operation that is compatible with the module structure. This includes distributivity over addition and linearity with respect to multiplication by elements of R. The theory of an algebra over a field is especially well developed.

- Associative algebra: an algebra over a ring such that the multiplication is associative.

- Nonassociative algebra: a module over a commutative ring, equipped with a ring multiplication operation that is not necessarily associative. Often associativity is replaced with a different identity, such as alternation, the Jacobi identity, or the Jordan identity.

- Coalgebra: a vector space with a "comultiplication" defined dually to that of associative algebras.

- Lie algebra: a special type of nonassociative algebra whose product satisfies the Jacobi identity.

- Lie coalgebra: a vector space with a "comultiplication" defined dually to that of Lie algebras.

- Graded algebra: a graded vector space with an algebra structure compatible with the grading. The idea is that if the grades of two elements a and b are known, then the grade of ab is known, and so the location of the product ab is determined in the decomposition.

- Inner product space: an F vector space V with a bilinear binary operation from $V \times V \to F$.

Four or more binary operations:

- Bialgebra: an associative algebra with a compatible coalgebra structure.

- Lie bialgebra: a Lie algebra with a compatible bialgebra structure.

- Clifford algebra: a graded associative algebra equipped with an exterior product from which may be derived several possible inner products. Exterior algebras and geometric algebras are special cases of this construction.

10.3   Hybrid structures

Algebraic structures can also coexist with added structure of non-algebraic nature, such as partial order or a topology. The added structure must be compatible, in some sense, with the algebraic structure.

- Topological group: a group with a topology compatible with the group operation.

- Lie group: a topological group with a compatible smooth manifold structure.

- Ordered groups, ordered rings and ordered fields: each type of structure with a compatible partial order.

- Archimedean group: a linearly ordered group for which the Archimedean property holds.

- Topological vector space: a vector space whose M has a compatible topology.

- Normed vector space: a vector space with a compatible norm. If such a space is complete (as a metric space) then it is called a Banach space.

- Hilbert space: an inner product space over the real or complex numbers whose inner product gives rise to a Banach space structure.

- Vertex operator algebra

- Von Neumann algebra: a *-algebra of operators on a Hilbert space equipped with the weak operator topology.

10.4   Universal algebra

Algebraic structures are defined through different configurations of axioms. Universal algebra abstractly studies such objects. One major dichotomy is between structures that are axiomatized entirely by *identities* and structures that are not. If all axioms defining a class of algebras are identities, then the class of objects is a variety (not to be confused with algebraic variety in the sense of algebraic geometry).

Identities are equations formulated using only the operations the structure allows, and variables that are tacitly universally quantified over the relevant universe. Identities contain no connectives, existentially quantified variables, or relations of any kind other than the allowed operations. The study of varieties is an important part of universal algebra. An algebraic structure in a variety may be understood as the quotient algebra of term algebra (also called "absolutely free algebra") divided by the equivalence relations generated by a set of identities. So, a collection of functions with given signatures generate a free algebra, the term algebra T. Given a set of equational identities (the axioms), one may consider their symmetric, transitive closure E. The quotient algebra T/E is then the algebraic structure or variety. Thus, for example, groups have a signature containing two operators: the multiplication operator m, taking two arguments, and the inverse operator i, taking one argument, and the identity element e, a constant, which may be considered an operator that takes zero arguments. Given a (countable) set of variables x, y, z, etc. the term algebra is the collection of all possible terms involving m, i, e and the variables; so for example, $m(i(x), m(x,m(y,e)))$ would be an element of the term algebra. One of the axioms defining a group is the identity $m(x, i(x)) = e$; another is $m(x,e) = x$. The axioms can be represented as trees. These equations induce equivalence classes on the free algebra; the quotient algebra then has the algebraic structure of a group.

Several non-variety structures fail to be varieties, because either:

1. It is necessary that $0 \neq 1$, 0 being the additive identity element and 1 being a multiplicative identity element, but this is a nonidentity;

2. Structures such as fields have some axioms that hold only for nonzero members of S. For an algebraic structure to be a variety, its operations must be defined for *all* members of S; there can be no partial operations.

Structures whose axioms unavoidably include nonidentities are among the most important ones in mathematics, e.g., fields and hence also vector spaces and algebras. Although structures with nonidentities retain an undoubted algebraic flavor, they suffer from defects varieties do not have. For example, the product of two fields is not a field.

10.5   Category theory

Category theory is another tool for studying algebraic structures (see, for example, Mac Lane 1998). A category is a collection of *objects* with associated *morphisms*. Every algebraic structure has its own notion of homomorphism, namely any function compatible with the operation(s) defining the structure. In this way, every algebraic structure gives rise to a category. For example, the category of groups has all groups as objects and all group homomorphisms as morphisms. This concrete category may be seen as a category of sets with added category-theoretic structure. Likewise, the category of topological groups (whose morphisms are the continuous group homomorphisms) is a category of topological spaces with extra structure. A forgetful functor between categories of algebraic structures "forgets" a part of a structure.

There are various concepts in category theory that try to capture the algebraic character of a context, for instance

- algebraic category
- essentially algebraic category
- presentable category
- locally presentable category
- monadic functors and categories
- universal property.

10.6   See also

- Mathematical structure
- Structure (mathematical logic)
- List of algebraic structures
- Signature (logic)
- Free object

10.7   References

[1] P.M. Cohn. (1981) *Universal Algebra*, Springer, p. 41.

[2] Jonathan D. H. Smith. *An Introduction to Quasigroups and Their Representations*. Chapman & Hall. Retrieved 2012-08-02.

[3] Ringoids and lattices can be clearly distinguished despite both having two defining binary operations. In the case of ringoids, the two operations are linked by the distributive law; in the case of lattices, they are linked by the absorption law. Ringoids also tend to have numerical models, while lattices tend to have set-theoretic models.

- Mac Lane, Saunders; Birkhoff, Garrett (1999), *Algebra* (2nd ed.), AMS Chelsea, ISBN 978-0-8218-1646-2
- Michel, Anthony N.; Herget, Charles J. (1993), *Applied Algebra and Functional Analysis*, New York: Dover Publications, ISBN 978-0-486-67598-5
- Burris, Stanley N.; Sankappanavar, H. P. (1981), *A Course in Universal Algebra*, Berlin, New York: Springer-Verlag, ISBN 978-3-540-90578-3

Category theory

- Mac Lane, Saunders (1998), *Categories for the Working Mathematician* (2nd ed.), Berlin, New York: Springer-Verlag, ISBN 978-0-387-98403-2
- Taylor, Paul (1999), *Practical foundations of mathematics*, Cambridge University Press, ISBN 978-0-521-63107-5

10.8   External links

- Jipsen's algebra structures. Includes many structures not mentioned here.
- Mathworld page on abstract algebra.
- Stanford Encyclopedia of Philosophy: Algebra by Vaughan Pratt.

Chapter 11

Number theory

Not to be confused with Numerology.

Number theory (or **arithmetic**[note 1]) is a branch of pure mathematics devoted primarily to the study of the integers. It is sometimes called "The Queen of Mathematics" because of its foundational place in the discipline.[1] Number theorists study prime numbers as well as the properties of objects made out of integers (e.g., rational numbers) or defined as generalizations of the integers (e.g., algebraic integers).

Integers can be considered either in themselves or as solutions to equations (Diophantine geometry). Questions in number theory are often best understood through the study of analytical objects (e.g., the Riemann zeta function) that encode properties of the integers, primes or other number-theoretic objects in some fashion (analytic number theory). One may also study real numbers in relation to rational numbers, e.g., as approximated by the latter (Diophantine approximation).

The older term for number theory is *arithmetic*. By the early twentieth century, it had been superseded by "number theory".[note 2] (The word "arithmetic" is used by the general public to mean "elementary calculations"; it has also acquired other meanings in mathematical logic, as in *Peano arithmetic*, and computer science, as in *floating point arithmetic*.) The use of the term *arithmetic* for *number theory* regained some ground in the second half of the 20th century, arguably in part due to French influence.[note 3] In particular, *arithmetical* is preferred as an adjective to *number-theoretic*.

11.1 History

11.1.1 Origins

Dawn of arithmetic

The first historical find of an arithmetical nature is a fragment of a table: the broken clay tablet Plimpton 322 (Larsa, Mesopotamia, ca. 1800 BCE) contains a list of "Pythagorean triples", i.e., integers (a, b, c) such that $a^2 + b^2 = c^2$. The triples are too many and too large to have been obtained by brute force. The heading over the first column reads: "The *takiltum* of the diagonal which has been subtracted such that the width..."[2]

The table's layout suggests[3] that it was constructed by means of what amounts, in modern language, to the identity

$$\left(\tfrac{1}{2} \left(x - \tfrac{1}{x} \right) \right)^2 + 1 = \left(\tfrac{1}{2} \left(x + \tfrac{1}{x} \right) \right)^2,$$

which is implicit in routine Old Babylonian exercises.[4] If some other method was used,[5] the triples were first constructed and then reordered by c/a, presumably for actual use as a "table", i.e., with a view to applications.

It is not known what these applications may have been, or whether there could have been any; Babylonian astronomy, for example, truly flowered only later. It has been suggested instead that the table was a source of numerical examples for school problems.[6][note 4]

While Babylonian number theory—or what survives of Babylonian mathematics that can be called thus—consists of this single, striking fragment, Babylonian algebra (in the secondary-school sense of "algebra") was exceptionally well developed.[7] Late Neoplatonic sources[8] state that Pythagoras learned mathematics from the Babylonians. Much earlier sources[9] state that Thales and Pythagoras traveled and studied in Egypt.

Euclid IX 21—34 is very probably Pythagorean;[10] it is very simple material ("odd times even is even", "if an odd number measures [= divides] an even number, then it also measures [= divides] half of it"), but it is all that is needed to prove that $\sqrt{2}$ is irrational.[11] Pythagorean mystics gave great importance to the odd and the even.[12] The discovery that $\sqrt{2}$ is irrational is credited to the early Pythagoreans (pre-Theodorus).[13] By revealing (in modern terms) that numbers could be irrational, this discovery seems to have provoked the first foundational crisis in mathematical history; its proof or its divulgation are sometimes credited to Hippasus, who was expelled or split from the Pythagorean sect.[14] This forced a distinction between *numbers* (integers and the rationals—the subjects of arithmetic), on the one hand, and *lengths* and *proportions* (which we would identify with real numbers, whether rational or not), on the other hand.

The Pythagorean tradition spoke also of so-called polygonal or figurate numbers.[15] While square numbers, cubic numbers, etc., are seen now as more natural than triangular numbers, pentagonal numbers, etc., the study of the sums of triangular and pentagonal numbers would prove fruitful in the early modern period (17th to early 19th century).

We know of no clearly arithmetical material in ancient Egyptian or Vedic sources, though there is some algebra in both. The Chinese remainder theorem appears as an exercise [16] in Sun Zi's *Suan Ching*, also known as *The Mathematical Classic of Sun Zi* (3rd, 4th or 5th century CE.)[17] (There is one important step glossed over in Sun Zi's solution:[note 5] it is the problem that was later solved by Āryabhaṭa's kuṭṭaka – see below.)

There is also some numerical mysticism in Chinese mathematics,[note 6] but, unlike that of the Pythagoreans, it seems to have led nowhere. Like the Pythagoreans' perfect numbers, magic squares have passed from superstition into recreation.

Classical Greece and the early Hellenistic period

Aside from a few fragments, the mathematics of Classical Greece is known to us either through the reports of contemporary non-mathematicians or through mathematical works from the early Hellenistic period.[18] In the case of number theory, this means, by and large, *Plato* and *Euclid*, respectively.

Plato had a keen interest in mathematics, and distinguished clearly between arithmetic and calculation. (By *arithmetic* he meant, in part, theorising on number, rather than what *arithmetic* or *number theory* have come to mean.) It is through one of Plato's dialogues—namely, *Theaetetus*—that we know that Theodorus had proven that $\sqrt{3}, \sqrt{5}, ..., \sqrt{17}$ are irrational. Theaetetus was, like Plato, a disciple of Theodorus's; he worked on distinguishing different kinds of incommensurables, and was thus arguably a pioneer in the study of number systems. (Book X of Euclid's Elements is described by Pappus as being largely based on Theaetetus's work.)

Euclid devoted part of his *Elements* to prime numbers and divisibility, topics that belong unambiguously to number theory and are basic to it (Books VII to IX of Euclid's Elements). In particular, he gave an algorithm for computing the greatest common divisor of two numbers (the Euclidean algorithm; *Elements*, Prop. VII.2) and the first known proof of the infinitude of primes (*Elements*, Prop. IX.20).

In 1773, Lessing published an epigram he had found in a manuscript during his work as a librarian; it claimed to be a letter sent by Archimedes to Eratosthenes.[19][20] The epigram proposed what has become known as Archimedes' cattle problem; its solution (absent from the manuscript) requires solving an indeterminate quadratic equation (which reduces to what would later be misnamed Pell's equation). As far as we know, such equations were first successfully treated by the Indian school. It is not known whether Archimedes himself had a method of solution.

Diophantus

Very little is known about Diophantus of Alexandria; he probably lived in the third century CE, that is, about five hundred years after Euclid. Six out of the thirteen books of Diophantus's *Arithmetica* survive in the original Greek; four more books survive in an Arabic translation. The *Arithmetica* is a collection of worked-out problems where the task is invariably to find rational solutions to a system of polynomial equations, usually of the form $f(x,y)=z^2$ or $f(x,y,z)=w^2$. Thus, nowadays, we speak of *Diophantine equations* when we speak of polynomial equations to which rational or integer solutions must be found.

One may say that Diophantus was studying rational points — i.e., points whose coordinates are rational — on curves and algebraic varieties; however, unlike the Greeks of the Classical period, who did what we would now call basic algebra in geometrical terms, Diophantus did what we would now call basic algebraic geometry in purely algebraic terms. In modern language, what Diophantus did was to find rational parametrizations of varieties; that is, given an equation of the form (say) $f(x_1, x_2, x_3) = 0$, his aim was to find (in essence) three rational functions g_1, g_2, g_3 such that, for all values of r and s, setting $x_i = g_i(r, s)$ for $i = 1, 2, 3$ gives a solution to $f(x_1, x_2, x_3) = 0$.

Diophantus also studied the equations of some non-rational curves, for which no rational parametrisation is possible. He managed to find some rational points on these curves (elliptic curves, as it happens, in what seems to be their first known occurrence) by means of what amounts to a tangent construction: translated into coordinate geometry (which did not exist in Diophantus's time), his method would be visualised as drawing a tangent to a curve at a known rational point, and then finding the other point of intersection of the tangent with the curve; that other point is a new rational point. (Diophantus also resorted to what could be called a special case of a secant construction.)

While Diophantus was concerned largely with rational solutions, he assumed some results on integer numbers, in particular that every integer is the sum of four squares (though he never stated as much explicitly).

Āryabhaṭa, Brahmagupta, Bhāskara

While Greek astronomy probably influenced Indian learning, to the point of introducing trigonometry,[21] it seems to be the case that Indian mathematics is otherwise an indigenous tradition;[22] in particular, there is no evidence that Euclid's Elements reached India before the 18th century.[23]

Āryabhaṭa (476–550 CE) showed that pairs of simultaneous congruences $n \equiv a_1 \pmod{m_1}$, $n \equiv a_2 \pmod{m_2}$ could be solved by a method he called kuttaka, or pulveriser;[24] this is a procedure close to (a generalisation of) the Euclidean algorithm, which was probably discovered independently in India.[25] Āryabhaṭa seems to have had in mind applications to astronomical calculations.[21]

Brahmagupta (628 CE) started the systematic study of indefinite quadratic equations—in particular, the misnamed Pell equation, in which Archimedes may have first been interested, and which did not start to be solved in the West until the time of Fermat and Euler. Later Sanskrit authors would follow, using Brahmagupta's technical terminology. A general procedure (the chakravala, or "cyclic method") for solving Pell's equation was finally found by Jayadeva (cited in the eleventh century; his work is otherwise lost); the earliest surviving exposition appears in Bhāskara II's Bīja-gaṇita (twelfth century).[26]

Unfortunately, Indian mathematics remained largely unknown in the West until the late eighteenth century;[27] Brahmagupta and Bhāskara's work was translated into English in 1817 by Henry Colebrooke.[28]

Arithmetic in the Islamic golden age

In the early ninth century, the caliph Al-Ma'mun ordered translations of many Greek mathematical works and at least one Sanskrit work (the Sindhind, which may [29] or may not[30] be Brahmagupta's Brāhmasphuṭasiddhānta). Diophantus's main work, the Arithmetica, was translated into Arabic by Qusta ibn Luqa (820–912). Part of the treatise al-Fakhri (by al-Karajī, 953 – ca. 1029) builds on it to some extent. According to Rashed Roshdi, Al-Karajī's contemporary Ibn al-Haytham knew[31] what would later be called Wilson's theorem.

Western Europe in the Middle Ages

Other than a treatise on squares in arithmetic progression by Fibonacci — who lived and studied in north Africa and Constantinople during his formative years, ca. 1175–1200 — no number theory to speak of was done in western Europe during the Middle Ages. Matters started to change in Europe in the late Renaissance, thanks to a renewed study of the works of Greek antiquity. A catalyst was the textual emendation and translation into Latin of Diophantus's Arithmetica (Bachet, 1621, following a first attempt by Xylander, 1575).

11.1.2   Early modern number theory

Fermat

Pierre de Fermat (1601–1665) never published his writings; in particular, his work on number theory is contained almost entirely in letters to mathematicians and in private marginal notes.[32] He wrote down nearly no proofs in number theory; he had no models in the area.[33] He did make repeated use of mathematical induction, introducing the method of infinite descent.

One of Fermat's first interests was perfect numbers (which appear in Euclid, *Elements* IX) and amicable numbers;[note 7] this led him to work on integer divisors, which were from the beginning among the subjects of the correspondence (1636 onwards) that put him in touch with the mathematical community of the day.[34] He had already studied Bachet's edition of Diophantus carefully;[35] by 1643, his interests had shifted largely to Diophantine problems and sums of squares[36] (also treated by Diophantus).

Fermat's achievements in arithmetic include:

- Fermat's little theorem (1640),[37] stating that, if a is not divisible by a prime p, then $a^{p-1} \equiv 1 \pmod{p}$.[note 8]

- If a and b are coprime, then $a^2 + b^2$ is not divisible by any prime congruent to −1 modulo 4;[38] *and* Every prime congruent to 1 modulo 4 can be written in the form $a^2 + b^2$.[39] These two statements also date from 1640; in 1659, Fermat stated to Huygens that he had proven the latter statement by the method of infinite descent.[40] Fermat and Frenicle also did some work (some of it erroneous)[41] on other quadratic forms.

- Fermat posed the problem of solving $x^2 - N y^2 = 1$ as a challenge to English mathematicians (1657). The problem was solved in a few months by Wallis and Brouncker.[42] Fermat considered their solution valid, but pointed out they had provided an algorithm without a proof (as had Jayadeva and Bhaskara, though Fermat would never know this.) He states that a proof can be found by descent.

- Fermat developed methods for (doing what in our terms amounts to) finding points on curves of genus 0 and 1. As in Diophantus, there are many special procedures and what amounts to a tangent construction, but no use of a secant construction.[43]

- Fermat states and proves (by descent) in the appendix to *Observations on Diophantus* (Obs. XLV)[44] that $x^4 + y^4 = z^4$ has no non-trivial solutions in the integers. Fermat also mentioned to his correspondents that $x^3 + y^3 = z^3$ has no non-trivial solutions, and that this could be proven by descent.[45] The first known proof is due to Euler (1753; indeed by descent).[46]

Fermat's claim ("Fermat's last theorem") to have shown there are no solutions to $x^n + y^n = z^n$ for all $n \geq 3$ (a fact the only known proof of which is beyond his methods) appears only in his annotations on the margin of his copy of Diophantus; he never claimed this to others[47] and thus would have had no need to retract it if he found any mistake in his supposed proof.

Euler

The interest of Leonhard Euler (1707–1783) in number theory was first spurred in 1729, when a friend of his, the amateur[note 9] Goldbach, pointed him towards some of Fermat's work on the subject.[48][49] This has been called the "rebirth" of modern number theory,[35] after Fermat's relative lack of success in getting his contemporaries' attention for the subject.[50] Euler's work on number theory includes the following:[51]

- *Proofs for Fermat's statements.* This includes Fermat's little theorem (generalised by Euler to non-prime moduli); the fact that $p = x^2 + y^2$ if and only if $p \equiv 1 \bmod 4$; initial work towards a proof that every integer is the sum of four squares (the first complete proof is by Joseph-Louis Lagrange (1770), soon improved by Euler himself[52]); the lack of non-zero integer solutions to $x^4 + y^4 = z^2$ (implying the case $n=4$ of Fermat's last theorem, the case $n=3$ of which Euler also proved by a related method).

- *Pell's equation*, first misnamed by Euler.[53] He wrote on the link between continued fractions and Pell's equation.[54]

- *First steps towards analytic number theory.* In his work of sums of four squares, partitions, pentagonal numbers, and the distribution of prime numbers, Euler pioneered the use of what can be seen as analysis (in particular, infinite series) in number theory. Since he lived before the development of complex analysis, most of his work is restricted to the formal manipulation of power series. He did, however, do some very notable (though not fully rigorous) early work on what would later be called the Riemann zeta function.[55]

- *Quadratic forms.* Following Fermat's lead, Euler did further research on the question of which primes can be expressed in the form $x^2 + Ny^2$, some of it prefiguring quadratic reciprocity.[56] [57][58]

- *Diophantine equations.* Euler worked on some Diophantine equations of genus 0 and 1.[59][60] In particular, he studied Diophantus's work; he tried to systematise it, but the time was not yet ripe for such an endeavour – algebraic geometry was still in its infancy.[61] He did notice there was a connection between Diophantine problems and elliptic integrals,[61] whose study he had himself initiated.

Lagrange, Legendre and Gauss

Joseph-Louis Lagrange (1736–1813) was the first to give full proofs of some of Fermat's and Euler's work and observations - for instance, the four-square theorem and the basic theory of the misnamed "Pell's equation" (for which an algorithmic solution was found by Fermat and his contemporaries, and also by Jayadeva and Bhaskara II before them.) He also studied quadratic forms in full generality (as opposed to $mx^2 + ny^2$) — defining their equivalence relation, showing how to put them in reduced form, etc.

Adrien-Marie Legendre (1752–1833) was the first to state the law of quadratic reciprocity. He also conjectured what amounts to the prime number theorem and Dirichlet's theorem on arithmetic progressions. He gave a full treatment of the equation $ax^2 + by^2 + cz^2 = 0$ [62] and worked on quadratic forms along the lines later developed fully by Gauss.[63] In his old age, he was the first to prove "Fermat's last theorem" for $n = 5$ (completing work by Peter Gustav Lejeune Dirichlet, and crediting both him and Sophie Germain).[64]

In his *Disquisitiones Arithmeticae* (1798), Carl Friedrich Gauss (1777–1855) proved the law of quadratic reciprocity and developed the theory of quadratic forms (in particular, defining their composition). He also introduced some basic notation (congruences) and devoted a section to computational matters, including primality tests.[65] The last section of the *Disquisitiones* established a link between roots of unity and number theory:

> The theory of the division of the circle...which is treated in sec. 7 does not belong by itself to arithmetic, but its principles can only be drawn from higher arithmetic.[66]

In this way, Gauss arguably made a first foray towards both Évariste Galois's work and algebraic number theory.

11.1.3   Maturity and division into subfields

Starting early in the nineteenth century, the following developments gradually took place:

- The rise to self-consciousness of number theory (or *higher arithmetic*) as a field of study.[67]

- The development of much of modern mathematics necessary for basic modern number theory: complex analysis, group theory, Galois theory—accompanied by greater rigor in analysis and abstraction in algebra.

- The rough subdivision of number theory into its modern subfields—in particular, analytic and algebraic number theory.

Algebraic number theory may be said to start with the study of reciprocity and cyclotomy, but truly came into its own with the development of abstract algebra and early ideal theory and valuation theory; see below. A conventional starting point for analytic number theory is Dirichlet's theorem on arithmetic progressions (1837),[68] [69] whose proof introduced

L-functions and involved some asymptotic analysis and a limiting process on a real variable.[70] The first use of analytic ideas in number theory actually goes back to Euler (1730s),[71] [72] who used formal power series and non-rigorous (or implicit) limiting arguments. The use of *complex* analysis in number theory comes later: the work of Bernhard Riemann (1859) on the zeta function is the canonical starting point;[73] Jacobi's four-square theorem (1839), which predates it, belongs to an initially different strand that has by now taken a leading role in analytic number theory (modular forms).[74]

The history of each subfield is briefly addressed in its own section below; see the main article of each subfield for fuller treatments. Many of the most interesting questions in each area remain open and are being actively worked on.

11.2 Main subdivisions

11.2.1 Elementary tools

The term *elementary* generally denotes a method that does not use complex analysis. For example, the prime number theorem was first proven using complex analysis in 1896, but an elementary proof was found only in 1949 by Erdős and Selberg.[75] The term is somewhat ambiguous: for example, proofs based on complex Tauberian theorems (e.g. Wiener–Ikehara) are often seen as quite enlightening but not elementary, in spite of using Fourier analysis, rather than complex analysis as such. Here as elsewhere, an *elementary* proof may be longer and more difficult for most readers than a non-elementary one.

Number theory has the reputation of being a field many of whose results can be stated to the layperson. At the same time, the proofs of these results are not particularly accessible, in part because the range of tools they use is, if anything, unusually broad within mathematics.[76]

11.2.2 Analytic number theory

Main article: Analytic number theory
Analytic number theory may be defined

- in terms of its tools, as the study of the integers by means of tools from real and complex analysis;[68] or

- in terms of its concerns, as the study within number theory of estimates on size and density, as opposed to identities.[77]

Some subjects generally considered to be part of analytic number theory, e.g., sieve theory,[note 10] are better covered by the second rather than the first definition: some of sieve theory, for instance, uses little analysis,[note 11] yet it does belong to analytic number theory.

The following are examples of problems in analytic number theory: the prime number theorem, the Goldbach conjecture (or the twin prime conjecture, or the Hardy–Littlewood conjectures), the Waring problem and the Riemann Hypothesis. Some of the most important tools of analytic number theory are the circle method, sieve methods and L-functions (or, rather, the study of their properties). The theory of modular forms (and, more generally, automorphic forms) also occupies an increasingly central place in the toolbox of analytic number theory.[78]

One may ask analytic questions about algebraic numbers, and use analytic means to answer such questions; it is thus that algebraic and analytic number theory intersect. For example, one may define prime ideals (generalizations of prime numbers in the field of algebraic numbers) and ask how many prime ideals there are up to a certain size. This question can be answered by means of an examination of Dedekind zeta functions, which are generalizations of the Riemann zeta function, a key analytic object at the roots of the subject.[79] This is an example of a general procedure in analytic number theory: deriving information about the distribution of a sequence (here, prime ideals or prime numbers) from the analytic behavior of an appropriately constructed complex-valued function.[80]

11.2.3 Algebraic number theory

An *algebraic number* is any complex number that is a solution to some polynomial equation $f(x) = 0$ with rational coefficients; for example, every solution x of $x^5 + (11/2)x^3 - 7x^2 + 9 = 0$ (say) is an algebraic number. Fields of algebraic numbers are also called *algebraic number fields*, or shortly *number fields*. Algebraic number theory studies algebraic number fields.[81] Thus, analytic and algebraic number theory can and do overlap: the former is defined by its methods, the latter by its objects of study.

It could be argued that the simplest kind of number fields (viz., quadratic fields) were already studied by Gauss, as the discussion of quadratic forms in *Disquisitiones arithmeticae* can be restated in terms of ideals and norms in quadratic fields. (A *quadratic field* consists of all numbers of the form $a + b\sqrt{d}$, where a and b are rational numbers and d is a fixed rational number whose square root is not rational.) For that matter, the 11th-century chakravala method amounts—in modern terms—to an algorithm for finding the units of a real quadratic number field. However, neither Bhāskara nor Gauss knew of number fields as such.

The grounds of the subject as we know it were set in the late nineteenth century, when *ideal numbers*, the *theory of ideals* and *valuation theory* were developed; these are three complementary ways of dealing with the lack of unique factorisation in algebraic number fields. (For example, in the field generated by the rationals and sqrt{-5}, the number 6 can be factorised both as $6 = 2 \cdot 3$ and $6 = (1 + \sqrt{-5})(1 - \sqrt{-5})$; all of 2, 3, $1 + \sqrt{-5}$ and $1 - \sqrt{-5}$ are irreducible, and thus, in a naïve sense, analogous to primes among the integers.) The initial impetus for the development of ideal numbers (by Kummer) seems to have come from the study of higher reciprocity laws,[82] i.e., generalisations of quadratic reciprocity.

Number fields are often studied as extensions of smaller number fields: a field L is said to be an *extension* of a field K if L contains K. (For example, the complex numbers C are an extension of the reals R, and the reals R are an extension of the rationals Q.) Classifying the possible extensions of a given number field is a difficult and partially open problem. Abelian extensions—that is, extensions L of K such that the Galois group[note 12] Gal(L/K) of L over K is an abelian group—are relatively well understood. Their classification was the object of the programme of class field theory, which was initiated in the late 19th century (partly by Kronecker and Eisenstein) and carried out largely in 1900—1950.

An example of an active area of research in algebraic number theory is Iwasawa theory. The Langlands program, one of the main current large-scale research plans in mathematics, is sometimes described as an attempt to generalise class field theory to non-abelian extensions of number fields.

11.2.4 Diophantine geometry

Main articles: Diophantine geometry and Glossary of arithmetic and Diophantine geometry

The central problem of *Diophantine geometry* is to determine when a Diophantine equation has solutions, and if it does, how many. The approach taken is to think of the solutions of an equation as a geometric object.

For example, an equation in two variables defines a curve in the plane. More generally, an equation, or system of equations, in two or more variables defines a curve, a surface or some other such object in n-dimensional space. In Diophantine geometry, one asks whether there are any *rational points* (points all of whose coordinates are rationals) or *integral points* (points all of whose coordinates are integers) on the curve or surface. If there are any such points, the next step is to ask how many there are and how they are distributed. A basic question in this direction is: are there finitely or infinitely many rational points on a given curve (or surface)? What about integer points?

An example here may be helpful. Consider the Pythagorean equation $x^2 + y^2 = 1$; we would like to study its rational solutions, i.e., its solutions (x, y) such that x and y are both rational. This is the same as asking for all integer solutions to $a^2 + b^2 = c^2$; any solution to the latter equation gives us a solution $x = a/c$, $y = b/c$ to the former. It is also the same as asking for all points with rational coordinates on the curve described by $x^2 + y^2 = 1$. (This curve happens to be a circle of radius 1 around the origin.)

The rephrasing of questions on equations in terms of points on curves turns out to be felicitous. The finiteness or not of the number of rational or integer points on an algebraic curve—that is, rational or integer solutions to an equation $f(x, y) = 0$, where f is a polynomial in two variables—turns out to depend crucially on the *genus* of the curve. The *genus* can be defined as follows:[note 13] allow the variables in $f(x, y) = 0$ to be complex numbers; then $f(x, y) = 0$ defines a 2-dimensional surface in (projective) 4-dimensional space (since two complex variables can be decomposed into four real variables, i.e., four dimensions). Count the number of (doughnut) holes in the surface; call this number the *genus* of $f(x, y) = 0$. Other geometrical notions turn out to be just as crucial.

There is also the closely linked area of Diophantine approximations: given a number x, how well can it be approximated by rationals? (We are looking for approximations that are good relative to the amount of space that it takes to write the rational: call a/q (with $\gcd(a, q) = 1$) a good approximation to x if $|x - a/q| < \frac{1}{q^c}$, where c is large.) This question is of special interest if x is an algebraic number. If x cannot be well approximated, then some equations do not have integer or rational solutions. Moreover, several concepts (especially that of height) turn out to be crucial both in Diophantine geometry and in the study of Diophantine approximations. This question is also of special interest in transcendence theory: if a number can be better approximated than any algebraic number, then it is a transcendental number. It is by this argument that π and e have been shown to be transcendental.

Diophantine geometry should not be confused with the geometry of numbers, which is a collection of graphical methods for answering certain questions in algebraic number theory. *Arithmetic geometry*, on the other hand, is a contemporary term for much the same domain as that covered by the term *Diophantine geometry*. The term *arithmetic geometry* is arguably used most often when one wishes to emphasise the connections to modern algebraic geometry (as in, for instance, Faltings' theorem) rather than to techniques in Diophantine approximations.

11.3 Recent approaches and subfields

The areas below date as such from no earlier than the mid-twentieth century, even if they are based on older material. For example, as is explained below, the matter of algorithms in number theory is very old, in some sense older than the concept of proof; at the same time, the modern study of computability dates only from the 1930s and 1940s, and computational complexity theory from the 1970s.

11.3.1 Probabilistic number theory

Main article: Probabilistic number theory

Take a number at random between one and a million. How likely is it to be prime? This is just another way of asking how many primes there are between one and a million. Further: how many prime divisors will it have, on average? How many divisors will it have altogether, and with what likelihood? What is the probability that it have many more or many fewer divisors or prime divisors than the average?

Much of probabilistic number theory can be seen as an important special case of the study of variables that are almost, but not quite, mutually independent. For example, the event that a random integer between one and a million be divisible by two and the event that it be divisible by three are almost independent, but not quite.

It is sometimes said that probabilistic combinatorics uses the fact that whatever happens with probability greater than 0 must happen sometimes; one may say with equal justice that many applications of probabilistic number theory hinge on the fact that whatever is unusual must be rare. If certain algebraic objects (say, rational or integer solutions to certain equations) can be shown to be in the tail of certain sensibly defined distributions, it follows that there must be few of them; this is a very concrete non-probabilistic statement following from a probabilistic one.

At times, a non-rigorous, probabilistic approach leads to a number of heuristic algorithms and open problems, notably Cramér's conjecture.

11.3.2   Arithmetic combinatorics

Main articles: Arithmetic combinatorics and Additive number theory

Let A be a set of N integers. Consider the set $A + A = \{\, m + n \mid m, n \in A \,\}$ consisting of all sums of two elements of A. Is $A + A$ much larger than A? Barely larger? If $A + A$ is barely larger than A, must A have plenty of arithmetic structure, for example, does A resemble an arithmetic progression?

If we begin from a fairly "thick" infinite set A, does it contain many elements in arithmetic progression: $a, a + b, a + 2b$, $a + 3b, \ldots, a + 10b$, say? Should it be possible to write large integers as sums of elements of A?

These questions are characteristic of *arithmetic combinatorics*. This is a presently coalescing field; it subsumes *additive number theory* (which concerns itself with certain very specific sets A of arithmetic significance, such as the primes or the squares) and, arguably, some of the *geometry of numbers*, together with some rapidly developing new material. Its focus on issues of growth and distribution accounts in part for its developing links with ergodic theory, finite group theory, model theory, and other fields. The term *additive combinatorics* is also used; however, the sets A being studied need not be sets of integers, but rather subsets of non-commutative groups, for which the multiplication symbol, not the addition symbol, is traditionally used; they can also be subsets of rings, in which case the growth of $A + A$ and $A \cdot A$ may be compared.

11.3.3   Computations in number theory

Main article: Computational number theory

While the word *algorithm* goes back only to certain readers of al-Khwārizmī, careful descriptions of methods of solution are older than proofs: such methods (that is, algorithms) are as old as any recognisable mathematics—ancient Egyptian, Babylonian, Vedic, Chinese—whereas proofs appeared only with the Greeks of the classical period. An interesting early case is that of what we now call the Euclidean algorithm. In its basic form (namely, as an algorithm for computing the greatest common divisor) it appears as Proposition 2 of Book VII in *Elements*, together with a proof of correctness. However, in the form that is often used in number theory (namely, as an algorithm for finding integer solutions to an equation $ax + by = c$, or, what is the same, for finding the quantities whose existence is assured by the Chinese remainder theorem) it first appears in the works of Āryabhaṭa (5th–6th century CE) as an algorithm called *kuṭṭaka* ("pulveriser"), without a proof of correctness.

There are two main questions: "can we compute this?" and "can we compute it rapidly?". Anybody can test whether a number is prime or, if it is not, split it into prime factors; doing so rapidly is another matter. We now know fast algorithms for testing primality, but, in spite of much work (both theoretical and practical), no truly fast algorithm for factoring.

The difficulty of a computation can be useful: modern protocols for encrypting messages (e.g., RSA) depend on functions that are known to all, but whose inverses (a) are known only to a chosen few, and (b) would take one too long a time to figure out on one's own. For example, these functions can be such that their inverses can be computed only if certain large integers are factorized. While many difficult computational problems outside number theory are known, most working encryption protocols nowadays are based on the difficulty of a few number-theoretical problems.

On a different note — some things may not be computable at all; in fact, this can be proven in some instances. For instance, in 1970, it was proven, as a solution to Hilbert's 10th problem, that there is no Turing machine which can solve all Diophantine equations.[83] In particular, this means that, given a computably enumerable set of axioms, there are Diophantine equations for which there is no proof, starting from the axioms, of whether the set of equations has or does not have integer solutions. (We would necessarily be speaking of Diophantine equations for which there are no integer solutions, since, given a Diophantine equation with at least one solution, the solution itself provides a proof of the fact that a solution exists. We cannot prove, of course, that a particular Diophantine equation is of this kind, since this would imply that it has no solutions.)

11.4 Applications

The number-theorist Leonard Dickson (1874-1954) said "Thank God that number theory is unsullied by any application". Such a view is no longer applicable to number theory.[84] In 1974, Donald Knuth said "...virtually every theorem in elementary number theory arises in a natural, motivated way in connection with the problem of making computers do high-speed numerical calculations".[85] Elementary number theory is taught in discrete mathematics courses for computer scientists; and, on the other hand, number theory also has applications to the continuous in numerical analysis.[86] As well as the well-known applications to cryptography, there are also applications to many other areas of mathematics.[87][88]

11.5 Literature

Two of the most popular introductions to the subject are:

- G. H. Hardy; E. M. Wright (2008) [1938]. *An introduction to the theory of numbers* (rev. by D. R. Heath-Brown and J. H. Silverman, 6th ed.). Oxford University Press. ISBN 978-0-19-921986-5.

- Vinogradov, I. M. (2003) [1954]. *Elements of Number Theory* (reprint of the 1954 ed.). Mineola, NY: Dover Publications.

Hardy and Wright's book is a comprehensive classic, though its clarity sometimes suffers due to the authors' insistence on elementary methods.[89] Vinogradov's main attraction consists in its set of problems, which quickly lead to Vinogradov's own research interests; the text itself is very basic and close to minimal. Other popular first introductions are:

- Ivan M. Niven; Herbert S. Zuckerman; Hugh L. Montgomery (2008) [1960]. *An introduction to the theory of numbers* (reprint of the 5th edition 1991 ed.). John Wiley & Sons. ISBN 978-8-12-651811-1.

- Kenneth H. Rosen (2010). *Elementary Number Theory* (6th ed.). Pearson Education. ISBN 978-0-32-171775-7.

Popular choices for a second textbook include:

- Borevich, A. I.; Shafarevich, Igor R. (1966). *Number theory*. Pure and Applied Mathematics 20. Boston, MA: Academic Press. ISBN 978-0-12-117850-5. MR 0195803.

- Serre, Jean-Pierre (1996) [1973]. *A course in arithmetic*. Graduate texts in mathematics 7. Springer. ISBN 978-0-387-90040-7.

11.6 Prizes

The American Mathematical Society awards the *Cole Prize in Number Theory*. Moreover number theory is one of the three mathematical subdisciplines rewarded by the *Fermat Prize*.

11.7 See also

- Algebraic function field
- Finite field
- p-adic number

11.8   Notes

[1] Especially in older sources; see two following notes.

[2] Already in 1921, T. L. Heath had to explain: "By arithmetic, Plato meant, not arithmetic in our sense, but the science which considers numbers in themselves, in other words, what we mean by the Theory of Numbers." (Heath 1921, p. 13)

[3] Take, e.g. Serre 1973. In 1952, Davenport still had to specify that he meant *The Higher Arithmetic*. Hardy and Wright wrote in the introduction to *An Introduction to the Theory of Numbers* (1938): "We proposed at one time to change [the title] to *An introduction to arithmetic*, a more novel and in some ways a more appropriate title; but it was pointed out that this might lead to misunderstandings about the content of the book." (Hardy & Wright 2008)

[4] Robson 2001, p. 201. This is controversial. See Plimpton 322. Robson's article is written polemically (Robson 2001, p. 202) with a view to "perhaps [...] knocking [Plimpton 322] off its pedestal" (Robson 2001, p. 167); at the same time, it settles to the conclusion that

> [...] the question "how was the tablet calculated?" does not have to have the same answer as the question "what problems does the tablet set?" The first can be answered most satisfactorily by reciprocal pairs, as first suggested half a century ago, and the second by some sort of right-triangle problems (Robson 2001, p. 202).

Robson takes issue with the notion that the scribe who produced Plimpton 322 (who had to "work for a living", and would not have belonged to a "leisured middle class") could have been motivated by his own "idle curiosity" in the absence of a "market for new mathematics".(Robson 2001, pp. 199–200)

[5] Sun Zi, *Suan Ching*, Ch. 3, Problem 26, in Lam & Ang 2004, pp. 219–220:

> [26] Now there are an unknown number of things. If we count by threes, there is a remainder 2; if we count by fives, there is a remainder 3; if we count by sevens, there is a remainder 2. Find the number of things. *Answer*: 23.
> *Method*: If we count by threes and there is a remainder 2, put down 140. If we count by fives and there is a remainder 3, put down 63. If we count by sevens and there is a remainder 2, put down 30. Add them to obtain 233 and subtract 210 to get the answer. If we count by threes and there is a remainder 1, put down 70. If we count by fives and there is a remainder 1, put down 21. If we count by sevens and there is a remainder 1, put down 15. When [a number] exceeds 106, the result is obtained by subtracting 105.

[6] See, e.g., Sun Zi, *Suan Ching*, Ch. 3, Problem 36, in Lam & Ang 2004, pp. 223–224:

> [36] Now there is a pregnant woman whose age is 29. If the gestation period is 9 months, determine the sex of the unborn child. *Answer*: Male.
> *Method*: Put down 49, add the gestation period and subtract the age. From the remainder take away 1 representing the heaven, 2 the earth, 3 the man, 4 the four seasons, 5 the five phases, 6 the six pitch-pipes, 7 the seven stars [of the Dipper], 8 the eight winds, and 9 the nine divisions [of China under Yu the Great]. If the remainder is odd, [the sex] is male and if the remainder is even, [the sex] is female.

This is the last problem in Sun Zi's otherwise matter-of-fact treatise.

[7] Perfect and especially amicable numbers are of little or no interest nowadays. The same was not true in medieval times – whether in the West or the Arab-speaking world – due in part to the importance given to them by the Neopythagorean (and hence mystical) Nicomachus (ca. 100 CE), who wrote a primitive but influential "Introduction to Arithmetic". See van der Waerden 1961, Ch. IV.

[8] Here, as usual, given two integers a and b and a non-zero integer m, we write $a \equiv b \pmod{m}$ (read "a is congruent to b modulo m") to mean that m divides $a - b$, or, what is the same, a and b leave the same residue when divided by m. This notation is actually much later than Fermat's; it first appears in section 1 of Gauss's Disquisitiones Arithmeticae. Fermat's little theorem is a consequence of the fact that the order of an element of a group divides the order of the group. The modern proof would have been within Fermat's means (and was indeed given later by Euler), even though the modern concept of a group came long after Fermat or Euler. (It helps to know that inverses exist modulo p (i.e., given a not divisible by a prime p, there is an integer x such that $xa \equiv 1 \pmod{p}$); this fact (which, in modern language, makes the residues mod p into a group, and which was already known to Āryabhata; see above) was familiar to Fermat thanks to its rediscovery by Bachet (Weil 1984, p. 7). Weil goes on to say that Fermat would have recognised that Bachet's argument is essentially Euclid's algorithm.

[9] Up to the second half of the seventeenth century, academic positions were very rare, and most mathematicians and scientists earned their living in some other way (Weil 1984, pp. 159, 161). (There were already some recognisable features of professional *practice*, viz., seeking correspondents, visiting foreign colleagues, building private libraries (Weil 1984, pp. 160–161). Matters started to shift in the late 17th century (Weil 1984, p. 161); scientific academies were founded in England (the Royal Society, 1662) and France (the Académie des sciences, 1666) and Russia (1724). Euler was offered a position at this last one in 1726; he accepted, arriving in St. Petersburg in 1727 (Weil 1984, p. 163 and Varadarajan 2006, p. 7). In this context, the term *amateur* usually applied to Goldbach is well-defined and makes some sense: he has been described as a man of letters who earned a living as a spy (Truesdell 1984, p. xv); cited in Varadarajan 2006, p. 9). Notice, however, that Goldbach published some works on mathematics and sometimes held academic positions.

[10] Sieve theory figures as one of the main subareas of analytic number theory in many standard treatments; see, for instance, Iwaniec & Kowalski 2004 or Montgomery & Vaughan 2007

[11] This is the case for small sieves (in particular, some combinatorial sieves such as the Brun sieve) rather than for large sieves; the study of the latter now includes ideas from harmonic and functional analysis.

[12] The Galois group of an extension K/L consists of the operations (isomorphisms) that send elements of L to other elements of L while leaving all elements of K fixed. Thus, for instance, $Gal(C/R)$ consists of two elements: the identity element (taking every element $x + iy$ of C to itself) and complex conjugation (the map taking each element $x + iy$ to $x - iy$). The Galois group of an extension tells us many of its crucial properties. The study of Galois groups started with Évariste Galois; in modern language, the main outcome of his work is that an equation $f(x) = 0$ can be solved by radicals (that is, x can be expressed in terms of the four basic operations together with square roots, cubic roots, etc.) if and only if the extension of the rationals by the roots of the equation $f(x) = 0$ has a Galois group that is solvable in the sense of group theory. ("Solvable", in the sense of group theory, is a simple property that can be checked easily for finite groups.)

[13] It may be useful to look at an example here. Say we want to study the curve $y^2 = x^3 + 7$. We allow x and y to be complex numbers: $(a + bi)^2 = (c + di)^3 + 7$. This is, in effect, a set of two equations on four variables, since both the real and the imaginary part on each side must match. As a result, we get a surface (two-dimensional) in four-dimensional space. After we choose a convenient hyperplane on which to project the surface (meaning that, say, we choose to ignore the coordinate a), we can plot the resulting projection, which is a surface in ordinary three-dimensional space. It then becomes clear that the result is a torus, i.e., the surface of a doughnut (somewhat stretched). A doughnut has one hole; hence the genus is 1.

11.9 References

[1] Long 1972, p. 1.

[2] Neugebauer & Sachs 1945, p. 40. The term *takiltum* is problematic. Robson prefers the rendering "The holding-square of the diagonal from which 1 is torn out, so that the short side comes up...".Robson 2001, p. 192

[3] Robson 2001, p. 189. Other sources give the modern formula $(p^2 - q^2, 2pq, p^2 + q^2)$. Van der Waerden gives both the modern formula and what amounts to the form preferred by Robson.(van der Waerden 1961, p. 79)

[4] van der Waerden 1961, p. 184.

[5] Neugebauer (Neugebauer 1969, pp. 36–40) discusses the table in detail and mentions in passing Euclid's method in modern notation (Neugebauer 1969, p. 39).

[6] Friberg 1981, p. 302.

[7] van der Waerden 1961, p. 43.

[8] Iamblichus, *Life of Pythagoras*,(trans. e.g. Guthrie 1987) cited in van der Waerden 1961, p. 108. See also Porphyry, *Life of Pythagoras*, paragraph 6, in Guthrie 1987 Van der Waerden (van der Waerden 1961, pp. 87–90) sustains the view that Thales knew Babylonian mathematics.

[9] Herodotus (II. 81) and Isocrates (*Busiris* 28), cited in: Huffman 2011. On Thales, see Eudemus ap. Proclus, 65.7, (e.g. Morrow 1992, p. 52) cited in: O'Grady 2004, p. 1. Proclus was using a work by Eudemus of Rhodes (now lost), the *Catalogue of Geometers*. See also introduction, Morrow 1992, p. xxx on Proclus' reliability.

[10] Becker 1936, p. 533, cited in: van der Waerden 1961, p. 108.

[11] Becker 1936.

[12] van der Waerden 1961, p. 109.

[13] Plato, *Theaetetus*, p. 147 B, (e.g. Jowett 1871), cited in von Fritz 2004, p. 212: "Theodorus was writing out for us something about roots, such as the roots of three or five, showing that they are incommensurable by the unit;..." *See also* Spiral of Theodorus.

[14] von Fritz 2004.

[15] Heath 1921, p. 76.

[16] Sun Zi, *Suan Ching*, Chapter 3, Problem 26. This can be found in Lam & Ang 2004, pp. 219–220, which contains a full translation of the *Suan Ching* (based on Qian 1963). See also the discussion in Lam & Ang 2004, pp. 138–140.

[17] The date of the text has been narrowed down to 220–420 AD (Yan Dunjie) or 280–473 AD (Wang Ling) through internal evidence (= taxation systems assumed in the text). See Lam & Ang 2004, pp. 27–28.

[18] Boyer & Merzbach 1991, p. 82.

[19] Vardi 1998, p. 305-319.

[20] Weil 1984, pp. 17–24.

[21] Plofker 2008, p. 119.

[22] Any early contact between Babylonian and Indian mathematics remains conjectural (Plofker 2008, p. 42).

[23] Mumford 2010, p. 387.

[24] Āryabhaṭa, Āryabhaṭīya, Chapter 2, verses 32–33, cited in: Plofker 2008, pp. 134–140. See also Clark 1930, pp. 42–50. A slightly more explicit description of the kuttaka was later given in Brahmagupta, *Brāhmasphuṭasiddhānta*, XVIII, 3–5 (in Colebrooke 1817, p. 325, cited in Clark 1930, p. 42).

[25] Mumford 2010, p. 388.

[26] Plofker 2008, p. 194.

[27] Plofker 2008, p. 283.

[28] Colebrooke 1817.

[29] Colebrooke 1817, p. lxv, cited in Hopkins 1990, p. 302. See also the preface in Sachau 1888 cited in Smith 1958, pp. 168

[30] Pingree 1968, pp. 97–125, and Pingree 1970, pp. 103–123, cited in Plofker 2008, p. 256.

[31] Rashed 1980, p. 305-321.

[32] Weil 1984, pp. 45–46.

[33] Weil 1984, p. 118. This was more so in number theory than in other areas (remark in Mahoney 1994, p. 284). Bachet's own proofs were "ludicrously clumsy" (Weil 1984, p. 33).

[34] Mahoney 1994, pp. 48, 53–54. The initial subjects of Fermat's correspondence included divisors ("aliquot parts") and many subjects outside number theory; see the list in the letter from Fermat to Roberval, 22.IX.1636, Tannery & Henry 1891, Vol. II, pp. 72, 74, cited in Mahoney 1994, p. 54.

[35] Weil 1984, pp. 1–2.

[36] Weil 1984, p. 53.

[37] Tannery & Henry 1891, Vol. II, p. 209, Letter XLVI from Fermat to Frenicle, 1640, cited in Weil 1984, p. 56

[38] Tannery & Henry 1891, Vol. II, p. 204, cited in Weil 1984, p. 63. All of the following citations from Fermat's *Varia Opera* are taken from Weil 1984, Chap. II. The standard Tannery & Henry work includes a revision of Fermat's posthumous *Varia Opera Mathematica* originally prepared by his son (Fermat 1679).

[39] Tannery & Henry 1891, Vol. II, p. 213.

[40] Tannery & Henry 1891, Vol. II, p. 423.

[41] Weil 1984, pp. 80, 91–92.

[42] Weil 1984, p. 92.

[43] Weil 1984, Ch. II, sect. XV and XVI.

[44] Tannery & Henry 1891, Vol. I, pp. 340–341.

[45] Weil 1984, p. 115.

[46] Weil 1984, pp. 115–116.

[47] Weil 1984, p. 104.

[48] Weil 1984, pp. 2, 172.

[49] Varadarajan 2006, p. 9.

[50] Weil 1984, p. 2 and Varadarajan 2006, p. 37

[51] Varadarajan 2006, p. 39 and Weil 1984, pp. 176–189

[52] Weil 1984, pp. 178–179.

[53] Weil 1984, p. 174. Euler was generous in giving credit to others (Varadarajan 2006, p. 14), not always correctly.

[54] Weil 1984, p. 183.

[55] Varadarajan 2006, pp. 45–55; see also chapter III.

[56] Varadarajan 2006, pp. 44–47.

[57] Weil 1984, pp. 177–179.

[58] Edwards 1983, pp. 285–291.

[59] Varadarajan 2006, pp. 55–56.

[60] Weil 1984, pp. 179–181.

[61] Weil 1984, p. 181.

[62] Weil 1984, pp. 327–328.

[63] Weil 1984, pp. 332–334.

[64] Weil 1984, pp. 337–338.

[65] Goldstein & Schappacher 2007, p. 14.

[66] From the preface of *Disquisitiones Arithmeticae*; the translation is taken from Goldstein & Schappacher 2007, p. 16

[67] See the discussion in section 5 of Goldstein & Schappacher 2007. Early signs of self-consciousness are present already in letters by Fermat: thus his remarks on what number theory is, and how "Diophantus's work [...] does not really belong to [it]" (quoted in Weil 1984, p. 25).

[68] Apostol 1976, p. 7.

[69] Davenport & Montgomery 2000, p. 1.

[70] See the proof in Davenport & Montgomery 2000, section 1

[71] Iwaniec & Kowalski 2004, p. 1.

[72] Varadarajan 2006, sections 2.5, 3.1 and 6.1.

[73] Granville 2008, pp. 322–348.

[74] See the comment on the importance of modularity in Iwaniec & Kowalski 2004, p. 1

[75] Goldfeld 2003.

[76] See, e.g., the initial comment in Iwaniec & Kowalski 2004, p. 1.

[77] Granville 2008, section 1: "The main difference is that in algebraic number theory [...] one typically considers questions with answers that are given by exact formulas, whereas in analytic number theory [...] one looks for *good approximations.*"

[78] See the remarks in the introduction to Iwaniec & Kowalski 2004, p. 1: "However much stronger...".

[79] Granville 2008, section 3: "[Riemann] defined what we now call the Riemann zeta function [...] Riemann's deep work gave birth to our subject [...]"

[80] See, e.g., Montgomery & Vaughan 2007, p. 1.

[81] CITEREFMilne2014, p. 2.

[82] Edwards 2000, p. 79.

[83] Davis, Martin; Matiyasevich, Yuri; Robinson, Julia (1976). "Hilbert's Tenth Problem: Diophantine Equations: Positive Aspects of a Negative Solution". In Felix E. Browder. *Mathematical Developments Arising from Hilbert Problems.* Proceedings of Symposia in Pure Mathematics. XXVIII.2. American Mathematical Society. pp. 323–378. ISBN 0-8218-1428-1. Zbl 0346.02026. Reprinted in *The Collected Works of Julia Robinson,* Solomon Feferman, editor, pp.269–378, American Mathematical Society 1996.

[84] "The Unreasonable Effectiveness of Number Theory", Stefan Andrus Burr, George E. Andrews, American Mathematical Soc., 1992, ISBN 9780821855010

[85] Computer science and its relation to mathematics" DF Knuth - The American Mathematical Monthly, 1974

[86] "Applications of number theory to numerical analysis", Lo-keng Hua, Luogeng Hua, Yuan Wang, Springer-Verlag, 1981, ISBN 978-3-540-10382-0

[87] "Practical applications of algebraic number theory". Mathoverflow.net. Retrieved 2012-05-18.

[88] "Where is number theory used in the rest of mathematics?". Mathoverflow.net. 2008-09-23. Retrieved 2012-05-18.

[89] Apostol n.d..

11.10   Sources

- Apostol, Tom M. (1976). *Introduction to analytic number theory.* Undergraduate Texts in Mathematics. Springer. ISBN 978-0-387-90163-3.

- Apostol, Tom M. (n.d.). "An Introduction to the Theory of Numbers". (Review of Hardy & Wright.) Mathematical Reviews (MathSciNet) MR0568909. American Mathematical Society. (Subscription needed)

- Becker, Oskar (1936). "Die Lehre von Geraden und Ungeraden im neunten Buch der euklidischen Elemente". *Quellen und Studien zur Geschichte der Mathematik, Astronomie und Physik.* Abteilung B:Studien (in German) (Berlin: J. Springer Verlag) 3: 533–53.

- Boyer, Carl Benjamin; Merzbach, Uta C. (1991) [1968]. *A History of Mathematics* (2nd ed.). New York: Wiley. ISBN 978-0-471-54397-8. 1968 edition at archive.org

- Clark, Walter Eugene (trans.) (1930). *The Āryabhaṭīya of Āryabhaṭa: An ancient Indian work on Mathematics and Astronomy.* University of Chicago Press.

- Colebrooke, Henry Thomas (1817). *Algebra, with Arithmetic and Mensuration, from the Sanscrit of Brahmegupta and Bháscara.* London: J. Murray.

- Davenport, Harold; Montgomery, Hugh L. (2000). *Multiplicative Number Theory*. Graduate texts in mathematics 74 (revised 3rd ed.). Springer. ISBN 978-0-387-95097-6.

- Edwards, Harold M. (November 1983). "Euler and Quadratic Reciprocity". *Mathematics Magazine* (Mathematical Association of America) 56 (5): 285–291. doi:10.2307/2690368. JSTOR 2690368.

- Edwards, Harold M. (2000) [1977]. *Fermat's Last Theorem: a Genetic Introduction to Algebraic Number Theory*. Graduate Texts in Mathematics 50 (reprint of 1977 ed.). Springer Verlag. ISBN 978-0-387-95002-0.

- Fermat, Pierre de (1679). *Varia Opera Mathematica* (in French and Latin). Toulouse: Joannis Pech.

- Friberg, Jöran (August 1981). "Methods and Traditions of Babylonian Mathematics: Plimpton 322, Pythagorean Triples and the Babylonian Triangle Parameter Equations". *Historia Mathematica* (Elsevier) 8 (3): 277–318. doi:10.1016/0315-0860(81)90069-0.

- von Fritz, Kurt (2004). "The Discovery of Incommensurability by Hippasus of Metapontum". In Christianidis, J. *Classics in the History of Greek Mathematics*. Berlin: Kluwer (Springer). ISBN 978-1-4020-0081-2.

- Gauss, Carl Friedrich; Waterhouse, William C. (trans.) (1966) [1801]. *Disquisitiones Arithmeticae*. Springer. ISBN 978-0-387-96254-2.

- Goldfeld, Dorian M. (2003). "Elementary Proof of the Prime Number Theorem: a Historical Perspective" (PDF).

- Goldstein, Catherine; Schappacher, Norbert (2007). "A book in search of a discipline". In Goldstein, C.; Schappacher, N.; Schwermer, Joachim. *The Shaping of Arithmetic after Gauss' "Disquisitiones Arithmeticae"*. Berlin & Heidelberg: Springer. pp. 3–66. ISBN 978-3-540-20441-1.

- Granville, Andrew (2008). "Analytic number theory". In Gowers, Timothy; Barrow-Green, June; Leader, Imre. *The Princeton Companion to Mathematics*. Princeton University Press. ISBN 978-0-691-11880-2.

- Porphyry; Guthrie, K. S. (trans.) (1920). *Life of Pythagoras*. Alpine, New Jersey: Platonist Press.

- Guthrie, Kenneth Sylvan (1987). *The Pythagorean Sourcebook and Library*. Grand Rapids, Michigan: Phanes Press. ISBN 978-0-933999-51-0.

- Hardy, Godfrey Harold; Wright, E. M. (2008) [1938]. *An Introduction to the Theory of Numbers* (Sixth ed.). Oxford University Press. ISBN 978-0-19-921986-5. MR 2445243.

- Heath, Thomas L. (1921). *A History of Greek Mathematics, Volume 1: From Thales to Euclid*. Oxford: Clarendon Press.

- Hopkins, J. F. P. (1990). "Geographical and Navigational Literature". In Young, M. J. L.; Latham, J. D.; Serjeant, R. B. *Religion, Learning and Science in the `Abbasid Period*. The Cambridge history of Arabic literature. Cambridge University Press. ISBN 978-0-521-32763-3.

- Huffman, Carl A. (8 August 2011). Zalta, Edward N., ed. "Pythagoras". *Stanford Encyclopaedia of Philosophy* (Fall 2011 ed.). Retrieved 7 February 2012.

- Iwaniec, Henryk; Kowalski, Emmanuel (2004). *Analytic Number Theory*. American Mathematical Society Colloquium Publications 53. Providence, RI,: American Mathematical Society. ISBN 0-8218-3633-1.

- Plato; Jowett, Benjamin (trans.) (1871). *Theaetetus*.

- Lam, Lay Yong; Ang, Tian Se (2004). *Fleeting Footsteps: Tracing the Conception of Arithmetic and Algebra in Ancient China* (revised ed.). Singapore: World Scientific. ISBN 978-981-238-696-0.

- Long, Calvin T. (1972). *Elementary Introduction to Number Theory* (2nd ed.). Lexington, VA: D. C. Heath and Company. LCCN 77171950.

- Mahoney, M. S. (1994). *The Mathematical Career of Pierre de Fermat, 1601–1665* (Reprint, 2nd ed.). Princeton University Press. ISBN 978-0-691-03666-3.

- Milne, J. S. (2014). "Algebraic Number Theory". Available at www.jmilne.org/math.

- Montgomery, Hugh L.; Vaughan, Robert C. (2007). *Multiplicative Number Theory: I, Classical Theory,*. Cambridge University Press. ISBN 978-0-521-84903-6.

- Morrow, Glenn Raymond (trans., ed.); Proclus (1992). *A Commentary on Book 1 of Euclid's Elements*. Princeton University Press. ISBN 978-0-691-02090-7.

- Mumford, David (March 2010). "Mathematics in India: reviewed by David Mumford" (PDF). *Notices of the American Mathematical Society* 57 (3): 387. ISSN 1088-9477.

- Neugebauer, Otto E. (1969). *The Exact Sciences in Antiquity* (corrected reprint of the 1957 ed.). New York: Dover Publications. ISBN 978-0-486-22332-2.

- Neugebauer, Otto E.; Sachs, Abraham Joseph; Götze, Albrecht (1945). *Mathematical Cuneiform Texts*. American Oriental Series 29. American Oriental Society etc.

- O'Grady, Patricia (September 2004). "Thales of Miletus". The Internet Encyclopaedia of Philosophy. Retrieved 7 February 2012.

- Pingree, David; Ya'qub, ibn Tariq (1968). "The Fragments of the Works of Ya'qub ibn Tariq". *Journal of Near Eastern Studies* (University of Chicago Press) 26.

- Pingree, D.; al-Fazari (1970). "The Fragments of the Works of al-Fazari". *Journal of Near Eastern Studies* (University of Chicago Press) 28.

- Plofker, Kim (2008). *Mathematics in India*. Princeton University Press. ISBN 978-0-691-12067-6.

- Qian, Baocong, ed. (1963). *Suanjing shi shu (Ten Mathematical Classics)* (in Chinese). Beijing: Zhonghua shuju.

- Rashed, Roshdi (1980). "Ibn al-Haytham et le théorème de Wilson". *Archive for History of Exact Sciences* 22 (4): 305–321. doi:10.1007/BF00717654.

- Robson, Eleanor (2001). "Neither Sherlock Holmes nor Babylon: a Reassessment of Plimpton 322" (PDF). *Historia Mathematica* (Elsevier) 28 (28): 167–206. doi:10.1006/hmat.2001.2317.

- Sachau, Eduard; Bīrūnī, Muḥammad ibn Aḥmad (1888). *Alberuni's India: An Account of the Religion, Philosophy, Literature, Geography, Chronology, Astronomy and Astrology of India, Vol. 1*. London: Kegan, Paul, Trench, Trübner & Co.

- Serre, Jean-Pierre (1996) [1973]. *A Course in Arithmetic*. Graduate texts in mathematics 7. Springer. ISBN 978-0-387-90040-7.

- Smith, D. E. (1958). *History of Mathematics, Vol I*. New York: Dover Publications.

- Tannery, Paul; Henry, Charles (eds.); Fermat, Pierre de (1891). *Oeuvres de Fermat*. (4 Vols.) (in French and Latin). Paris: Imprimerie Gauthier-Villars et Fils. Volume 1 Volume 2 Volume 3 Volume 4 (1912)

- Iamblichus; Taylor, Thomas (trans.) (1818). *Life of Pythagoras or, Pythagoric Life*. London: J. M. Watkins. For other editions, see Iamblichus#List of editions and translations

- Truesdell, C. A. (1984). "Leonard Euler, Supreme Geometer". In Hewlett, John (trans.). *Leonard Euler, Elements of Algebra* (reprint of 1840 5th ed.). New York: Springer-Verlag. ISBN 978-0-387-96014-2. This Google books preview of *Elements of algebra* lacks Truesdell's intro, which is reprinted (slightly abridged) in the following book:

- Truesdell, C. A. (2007). "Leonard Euler, Supreme Geometer". In Dunham, William. *The Genius of Euler: reflections on his life and work*. Volume 2 of MAA tercentenary Euler celebration. New York: Mathematical Association of America. ISBN 978-0-88385-558-4.

- Varadarajan, V. S. (2006). *Euler Through Time: A New Look at Old Themes*. American Mathematical Society. ISBN 978-0-8218-3580-7.

- Vardi, Ilan (April 1998). "Archimedes' Cattle Problem" (PDF). *American Mathematical Monthly* 105 (4): 305–319. doi:10.2307/2589706.

- van der Waerden, Bartel L.; Dresden, Arnold (trans) (1961). *Science Awakening*. Vol. 1 or Vol 2. New York: Oxford University Press.

- Weil, André (1984). *Number Theory: an Approach Through History – from Hammurapi to Legendre*. Boston: Birkhäuser. ISBN 978-0-8176-3141-3.

11.11 External links

- Hazewinkel, Michiel, ed. (2001), "Number theory", *Encyclopedia of Mathematics*, Springer, ISBN 978-1-55608-010-4

- Quotations related to Number theory at Wikiquote

- Number Theory Web

A Lehmer sieve, which is a primitive digital computer once used for finding primes and solving simple Diophantine equations.

The Plimpton 322 tablet

DIOPHANTI
ALEXANDRINI
ARITHMETICORVM
LIBRI SEX,
ET DE NVMERIS MVLTANGVLIS,
LIBER VNVS.

Nunc primùm Græcè & Latinè editi, atque absolutissimis
Commentariis illustrati.

AVCTORE CLAVDIO GASPARE BACHETO
MEZIRIACO SEBVSIANO V.C.

LVTETIAE PARISIORVM,
Sumptibus Sebastiani Cramoisy, via
Iacobæa, sub Ciconiis.
M. DC. XXI.
CVM PRIVILEGIO REGIS.

Leonhard Euler

DISQVISITIONES

ARITHMETICAE

AVCTORE

D. CAROLO FRIDERICO GAVSS

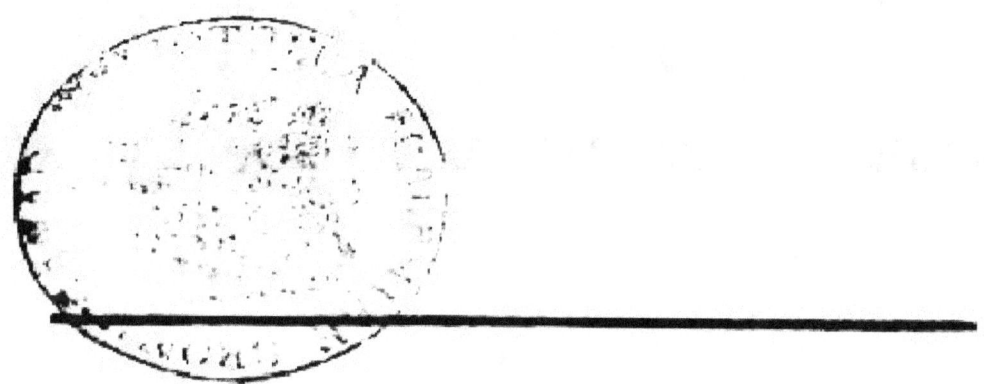

LIPSIAE

IN COMMISSIS APVD GERH. FLEISCHER, JVN.

1 8 0 1.

Carl Friedrich Gauss

Ernst Kummer

Peter Gustav Lejeune Dirichlet

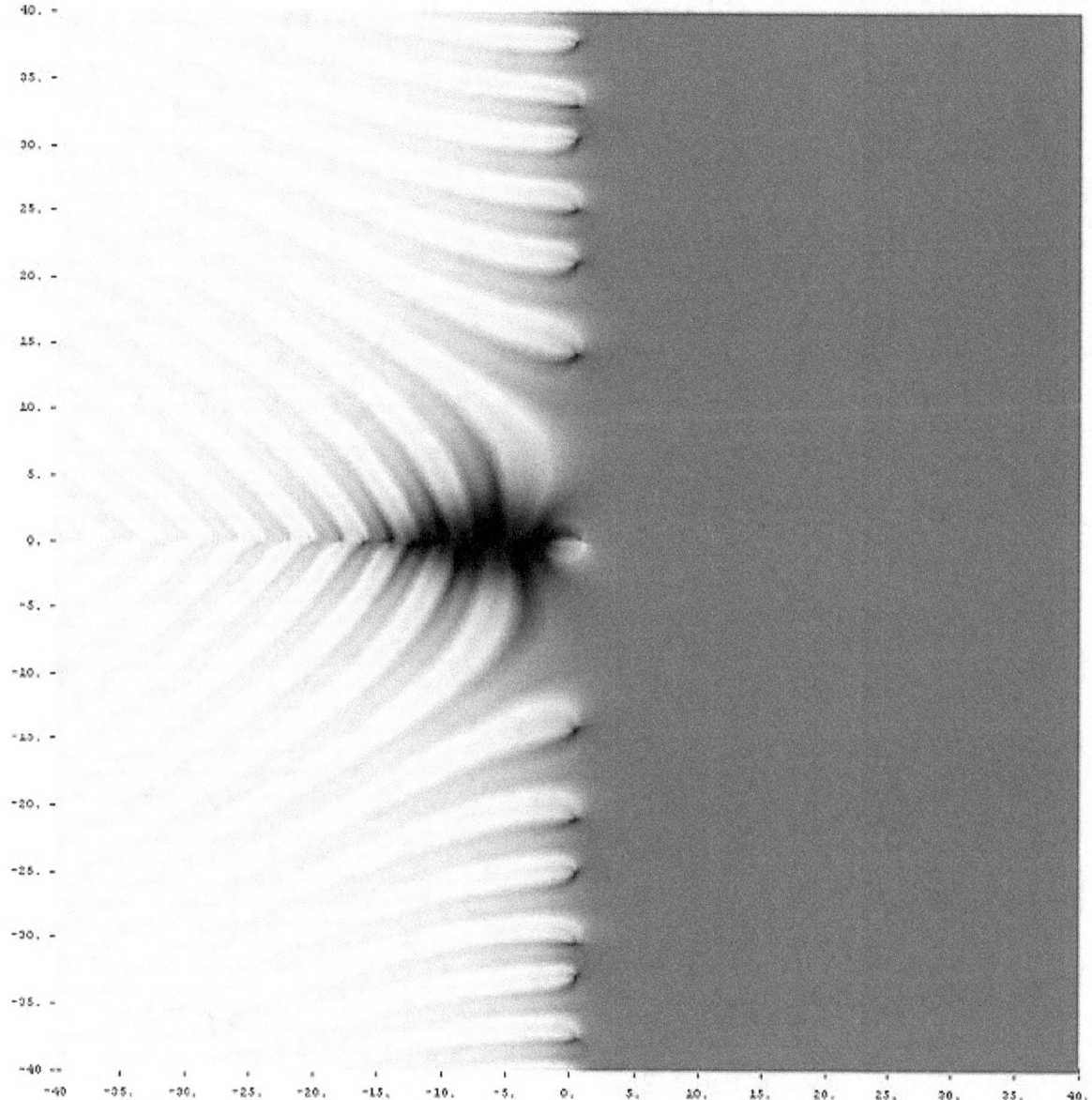

Riemann zeta function ζ(s) in the complex plane. The color of a point s gives the value of ζ(s): dark colors denote values close to zero and hue gives the value's argument.

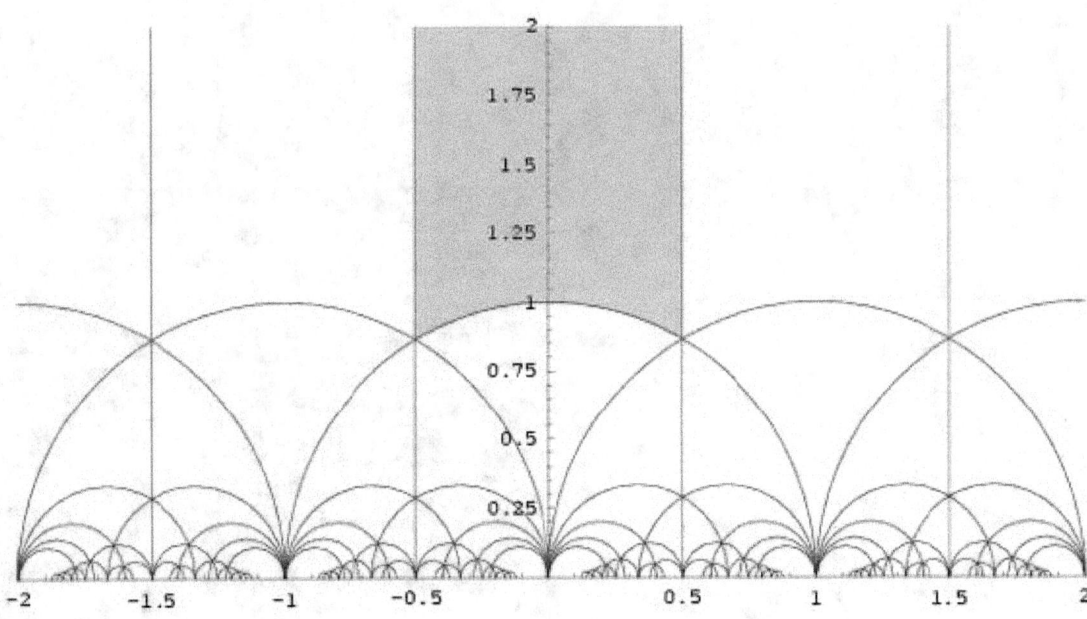

The action of the modular group on the upper half plane. The region in grey is the standard fundamental domain.

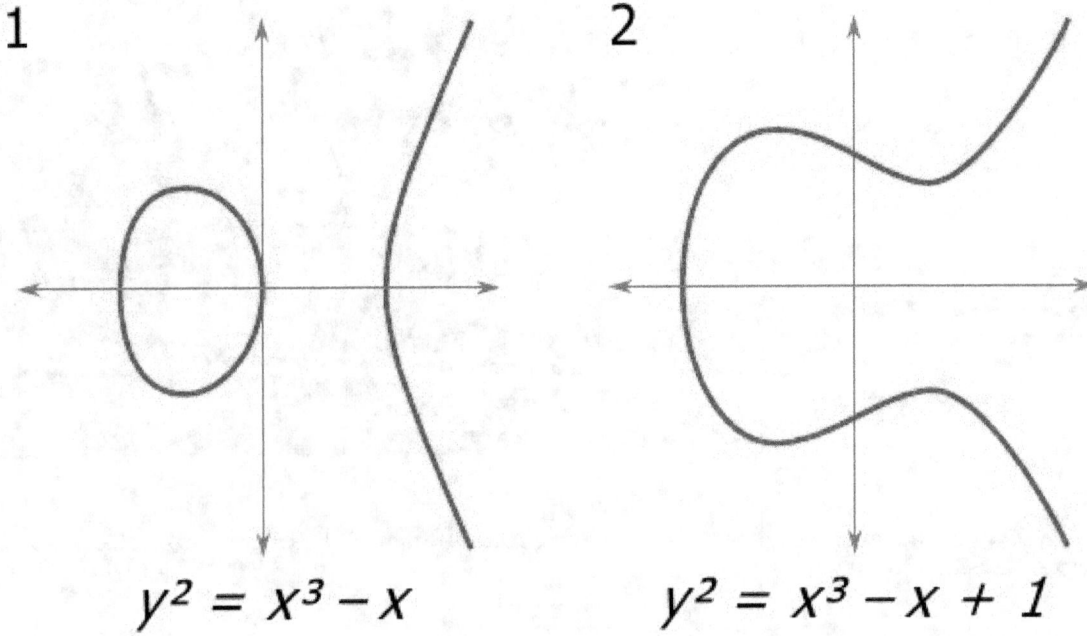

Two examples of an elliptic curve, i.e., a curve of genus 1 having at least one rational point. (Either graph can be seen as a slice of a torus in four-dimensional space.)

Chapter 12

Field (mathematics)

This article is about fields in algebra. For fields in geometry, see Vector field. For other uses, see Field (disambiguation).

In abstract algebra, a **field** is a nonzero commutative division ring, or equivalently a ring whose nonzero elements form an abelian group under multiplication. As such it is an algebraic structure with notions of addition, subtraction, multiplication, and division satisfying the appropriate abelian group equations and distributive law. The most commonly used fields are the field of real numbers, the field of complex numbers, and the field of rational numbers, but there are also finite fields, fields of functions, algebraic number fields, p-adic fields, and so forth.

Any field may be used as the scalars for a vector space, which is the standard general context for linear algebra. The theory of field extensions (including Galois theory) involves the roots of polynomials with coefficients in a field; among other results, this theory leads to impossibility proofs for the classical problems of angle trisection and squaring the circle with a compass and straightedge, as well as a proof of the Abel–Ruffini theorem on the algebraic insolubility of quintic equations. In modern mathematics, the theory of fields (or **field theory**) plays an essential role in number theory and algebraic geometry.

As an algebraic structure, every field is a ring, but not every ring is a field. The most important difference is that fields allow for division (though not division by zero), while a ring need not possess multiplicative inverses; for example the integers form a ring, but $2x = 1$ has no solution in integers. Also, the multiplication operation in a field is required to be commutative. A ring in which division is possible but commutativity is not assumed (such as the quaternions) is called a *division ring* or *skew field*. (Historically, division rings were sometimes referred to as fields, while fields were called *commutative fields*.)

As a ring, a field may be classified as a specific type of integral domain, and can be characterized by the following (not exhaustive) chain of class inclusions:

> **Commutative rings** ⊃ **integral domains** ⊃ **integrally closed domains** ⊃ **unique factorization domains**
> ⊃ **principal ideal domains** ⊃ **Euclidean domains** ⊃ **fields** ⊃ **finite fields**

12.1 Definition and illustration

Intuitively, a field is a set F that is a commutative group with respect to two compatible operations, addition and multiplication (the latter excluding zero), with "compatible" being formalized by *distributivity*, and the caveat that the additive and the multiplicative identities are distinct ($0 \neq 1$).

The most common way to formalize this is by defining a *field* as a set together with two operations, usually called *addition* and *multiplication*, and denoted by + and ·, respectively, such that the following axioms hold; *subtraction* and *division* are defined in terms of the inverse operations of addition and multiplication:[note 1]

Closure of **F** under addition and multiplication For all a, b in F, both $a + b$ and $a \cdot b$ are in F (or more formally, +

and $\cdot$ are binary operations on F).

Associativity of addition and multiplication For all a, b, and c in F, the following equalities hold: $a + (b + c) = (a + b) + c$ and $a \cdot (b \cdot c) = (a \cdot b) \cdot c$.

Commutativity of addition and multiplication For all a and b in F, the following equalities hold: $a + b = b + a$ and $a \cdot b = b \cdot a$.

Existence of additive and multiplicative identity elements There exists an element of F, called the *additive identity* element and denoted by 0, such that for all a in F, $a + 0 = a$. Likewise, there is an element, called the *multiplicative identity* element and denoted by 1, such that for all a in F, $a \cdot 1 = a$. To exclude the trivial ring, the additive identity and the multiplicative identity are required to be distinct.

Existence of additive inverses and multiplicative inverses For every a in F, there exists an element $-a$ in F, such that $a + (-a) = 0$. Similarly, for any a in F other than 0, there exists an element a^{-1} in F, such that $a \cdot a^{-1} = 1$. (The elements $a + (-b)$ and $a \cdot b^{-1}$ are also denoted $a - b$ and a/b, respectively.) In other words, *subtraction* and *division* operations exist.

Distributivity of multiplication over addition For all a, b and c in F, the following equality holds: $a \cdot (b + c) = (a \cdot b) + (a \cdot c)$.

A field is therefore an algebraic structure $\langle F, +, \cdot, -, ^{-1}, 0, 1 \rangle$; of type $\langle 2, 2, 1, 1, 0, 0 \rangle$, consisting of two abelian groups:

- F under $+$, $-$, and 0;

- $F \setminus \{0\}$ under $\cdot$, $^{-1}$, and 1, with $0 \neq 1$,

with $\cdot$ distributing over $+$.[1]

12.1.1   First example: rational numbers

A simple example of a field is the field of rational numbers, consisting of numbers which can be written as fractions a/b, where a and b are integers, and $b \neq 0$. The additive inverse of such a fraction is simply $-a/b$, and the multiplicative inverse (provided that $a \neq 0$) is b/a. To see the latter, note that

$$\frac{b}{a} \cdot \frac{a}{b} = \frac{ba}{ab} = 1.$$

The abstractly required field axioms reduce to standard properties of rational numbers, such as the law of distributivity

$$\frac{a}{b} \cdot \left(\frac{c}{d} + \frac{e}{f} \right)$$

$$= \frac{a}{b} \cdot \left(\frac{c}{d} \cdot \frac{f}{f} + \frac{e}{f} \cdot \frac{d}{d} \right) = \frac{a}{b} \cdot \frac{cf + ed}{df}$$

$$= \frac{a(cf + ed)}{bdf} = \frac{acf}{bdf} + \frac{aed}{bdf} = \frac{ac}{bd} + \frac{ae}{bf}$$

$$= \frac{a}{b} \cdot \frac{c}{d} + \frac{a}{b} \cdot \frac{e}{f}$$

ae bf

or the law of commutativity and law of associativity.

ae bf

or the law of commutativity and law of associativity.

12.1.2   Second example: a field with four elements

In addition to familiar number systems such as the rationals, there are other, less immediate examples of fields. The following example is a field consisting of four elements called O, I, A and B. The notation is chosen such that O plays the role of the additive identity element (denoted 0 in the axioms), and I is the multiplicative identity (denoted 1 above). One can check that all field axioms are satisfied. For example:

A $\cdot$ (B + A) = A $\cdot$ I = A, which equals A $\cdot$ B + A $\cdot$ A = I + B = A, as required by the distributivity.

The above field is called a finite field with four elements, and can be denoted **F**4. Field theory is concerned with understanding the reasons for the existence of this field, defined in a fairly ad-hoc manner, and describing its inner structure. For example, from a glance at the multiplication table, it can be seen that any non-zero element (i.e., I, A, and B) is a power of A: A = A^1, B = A^2 = A $\cdot$ A, and finally I = A^3 = A $\cdot$ A $\cdot$ A. This is not a coincidence, but rather one of the starting points of a deeper understanding of (finite) fields.

12.1.3   Alternative axiomatizations

As with other algebraic structures, there exist alternative axiomatizations. Because of the relations between the operations, one can alternatively axiomatize a field by explicitly assuming that there are four binary operations (add, subtract, multiply, divide) with axioms relating these, or (by functional decomposition) in terms of two binary operations (add and multiply) and two unary operations (additive inverse and multiplicative inverse), or other variants.

The usual axiomatization in terms of the two operations of addition and multiplication is brief and allows the other operations to be defined in terms of these basic ones, but in other contexts, such as topology and category theory, it is important to include all operations as explicitly given, rather than implicitly defined (compare topological group). This is because without further assumptions, the implicitly defined inverses may not be continuous (in topology), or may not be able to be defined (in category theory). Defining an inverse requires that one is working with a set, not a more general object.

For a very economical axiomatization of the field of real numbers, whose primitives are merely a set **R** with 1 $\in$ **R**, addition, and a binary relation, "<". See Tarski's axiomatization of the reals.

12.2   Related algebraic structures

The axioms imposed above resemble the ones familiar from other algebraic structures. For example, the existence of the binary operation "$\cdot$", together with its commutativity, associativity, (multiplicative) identity element and inverses are precisely the axioms for an abelian group. In other words, for any field, the subset of nonzero elements $F \setminus \{0\}$, also often denoted F^*, is an abelian group (F^*, $\cdot$) usually called multiplicative group of the field. Likewise (F, +) is an abelian group. The structure of a field is hence the same as specifying such two group structures (on the same set), obeying the distributivity.

Important other algebraic structures such as rings arise when requiring only part of the above axioms. For example, if the requirement of commutativity of the multiplication operation $\cdot$ is dropped, one gets structures usually called division rings or *skew fields*.

12.2.1   Remarks

By elementary group theory, applied to the abelian groups (F^*, $\cdot$), and (F, +), the additive inverse $-a$ and the multiplicative inverse a^{-1} are uniquely determined by a.

Similar direct consequences from the field axioms include

$-(a \cdot b) = (-a) \cdot b = a \cdot (-b)$, in particular $-a = (-1) \cdot a$

as well as

$$a \cdot 0 = 0.$$

Both can be shown by replacing *b* or *c* with 0 in the distributive property.

12.3 History

The concept of *field* was used implicitly by Niels Henrik Abel and Évariste Galois in their work on the solvability of polynomial equations with rational coefficients of degree five or higher.

In 1857, Karl von Staudt published his Algebra of Throws which provided a geometric model satisfying the axioms of a field.[2] This construction has been frequently recalled as a contribution to the foundations of mathematics.

In 1871, Richard Dedekind introduced, for a set of real or complex numbers which is closed under the four arithmetic operations, the German word *Körper*, which means "body" or "corpus" (to suggest an organically closed entity),[3] hence the common use of the letter *K* to denote a field. He also defined rings (then called *order* or *order-modul*), but the term "*a ring*" (*Zahlring*) was invented by Hilbert.[4] In 1893, Eliakim Hastings Moore called the concept "field" in English.[5][6]

In 1881, Leopold Kronecker defined what he called a "domain of rationality", which is indeed a field of polynomials in modern terms. In 1893, Heinrich M. Weber gave the first clear definition of an abstract field.[7] In 1910, Ernst Steinitz published the very influential paper *Algebraische Theorie der Körper* (English: Algebraic Theory of Fields).[8] In this paper he axiomatically studies the properties of fields and defines many important field theoretic concepts like prime field, perfect field and the transcendence degree of a field extension.

Emil Artin developed the relationship between groups and fields in great detail from 1928 through 1942.

12.4 Examples

12.4.1 Rationals and algebraic numbers

The field of rational numbers $\mathbf{Q}$ has been introduced above. A related class of fields very important in number theory are algebraic number fields. We will first give an example, namely the field $\mathbf{Q}(\zeta)$ consisting of numbers of the form

$$a + b\zeta,$$

with $a, b: \mathbf{Q}$, where ζ is a primitive third root of unity, i.e., a complex number satisfying $\zeta^3 = 1, \zeta \neq 1$. This field extension can be used to prove a special case of Fermat's last theorem, which asserts the non-existence of rational nonzero solutions to the equation

$$x^3 + y^3 = z^3.$$

In the language of field extensions detailed below, $\mathbf{Q}(\zeta)$ is a field extension of degree 2. Algebraic number fields are by definition finite field extensions of $\mathbf{Q}$, that is, fields containing $\mathbf{Q}$ having finite dimension as a $\mathbf{Q}$-vector space.

12.4.2 Reals, complex numbers, and *p*-adic numbers

Take the real numbers $\mathbf{R}$, under the usual operations of addition and multiplication. When the real numbers are given the usual ordering, they form a *complete ordered field*; it is this structure which provides the foundation for most formal treatments of calculus.

The complex numbers $\mathbf{C}$ consist of expressions

$a + bi$

where i is the imaginary unit, i.e., a (non-real) number satisfying $i^2 = -1$. Addition and multiplication of real numbers are defined in such a way that all field axioms hold for **C**. For example, the distributive law enforces

$(a + bi) \cdot (c + di) = ac + bdi + adi + bdi^2$, which equals $ac - bd + (bc + ad)i$.

The real numbers can be constructed by completing the rational numbers, i.e., filling the "gaps": for example $\sqrt{2}$ is such a gap. By a formally very similar procedure, another important class of fields, the field of *p*-adic numbers $\mathbf{Q}_p$ is built. It is used in number theory and *p*-adic analysis.

Hyperreal numbers and superreal numbers extend the real numbers with the addition of infinitesimal and infinite numbers.

12.4.3   Constructible numbers

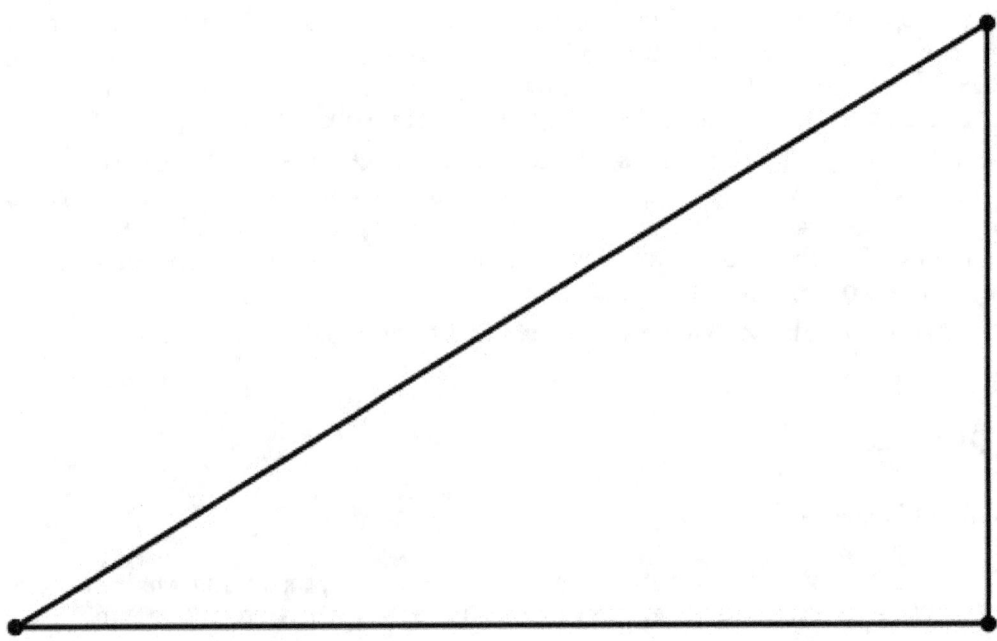

Given 0, 1, r_1 and r_2, the construction yields $r_1 r_2$

In antiquity, several geometric problems concerned the (in)feasibility of constructing certain numbers with compass and straightedge. For example it was unknown to the Greeks that it is in general impossible to trisect a given angle. Using the field notion and field theory allows these problems to be settled. To do so, the field of constructible numbers is considered. It contains, on the plane, the points 0 and 1, and all complex numbers that can be constructed from these two by a finite number of construction steps using only compass and straightedge. This set, endowed with the usual addition and multiplication of complex numbers does form a field. For example, multiplying two (real) numbers r_1 and r_2 that have already been constructed can be done using construction at the right, based on the intercept theorem. This way, the obtained field F contains all rational numbers, but is bigger than **Q**, because for any $f \in F$, the square root of f is also a constructible number.

A closely related concept is that of a Euclidean field, namely an ordered field whose positive elements are closed under square root. The real constructible numbers form the least Euclidean field, and the Euclidean fields are precisely the ordered extensions thereof.

12.4.4 Finite fields

Main article: Finite field

Finite fields (also called *Galois fields*) are fields with finitely many elements. The above introductory example F_4 is a field with four elements. F_2 consists of two elements, 0 and 1. This is the smallest field, because by definition a field has at least two distinct elements $1 \neq 0$. Interpreting the addition and multiplication in this latter field as XOR and AND operations, this field finds applications in computer science, especially in cryptography and coding theory.

In a finite field there is necessarily an integer n such that $1 + 1 + \cdots + 1$ (n repeated terms) equals 0. It can be shown that the smallest such n must be a prime number, called the *characteristic* of the field. If a (necessarily infinite) field has the property that $1 + 1 + \cdots + 1$ is never zero, for any number of summands, such as in Q, for example, the characteristic is said to be zero.

A basic class of finite fields are the fields F_p with p elements (p a prime number):

$$F_p = Z/pZ = \{0, 1, ..., p - 1\},$$

where the operations are defined by performing the operation in the set of integers Z, dividing by p and taking the remainder; see modular arithmetic. A field K of characteristic p necessarily contains F_p[9] and therefore may be viewed as a vector space over F_p, of finite dimension if K is finite. Thus a finite field K has prime power order, i.e., K has $q = p^n$ elements (where $n > 0$ is the number of elements in a basis of K over F_p). By developing more field theory, in particular the notion of the splitting field of a polynomial f over a field K, which is the smallest field containing K and all roots of f, one can show that two finite fields with the same number of elements are isomorphic, i.e., there is a one-to-one mapping of one field onto the other that preserves multiplication and addition. Thus we may speak of *the* finite field with q elements, usually denoted by F_q or GF(q).

12.4.5 Archimedean fields

Main article: Archimedean field

An Archimedean field is an ordered field such that for each element there exists a finite expression $1 + 1 + \cdots + 1$ whose value is greater than that element, that is, there are no infinite elements. Equivalently, the field contains no infinitesimals; or, the field is isomorphic to a subfield of the reals. A necessary condition for an ordered field to be complete is that it be Archimedean, since in any non-Archimedean field there is neither a greatest infinitesimal nor a least positive rational, whence the sequence 1/2, 1/3, 1/4, ..., every element of which is greater than every infinitesimal, has no limit. (And since every proper subfield of the reals also contains such gaps, up to isomorphism the reals form the unique complete ordered field.)

12.4.6 Field of functions

Given a geometric object X, one can consider functions on such objects. Adding and multiplying them pointwise, i.e., $(f \cdot g)(x) = f(x) \cdot g(x)$ this leads to a field. However, for having multiplicative inverses, one has to consider partial functions, which, almost everywhere, are defined and have a non-zero value.

If X is an algebraic variety over a field F, then the rational functions $X \to F$ form a field, the function field of X. This field consists of the functions that are defined and are the quotient of two polynomial functions outside some subvariety. Likewise, if X is a Riemann surface, then the meromorphic functions $S \to C$ form a field. Under certain circumstances, namely when S is compact, S can be reconstructed from this field.

12.4.7   Local and global fields

Another important distinction in the realm of fields, especially with regard to number theory, are local fields and global fields. Local fields are completions of global fields at a given place. For example, $\mathbf{Q}$ is a global field, and the attached local fields are $\mathbf{Q}_p$ and $\mathbf{R}$ (Ostrowski's theorem). Algebraic number fields and function fields over $\mathbf{F}_q$ are further global fields. Studying arithmetic questions in global fields may sometimes be done by looking at the corresponding questions locally—this technique is called local-global principle.

12.5   Some first theorems

- Every finite subgroup of the multiplicative group $F^\times$ is cyclic. This applies in particular to $\mathbf{F}_q^\times$, it is cyclic of order $q-1$. In the introductory example, a generator of $\mathbf{F}_4^\times$ is the element A.

- A integral domain is a field if and only if it has no ideals except $\{0\}$ and itself. Equivalently, an integral domain is a field if and only if its Krull dimension is 0.

- Isomorphism extension theorem

12.6   Constructing fields

12.6.1   Closure operations

Assuming the axiom of choice, for every field F, there exists a field $\bar{F}$, called the algebraic closure of F, which contains F, is algebraic over F, which means that any element x of $\bar{F}$ satisfies a polynomial equation

$$f_n x^n + f_{n-1} x^{n-1} + \cdots + f_1 x + f_0 = 0, \text{ with coefficients } f_n, ..., f_0 \in F,$$

and is algebraically closed, i.e., any such polynomial does have at least one solution in $\bar{F}$. The algebraic closure is unique up to isomorphism inducing the identity on F. However, in many circumstances in mathematics, it is not appropriate to treat $\bar{F}$ as being uniquely determined by F, since the isomorphism above is not itself unique. In these cases, one refers to such a $\bar{F}$ as *an* algebraic closure of F. A similar concept is the separable closure, containing all roots of separable polynomials, instead of all polynomials.

For example, if $F = \mathbf{Q}$, the algebraic closure $\bar{\mathbf{Q}}$ is also called *field of algebraic numbers*. The field of algebraic numbers is an example of an algebraically closed field of characteristic zero; as such it satisfies the same first-order sentences as the field of complex numbers $\mathbf{C}$.

In general, all algebraic closures of a field are isomorphic. However, there is in general no preferable isomorphism between two closures. Likewise for separable closures.

12.6.2   Subfields and field extensions

A *subfield* is, informally, a small field contained in a bigger one. Formally, a subfield E of a field F is a subset containing 0 and 1, closed under the operations $+$, $-$, $\cdot$ and multiplicative inverses and with its own operations defined by restriction. For example, the real numbers contain several interesting subfields: the real algebraic numbers, the computable numbers and the rational numbers are examples.

The notion of field extension lies at the heart of field theory, and is crucial to many other algebraic domains. A field extension F / E is simply a field F and a subfield $E \subset F$. Constructing such a field extension F / E can be done by "adding new elements" or *adjoining elements* to the field E. For example, given a field E, the set $F = E(X)$ of rational functions, i.e., equivalence classes of expressions of the kind

$$\frac{p(X)}{q(X)},$$

where $p(X)$ and $q(X)$ are polynomials with coefficients in E, and q is not the zero polynomial, forms a field. This is the simplest example of a transcendental extension of E. It also is an example of a domain (the ring of polynomials E in this case) being embedded into its field of fractions $E(x)$.

The ring of formal power series $E[[X]]$ is also a domain, and again the (equivalence classes of) fractions of the form $p(X)/q(X)$ where p and q are elements of $E[[X]]$ form the field of fractions for $E[[X]]$. This field is actually the ring of Laurent series over the field E, denoted $E((x))$.

In the above two cases, the added symbol X and its powers did not interact with elements of E. It is possible however that the adjoined symbol may interact with E. This idea will be illustrated by adjoining an element to the field of real numbers $\mathbf{R}$. As explained above, $\mathbf{C}$ is an extension of $\mathbf{R}$. $\mathbf{C}$ can be obtained from $\mathbf{R}$ by adjoining the imaginary symbol i which satisfies $i^2 = -1$. The result is that $\mathbf{R}[i]=\mathbf{C}$. This is different from adjoining the symbol X to $\mathbf{R}$, because in that case, the powers of X are all distinct objects, but here, $i^2=-1$ is actually an element of $\mathbf{R}$.

Another way to view this last example is to note that i is a zero of the polynomial $p(X) = X^2 + 1$. The quotient ring $\mathbf{R}[X]/(X^2 + 1)$ can be mapped onto $\mathbf{C}$ using the map $a + bX \mapsto a + ib$. Since the ideal $(X^2 +1)$ is generated by a polynomial irreducible over $\mathbf{R}$, the ideal is maximal, hence the quotient ring is a field. This nonzero ring map from the quotient to $\mathbf{C}$ is necessarily an isomorphism of rings.

The above construction generalises to any irreducible polynomial in the polynomial ring $E[X]$, i.e., a polynomial $p(X)$ that cannot be written as a product of non-constant polynomials. The quotient ring $F = E[X] / (p(X))$, is again a field.

Alternatively, constructing such field extensions can also be done, if a bigger container is already given. Suppose given a field E, and a field G containing E as a subfield, for example G could be the algebraic closure of E. Let x be an element of G not in E. Then there is a smallest subfield of G containing E and x, denoted $F = E(x)$ and called *field extension F / E generated by x in G*.[10] Such extensions are also called *simple extensions*. Many extensions are of this type; see the primitive element theorem. For instance, $\mathbf{Q}(i)$ is the subfield of $\mathbf{C}$ consisting of all numbers of the form $a + bi$ where both a and b are rational numbers.

One distinguishes between extensions having various qualities. For example, an extension K of a field k is called *algebraic*, if every element of K is a root of some polynomial with coefficients in k. Otherwise, the extension is called *transcendental*. The aim of Galois theory is the study of *algebraic extensions* of a field.

12.6.3   Rings vs fields

Adding multiplicative inverses to an integral domain R yields the field of fractions of R. For example, the field of fractions of the integers $\mathbf{Z}$ is just $\mathbf{Q}$. Also, the field $F(X)$ is the quotient field of the ring of polynomials $F[X]$.

Another method to obtain a field from a commutative ring R is taking the quotient R / m, where m is any maximal ideal of R. The above construction of $F = E[X] / (p(X))$, is an example, because the irreducibility of the polynomial $p(X)$ is equivalent to the maximality of the ideal generated by this polynomial. Another example are the finite fields $\mathbf{F}p= \mathbf{Z} / p\mathbf{Z}$.

12.6.4   Ultraproducts

If I is an index set, U is an ultrafilter on I, and Fi is a field for every i in I, the ultraproduct of the Fi with respect to U is a field.

For example, a non-principal ultraproduct of finite fields is a pseudo finite field; i.e., a PAC field having exactly one extension of any degree.

12.7   Galois theory

Main article: Galois theory

Galois theory aims to study the algebraic extensions of a field by studying the symmetry in the arithmetic operations of addition and multiplication. The fundamental theorem of Galois theory shows that there is a strong relation between the structure of the symmetry group and the set of algebraic extensions.

In the case where F / E is a finite (Galois) extension, Galois theory studies the algebraic extensions of E that are subfields of F. Such fields are called intermediate extensions. Specifically, the Galois group of F over E, denoted $\mathrm{Gal}(F/E)$, is the group of field automorphisms of F that are trivial on E (i.e., the bijections $\sigma : F \to F$ that preserve addition and multiplication and that send elements of E to themselves), and the fundamental theorem of Galois theory states that there is a one-to-one correspondence between subgroups of $\mathrm{Gal}(F/E)$ and the set of intermediate extensions of the extension F/E. The theorem, in fact, gives an explicit correspondence and further properties.

To study all (separable) algebraic extensions of E at once, one must consider the absolute Galois group of E, defined as the Galois group of the separable closure, E^{sep}, of E over E (i.e., $\mathrm{Gal}(E^{sep}/E)$). It is possible that the degree of this extension is infinite (as in the case of $E = \mathbf{Q}$). It is thus necessary to have a notion of Galois group for an infinite algebraic extension. The Galois group in this case is obtained as a "limit" (specifically an inverse limit) of the Galois groups of the finite Galois extensions of E. In this way, it acquires a topology.[note 2] The fundamental theorem of Galois theory can be generalized to the case of infinite Galois extensions by taking into consideration the topology of the Galois group, and in the case of E^{sep}/E it states that there this a one-to-one correspondence between *closed* subgroups of $\mathrm{Gal}(E^{sep}/E)$ and the set of all separable algebraic extensions of E (technically, one only obtains those separable algebraic extensions of E that occur as subfields of the *chosen* separable closure E^{sep}, but since all separable closures of E are isomorphic, choosing a different separable closure would give the same Galois group and thus an "equivalent" set of algebraic extensions).

12.8   Generalizations

There are also proper classes with field structure, which are sometimes called **Fields**, with a capital F:

- The surreal numbers form a Field containing the reals, and would be a field except for the fact that they are a proper class, not a set.

- The nimbers form a Field. The set of nimbers with birthday smaller than 2^{2^n}, the nimbers with birthday smaller than any infinite cardinal are all examples of fields.

In a different direction, differential fields are fields equipped with a derivation. For example, the field $\mathbf{R}(X)$, together with the standard derivative of polynomials forms a differential field. These fields are central to differential Galois theory. Exponential fields, meanwhile, are fields equipped with an exponential function that provides a homomorphism between the additive and multiplicative groups within the field. The usual exponential function makes the real and complex numbers exponential fields, denoted $\mathbf{R}_{ex}$ and $\mathbf{C}_{ex}$ respectively.

Generalizing in a more categorical direction yields the field with one element and related objects.

12.8.1   Exponentiation

One does not in general study generalizations of fields with *three* binary operations. The familiar addition/subtraction, multiplication/division, exponentiation/root-extraction/logarithm operations from the natural numbers to the reals, each built up in terms of iteration of the last, mean that generalizing exponentiation as a binary operation is tempting, but has generally not proven fruitful; instead, an exponential field assumes a unary exponential function from the additive group to the multiplicative group, not a partially defined binary function. Note that the exponential operation of a^b is neither associative nor commutative, nor has a unique inverse (± 2 are both square roots of 4, for instance), unlike addition and multiplication, and further is not defined for many pairs—for example, $(-1)^{1/2} = \sqrt{-1}$ does not define a single number.

These all show that even for rational numbers exponentiation is not nearly as well-behaved as addition and multiplication, which is why one does not in general axiomatize exponentiation.

12.9 Applications

The concept of a field is of use, for example, in defining vectors and matrices, two structures in linear algebra whose components can be elements of an arbitrary field.

Finite fields are used in number theory, Galois theory, cryptography, coding theory and combinatorics; and again the notion of algebraic extension is an important tool.

12.10 See also

- Category of fields
- Glossary of field theory for more definitions in field theory.
- Heyting field
- Lefschetz principle
- Puiseux series
- Ring
- Vector space
- Vector spaces without fields

12.11 Notes

[1] That is, the axiom for addition only assumes a binary operation $+: F \times F \to F, a, b \mapsto a + b$. The axiom of inverse allows one to define a unary operation $-: F \to F a \mapsto -a$ that sends an element to its negative (its additive inverse); this is not taken as given, but is implicitly defined in terms of addition as " $-a$ is the unique b such that $a + b = 0$ ", "implicitly" because it is defined in terms of solving an equation—and one then defines the binary operation of subtraction, also denoted by "−", as $-: F \times F \to F, a, b \mapsto a - b := a + (-b)$ in terms of addition and additive inverse. In the same way, one defines the binary operation of division $\div$ in terms of the assumed binary operation of multiplication and the implicitly defined operation of "reciprocal" (multiplicative inverse).

[2] As an inverse limit of finite discrete groups, it is equipped with the profinite topology, making it a profinite topological group

12.12 References

[1] Wallace, D A R (1998) *Groups, Rings, and Fields*, SUMS. Springer-Verlag: 151, Th. 2.

[2] Karl Georg Christian v. Staudt, *Beiträge zur Geometrie der Lage* (Contributions to the Geometry of Position), volume 2 (Nürnberg, (Germany): Bauer and Raspe, 1857). See: "Summen von Würfen" (sums of throws), pp. 166-171 ; "Produckte aus Würfen" (products of throws), pp. 171-176 ; "Potenzen von Würfen" (powers of throws), pp. 176-182.

[3] Peter Gustav Lejeune Dirichlet with R. Dedekind, *Vorlesungen über Zahlentheorie von P. G. Lejeune Dirichlet* (Lectures on Number Theory by P.G. Lejeune Dirichlet), 2nd ed., volume 1 (Braunschweig, Germany: Friedrich Vieweg und Sohn, 1871), p. 424. From page 424: *"Unter einem Körper wollen wir jedes System von unendlich vielen reellen oder complexen Zahlen verstehen, welches in sich so abgeschlossen und vollständig ist, dass die Addition, Subtraction, Multiplication und Division von je zwei dieser Zahlen immer wieder eine Zahl desselben Systems hervorbringt."* (By a "field" we will understand any system of

infinitely many real or complex numbers, which is so closed and complete that the addition, subtraction, multiplication, and division of any two of these numbers always again produces a number of the same system.)

[4] J J O'Connor and E F Robertson, *The development of Ring Theory*, September 2004.

[5] Moore, E. Hastings (1893), "A doubly-infinite system of simple groups", *Bulletin of the New York Mathematical Society* 3 (3): 73–78, doi:10.1090/S0002-9904-1893-00178-X, JFM 25.0198.01. From page 75: "Such a system of *s* marks [i.e., a finite field with *s* elements] we call a *field of order s*."

[6] *Earliest Known Uses of Some of the Words of Mathematics (F)*

[7] Fricke, Robert; Weber, Heinrich Martin (1924), *Lehrbuch der Algebra*, Vieweg, JFM 50.0042.03

[8] Steinitz, Ernst (1910), "Algebraische Theorie der Körper", *Journal für die reine und angewandte Mathematik* 137: 167–309, doi:10.1515/crll.1910.137.167, ISSN 0075-4102, JFM 41.0445.03

[9] Jacobson (2009), p. 213

[10] Jacobson (2009), p. 213

12.13   Sources

• Artin, Michael (1991), *Algebra*, Prentice Hall, ISBN 978-0-13-004763-2, especially Chapter 13

• Allenby, R.B.J.T. (1991), *Rings, Fields and Groups*, Butterworth-Heinemann, ISBN 978-0-340-54440-2

• Blyth, T.S.; Robertson, E. F. (1985), *Groups, rings and fields: Algebra through practice*, Cambridge University Press. See especially Book 3 (ISBN 0-521-27288-2) and Book 6 (ISBN 0-521-27291-2).

• Jacobson, Nathan (2009), *Basic algebra* 1 (2nd ed.), Dover, ISBN 978-0-486-47189-1

• James Ax (1968), *The elementary theory of finite fields*, Ann. of Math. (2), 88, 239–271

12.14   External links

• Hazewinkel, Michiel, ed. (2001), "Field", *Encyclopedia of Mathematics*, Springer, ISBN 978-1-55608-010-4

• Field Theory Q&A

• Fields at ProvenMath definition and basic properties.

• Field at PlanetMath.org.

Chapter 13

Commutative ring

In ring theory, a branch of abstract algebra, a **commutative ring** is a ring in which the multiplication operation is commutative. The study of commutative rings is called commutative algebra.

Some specific kinds of commutative rings are given with the following chain of class inclusions:

> **Commutative rings ⊃ integral domains ⊃ integrally closed domains ⊃ unique factorization domains ⊃ principal ideal domains ⊃ Euclidean domains ⊃ fields ⊃ finite fields**

13.1 Definition and first examples

13.1.1 Definition

For more details on the definition of rings, see Ring (mathematics).

A *ring* is a set R equipped with two binary operations, i.e. operations combining any two elements of the ring to a third. They are called *addition* and *multiplication* and commonly denoted by "+" and "·"; e.g. $a + b$ and $a \cdot b$. To form a ring these two operations have to satisfy a number of properties: the ring has to be an abelian group under addition as well as a monoid under multiplication, where multiplication distributes over addition; i.e., $a \cdot (b + c) = (a \cdot b) + (a \cdot c)$. The identity elements for addition and multiplication are denoted 0 and 1, respectively.

If the multiplication is commutative, i.e.

$$a \cdot b = b \cdot a,$$

then the ring R is called *commutative*. In the remainder of this article, all rings will be commutative, unless explicitly stated otherwise.

13.1.2 First examples

An important example, and in some sense crucial, is the ring of integers $\mathbb{Z}$ with the two operations of addition and multiplication. As the multiplication of integers is a commutative operation, this is a commutative ring. It is usually denoted $\mathbb{Z}$ as an abbreviation of the German word *Zahlen* (numbers).

A field is a commutative ring where every non-zero element a is invertible; i.e., has a multiplicative inverse b such that $a \cdot b = 1$. Therefore, by definition, any field is a commutative ring. The rational, real and complex numbers form fields.

The ring of 2×2 matrices is *not* commutative, since matrix multiplication fails to be commutative, as the following example shows:

$$\begin{bmatrix} 1 & 1 \\ 0 & 1 \end{bmatrix} \cdot \begin{bmatrix} 1 & 1 \\ 1 & 0 \end{bmatrix} = \begin{bmatrix} 2 & 1 \\ 1 & 0 \end{bmatrix}$$

$$\begin{bmatrix} 1 & 1 \\ 1 & 0 \end{bmatrix} \cdot \begin{bmatrix} 1 & 1 \\ 0 & 1 \end{bmatrix} = \begin{bmatrix} 1 & 2 \\ 1 & 1 \end{bmatrix}$$

However, matrices that can be diagonalized with the same similarity transformation do form a commutative ring. An example is the set of matrices of divided differences with respect to a fixed set of nodes.

If R is a given commutative ring, then the set of all polynomials in the variable X whose coefficients are in R forms the polynomial ring, denoted $R[X]$. The same holds true for several variables.

If V is some topological space, for example a subset of some $\mathbf{R}^n$, real- or complex-valued continuous functions on V form a commutative ring. The same is true for differentiable or holomorphic functions, when the two concepts are defined, such as for V a complex manifold.

13.2   Ideals and the spectrum

In the following, R denotes a commutative ring.

In contrast to fields, where every nonzero element is multiplicatively invertible, the theory of rings is more complicated. There are several notions to cope with that situation. First, an element a of ring R is called a unit if it possesses a multiplicative inverse. Another particular type of element is the zero divisors, i.e. a non-zero element a such that there exists a non-zero element b of the ring such that $ab = 0$. If R possesses no zero divisors, it is called an integral domain since it closely resembles the integers in some ways.

Many of the following notions also exist for not necessarily commutative rings, but the definitions and properties are usually more complicated. For example, all ideals in a commutative ring are automatically two-sided, which simplifies the situation considerably.

13.2.1   Ideals and factor rings

Main articles: Ideal and Factor ring

The inner structure of a commutative ring is determined by considering its ideals, i.e. nonempty subsets that are closed under multiplication with arbitrary ring elements and addition: for all r in R, i and j in I, both ri and $i + j$ are required to be in I. Given any subset $F = \{f_j\}_{j \in J}$ of R (where J is some index set), the ideal *generated by* F is the smallest ideal that contains F. Equivalently, it is given by finite linear combinations

$$r_1 f_1 + r_2 f_2 + \ldots + r_n f_n.$$

An ideal generated by one element is called a principal ideal. A ring all of whose ideals are principal is called a principal ideal ring; two important cases are $\mathbf{Z}$ and $k[X]$, the polynomial ring over a field k. Any ring has two ideals, namely the zero ideal $\{0\}$ and R, the whole ring. An ideal is *proper* if it is strictly smaller than the whole ring. An ideal that is not strictly contained in any proper ideal is called maximal. An ideal m is maximal if and only if R/m is a field. Except for the zero ring, any ring (with identity) possesses at least one maximal ideal; this follows from Zorn's lemma.

The definition of ideals is such that "dividing" I "out" gives another ring, the *factor ring* R/I: it is the set of cosets of I together with the operations

$$(a + I) + (b + I) = (a + b) + I \text{ and } (a + I)(b + I) = ab + I.$$

For example, the ring $\mathbf{Z}/n\mathbf{Z}$ (also denoted $\mathbf{Z}_n$), where n is an integer, is the ring of integers modulo n. It is the basis of modular arithmetic.

13.2.2   Localizations

Main article: Localization of a ring

The *localization* of a ring is the counterpart to factor rings insofar as in a factor ring R / I certain elements (namely the elements of I) become zero, whereas in the localization certain elements are rendered invertible, i.e. multiplicative inverses are added to the ring. Concretely, if S is a multiplicatively closed subset of R (i.e. whenever s, t: S then so is st) then the *localization* of R at S, or *ring of fractions* with denominators in S, usually denoted $S^{-1}R$ consists of symbols

$$\frac{r}{s}$$

subject to certain rules that mimick the cancellation familiar from rational numbers. Indeed, in this language $\mathbb{Q}$ is the localization of $\mathbb{Z}$ at all nonzero integers. This construction works for any integral domain R instead of $\mathbb{Z}$. The localization $(R \setminus \{0\})^{-1} R$ is called the quotient field of R. If S consists of the powers of one fixed element f, the localisation is written R_f.

13.2.3   Prime ideals and the spectrum

Main articles: Prime ideal and Spectrum of a ring

A particularly important type of ideals is *prime ideals*, often denoted p. This notion arose when algebraists (in the 19th century) realized that, unlike in $\mathbb{Z}$, in many rings there is no unique factorization into prime numbers. (Rings where it does hold are called unique factorization domains.) By definition, a prime ideal is a proper ideal such that, whenever the product ab of any two ring elements a and b is in p, at least one of the two elements is already in p. (The opposite conclusion holds for any ideal, by definition). Equivalently, the factor ring R / p is an integral domain. Yet another way of expressing the same is to say that the complement $R \setminus p$ is multiplicatively closed. The localisation $(R \setminus p)^{-1} R$ is important enough to have its own notation: R_p. This ring has only one maximal ideal, namely pR_p. Such rings are called local.

By the above, any maximal ideal is prime. Proving that an ideal is prime, or equivalently that a ring has no zero-divisors can be very difficult.

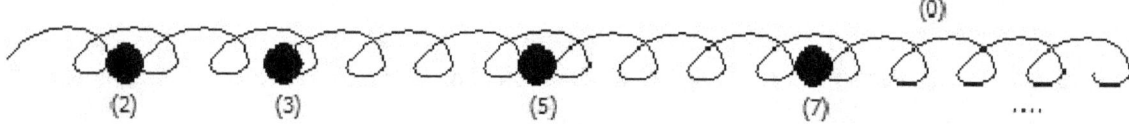

The spectrum of $\mathbb{Z}$.

Prime ideals are the key step in interpreting a ring *geometrically*, via the *spectrum of a ring* Spec R: it is the set of all prime ideals of R.[nb 1] As noted above, there is at least one prime ideal, therefore the spectrum is nonempty. If R is a field, the only prime ideal is the zero ideal, therefore the spectrum is just one point. The spectrum of $\mathbb{Z}$, however, contains one point for the zero ideal, and a point for any prime number p (which generates the prime ideal $p\mathbb{Z}$). The spectrum is endowed with a topology called the Zariski topology, which is determined by specifying that subsets $D(f) = \{p$: Spec R, $f \notin p\}$, where f is any ring element, be open. This topology tends to be different from those encountered in analysis or differential geometry; for example, there will generally be points which are not closed. The closure of the point corresponding to the zero ideal $0 \subset \mathbb{Z}$, for example, is the whole spectrum of $\mathbb{Z}$.

The notion of a spectrum is the common basis of commutative algebra and algebraic geometry. Algebraic geometry proceeds by endowing Spec R with a sheaf O (an entity that collects functions defined locally, i.e. on varying open subsets). The datum of the space and the sheaf is called an affine scheme. Given an affine scheme, the underlying ring R can be recovered as the global sections of O. Moreover, the established one-to-one correspondence between rings and affine schemes is also compatible with ring homomorphisms: any $f : R \to S$ gives rise to a continuous map in the opposite direction

> $Spec\,S \to Spec\,R,\ q \mapsto f^{-1}(q)$, i.e. any prime ideal of S is mapped to its preimage under f, which is a prime ideal of R.

The spectrum also makes precise the intuition that localisation and factor rings are complementary: the natural maps $R \to Rf$ and $R \to R\,/\,fR$ correspond, after endowing the spectra of the rings in question with their Zariski topology, to complementary open and closed immersions respectively.

Altogether the equivalence of the two said categories is very apt to reflect algebraic properties of rings in a geometrical manner. Affine schemes are—much the same way as manifolds are locally given by open subsets of $\mathbf{R}^n$—local models for schemes, which are the object of study in algebraic geometry. Therefore, many notions that apply to rings and homomorphisms stem from geometric intuition.

13.3   Ring homomorphisms

Main article: Ring homomorphism

As usual in algebra, a function f between two objects that respects the structures of the objects in question is called homomorphism. In the case of rings, a *ring homomorphism* is a map $f : R \to S$ such that

> $f(a + b) = f(a) + f(b),\ f(ab) = f(a)f(b)$ and $f(1) = 1$.

These conditions ensure $f(0) = 0$, but the requirement that the multiplicative identity element 1 is preserved under f would not follow from the two remaining properties. In such a situation S is also called an R-algebra, by understanding that s in S may be multiplied by some r of R, by setting

> $r \cdot s := f(r) \cdot s$.

The *kernel* and *image* of f are defined by $\ker (f) = \{r \in R,\ f(r) = 0\}$ and $im (f) = f(R) = \{f(r),\ r \in R\}$. The kernel is an ideal of R, and the image is a subring of S.

13.4   Modules

Main article: Modules

The outer structure of a commutative ring is determined by considering linear algebra over that ring, i.e., by investigating the theory of its modules, which are similar to vector spaces, except that the base is not necessarily a field, but can be any ring R. The theory of R-modules is significantly more difficult than linear algebra of vector spaces. Module theory has to grapple with difficulties such as modules not having bases, that the rank of a free module (i.e. the analog of the dimension of vector spaces) may not be well-defined and that submodules of finitely generated modules need not be finitely generated (unless R is Noetherian, see below).

Ideals within a ring R can be characterized as R-modules which are submodules of R. On the one hand, a good understanding of R-modules necessitates enough information about R. Vice versa, however, many techniques in commutative algebra that study the structure of R, by examining its ideals, proceed by studying modules in general.

13.5   Noetherian rings

Main article: Noetherian ring

A ring is called *Noetherian* (in honor of Emmy Noether, who developed this concept) if every ascending chain of ideals

$$0 \subseteq I_0 \subseteq I_1 \ldots \subseteq I_n \subseteq I_{n+1} \subseteq \ldots$$

becomes stationary, i.e. becomes constant beyond some index n. Equivalently, any ideal is generated by finitely many elements, or, yet equivalent, submodules of finitely generated modules are finitely generated. A ring is called Artinian (after Emil Artin), if every descending chain of ideals

$$R \supseteq I_0 \supseteq I_1 \ldots \supseteq I_n \supseteq I_{n+1} \supseteq \ldots$$

becomes stationary eventually. Despite the two conditions appearing symmetric, Noetherian rings are much more general than Artinian rings. For example, $\mathbb{Z}$ is Noetherian, since every ideal can be generated by one element, but is not Artinian, as the chain

$$\mathbb{Z} \supsetneq 2\mathbb{Z} \supsetneq 4\mathbb{Z} \supsetneq 8\mathbb{Z} \supsetneq \ldots$$

shows. In fact, by the Hopkins–Levitzki theorem, every Artinian ring is Noetherian.

Being Noetherian is an extremely important finiteness condition. The condition is preserved under many operations that occur frequently in geometry: if R is Noetherian, then so is the polynomial ring $R[X_1, X_2, ..., X_n]$ (by Hilbert's basis theorem), any localization $S^{-1}R$, factor rings R/I.

13.6   Dimension

Main article: Krull dimension

The *Krull dimension* (or simply dimension) *dim R* of a ring R is a notion to measure the "size" of a ring, very roughly by the counting independent elements in R. Precisely, it is defined as the supremum of lengths n of chains of prime ideals

$$p_0 \subsetneq p_1 \subsetneq \cdots \subsetneq p_n$$

For example, a field is zero-dimensional, since the only prime ideal is the zero ideal. It is also known that a commutative ring is Artinian if and only if it is Noetherian and zero-dimensional (i.e., all its prime ideals are maximal). The integers are one-dimensional: any chain of prime ideals is of the form

$$0 = p_0 \subsetneq p\mathbb{Z} = p_1 \text{ , where } p \text{ is a prime number}$$

since any ideal in $\mathbb{Z}$ is principal.

The dimension behaves well if the rings in question are Noetherian: the expected equality

$$\dim R[X] = \dim R + 1$$

holds in this case (in general, one has only $\dim R + 1 \le \dim R[X] \le 2 \cdot \dim R + 1$). Furthermore, since the dimension depends only on one maximal chain, the dimension of R is the supremum of all dimensions of its localisations R_p, where p is an arbitrary prime ideal. Intuitively, the dimension of R is a local property of the spectrum of R. Therefore, the dimension is often considered for local rings only, also since general Noetherian rings may still be infinite, despite all their localisations being finite-dimensional.

Determining the dimension of, say,

$k[X_1, X_2, ..., X_n] / (f_1, f_2, ..., f_m)$, where k is a field and the f_i are some polynomials in n variables,

is generally not easy. For R Noetherian, the dimension of R / I is, by Krull's principal ideal theorem, at least dim $R - n$, if I is generated by n elements. If the dimension does drops as much as possible, i.e. dim R / I = dim $R - n$, the R / I is called a complete intersection.

A local ring R, i.e. one with only one maximal ideal m, is called regular, if the (Krull) dimension of R equals the dimension (as a vector space over the field R / m) of the cotangent space m / m^2.

13.7   Constructing commutative rings

There are several ways to construct new rings out of given ones. The aim of such constructions is often to improve certain properties of the ring so as to make it more readily understandable. For example, an integral domain that is integrally closed in its field of fractions is called normal. This is a desirable property, for example any normal one-dimensional ring is necessarily regular. Rendering a ring normal is known as *normalization*.

13.7.1   Completions

If I is an ideal in a commutative ring R, the powers of I form topological neighborhoods of 0 which allow R to be viewed as a topological ring. This topology is called the I-adic topology. R can then be completed with respect to this topology. Formally, the I-adic completion is the inverse limit of the rings R/I^n. For example, if k is a field, $k[[X]]$, the formal power series ring in one variable over k, is the I-adic completion of $k[X]$ where I is the principal ideal generated by X. Analogously, the ring of p-adic integers is the I-adic completion of Z where I is the principal ideal generated by p. Any ring that is isomorphic to its own completion, is called complete.

13.8   Properties

By Wedderburn's theorem, every finite division ring is commutative, and therefore a finite field. Another condition ensuring commutativity of a ring, due to Jacobson, is the following: for every element r of R there exists an integer n > 1 such that $r^n = r$.[1] If, $r^2 = r$ for every r, the ring is called Boolean ring. More general conditions which guarantee commutativity of a ring are also known.[2]

13.9   See also

- Graded ring

- Almost commutative ring

- Almost ring, a certain generalization of a commutative ring.

- Simplicial commutative ring, a simplicial object in the category of commutative rings.

13.10   Notes

[1] This notion can be related to the spectrum of a linear operator, see Spectrum of a C*-algebra and Gelfand representation.

13.10.1 Citations

[1] Jacobson 1945

[2] Pinter-Lucke 2007

13.11 References

• Atiyah, Michael; Macdonald, I. G. (1969), *Introduction to commutative algebra*, Addison-Wesley Publishing Co.

• Balcerzyk, Stanisław; Józefiak, Tadeusz (1989), *Commutative Noetherian and Krull rings*, Ellis Horwood Series: Mathematics and its Applications, Chichester: Ellis Horwood Ltd., ISBN 978-0-13-155615-7

• Balcerzyk, Stanisław; Józefiak, Tadeusz (1989), *Dimension, multiplicity and homological methods*, Ellis Horwood Series: Mathematics and its Applications., Chichester: Ellis Horwood Ltd., ISBN 978-0-13-155623-2

• Eisenbud, David (1995), *Commutative algebra. With a view toward algebraic geometry*, Graduate Texts in Mathematics 150, Berlin, New York: Springer-Verlag, ISBN 978-0-387-94268-1, MR 1322960

• Jacobson, Nathan (1945), "Structure theory of algebraic algebras of bounded degree", *Annals of Mathematics* 46 (4): 695–707, doi:10.2307/1969205, ISSN 0003-486X, JSTOR 1969205

• Kaplansky, Irving (1974), *Commutative rings* (Revised ed.), University of Chicago Press, MR 0345945

• Matsumura, Hideyuki (1989), *Commutative Ring Theory*, Cambridge Studies in Advanced Mathematics (2nd ed.), Cambridge University Press, ISBN 978-0-521-36764-6

• Nagata, Masayoshi (1975) [1962], *Local rings*, Interscience Tracts in Pure and Applied Mathematics 13, Interscience Publishers, pp. xiii+234, ISBN 978-0-88275-228-0, MR 0155856

• Pinter-Lucke, James (2007), "Commutativity conditions for rings: 1950–2005", *Expositiones Mathematicae* 25 (2): 165–174, doi:10.1016/j.exmath.2006.07.001, ISSN 0723-0869

• Zariski, Oscar; Samuel, Pierre (1958–60), *Commutative Algebra I, II*, University series in Higher Mathematics, Princeton, N.J.: D. van Nostrand, Inc. *(Reprinted 1975-76 by Springer as volumes 28-29 of Graduate Texts in Mathematics.)*

Chapter 14

Commutative algebra

This article is about the branch of algebra that studies commutative rings. For algebras that are commutative, see Commutative algebra (structure).

Commutative algebra is the branch of algebra that studies commutative rings, their ideals, and modules over such rings. Both algebraic geometry and algebraic number theory build on commutative algebra. Prominent examples of commutative rings include polynomial rings, rings of algebraic integers, including the ordinary integers $\mathbb{Z}$, and p-adic integers.[1]

Commutative algebra is the main technical tool in the local study of schemes.

The study of rings which are not necessarily commutative is known as noncommutative algebra; it includes ring theory, representation theory, and the theory of Banach algebras.

14.1 Overview

Commutative algebra is essentially the study of the rings occurring in algebraic number theory and algebraic geometry.

In algebraic number theory, the rings of algebraic integers are Dedekind rings, which constitute therefore an important class of commutative rings. Considerations related to modular arithmetic have led to the notion of valuation ring. The restriction of algebraic field extensions to subrings has led to the notions of integral extensions and integrally closed domains as well as the notion of ramification of an extension of valuation rings.

The notion of localization of a ring (in particular the localization with respect to a prime ideal, the localization consisting in inverting a single element and the total quotient ring) is one of the main differences between commutative algebra and the theory of non-commutative rings. It leads to an important class of commutative rings, the local rings that have only one maximal ideal. The set of the prime ideals of a commutative ring is naturally equipped with a topology, the Zariski topology. All these notions are widely used in algebraic geometry and are the basic technical tools for the definition of scheme theory, a generalization of algebraic geometry introduced by Grothendieck.

Many other notions of commutative algebra are counterparts of geometrical notions occurring in algebraic geometry. This is the case of Krull dimension, primary decomposition, regular rings, Cohen–Macaulay rings, Gorenstein rings and many other notions.

14.2 History

The subject, first known as ideal theory, began with Richard Dedekind's work on ideals, itself based on the earlier work of Ernst Kummer and Leopold Kronecker. Later, David Hilbert introduced the term *ring* to generalize the earlier term *number ring*. Hilbert introduced a more abstract approach to replace the more concrete and computationally oriented methods grounded in such things as complex analysis and classical invariant theory. In turn, Hilbert strongly influenced

A 1915 postcard from one of the pioneers of commutative algebra, Emmy Noether, to E. Fischer, discussing her work in commutative algebra.

Emmy Noether, who recast many earlier results in terms of an ascending chain condition, now known as the Noetherian condition. Another important milestone was the work of Hilbert's student Emanuel Lasker, who introduced primary ideals and proved the first version of the Lasker–Noether theorem.

The main figure responsible for the birth of commutative algebra as a mature subject was Wolfgang Krull, who introduced the fundamental notions of localization and completion of a ring, as well as that of regular local rings. He established the concept of the Krull dimension of a ring, first for Noetherian rings before moving on to expand his theory to cover general valuation rings and Krull rings. To this day, Krull's principal ideal theorem is widely considered the single most important foundational theorem in commutative algebra. These results paved the way for the introduction of commutative algebra into algebraic geometry, an idea which would revolutionize the latter subject.

Much of the modern development of commutative algebra emphasizes modules. Both ideals of a ring R and R-algebras are special cases of R-modules, so module theory encompasses both ideal theory and the theory of ring extensions. Though it was already incipient in Kronecker's work, the modern approach to commutative algebra using module theory is usually credited to Krull and Noether.

14.3   Main tools and results

14.3.1   Noetherian rings

Main article: Noetherian ring

In mathematics, more specifically in the area of modern algebra known as ring theory, a **Noetherian ring**, named after Emmy Noether, is a ring in which every non-empty set of ideals has a maximal element. Equivalently, a ring is Noetherian if it satisfies the ascending chain condition on ideals; that is, given any chain:

$$I_1 \subseteq \cdots I_{k-1} \subseteq I_k \subseteq I_{k+1} \subseteq \cdots$$

there exists an n such that:

$$I_n = I_{n+1} = \cdots$$

For a commutative ring to be Noetherian it suffices that every prime ideal of the ring is finitely generated. (The result is due to I. S. Cohen.)

The notion of a Noetherian ring is of fundamental importance in both commutative and noncommutative ring theory, due to the role it plays in simplifying the ideal structure of a ring. For instance, the ring of integers and the polynomial ring over a field are both Noetherian rings, and consequently, such theorems as the Lasker–Noether theorem, the Krull intersection theorem, and the Hilbert's basis theorem hold for them. Furthermore, if a ring is Noetherian, then it satisfies the descending chain condition on *prime ideals*. This property suggests a deep theory of dimension for Noetherian rings beginning with the notion of the Krull dimension.

14.3.2   Hilbert's basis theorem

Main article: Hilbert's basis theorem

> **Theorem.** If R is a left (resp. right) Noetherian ring, then the polynomial ring $R[X]$ is also a left (resp. right) Noetherian ring.

Hilbert's basis theorem has some immediate corollaries:

1. By induction we see that $R[X_0, \ldots, X_{n-1}]$ will also be Noetherian.

2. Since any affine variety over R^n (i.e. a locus-set of a collection of polynomials) may be written as the locus of an ideal a $\subset R[X_0, \ldots, X_{n-1}]$ and further as the locus of its generators, it follows that every affine variety is the locus of finitely many polynomials — i.e. the intersection of finitely many hypersurfaces.

3. If A is a finitely-generated R-algebra, then we know that $A \simeq R[X_0, \ldots, X_{n-1}]/a$, where a is an ideal. The basis theorem implies that a must be finitely generated, say a $= (p_0, \ldots, p_{N-1})$, i.e. A is finitely presented.

14.3.3   Primary decomposition

Main article: Primary decomposition

An ideal Q of a ring is said to be *primary* if Q is proper and whenever $xy \in Q$, either $x \in Q$ or $y^n \in Q$ for some positive integer n. In $\mathbf{Z}$, the primary ideals are precisely the ideals of the form (p^e) where p is prime and e is a positive integer. Thus, a primary decomposition of (n) corresponds to representing (n) as the intersection of finitely many primary ideals. The *Lasker–Noether theorem*, given here, may be seen as a certain generalization of the fundamental theorem of arithmetic:

> **Lasker-Noether Theorem.** Let R be a commutative Noetherian ring and let I be an ideal of R. Then I may be written as the intersection of finitely many primary ideals with distinct radicals; that is:

$$I = \bigcap_{i=1}^{t} Q_i$$

> with Q_i primary for all i and $\text{Rad}(Q_i) \neq \text{Rad}(Q_j)$ for $i \neq j$. Furthermore, if:

$$I = \bigcap_{i=1}^{k} P_i$$

> is decomposition of I with $\text{Rad}(P_i) \neq \text{Rad}(P_j)$ for $i \neq j$, and both decompositions of I are *irredundant* (meaning that no proper subset of either $\{Q_1, \ldots, Q_t\}$ or $\{P_1, \ldots, P_k\}$ yields an intersection equal to I), $t = k$ and (after possibly renumbering the Q_i) $\text{Rad}(Q_i) = \text{Rad}(P_i)$ for all i.

For any primary decomposition of I, the set of all radicals, that is, the set $\{\text{Rad}(Q_1), \ldots, \text{Rad}(Q_t)\}$ remains the same by the Lasker–Noether theorem. In fact, it turns out that (for a Noetherian ring) the set is precisely the assassinator of the module R/I, that is, the set of all annihilators of R/I (viewed as a module over R) that are prime.

14.3.4   Localization

Main article: Localization (algebra)

The localization is a formal way to introduce the "denominators" to a given ring or a module. That is, it introduces a new ring/module out of an existing one so that it consists of fractions

$$\frac{m}{s}$$

where the denominators s range in a given subset S of R. The archetypal example is the construction of the ring $\mathbf{Q}$ of rational numbers from the ring $\mathbf{Z}$ of integers.

14.3.5   Completion

Main article: Completion (ring theory)

A completion is any of several related functors on rings and modules that result in complete topological rings and modules. Completion is similar to localization, and together they are among the most basic tools in analysing commutative rings. Complete commutative rings have simpler structure than the general ones and Hensel's lemma applies to them.

14.3.6   Zariski topology on prime ideals

Main article: Zariski topology

The Zariski topology defines a topology on the spectrum of a ring (the set of prime ideals).[2] In this formulation, the Zariski-closed sets are taken to be the sets

$$V(I) = \{ P \in \operatorname{Spec}(A) \mid I \subseteq P \}$$

where A is a fixed commutative ring and I is an ideal. This is defined in analogy with the classical Zariski topology, where closed sets in affine space are those defined by polynomial equations . To see the connection with the classical picture, note that for any set S of polynomials (over an algebraically closed field), it follows from Hilbert's Nullstellensatz that the points of $V(S)$ (in the old sense) are exactly the tuples $(a_1, ..., a_n)$ such that $(x_1 - a_1, ..., x_n - a_n)$ contains S; moreover, these are maximal ideals and by the "weak" Nullstellensatz, an ideal of any affine coordinate ring is maximal if and only if it is of this form. Thus, $V(S)$ is "the same as" the maximal ideals containing S. Grothendieck's innovation in defining Spec was to replace maximal ideals with all prime ideals; in this formulation it is natural to simply generalize this observation to the definition of a closed set in the spectrum of a ring.

14.4   Examples

The fundamental example in commutative algebra is the ring of integers $\mathbb{Z}$. The existence of primes and the unique factorization theorem laid the foundations for concepts such as Noetherian rings and the primary decomposition.

Other important examples are:

- Polynomial rings $R[x_1, ..., x_n]$
- The p-adic integers
- Rings of algebraic integers.

14.5   Connections with algebraic geometry

Commutative algebra (in the form of polynomial rings and their quotients, used in the definition of algebraic varieties) has always been a part of algebraic geometry. However, in late 1950s, algebraic varieties were subsumed into Alexander Grothendieck's concept of a scheme. Their local objects are affine schemes or prime spectra which are locally ringed spaces which form a category which is antiequivalent (dual) to the category of commutative unital rings, extending the duality between the category of affine algebraic varieties over a field k, and the category of finitely generated reduced k-algebras. The gluing is along Zariski topology; one can glue within the category of locally ringed spaces, but also, using the Yoneda embedding, within the more abstract category of presheaves of sets over the category of affine schemes. The Zariski topology in the set theoretic sense is then replaced by a Zariski topology in the sense of Grothendieck topology. Grothendieck introduced Grothendieck topologies having in mind more exotic but geometrically finer and more sensitive examples than the crude Zariski topology, namely the étale topology, and the two flat Grothendieck topologies: fppf and fpqc; nowadays some other examples became prominent including Nisnevich topology. Sheaves can be furthermore generalized to stacks in the sense of Grothendieck, usually with some additional representability conditions leading to Artin stacks and, even finer, Deligne-Mumford stacks, both often called algebraic stacks.

14.6   See also

- List of commutative algebra topics
- Glossary of commutative algebra
- Combinatorial commutative algebra
- Gröbner basis
- Homological algebra

14.7   References

[1] Atiyah and Macdonald, 1969, Chapter 1

[2] Dummit, D. S.; Foote, R. (2004). *Abstract Algebra* (3 ed.). Wiley. pp. 71–72. ISBN 9780471433347.

- Michael Atiyah & Ian G. Macdonald, *Introduction to Commutative Algebra*, Massachusetts : Addison-Wesley Publishing, 1969.

- Bourbaki, Nicolas, *Commutative algebra. Chapters 1-–7*. Translated from the French. Reprint of the 1989 English translation. Elements of Mathematics (Berlin). Springer-Verlag, Berlin, 1998. xxiv+625 pp. ISBN 3-540-64239-0

- Bourbaki, Nicolas, *Éléments de mathématique. Algèbre commutative. Chapitres 8 et 9.* (Elements of mathematics. Commutative algebra. Chapters 8 and 9) Reprint of the 1983 original. Springer, Berlin, 2006. ii+200 pp. ISBN 978-3-540-33942-7

- David Eisenbud, *Commutative Algebra With a View Toward Algebraic Geometry*, New York : Springer-Verlag, 1999.

- Rémi Goblot, "Algèbre commutative, cours et exercices corrigés", 2e édition, Dunod 2001, ISBN 2-10-005779-0

- Ernst Kunz, "Introduction to Commutative algebra and algebraic geometry", Birkhauser 1985, ISBN 0-8176-3065-1

- Matsumura, Hideyuki, *Commutative algebra*. Second edition. Mathematics Lecture Note Series, 56. Benjamin/Cgs Publishing Co., Inc., Reading, Mass., 1980. xv+313 pp. ISBN 0-8053-7026-9

- Matsumura, Hideyuki, *Commutative Ring Theory*. Second edition. Translated from the Japanese. Cambridge Studies in Advanced Mathematics, Cambridge, UK : Cambridge University Press, 1989. ISBN 0-521-36764-6

- Nagata, Masayoshi, *Local rings*. Interscience Tracts in Pure and Applied Mathematics, No. 13. Interscience Publishers a division of John Wiley and Sons, New York-London 1962 xiii+234 pp.

- Miles Reid, *Undergraduate Commutative Algebra (London Mathematical Society Student Texts)*, Cambridge, UK : Cambridge University Press, 1996.

- Jean-Pierre Serre, *Local algebra*. Translated from the French by CheeWhye Chin and revised by the author. (Original title: *Algèbre locale, multiplicités*) Springer Monographs in Mathematics. Springer-Verlag, Berlin, 2000. xiv+128 pp. ISBN 3-540-66641-9

- Sharp, R. Y., *Steps in commutative algebra*. Second edition. London Mathematical Society Student Texts, 51. Cambridge University Press, Cambridge, 2000. xii+355 pp. ISBN 0-521-64623-5

- Zariski, Oscar; Samuel, Pierre, *Commutative algebra*. Vol. 1, 2. With the cooperation of I. S. Cohen. Corrected reprinting of the 1958, 1960 edition. Graduate Texts in Mathematics, No. 28, 29. Springer-Verlag, New York-Heidelberg-Berlin, 1975.

- Zeidler, A. Bernhard (2014), *Abstract Algebra* (PDF) A web-book on algebra and commutative algebra. Warning: work in progress! Free downloadable PDF under Open Publication License.

Chapter 15

Mathematical analysis

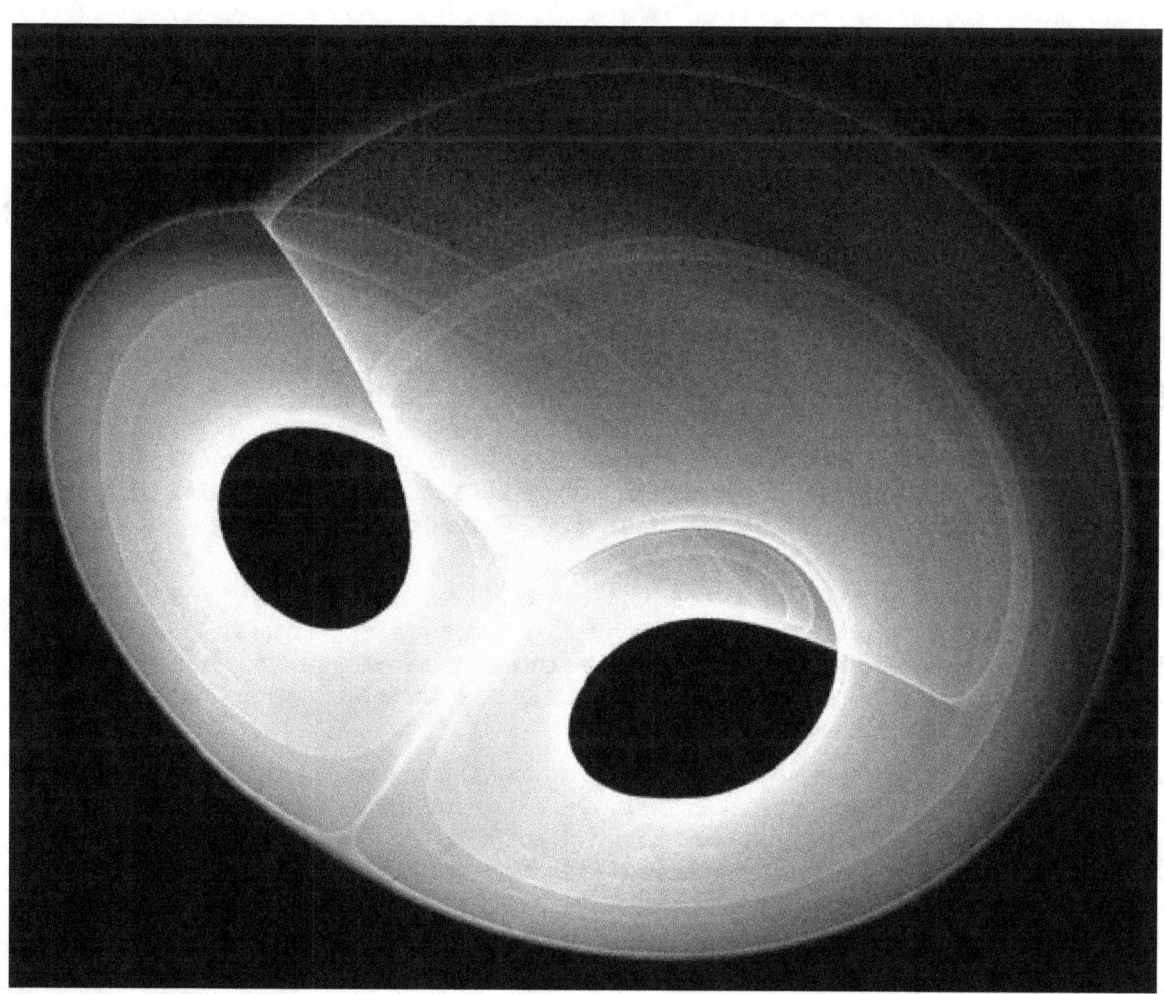

A strange attractor arising from a differential equation. Differential equations are an important area of mathematical analysis with many applications to science and engineering.

Mathematical analysis is a branch of mathematics that includes the theories of differentiation, integration, measure, limits, infinite series, and analytic functions.[1]

These theories are usually studied in the context of real and complex numbers and functions. Analysis evolved from calculus, which involves the elementary concepts and techniques of analysis. Analysis may be distinguished from geometry; however, it can be applied to any space of mathematical objects that has a definition of nearness (a topological space) or specific distances between objects (a metric space).

15.1   History

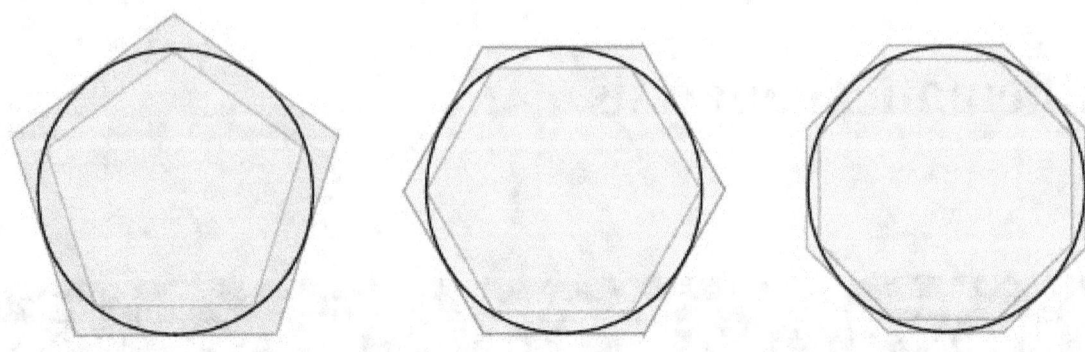

Archimedes used the method of exhaustion to compute the area inside a circle by finding the area of regular polygons with more and more sides. This was an early but informal example of a limit, one of the most basic concepts in mathematical analysis.

Mathematical analysis formally developed in the 17th century during the Scientific Revolution,[2] but many of its ideas can be traced back to earlier mathematicians. Early results in analysis were implicitly present in the early days of ancient Greek mathematics. For instance, an infinite geometric sum is implicit in Zeno's paradox of the dichotomy.[3] Later, Greek mathematicians such as Eudoxus and Archimedes made more explicit, but informal, use of the concepts of limits and convergence when they used the method of exhaustion to compute the area and volume of regions and solids.[4] The explicit use of infinitesimals appears in Archimedes' *The Method of Mechanical Theorems*, a work rediscovered in the 20th century.[5] In Asia, the Chinese mathematician Liu Hui used the method of exhaustion in the 3rd century AD to find the area of a circle.[6] Zu Chongzhi established a method that would later be called Cavalieri's principle to find the volume of a sphere in the 5th century.[7] The Indian mathematician Bhāskara II gave examples of the derivative and used what is now known as Rolle's theorem in the 12th century.[8]

In the 14th century, Madhava of Sangamagrama developed infinite series expansions, like the power series and the Taylor series, of functions such as sine, cosine, tangent and arctangent.[9] Alongside his development of the Taylor series of the trigonometric functions, he also estimated the magnitude of the error terms created by truncating these series and gave a rational approximation of an infinite series. His followers at the Kerala school of astronomy and mathematics further expanded his works, up to the 16th century.

The modern foundations of mathematical analysis were established in 17th century Europe.[2] Descartes and Fermat independently developed analytic geometry, and a few decades later Newton and Leibniz independently developed infinitesimal calculus, which grew, with the stimulus of applied work that continued through the 18th century, into analysis topics such as the calculus of variations, ordinary and partial differential equations, Fourier analysis, and generating functions. During this period, calculus techniques were applied to approximate discrete problems by continuous ones.

In the 18th century, Euler introduced the notion of mathematical function.[10] Real analysis began to emerge as an independent subject when Bernard Bolzano introduced the modern definition of continuity in 1816,[11] but Bolzano's work did not become widely known until the 1870s. In 1821, Cauchy began to put calculus on a firm logical foundation by rejecting the principle of the generality of algebra widely used in earlier work, particularly by Euler. Instead, Cauchy formulated calculus in terms of geometric ideas and infinitesimals. Thus, his definition of continuity required an infinitesimal change in x to correspond to an infinitesimal change in y. He also introduced the concept of the Cauchy sequence, and started the formal theory of complex analysis. Poisson, Liouville, Fourier and others studied partial differential equations and harmonic analysis. The contributions of these mathematicians and others, such as Weierstrass, developed the (ε, δ)-definition of limit approach, thus founding the modern field of mathematical analysis.

In the middle of the 19th century Riemann introduced his theory of integration. The last third of the century saw the arithmetization of analysis by Weierstrass, who thought that geometric reasoning was inherently misleading, and introduced the "epsilon-delta" definition of limit. Then, mathematicians started worrying that they were assuming the existence of a continuum of real numbers without proof. Dedekind then constructed the real numbers by Dedekind cuts, in which irrational numbers are formally defined, which serve to fill the "gaps" between rational numbers, thereby creating a complete set: the continuum of real numbers, which had already been developed by Simon Stevin in terms of decimal expansions. Around that time, the attempts to refine the theorems of Riemann integration led to the study of the "size" of the set of discontinuities of real functions.

Also, "monsters" (nowhere continuous functions, continuous but nowhere differentiable functions, space-filling curves) began to be investigated. In this context, Jordan developed his theory of measure, Cantor developed what is now called naive set theory, and Baire proved the Baire category theorem. In the early 20th century, calculus was formalized using an axiomatic set theory. Lebesgue solved the problem of measure, and Hilbert introduced Hilbert spaces to solve integral equations. The idea of normed vector space was in the air, and in the 1920s Banach created functional analysis.

15.2 Important concepts

15.2.1 Metric spaces

Main article: Metric space

In mathematics, a **metric space** is a set where a notion of distance (called a metric) between elements of the set is defined.

Much of analysis happens in some metric space; the most commonly used are the real line, the complex plane, Euclidean space, other vector spaces, and the integers. Examples of analysis without a metric include measure theory (which describes size rather than distance) and functional analysis (which studies topological vector spaces that need not have any sense of distance).

Formally, A **metric space** is an ordered pair (M, d) where M is a set and d is a metric on M, i.e., a function

$$d: M \times M \to \mathbb{R}$$

such that for any $x, y, z \in M$, the following holds:

1. $d(x, y) = 0$ if and only if $x = y$ (*identity of indiscernibles*),

2. $d(x, y) = d(y, x)$ (*symmetry*) and

3. $d(x, z) \leq d(x, y) + d(y, z)$ (*triangle inequality*).

By taking the third property and letting $z = x$, it can be shown that $d(x, y) \geq 0$ (*non-negative*).

15.2.2 Sequences and limits

Main article: Sequence

A **sequence** is an ordered list. Like a set, it contains members (also called *elements* or *terms*). Unlike a set, order matters, and exactly the same elements can appear multiple times at different positions in the sequence. Most precisely, a sequence can be defined as a function whose domain is a countable totally ordered set, such as the natural numbers.

One of the most important properties of a sequence is *convergence*. Informally, a sequence converges if it has a *limit*. Continuing informally, a (singly-infinite) sequence has a limit if it approaches some point x, called the limit, as n becomes very large. That is, for an abstract sequence (a_n) (with n running from 1 to infinity understood) the distance between a_n and x approaches 0 as $n \to \infty$, denoted

$$\lim_{n \to \infty} a_n = x.$$

15.3   Main branches

15.3.1   Real analysis

Main article: Real analysis

Real analysis (traditionally, **the theory of functions of a real variable**) is a branch of mathematical analysis dealing with the real numbers and real-valued functions of a real variable.[12][13] In particular, it deals with the analytic properties of real functions and sequences, including convergence and limits of sequences of real numbers, the calculus of the real numbers, and continuity, smoothness and related properties of real-valued functions.

15.3.2   Complex analysis

Main article: Complex analysis

Complex analysis, traditionally known as **the theory of functions of a complex variable**, is the branch of mathematical analysis that investigates functions of complex numbers.[14] It is useful in many branches of mathematics, including algebraic geometry, number theory, applied mathematics; as well as in physics, including hydrodynamics, thermodynamics, mechanical engineering, electrical engineering, and particularly, quantum field theory.

Complex analysis is particularly concerned with the analytic functions of complex variables (or, more generally, meromorphic functions). Because the separate real and imaginary parts of any analytic function must satisfy Laplace's equation, complex analysis is widely applicable to two-dimensional problems in physics.

15.3.3   Functional analysis

Main article: Functional analysis

Functional analysis is a branch of mathematical analysis, the core of which is formed by the study of vector spaces endowed with some kind of limit-related structure (e.g. inner product, norm, topology, etc.) and the linear operators acting upon these spaces and respecting these structures in a suitable sense.[15][16] The historical roots of functional analysis lie in the study of spaces of functions and the formulation of properties of transformations of functions such as the Fourier transform as transformations defining continuous, unitary etc. operators between function spaces. This point of view turned out to be particularly useful for the study of differential and integral equations.

15.3.4   Diff erential equations

Main article: Differential equations

A **diff erential equation** is a mathematical equation for an unknown function of one or several variables that relates the values of the function itself and its derivatives of various orders.[17][18][19] Differential equations play a prominent role in engineering, physics, economics, biology, and other disciplines.

Differential equations arise in many areas of science and technology, specifically whenever a deterministic relation involving some continuously varying quantities (modeled by functions) and their rates of change in space and/or time (expressed as derivatives) is known or postulated. This is illustrated in classical mechanics, where the motion of a body is described by its position and velocity as the time value varies. Newton's laws allow one (given the position, velocity, acceleration and various forces acting on the body) to express these variables dynamically as a differential equation for the unknown position of the body as a function of time. In some cases, this differential equation (called an equation of motion) may be solved explicitly.

15.3.5 Measure theory

Main article: Measure (mathematics)

A **measure** on a set is a systematic way to assign a number to each suitable subset of that set, intuitively interpreted as its size.[20] In this sense, a measure is a generalization of the concepts of length, area, and volume. A particularly important example is the Lebesgue measure on a Euclidean space, which assigns the conventional length, area, and volume of Euclidean geometry to suitable subsets of the n-dimensional Euclidean space R^n. For instance, the Lebesgue measure of the interval $[0, 1]$ in the real numbers is its length in the everyday sense of the word — specifically, 1.

Technically, a measure is a function that assigns a non-negative real number or $+\infty$ to (certain) subsets of a set X. It must assign 0 to the empty set and be (countably) additive: the measure of a 'large' subset that can be decomposed into a finite (or countable) number of 'smaller' disjoint subsets, is the sum of the measures of the "smaller" subsets. In general, if one wants to associate a *consistent* size to *each* subset of a given set while satisfying the other axioms of a measure, one only finds trivial examples like the counting measure. This problem was resolved by defining measure only on a sub-collection of all subsets; the so-called *measurable* subsets, which are required to form a σ-algebra. This means that countable unions, countable intersections and complements of measurable subsets are measurable. Non-measurable sets in a Euclidean space, on which the Lebesgue measure cannot be defined consistently, are necessarily complicated in the sense of being badly mixed up with their complement. Indeed, their existence is a non-trivial consequence of the axiom of choice.

15.3.6 Numerical analysis

Main article: Numerical analysis

Numerical analysis is the study of algorithms that use numerical approximation (as opposed to general symbolic manipulations) for the problems of mathematical analysis (as distinguished from discrete mathematics).[21]

Modern numerical analysis does not seek exact answers, because exact answers are often impossible to obtain in practice. Instead, much of numerical analysis is concerned with obtaining approximate solutions while maintaining reasonable bounds on errors.

Numerical analysis naturally finds applications in all fields of engineering and the physical sciences, but in the 21st century, the life sciences and even the arts have adopted elements of scientific computations. Ordinary differential equations appear in celestial mechanics (planets, stars and galaxies); numerical linear algebra is important for data analysis; stochastic differential equations and Markov chains are essential in simulating living cells for medicine and biology.

15.4 Other topics in mathematical analysis

- Calculus of variations deals with extremizing functionals, as opposed to ordinary calculus which deals with functions.

- Harmonic analysis deals with Fourier series and their abstractions.

- Geometric analysis involves the use of geometrical methods in the study of partial differential equations and the application of the theory of partial differential equations to geometry.

- Clifford analysis, the study of Clifford valued functions that are annihilated by Dirac or Dirac-like operators, termed in general as monogenic or Clifford analytic functions.

- p-adic analysis, the study of analysis within the context of p-adic numbers, which differs in some interesting and surprising ways from its real and complex counterparts.

- Non-standard analysis, which investigates the hyperreal numbers and their functions and gives a rigorous treatment of infinitesimals and infinitely large numbers.

- Computable analysis, the study of which parts of analysis can be carried out in a computable manner.

- Stochastic calculus – analytical notions developed for stochastic processes.

- Set-valued analysis – applies ideas from analysis and topology to set-valued functions.

- Convex analysis, the study of convex sets and functions.

- Tropical analysis (or idempotent analysis) – analysis in the context of the semiring of the max-plus algebra where the lack of an additive inverse is compensated somewhat by the idempotent rule $A + A = A$. When transferred to the tropical setting, many nonlinear problems become linear.[22]

15.5   Applications

Techniques from analysis are also found in other areas such as:

15.5.1   Physical sciences

The vast majority of classical mechanics, relativity, and quantum mechanics is based on applied analysis, and differential equations in particular. Examples of important differential equations include Newton's second law, the Schrödinger equation, and the Einstein field equations.

Functional analysis is also a major factor in quantum mechanics.

15.5.2   Signal processing

When processing signals, such as audio, radio waves, light waves, seismic waves, and even images, Fourier analysis can isolate individual components of a compound waveform, concentrating them for easier detection and/or removal. A large family of signal processing techniques consist of Fourier-transforming a signal, manipulating the Fourier-transformed data in a simple way, and reversing the transformation.[23]

15.5.3   Other areas of mathematics

Techniques from analysis are used in many areas of mathematics, including:

- Analytic number theory
- Analytic combinatorics
- Continuous probability
- Differential entropy in information theory
- Differential games

- Differential geometry, the application of calculus to specific mathematical spaces known as manifolds that possess a complicated internal structure but behave in a simple manner locally.

- Differential topology

- Mathematical finance

15.6   See also

- Constructive analysis

- History of calculus

- Non-classical analysis

- Paraconsistent mathematics

- Smooth infinitesimal analysis

- Timeline of calculus and mathematical analysis

15.7   Notes

[1] Edwin Hewitt and Karl Stromberg, "Real and Abstract Analysis", Springer-Verlag, 1965

[2] Jahnke, Hans Niels (2003). *A History of Analysis*. American Mathematical Society. p. 7. ISBN 978-0-8218-2623-2.

[3] Stillwell (2004). "Infinite Series". Mathematics and its History (2nd ed.). Springer Science + Business Media Inc. p. 170. ISBN 0-387-95336-1. Infinite series were present in Greek mathematics, [...] There is no question that Zeno's paradox of the dichotomy (Section 4.1), for example, concerns the decomposition of the number 1 into the infinite series $\frac{1}{2} + \frac{1}{2}^2 + \frac{1}{2}^3 + \frac{1}{2}^4 + \ldots$ and that Archimedes found the area of the parabolic segment (Section 4.4) essentially by summing the infinite series $1 + \frac{1}{4} + \frac{1}{4}^2 + \frac{1}{4}^3 + \ldots = \frac{4}{3}$. Both these examples are special cases of the result we express as summation of a geometric series

[4] (Smith, 1958)

[5] Pinto, J. Sousa (2004). *Infinitesimal Methods of Mathematical Analysis*. Horwood Publishing. p. 8. ISBN 978-1-898563-99-0.

[6] Dun, Liu; Fan, Dainian; Cohen, Robert Sonné (1966). "A comparison of Archimdes' and Liu Hui's studies of circles". Chinese studies in the history and philosophy of science and technology 130. Springer. p. 279. ISBN 0-7923-3463-9., Chapter , p. 279

[7] Zill, Dennis G.; Wright, Scott; Wright, Warren S. (2009). *Calculus: Early Transcendentals* (3 ed.). Jones & Bartlett Learning. p. xxvii. ISBN 0-7637-5995-3., Extract of page 27

[8] Seal, Sir Brajendranath (1915), *The positive sciences of the ancient Hindus*, Longmans, Green and co.

[9] C. T. Rajagopal and M. S. Rangachari (June 1978). "On an untapped source of medieval Keralese Mathematics". *Archive for History of Exact Sciences* 18 (2): 89–102. doi:10.1007/BF00348142.

[10] Dunham, William (1999). *Euler: The Master of Us All*. The Mathematical Association of America. p. 17.

[11]   • Cooke, Roger (1997). "Beyond the Calculus". *The History of Mathematics: A Brief Course*. Wiley-Interscience. p. 379. ISBN 0-471-18082-3. Real analysis began its growth as an independent subject with the introduction of the modern definition of continuity in 1816 by the Czech mathematician Bernard Bolzano (1781–1848)

[12] Rudin, Walter. *Principles of Mathematical Analysis*. Walter Rudin Student Series in Advanced Mathematics (3rd ed.). McGraw–Hill. ISBN 978-0-07-054235-8.

[13] Abbott, Stephen (2001). *Understanding Analysis*. Undergradutate Texts in Mathematics. New York: Springer-Verlag. ISBN 0-387-95060-5.

[14] Ahlfors., *Complex Analysis* (McGraw-Hill)

[15] Rudin, W.: *Functional Analysis*, McGraw-Hill Science, 1991

[16] Conway, J. B.: *A Course in Functional Analysis*, 2nd edition, Springer-Verlag, 1994, ISBN 0-387-97245-5

[17] E. L. Ince, *Ordinary Differential Equations*, Dover Publications, 1958, ISBN 0-486-60349-0

[18] Witold Hurewicz, *Lectures on Ordinary Differential Equations*, Dover Publications, ISBN 0-486-49510-8

[19] Evans, L. C. (1998), *Partial Differential Equations*, Providence: American Mathematical Society, ISBN 0-8218-0772-2

[20] Terence Tao, 2011. *An Introduction to Measure Theory*. American Mathematical Society.

[21] Hildebrand, F. B. (1974). *Introduction to Numerical Analysis* (2nd edition ed.). McGraw-Hill. ISBN 0-07-028761-9.

[22] THE MASLOV DEQUANTIZATION, IDEMPOTENT AND TROPICAL MATHEMATICS: A BRIEF INTRODUCTION

[23] Theory and application of digital signal processing Rabiner, L. R.; Gold, B. Englewood Cliffs. N.J., Prentice-Hall, Inc., 1975.

15.8   References

- Aleksandrov, A. D., Kolmogorov, A. N., Lavrent'ev, M. A. (eds.). 1984. *Mathematics, its Content, Methods, and Meaning*. 2nd ed. Translated by S. H. Gould, K. A. Hirsch and T. Bartha; translation edited by S. H. Gould. MIT Press; published in cooperation with the American Mathematical Society.

- Apostol, Tom M. 1974. *Mathematical Analysis*. 2nd ed. Addison–Wesley. ISBN 978-0-201-00288-1.

- Binmore, K.G. 1980–1981. *The foundations of analysis: a straightforward introduction*. 2 volumes. Cambridge University Press.

- Johnsonbaugh, Richard, & W. E. Pfaffenberger. 1981. *Foundations of mathematical analysis*. New York: M. Dekker.

- Nikol'skii, S. M. 2002. "Mathematical analysis". In *Encyclopaedia of Mathematics*, Michiel Hazewinkel (editor). Springer-Verlag. ISBN 1-4020-0609-8.

- Rombaldi, Jean-Étienne. 2004. *Éléments d'analyse réelle : CAPES et agrégation interne de mathématiques*. EDP Sciences. ISBN 2-86883-681-X.

- Rudin, Walter. 1976. *Principles of Mathematical Analysis*. McGraw–Hill Publishing Co.; 3rd revised edition (September 1, 1976), ISBN 978-0-07-085613-4.

- Smith, David E. 1958. *History of Mathematics*. Dover Publications. ISBN 0-486-20430-8.

- Whittaker, E. T. and Watson, G. N.. 1927. *A Course of Modern Analysis*. 4th edition. Cambridge University Press. ISBN 0-521-58807-3.

- Real Analysis - Course Notes

15.9   External links

- Earliest Known Uses of Some of the Words of Mathematics: Calculus & Analysis

- Basic Analysis: Introduction to Real Analysis by Jiri Lebl (Creative Commons BY-NC-SA)

Chapter 16

Cryptography

"Secret code" redirects here. For the Aya Kamiki album, see Secret Code.
"Cryptology" redirects here. For the David S. Ware album, see Cryptology (album).

Cryptography or **cryptology**; from Greek κρυπτός, *kryptós*, "hidden, secret"; and γράφειν *graphein*, "writing", or

German Lorenz cipher machine, used in World War II to encrypt very-high-level general staff messages

-λογία *-logia*, "study", respectively[1] is the practice and study of techniques for secure communication in the presence of third parties (called adversaries).[2] More generally, it is about constructing and analyzing protocols that block adversaries;[3] various aspects in information security such as data confidentiality, data integrity, authentication, and non-repudiation[4] are central to modern cryptography. Modern cryptography exists at the intersection of the disciplines of mathematics, computer science, and electrical engineering. Applications of cryptography include ATM cards, computer passwords, and electronic commerce.

Cryptography prior to the modern age was effectively synonymous with *encryption*, the conversion of information from a readable state to apparent nonsense. The originator of an encrypted message shared the decoding technique needed to recover the original information only with intended recipients, thereby precluding unwanted persons from doing the same. Since World War I and the advent of the computer, the methods used to carry out cryptology have become increasingly complex and its application more widespread.

Modern cryptography is heavily based on mathematical theory and computer science practice; cryptographic algorithms are designed around computational hardness assumptions, making such algorithms hard to break in practice by any adversary. It is theoretically possible to break such a system, but it is infeasible to do so by any known practical means. These schemes are therefore termed computationally secure; theoretical advances, e.g., improvements in integer factorization algorithms, and faster computing technology require these solutions to be continually adapted. There exist information-theoretically secure schemes that provably cannot be broken even with unlimited computing power—an example is the one-time pad—but these schemes are more difficult to implement than the best theoretically breakable but computationally secure mechanisms.

The growth of cryptographic technology has raised a number of legal issues in the information age. Cryptography's potential for use as a tool for espionage and sedition has led many governments to classify it as a weapon and to limit or even prohibit its use and export.[5] In some jurisdictions where the use of cryptography is legal, laws permit investigators to compel the disclosure of encryption keys for documents relevant to an investigation.[6] Cryptography also plays a major role in digital rights management and piracy of digital media.[7]

16.1   Terminology

Until modern times cryptography referred almost exclusively to *encryption*, which is the process of converting ordinary information (called plaintext) into unintelligible text (called ciphertext).[8] Decryption is the reverse, in other words, moving from the unintelligible ciphertext back to plaintext. A *cipher* (or *cypher*) is a pair of algorithms that create the encryption and the reversing decryption. The detailed operation of a cipher is controlled both by the algorithm and in each instance by a "key". This is a secret (ideally known only to the communicants), usually a short string of characters, which is needed to decrypt the ciphertext. Formally, a "cryptosystem" is the ordered list of elements of finite possible plaintexts, finite possible cyphertexts, finite possible keys, and the encryption and decryption algorithms which correspond to each key. Keys are important both formally and in actual practice, as ciphers without variable keys can be trivially broken with only the knowledge of the cipher used and are therefore useless (or even counter-productive) for most purposes. Historically, ciphers were often used directly for encryption or decryption without additional procedures such as authentication or integrity checks.

In colloquial use, the term "code" is often used to mean any method of encryption or concealment of meaning. However, in cryptography, *code* has a more specific meaning. It means the replacement of a unit of plaintext (i.e., a meaningful word or phrase) with a code word (for example, "wallaby" replaces "attack at dawn"). Codes are no longer used in serious cryptography—except incidentally for such things as unit designations (e.g., Bronco Flight or Operation Overlord)—since properly chosen ciphers are both more practical and more secure than even the best codes and also are better adapted to computers.

Cryptanalysis is the term used for the study of methods for obtaining the meaning of encrypted information without access to the key normally required to do so; i.e., it is the study of how to crack encryption algorithms or their implementations.

Some use the terms *cryptography* and *cryptology* interchangeably in English, while others (including US military practice generally) use *cryptography* to refer specifically to the use and practice of cryptographic techniques and *cryptology* to refer to the combined study of cryptography and cryptanalysis.[9][10] English is more flexible than several other languages in which *cryptology* (done by cryptologists) is always used in the second sense above. RFC 2828 advises that steganography is sometimes included in cryptology.[11]

The study of characteristics of languages that have some application in cryptography or cryptology (e.g. frequency data, letter combinations, universal patterns, etc.) is called cryptolinguistics.

16.2 History of cryptography and cryptanalysis

Main article: History of cryptography

Before the modern era, cryptography was concerned solely with message confidentiality (i.e., encryption)—conversion of messages from a comprehensible form into an incomprehensible one and back again at the other end, rendering it unreadable by interceptors or eavesdroppers without secret knowledge (namely the key needed for decryption of that message). Encryption attempted to ensure secrecy in communications, such as those of spies, military leaders, and diplomats. In recent decades, the field has expanded beyond confidentiality concerns to include techniques for message integrity checking, sender/receiver identity authentication, digital signatures, interactive proofs and secure computation, among others.

16.2.1 Classic cryptography

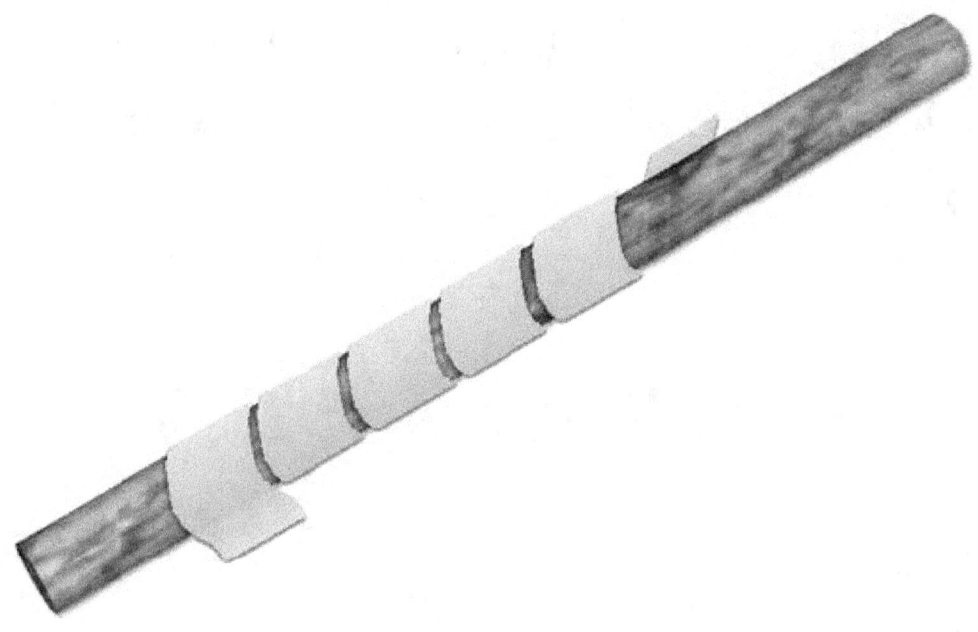

Reconstructed ancient Greek scytale, an early cipher device

The earliest forms of secret writing required little more than writing implements since most people could not read. More literacy, or literate opponents, required actual cryptography. The main classical cipher types are transposition ciphers, which rearrange the order of letters in a message (e.g., 'hello world' becomes 'ehlol owrdl' in a trivially simple rearrangement scheme), and substitution ciphers, which systematically replace letters or groups of letters with other letters or groups of letters (e.g., 'fly at once' becomes 'gmz bu podf' by replacing each letter with the one following it in the Latin alphabet). Simple versions of either have never offered much confidentiality from enterprising opponents. An early substitution cipher was the Caesar cipher, in which each letter in the plaintext was replaced by a letter some fixed number of positions further down the alphabet. Suetonius reports that Julius Caesar used it with a shift of three to communicate with his generals. Atbash is an example of an early Hebrew cipher. The earliest known use of cryptography is some carved ciphertext on stone in Egypt (ca 1900 BCE), but this may have been done for the amusement of literate observers rather than as a way of concealing information.

The Greeks of Classical times are said to have known of ciphers (e.g., the scytale transposition cipher claimed to have been used by the Spartan military).[12] Steganography (i.e., hiding even the existence of a message so as to keep it confidential) was also first developed in ancient times. An early example, from Herodotus, was a message tattooed on a slave's shaved head and concealed under the regrown hair.[8] More modern examples of steganography include the use of invisible ink, microdots, and digital watermarks to conceal information.

In India, the 2000-year old Kamasutra of Vātsyāyana speaks of two different kinds of ciphers called Kautiliyam and Mulavediya. In the Kautiliyam, the cipher letter substitutions are based on phonetic relations, such as vowels becoming consonants. In the Mulavediya, the cipher alphabet consists of pairing letters and using the reciprocal ones.[8]

First page of a book by Al-Kindi which discusses encryption of messages

Ciphertexts produced by a classical cipher (and some modern ciphers) always reveal statistical information about the plaintext, which can often be used to break them. After the discovery of frequency analysis, perhaps by the Arab mathematician and polymath Al-Kindi (also known as *Alkindus*) in the 9th century,[13] nearly all such ciphers became more or less readily breakable by any informed attacker. Such classical ciphers still enjoy popularity today, though mostly as puzzles (see cryptogram). Al-Kindi wrote a book on cryptography entitled *Risalah fi Istikhraj al-Mu'amma* (*Manuscript for the Deciphering Cryptographic Messages*), which described the first known use frequency analysis cryptanalysis techniques.[13][14]

16th-century book-shaped French cipher machine, with arms of Henri II of France

Essentially all ciphers remained vulnerable to cryptanalysis using the frequency analysis technique until the development of the polyalphabetic cipher, most clearly by Leon Battista Alberti around the year 1467, though there is some indication that it was already known to Al-Kindi.[14] Alberti's innovation was to use different ciphers (i.e., substitution alphabets) for various parts of a message (perhaps for each successive plaintext letter at the limit). He also invented what was probably the first automatic cipher device, a wheel which implemented a partial realization of his invention. In the polyalphabetic Vigenère cipher, encryption uses a *key word*, which controls letter substitution depending on which letter of the key word is used. In the mid-19th century Charles Babbage showed that the Vigenère cipher was vulnerable to Kasiski examination, but this was first published about ten years later by Friedrich Kasiski.[15]

Although frequency analysis can be a powerful and general technique against many ciphers, encryption has still often been effective in practice, as many a would-be cryptanalyst was unaware of the technique. Breaking a message without using frequency analysis essentially required knowledge of the cipher used and perhaps of the key involved, thus making espionage, bribery, burglary, defection, etc., more attractive approaches to the cryptanalytically uninformed. It was finally explicitly recognized in the 19th century that secrecy of a cipher's algorithm is not a sensible nor practical safeguard of message security; in fact, it was further realized that any adequate cryptographic scheme (including ciphers) should remain secure even if the adversary fully understands the cipher algorithm itself. Security of the key used should alone be sufficient for a good cipher to maintain confidentiality under an attack. This fundamental principle was first explicitly stated in 1883 by Auguste Kerckhoffs and is generally called Kerckhoff's Principle; alternatively and more bluntly, it was restated by Claude Shannon, the inventor of information theory and the fundamentals of theoretical cryptography, as *Shannon's Maxim*—'the enemy knows the system'.

Different physical devices and aids have been used to assist with ciphers. One of the earliest may have been the scytale

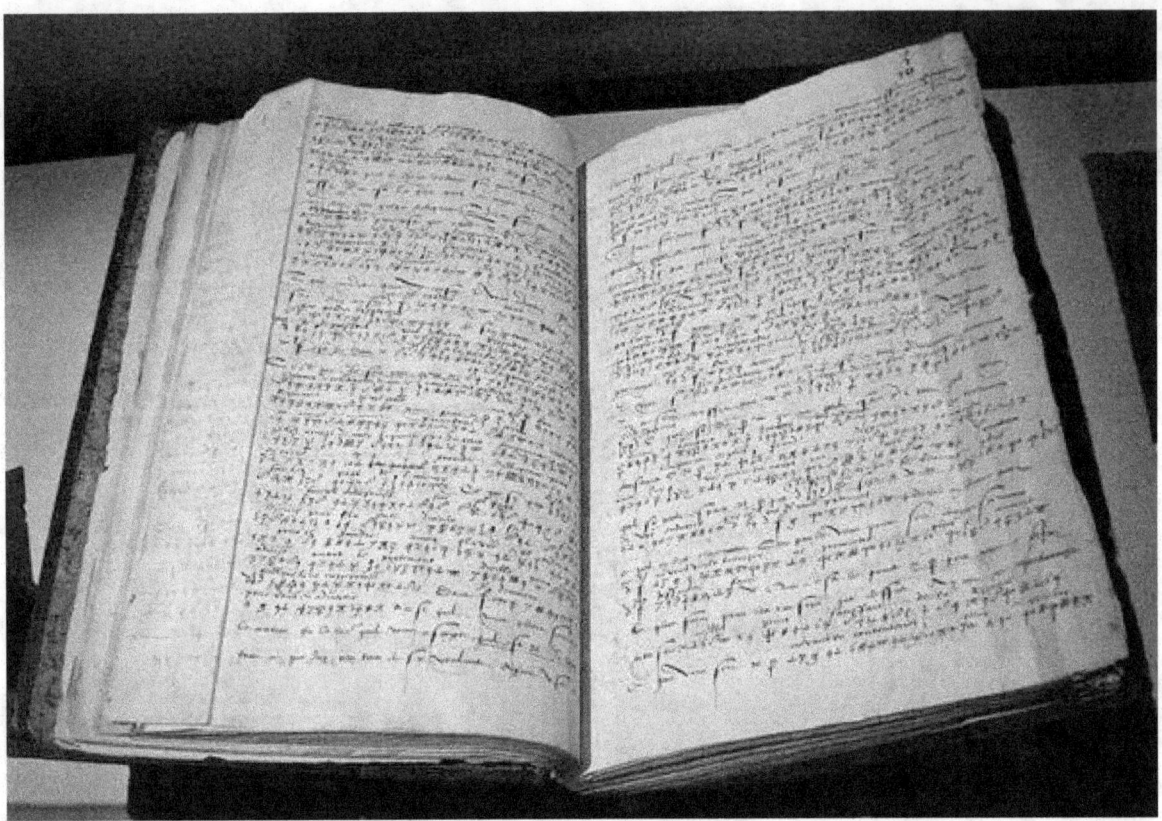

Enciphered letter from Gabriel de Luetz d'Aramon, French Ambassador to the Ottoman Empire, after 1546, with partial decipherment

of ancient Greece, a rod supposedly used by the Spartans as an aid for a transposition cipher (see image above). In medieval times, other aids were invented such as the cipher grille, which was also used for a kind of steganography. With the invention of polyalphabetic ciphers came more sophisticated aids such as Alberti's own cipher disk, Johannes Trithemius' tabula recta scheme, and Thomas Jefferson's multi cylinder (not publicly known, and reinvented independently by Bazeries around 1900). Many mechanical encryption/decryption devices were invented early in the 20th century, and several patented, among them rotor machines—famously including the Enigma machine used by the German government and military from the late 1920s and during World War II.[16] The ciphers implemented by better quality examples of these machine designs brought about a substantial increase in cryptanalytic difficulty after WWI.[17]

16.2.2   Computer era

Cryptanalysis of the new mechanical devices proved to be both difficult and laborious. In the United Kingdom, cryptanalytic efforts at Bletchley Park during WWII spurred the development of more efficient means for carrying out repetitious tasks. This culminated in the development of the Colossus, the world's first fully electronic, digital, programmable computer, which assisted in the decryption of ciphers generated by the German Army's Lorenz SZ40/42 machine.

Just as the development of digital computers and electronics helped in cryptanalysis, it made possible much more complex ciphers. Furthermore, computers allowed for the encryption of any kind of data representable in any binary format, unlike classical ciphers which only encrypted written language texts; this was new and significant. Computer use has thus supplanted linguistic cryptography, both for cipher design and cryptanalysis. Many computer ciphers can be characterized by their operation on binary bit sequences (sometimes in groups or blocks), unlike classical and mechanical schemes, which generally manipulate traditional characters (i.e., letters and digits) directly. However, computers have also assisted cryptanalysis, which has compensated to some extent for increased cipher complexity. Nonetheless, good modern ciphers have stayed ahead of cryptanalysis; it is typically the case that use of a quality cipher is very efficient (i.e., fast and requiring few resources, such as memory or CPU capability), while breaking it requires an effort many orders of magnitude larger, and vastly larger than that required for any classical cipher, making cryptanalysis so inefficient and impractical as to be effectively impossible.

Extensive open academic research into cryptography is relatively recent; it began only in the mid-1970s. In recent times, IBM personnel designed the algorithm that became the Federal (i.e., US) Data Encryption Standard; Whitfield Diffie and Martin Hellman published their key agreement algorithm;[18] and the RSA algorithm was published in Martin Gardner's *Scientific American* column. Since then, cryptography has become a widely used tool in communications, computer networks, and computer security generally. Some modern cryptographic techniques can only keep their keys secret if certain mathematical problems are intractable, such as the integer factorization or the discrete logarithm problems, so there are deep connections with abstract mathematics. There are very few cryptosytems that are proven to be unconditionally secure. The one-time pad is one. There are a few important ones that are proven secure under certain unproven assumptions. For example, the infeasibility of factoring extremely large integers is the basis for believing that RSA is secure, and some other systems, but even there, the proof is usually lost due to practical considerations. There are systems similar to RSA, such as one by Michael O. Rabin that is provably secure provided factoring n = pq is impossible, but the more practical system RSA has never been proved secure in this sense. The discrete logarithm problem is the basis for believing some other cryptosystems are secure, and again, there are related, less practical systems that are provably secure relative to the discrete log problem.[19]

As well as being aware of cryptographic history, cryptographic algorithm and system designers must also sensibly consider probable future developments while working on their designs. For instance, continuous improvements in computer processing power have increased the scope of brute-force attacks, so when specifying key lengths, the required key lengths are similarly advancing.[20] The potential effects of quantum computing are already being considered by some cryptographic system designers; the announced imminence of small implementations of these machines may be making the need for this preemptive caution rather more than merely speculative.[4]

Essentially, prior to the early 20th century, cryptography was chiefly concerned with linguistic and lexicographic patterns. Since then the emphasis has shifted, and cryptography now makes extensive use of mathematics, including aspects of information theory, computational complexity, statistics, combinatorics, abstract algebra, number theory, and finite mathematics generally. Cryptography is also a branch of engineering, but an unusual one since it deals with active, intelligent, and malevolent opposition (see cryptographic engineering and security engineering); other kinds of engineering (e.g., civil or chemical engineering) need deal only with neutral natural forces. There is also active research examining the relationship between cryptographic problems and quantum physics (see quantum cryptography and quantum computer).

16.3   Modern cryptography

The modern field of cryptography can be divided into several areas of study. The chief ones are discussed here; see Topics in Cryptography for more.

16.3.1   Symmetric-key cryptography

Main article: Symmetric-key algorithm
 Symmetric-key cryptography refers to encryption methods in which both the sender and receiver share the same key (or, less commonly, in which their keys are different, but related in an easily computable way). This was the only kind of encryption publicly known until June 1976.[18]

Symmetric key ciphers are implemented as either block ciphers or stream ciphers. A block cipher enciphers input in blocks of plaintext as opposed to individual characters, the input form used by a stream cipher.

The Data Encryption Standard (DES) and the Advanced Encryption Standard (AES) are block cipher designs which have been designated cryptography standards by the US government (though DES's designation was finally withdrawn after the AES was adopted).[21] Despite its deprecation as an official standard, DES (especially its still-approved and much more secure triple-DES variant) remains quite popular; it is used across a wide range of applications, from ATM encryption[22] to e-mail privacy[23] and secure remote access.[24] Many other block ciphers have been designed and released, with considerable variation in quality. Many have been thoroughly broken, such as FEAL.[4][25]

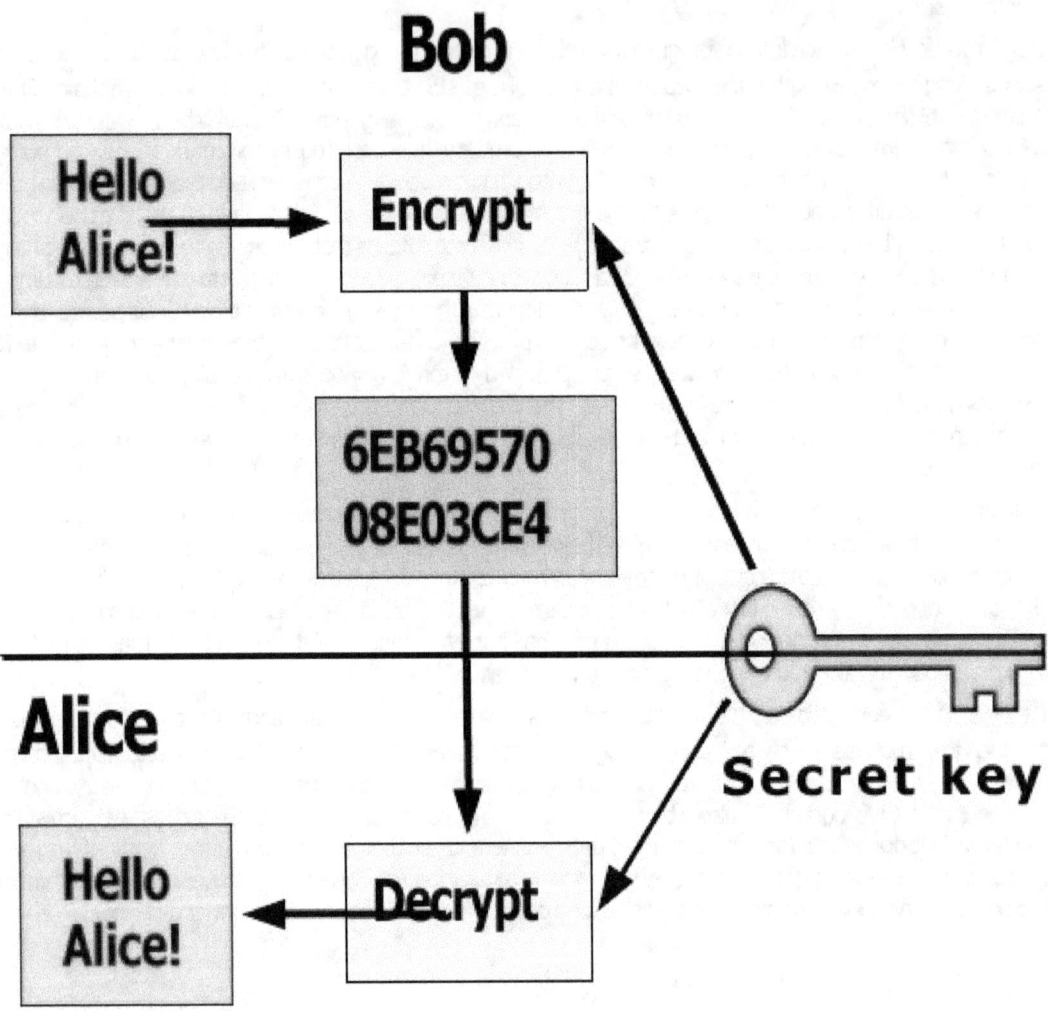

Symmetric-key cryptography, where a single key is used for encryption and decryption

Stream ciphers, in contrast to the 'block' type, create an arbitrarily long stream of key material, which is combined with the plaintext bit-by-bit or character-by-character, somewhat like the one-time pad. In a stream cipher, the output stream is created based on a hidden internal state which changes as the cipher operates. That internal state is initially set up using the secret key material. RC4 is a widely used stream cipher; see Category:Stream ciphers.[4] Block ciphers can be used as stream ciphers; see Block cipher modes of operation.

Cryptographic hash functions are a third type of cryptographic algorithm. They take a message of any length as input, and output a short, fixed length hash which can be used in (for example) a digital signature. For good hash functions, an attacker cannot find two messages that produce the same hash. MD4 is a long-used hash function which is now broken; MD5, a strengthened variant of MD4, is also widely used but broken in practice. The US National Security Agency developed the Secure Hash Algorithm series of MD5-like hash functions: SHA-0 was a flawed algorithm that the agency withdrew; SHA-1 is widely deployed and more secure than MD5, but cryptanalysts have identified attacks against it; the SHA-2 family improves on SHA-1, but it isn't yet widely deployed; and the US standards authority thought it "prudent" from a security perspective to develop a new standard to "significantly improve the robustness of NIST's overall hash algorithm toolkit."[26] Thus, a hash function design competition was meant to select a new U.S. national standard, to be

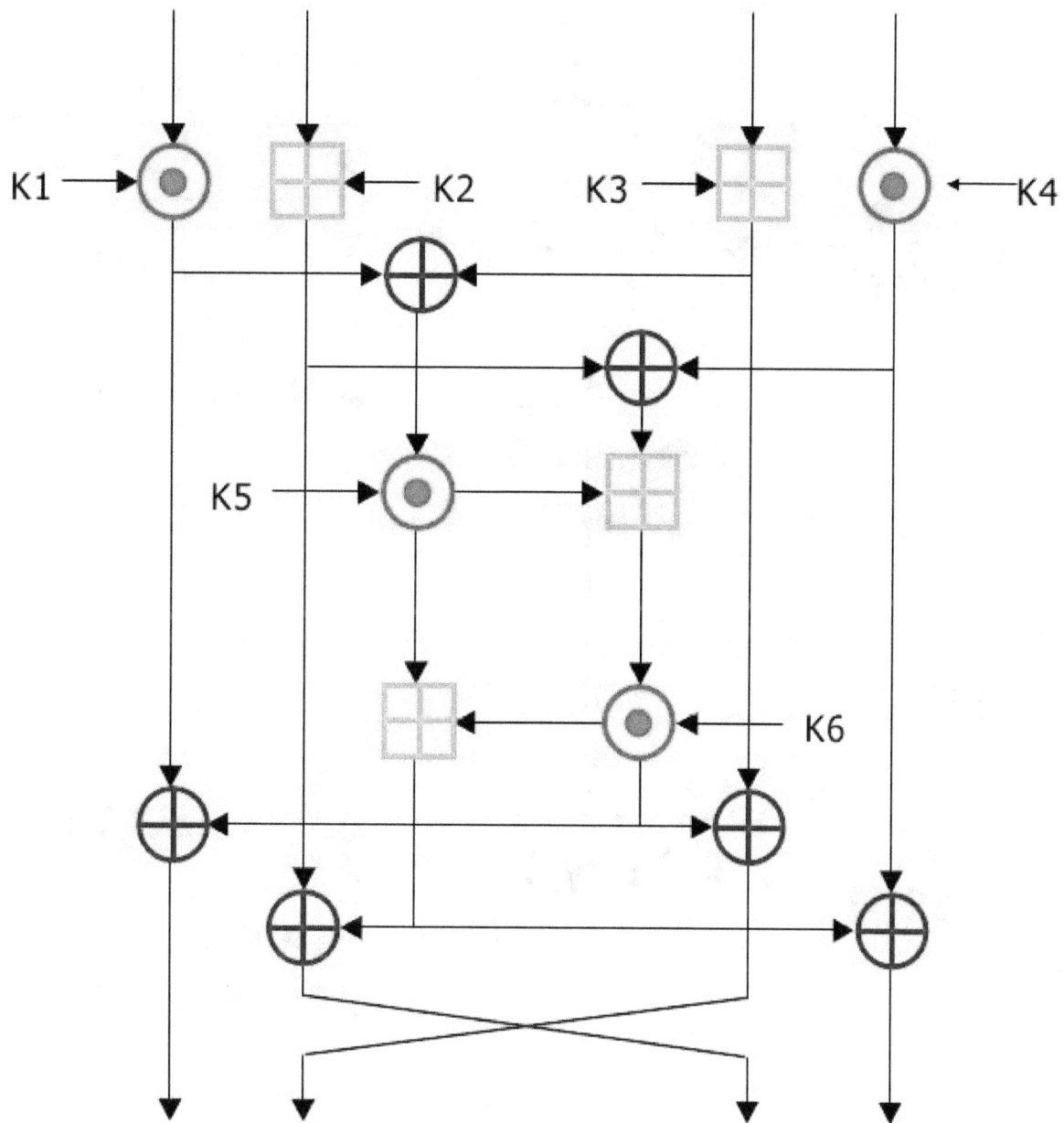

One round (out of 8.5) of the IDEA cipher, used in some versions of PGP for high-speed encryption of, for instance, e-mail

called SHA-3, by 2012. The competition ended on October 2, 2012 when the NIST announced that Keccak would be the new SHA-3 hash algorithm.[27]

Message authentication codes (MACs) are much like cryptographic hash functions, except that a secret key can be used to authenticate the hash value upon receipt;[4] this additional complication blocks an attack scheme against bare digest algorithms, and so has been thought worth the effort.

16.3.2   Public-key cryptography

Main article: Public-key cryptography

Symmetric-key cryptosystems use the same key for encryption and decryption of a message, though a message or group of messages may have a different key than others. A significant disadvantage of symmetric ciphers is the key management

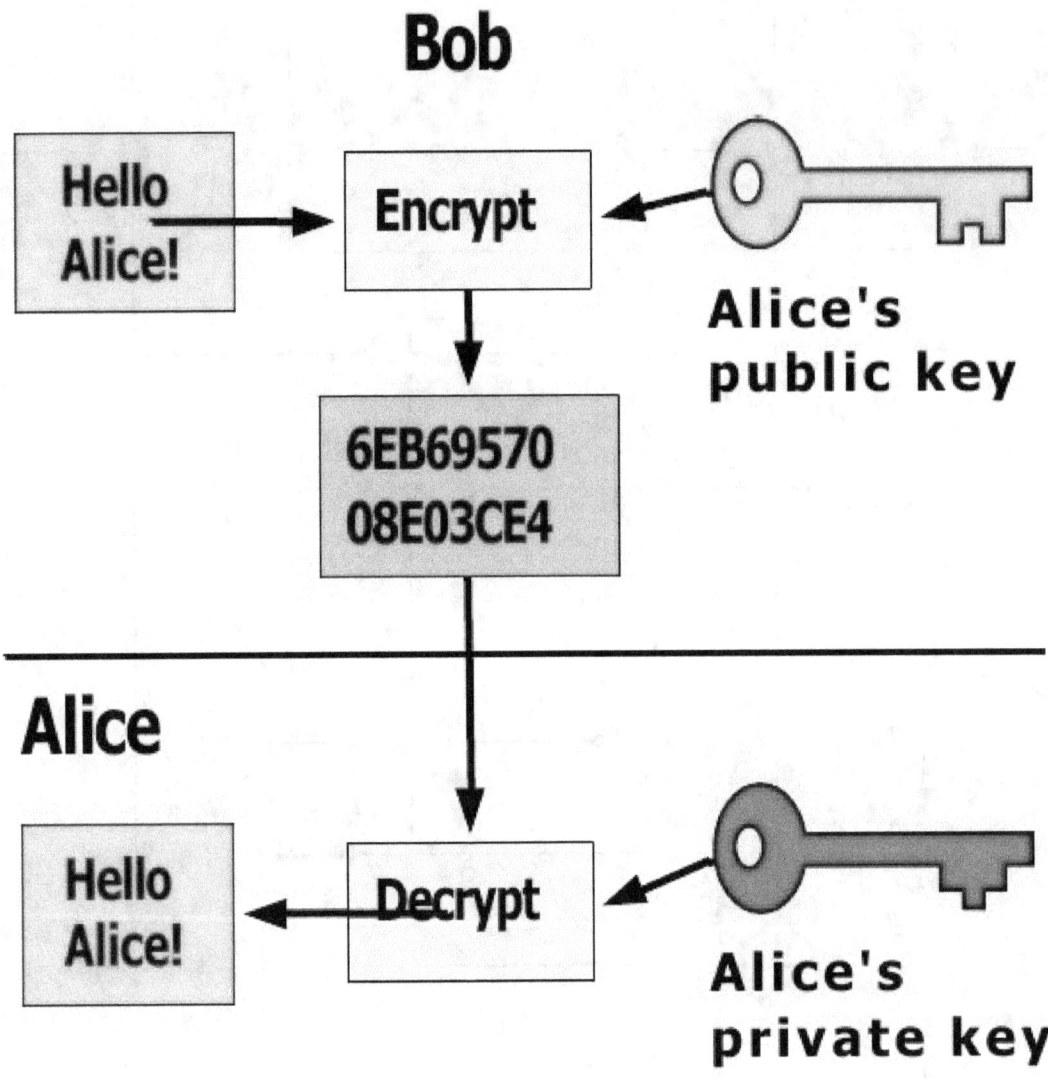

Public-key cryptography, where different keys are used for encryption and decryption

necessary to use them securely. Each distinct pair of communicating parties must, ideally, share a different key, and perhaps each ciphertext exchanged as well. The number of keys required increases as the square of the number of network members, which very quickly requires complex key management schemes to keep them all consistent and secret. The difficulty of securely establishing a secret key between two communicating parties, when a secure channel does not already exist between them, also presents a chicken-and-egg problem which is a considerable practical obstacle for cryptography users in the real world.

In a groundbreaking 1976 paper, Whitfield Diffie and Martin Hellman proposed the notion of *public-key* (also, more generally, called *asymmetric key*) cryptography in which two different but mathematically related keys are used—a *public* key and a *private* key.[28] A public key system is so constructed that calculation of one key (the 'private key') is computationally infeasible from the other (the 'public key'), even though they are necessarily related. Instead, both keys are generated secretly, as an interrelated pair.[29] The historian David Kahn described public-key cryptography as "the most revolutionary new concept in the field since polyalphabetic substitution emerged in the Renaissance".[30]

In public-key cryptosystems, the public key may be freely distributed, while its paired private key must remain secret. In

Whitfield Diffie and Martin Hellman, authors of the first published paper on public-key cryptography

a public-key encryption system, the *public key* is used for encryption, while the *private* or *secret key* is used for decryption. While Diffie and Hellman could not find such a system, they showed that public-key cryptography was indeed possible by presenting the Diffie–Hellman key exchange protocol, a solution that is now widely used in secure communications to allow two parties to secretly agree on a shared encryption key.[18]

Diffie and Hellman's publication sparked widespread academic efforts in finding a practical public-key encryption system. This race was finally won in 1978 by Ronald Rivest, Adi Shamir, and Len Adleman, whose solution has since become known as the RSA algorithm.[31]

The Diffie–Hellman and RSA algorithms, in addition to being the first publicly known examples of high quality public-key algorithms, have been among the most widely used. Others include the Cramer–Shoup cryptosystem, ElGamal encryption, and various elliptic curve techniques. See Category:Asymmetric-key cryptosystems.

To much surprise, a document published in 1997 by the Government Communications Headquarters (GCHQ), a British intelligence organization, revealed that cryptographers at GCHQ had anticipated several academic developments.[32] Reportedly, around 1970, James H. Ellis had conceived the principles of asymmetric key cryptography. In 1973, Clifford Cocks invented a solution that essentially resembles the RSA algorithm.[32][33] And in 1974, Malcolm J. Williamson is claimed to have developed the Diffie–Hellman key exchange.[34]

Padlock icon from the Firefox Web browser, which indicates that TLS, a public-key cryptography system, is in use.

Public-key cryptography can also be used for implementing digital signature schemes. A digital signature is reminiscent of an ordinary signature; they both have the characteristic of being easy for a user to produce, but difficult for anyone else to forge. Digital signatures can also be permanently tied to the content of the message being signed; they cannot then be 'moved' from one document to another, for any attempt will be detectable. In digital signature schemes, there are two algorithms: one for *signing*, in which a secret key is used to process the message (or a hash of the message, or both), and one for *verification*, in which the matching public key is used with the message to check the validity of the signature. RSA and DSA are two of the most popular digital signature schemes. Digital signatures are central to the operation of public key infrastructures and many network security schemes (e.g., SSL/TLS, many VPNs, etc.).[25]

Public-key algorithms are most often based on the computational complexity of "hard" problems, often from number theory. For example, the hardness of RSA is related to the integer factorization problem, while Diffie–Hellman and DSA are related to the discrete logarithm problem. More recently, *elliptic curve cryptography* has developed, a system in which security is based on number theoretic problems involving elliptic curves. Because of the difficulty of the underlying problems, most public-key algorithms involve operations such as modular multiplication and exponentiation, which are much more computationally expensive than the techniques used in most block ciphers, especially with typical key sizes. As a result, public-key cryptosystems are commonly hybrid cryptosystems, in which a fast high-quality symmetric-key encryption algorithm is used for the message itself, while the relevant symmetric key is sent with the message, but encrypted using a public-key algorithm. Similarly, hybrid signature schemes are often used, in which a cryptographic hash function is computed, and only the resulting hash is digitally signed.[4]

16.3.3   Cryptanalysis

Main article: Cryptanalysis
The goal of cryptanalysis is to find some weakness or insecurity in a cryptographic scheme, thus permitting its subversion or evasion.

It is a common misconception that every encryption method can be broken. In connection with his WWII work at Bell Labs, Claude Shannon proved that the one-time pad cipher is unbreakable, provided the key material is truly random, never reused, kept secret from all possible attackers, and of equal or greater length than the message.[35] Most ciphers, apart from the one-time pad, can be broken with enough computational effort by brute force attack, but the amount of effort needed may be exponentially dependent on the key size, as compared to the effort needed to make use of the cipher. In such cases, effective security could be achieved if it is proven that the effort required (i.e., "work factor", in Shannon's terms) is beyond the ability of any adversary. This means it must be shown that no efficient method (as opposed to the time-consuming brute force method) can be found to break the cipher. Since no such proof has been found to date, the one-time-pad remains the only theoretically unbreakable cipher.

There are a wide variety of cryptanalytic attacks, and they can be classified in any of several ways. A common distinction turns on what an attacker knows and what capabilities are available. In a ciphertext-only attack, the cryptanalyst has access only to the ciphertext (good modern cryptosystems are usually effectively immune to ciphertext-only attacks). In a known-plaintext attack, the cryptanalyst has access to a ciphertext and its corresponding plaintext (or to many such pairs). In a chosen-plaintext attack, the cryptanalyst may choose a plaintext and learn its corresponding ciphertext (perhaps many times); an example is gardening, used by the British during WWII. Finally, in a chosen-ciphertext attack, the cryptanalyst may be able to *choose* ciphertexts and learn their corresponding plaintexts.[4] Also important, often overwhelmingly so, are mistakes (generally in the design or use of one of the protocols involved; see Cryptanalysis of the Enigma for some historical examples of this).

Cryptanalysis of symmetric-key ciphers typically involves looking for attacks against the block ciphers or stream ciphers that are more efficient than any attack that could be against a perfect cipher. For example, a simple brute force attack against DES requires one known plaintext and 2^{55} decryptions, trying approximately half of the possible keys, to reach a point at which chances are better than even that the key sought will have been found. But this may not be enough assurance; a linear cryptanalysis attack against DES requires 2^{43} known plaintexts and approximately 2^{43} DES operations.[36] This is a considerable improvement on brute force attacks.

Public-key algorithms are based on the computational difficulty of various problems. The most famous of these is integer factorization (e.g., the RSA algorithm is based on a problem related to integer factoring), but the discrete logarithm problem is also important. Much public-key cryptanalysis concerns numerical algorithms for solving these computational problems, or some of them, efficiently (i.e., in a practical time). For instance, the best known algorithms for solving the elliptic curve-based version of discrete logarithm are much more time-consuming than the best known algorithms for factoring, at least for problems of more or less equivalent size. Thus, other things being equal, to achieve an equivalent strength of attack resistance, factoring-based encryption techniques must use larger keys than elliptic curve techniques.

For this reason, public-key cryptosystems based on elliptic curves have become popular since their invention in the mid-1990s.

While pure cryptanalysis uses weaknesses in the algorithms themselves, other attacks on cryptosystems are based on actual use of the algorithms in real devices, and are called *side-channel attacks*. If a cryptanalyst has access to, for example, the amount of time the device took to encrypt a number of plaintexts or report an error in a password or PIN character, he may be able to use a timing attack to break a cipher that is otherwise resistant to analysis. An attacker might also study the pattern and length of messages to derive valuable information; this is known as traffic analysis[37] and can be quite useful to an alert adversary. Poor administration of a cryptosystem, such as permitting too short keys, will make any system vulnerable, regardless of other virtues. And, of course, social engineering, and other attacks against the personnel who work with cryptosystems or the messages they handle (e.g., bribery, extortion, blackmail, espionage, torture, ...) may be the most productive attacks of all.

16.3.4 Cryptographic primitives

Much of the theoretical work in cryptography concerns cryptographic *primitives*—algorithms with basic cryptographic properties—and their relationship to other cryptographic problems. More complicated cryptographic tools are then built from these basic primitives. These primitives provide fundamental properties, which are used to develop more complex tools called *cryptosystems* or *cryptographic protocols*, which guarantee one or more high-level security properties. Note however, that the distinction between cryptographic *primitives* and cryptosystems, is quite arbitrary; for example, the RSA algorithm is sometimes considered a cryptosystem, and sometimes a primitive. Typical examples of cryptographic primitives include pseudorandom functions, one-way functions, etc.

16.3.5 Cryptosystems

One or more cryptographic primitives are often used to develop a more complex algorithm, called a cryptographic system, or *cryptosystem*. Cryptosystems (e.g., El-Gamal encryption) are designed to provide particular functionality (e.g., public key encryption) while guaranteeing certain security properties (e.g., chosen-plaintext attack (CPA) security in the random oracle model). Cryptosystems use the properties of the underlying cryptographic primitives to support the system's security properties. Of course, as the distinction between primitives and cryptosystems is somewhat arbitrary, a sophisticated cryptosystem can be derived from a combination of several more primitive cryptosystems. In many cases, the cryptosystem's structure involves back and forth communication among two or more parties in space (e.g., between the sender of a secure message and its receiver) or across time (e.g., cryptographically protected backup data). Such cryptosystems are sometimes called *cryptographic protocols*.

Some widely known cryptosystems include RSA encryption, Schnorr signature, El-Gamal encryption, PGP, etc. More complex cryptosystems include electronic cash[38] systems, signcryption systems, etc. Some more 'theoretical' cryptosystems include interactive proof systems,[39] (like zero-knowledge proofs),[40] systems for secret sharing,[41][42] etc.

Until recently, most security properties of most cryptosystems were demonstrated using empirical techniques or using ad hoc reasoning. Recently, there has been considerable effort to develop formal techniques for establishing the security of cryptosystems; this has been generally called *provable security*. The general idea of provable security is to give arguments about the computational difficulty needed to compromise some security aspect of the cryptosystem (i.e., to any adversary).

The study of how best to implement and integrate cryptography in software applications is itself a distinct field (see Cryptographic engineering and Security engineering).

16.4 Legal issues

See also: Cryptography laws in different nations

16.4.1   Prohibitions

Cryptography has long been of interest to intelligence gathering and law enforcement agencies. Secret communications may be criminal or even treasonous. Because of its facilitation of privacy, and the diminution of privacy attendant on its prohibition, cryptography is also of considerable interest to civil rights supporters. Accordingly, there has been a history of controversial legal issues surrounding cryptography, especially since the advent of inexpensive computers has made widespread access to high quality cryptography possible.

In some countries, even the domestic use of cryptography is, or has been, restricted. Until 1999, France significantly restricted the use of cryptography domestically, though it has since relaxed many of these rules. In China and Iran, a license is still required to use cryptography.[5] Many countries have tight restrictions on the use of cryptography. Among the more restrictive are laws in Belarus, Kazakhstan, Mongolia, Pakistan, Singapore, Tunisia, and Vietnam.[43]

In the United States, cryptography is legal for domestic use, but there has been much conflict over legal issues related to cryptography. One particularly important issue has been the export of cryptography and cryptographic software and hardware. Probably because of the importance of cryptanalysis in World War II and an expectation that cryptography would continue to be important for national security, many Western governments have, at some point, strictly regulated export of cryptography. After World War II, it was illegal in the US to sell or distribute encryption technology overseas; in fact, encryption was designated as auxiliary military equipment and put on the United States Munitions List.[44] Until the development of the personal computer, asymmetric key algorithms (i.e., public key techniques), and the Internet, this was not especially problematic. However, as the Internet grew and computers became more widely available, high-quality encryption techniques became well known around the globe.

16.4.2   Export controls

Main article: Export of cryptography

In the 1990s, there were several challenges to US export regulation of cryptography. After the source code for Philip Zimmermann's Pretty Good Privacy (PGP) encryption program found its way onto the Internet in June 1991, a complaint by RSA Security (then called RSA Data Security, Inc.) resulted in a lengthy criminal investigation of Zimmermann by the US Customs Service and the FBI, though no charges were ever filed.[45][46] Daniel J. Bernstein, then a graduate student at UC Berkeley, brought a lawsuit against the US government challenging some aspects of the restrictions based on free speech grounds. The 1995 case Bernstein v. United States ultimately resulted in a 1999 decision that printed source code for cryptographic algorithms and systems was protected as free speech by the United States Constitution.[47]

In 1996, thirty-nine countries signed the Wassenaar Arrangement, an arms control treaty that deals with the export of arms and "dual-use" technologies such as cryptography. The treaty stipulated that the use of cryptography with short key-lengths (56-bit for symmetric encryption, 512-bit for RSA) would no longer be export-controlled.[48] Cryptography exports from the US became less strictly regulated as a consequence of a major relaxation in 2000;[49] there are no longer very many restrictions on key sizes in US-exported mass-market software. Since this relaxation in US export restrictions, and because most personal computers connected to the Internet include US-sourced web browsers such as Firefox or Internet Explorer, almost every Internet user worldwide has potential access to quality cryptography via their browsers (e.g., via Transport Layer Security). The Mozilla Thunderbird and Microsoft Outlook E-mail client programs similarly can transmit and receive emails via TLS, and can send and receive email encrypted with S/MIME. Many Internet users don't realize that their basic application software contains such extensive cryptosystems. These browsers and email programs are so ubiquitous that even governments whose intent is to regulate civilian use of cryptography generally don't find it practical to do much to control distribution or use of cryptography of this quality, so even when such laws are in force, actual enforcement is often effectively impossible.

16.4.3   NSA involvement

See also: Clipper chip

Another contentious issue connected to cryptography in the United States is the influence of the National Security Agency on cipher development and policy. The NSA was involved with the design of DES during its development at IBM and its consideration by the National Bureau of Standards as a possible Federal Standard for cryptography.[50] DES was designed to be resistant to differential cryptanalysis,[51] a powerful and general cryptanalytic technique known to the NSA and IBM, that became publicly known only when it was rediscovered in the late 1980s.[52] According to Steven Levy, IBM discovered differential cryptanalysis,[46] but kept the technique secret at the NSA's request. The technique became publicly known only when Biham and Shamir re-discovered and announced it some years later. The entire affair illustrates the difficulty of determining what resources and knowledge an attacker might actually have.

Another instance of the NSA's involvement was the 1993 Clipper chip affair, an encryption microchip intended to be part of the Capstone cryptography-control initiative. Clipper was widely criticized by cryptographers for two reasons. The cipher algorithm (called Skipjack) was then classified (declassified in 1998, long after the Clipper initiative lapsed). The classified cipher caused concerns that the NSA had deliberately made the cipher weak in order to assist its intelligence efforts. The whole initiative was also criticized based on its violation of Kerckhoff ss Principle, as the scheme included a special escrow key held by the government for use by law enforcement, for example in wiretaps.[46]

16.4.4   Digital rights management

Main article: Digital rights management

Cryptography is central to digital rights management (DRM), a group of techniques for technologically controlling use of copyrighted material, being widely implemented and deployed at the behest of some copyright holders. In 1998, U.S. President Bill Clinton signed the Digital Millennium Copyright Act (DMCA), which criminalized all production, dissemination, and use of certain cryptanalytic techniques and technology (now known or later discovered); specifically, those that could be used to circumvent DRM technological schemes.[53] This had a noticeable impact on the cryptography research community since an argument can be made that *any* cryptanalytic research violated, or might violate, the DMCA. Similar statutes have since been enacted in several countries and regions, including the implementation in the EU Copyright Directive. Similar restrictions are called for by treaties signed by World Intellectual Property Organization member-states.

The United States Department of Justice and FBI have not enforced the DMCA as rigorously as had been feared by some, but the law, nonetheless, remains a controversial one. Niels Ferguson, a well-respected cryptography researcher, has publicly stated that he will not release some of his research into an Intel security design for fear of prosecution under the DMCA.[54] Both Alan Cox (longtime number 2 in Linux kernel development) and Edward Felten (and some of his students at Princeton) have encountered problems related to the Act. Dmitry Sklyarov was arrested during a visit to the US from Russia, and jailed for five months pending trial for alleged violations of the DMCA arising from work he had done in Russia, where the work was legal. In 2007, the cryptographic keys responsible for Blu-ray and HD DVD content scrambling were discovered and released onto the Internet. In both cases, the MPAA sent out numerous DMCA takedown notices, and there was a massive Internet backlash[7] triggered by the perceived impact of such notices on fair use and free speech.

16.4.5   Forced disclosure of encryption keys

Main article: Key disclosure law

In the United Kingdom, the Regulation of Investigatory Powers Act gives UK police the powers to force suspects to decrypt files or hand over passwords that protect encryption keys. Failure to comply is an offense in its own right, punishable on conviction by a two-year jail sentence or up to five years in cases involving national security.[6] Successful prosecutions have occurred under the Act; the first, in 2009,[55] resulted in a term of 13 months' imprisonment.[56] Similar forced disclosure laws in Australia, Finland, France, and India compel individual suspects under investigation to hand over encryption keys or passwords during a criminal investigation.

In the United States, the federal criminal case of United States v. Fricosu addressed whether a search warrant can compel a person to reveal an encryption passphrase or password.[57] The Electronic Frontier Foundation (EFF) argued that this is a violation of the protection from self-incrimination given by the Fifth Amendment.[58] In 2012, the court ruled that under the All Writs Act, the defendant was required to produce an unencrypted hard drive for the court.[59]

In many jurisdictions, the legal status of forced disclosure remains unclear.

16.5   See also

- List of cryptographers
- Encyclopedia of Cryptography and Security
- List of important publications in cryptography
- List of multiple discoveries (see "RSA")
- List of unsolved problems in computer science
- Outline of cryptography
- Global surveillance
- Strong cryptography

16.6   References

[1] Liddell, Henry George; Scott, Robert; Jones, Henry Stuart; McKenzie, Roderick (1984). *A Greek-English Lexicon*. Oxford University Press.

[2] Rivest, Ronald L. (1990). "Cryptology". In J. Van Leeuwen. *Handbook of Theoretical Computer Science* 1. Elsevier.

[3] Bellare, Mihir; Rogaway, Phillip (21 September 2005). "Introduction". *Introduction to Modern Cryptography*. p. 10.

[4] Menezes, A. J.; van Oorschot, P. C.; Vanstone, S. A. *Handbook of Applied Cryptography*. ISBN 0-8493-8523-7.

[5] "Overview per country". *Crypto Law Survey*. February 2013. Retrieved 26 March 2015.

[6] "UK Data Encryption Disclosure Law Takes Effect". *PC World*. 1 October 2007. Retrieved 26 March 2015.

[7] Doctorow, Cory (2 May 2007). "Digg users revolt over AACS key". *Boing Boing*. Retrieved 26 March 2015.

[8] Kahn, David (1967). *The Codebreakers*. ISBN 0-684-83130-9.

[9] Oded Goldreich, *Foundations of Cryptography, Volume 1: Basic Tools*, Cambridge University Press, 2001, ISBN 0-521-79172-3

[10] "Cryptology (definition)". *Merriam-Webster's Collegiate Dictionary* (11th ed.). Merriam-Webster. Retrieved 26 March 2015.

[11] "RFC 2828 - Internet Security Glossary". *Internet Engineering Task Force*. May 2000. Retrieved 26 March 2015.

[12] IA'shchenko, V. V. (2002). *Cryptography: an introduction*. AMS Bookstore. p. 6. ISBN 0-8218-2986-6.

[13] Singh, Simon (2000). *The Code Book*. New York: Anchor Books. pp. 14–20. ISBN 9780385495325.

[14] Al-Kadi, Ibrahim A. (April 1992). "The origins of cryptology: The Arab contributions". *Cryptologia* 16 (2): 97–126.

[15] Schrödel, Tobias (October 2008). "Breaking Short Vigenère Ciphers". *Cryptologia* 32(4): 334–337. doi:10.1080/0161119080297.

[16] Hakim, Joy (1995). *A History of US: War, Peace and all that Jazz*. New York: Oxford University Press. ISBN 0-19-509514-6.

[17] Gannon, James (2001). *Stealing Secrets, Telling Lies: How Spies and Codebreakers Helped Shape the Twentieth Century*. Washington, D.C.: Brassey's. ISBN 1-57488-367-4.

[18] Diffie, Whitfield; Hellman, Martin (November 1976). "New Directions in Cryptography" (PDF). *IEEE Transactions on Information Theory*. IT-22: 644–654.

[19] *Cryptography: Theory and Practice*, Third Edition (Discrete Mathematics and Its Applications), 2005, by Douglas R. Stinson, Chapman and Hall/CRC

[20] Blaze, Matt; Diffie, Whitefield; Rivest, Ronald L.; Schneier, Bruce; Shimomura, Tsutomu; Thompson, Eric; Wiener, Michael (January 1996). "Minimal key lengths for symmetric ciphers to provide adequate commercial security". Fortify. Retrieved 26 March 2015.

[21] "FIPS PUB 197: The official Advanced Encryption Standard" (PDF). *Computer Security Resource Center*. National Institute of Standards and Technology. Retrieved 26 March 2015.

[22] "NCUA letter to credit unions" (PDF). *National Credit Union Administration*. July 2004. Retrieved 26 March 2015.

[23] "RFC 2440 - Open PGP Message Format". *Internet Engineering Task Force*. November 1998. Retrieved 26 March 2015.

[24] Golen, Pawel (19 July 2002). "SSH". *WindowSecurity*. Retrieved 26 March 2015.

[25] Schneier, Bruce (1996). *Applied Cryptography* (2nd ed.). Wiley. ISBN 0-471-11709-9.

[26] "Notices". *Federal Register* 72 (212). 2 November 2007.
Archived 28 February 2008 at the Wayback Machine

[27] "NIST Selects Winner of Secure Hash Algorithm (SHA-3) Competition". *Tech Beat*. National Institute of Standards and Technology. October 2, 2012. Retrieved 26 March 2015.

[28] Diffie, Whitfield; Hellman, Martin (8 June 1976). "Multi-user cryptographic techniques". *AFIPS Proceedings* 45: 109–112.

[29] Ralph Merkle was working on similar ideas at the time and encountered publication delays, and Hellman has suggested that the term used should be Diffi e–Hellman–Merkle aysmmetric key cryptography.

[30] Kahn, David (Fall 1979). "Cryptology Goes Public". *Foreign Affairs* 58 (1): 153.

[31] Rivest, Ronald L.; Shamir, A.; Adleman, L. (1978). "A Method for Obtaining Digital Signatures and Public-Key Cryptosystems". *Communications of the ACM* (Association for Computing Machinery) 21 (2): 120–126.
Archived November 16, 2001 at the Wayback Machine
Previously released as an MIT "Technical Memo" in April 1977, and published in Martin Gardner's *Scientific American* Mathematical recreations column

[32] Wayner, Peter (24 December 1997). "British Document Outlines Early Encryption Discovery". *New York Times*. Retrieved 26 March 2015.

[33] Cocks, Clifford (20 November 1973). "A Note on 'Non-Secret Encryption'" (PDF). *CESG Research Report*.

[34] Singh, Simon (1999). *The Code Book*. Doubleday. pp. 279–292.

[35] Shannon, Claude; Weaver, Warren (1963). *The Mathematical Theory of Communication*. University of Illinois Press. ISBN 0-252-72548-4.

[36] Junod, Pascal (2001). "On the Complexity of Matsui's Attack" (PDF). *Selected Areas in Cryptography*.

[37] Song, Dawn; Wagner, David A.; Tian, Xuqing (2001). "Timing Analysis of Keystrokes and Timing Attacks on SSH" (PDF). *Tenth USENIX Security Symposium*.

[38] Brands, S. (1994). "Untraceable Off-line Cash in Wallets with Observers". *Advances in Cryptology—Proceedings of CRYPTO* (Springer-Verlag).

[39] Babai, László (1985). "Trading group theory for randomness". *Proceedings of the Seventeenth Annual Symposium on the Theory of Computing* (Association for Computing Machinery).

[40] Goldwasser, S.; Micali, S.; Rackoff, C. (1989). "The Knowledge Complexity of Interactive Proof Systems". *SIAM Journal on Computing* 18 (1): 186–208.

[41] Blakley, G. (June 1979). "Safeguarding cryptographic keys". *Proceedings of AFIPS 1979* 48: 313–317.

[42] Shamir, A. (1979). "How to share a secret". *Communications of the ACM* (Association for Computing Machinery) 22: 612–613.

[43] "6.5.1 WHAT ARE THE CRYPTOGRAPHIC POLICIES OF SOME COUNTRIES?". RSA Laboratories. Retrieved 26 March 2015.

[44] Rosenoer, Jonathan (1995). "CRYPTOGRAPHY & SPEECH". *CyberLaw*. Archived December 1, 2005 at the Wayback Machine

[45] "Case Closed on Zimmermann PGP Investigation". *IEEE Computer Society's Technical Committee on Security and Privacy*. 14 February 1996. Retrieved 26 March 2015.

[46] Levy, Steven (2001). *Crypto: How the Code Rebels Beat the Government—Saving Privacy in the Digital Age*. Penguin Books. p. 56. ISBN 0-14-024432-8. OCLC 244148644 48066852 48846639.

[47] "Bernstein v USDOJ". *Electronic Privacy Information Center*. United States Court of Appeals for the Ninth Circuit. 6 May 1999. Retrieved 26 March 2015.

[48] "DUAL-USE LIST - CATEGORY 5 – PART 2 – "INFORMATION SECURITY"" (DOC). *Wassenaar Arrangement*. Retrieved 26 March 2015.

[49] "6.4 UNITED STATES CRYPTOGRAPHY EXPORT/IMPORT LAWS". *RSA Laboratories*. Retrieved 26 March 2015.

[50] Schneier, Bruce (15 June 2000). "The Data Encryption Standard (DES)". *Crypto-Gram*. Retrieved 26 March 2015.

[51] Coppersmith, D. (May 1994). "The Data Encryption Standard (DES) and its strength against attacks" (PDF). *IBM Journal of Research and Development* 38 (3): 243. doi:10.1147/rd.383.0243. Retrieved 26 March 2015.

[52] Biham, E.; Shamir, A. (1991). "Differential cryptanalysis of DES-like cryptosystems" (PDF). *Journal of Cryptology* (Springer-Verlag) 4 (1): 3–72. Retrieved 26 March 2015.

[53] "The Digital Millennium Copyright Act of 1998" (PDF). *United States Copyright Office*. Retrieved 26 March 2015.

[54] Ferguson, Niels (15 August 2001). "Censorship in action: why I don't publish my HDCP results". Archived December 1, 2001 at the Wayback Machine

[55] Williams, Christopher (11 August 2009). "Two convicted for refusal to decrypt data". *The Register*. Retrieved 26 March 2015.

[56] Williams, Christopher (24 November 2009). "UK jails schizophrenic for refusal to decrypt files". *The Register*. Retrieved 26 March 2015.

[57] Ingold, John (January 4, 2012). "Password case reframes Fifth Amendment rights in context of digital world". *The Denver Post*. Retrieved 26 March 2015.

[58] Leyden, John (13 July 2011). "US court test for rights not to hand over crypto keys". *The Register*. Retrieved 26 March 2015.

[59] "ORDER GRANTING APPLICATION UNDER THE ALL WRITS ACT REQUIRING DEFENDANT FRICOSU TO ASSIST IN THE EXECUTION OF PREVIOUSLY ISSUED SEARCH WARRANTS" (PDF). United States District Court for the District of Colorado. Retrieved 26 March 2015.

16.7   Further reading

Further information: Books on cryptography

- Becket, B (1988). *Introduction to Cryptology*. Blackwell Scientific Publications. ISBN 0-632-01836-4. OCLC 16832704. Excellent coverage of many classical ciphers and cryptography concepts and of the "modern" DES and RSA systems.

- *Cryptography and Mathematics* by Bernhard Esslinger, 200 pages, part of the free open-source package CrypTool, PDF download at the Wayback Machine (archived July 22, 2011). CrypTool is the most widespread e-learning program about cryptography and cryptanalysis, open source.

- *In Code: A Mathematical Journey* by Sarah Flannery (with David Flannery). Popular account of Sarah's award-winning project on public-key cryptography, co-written with her father.

- James Gannon, *Stealing Secrets, Telling Lies: How Spies and Codebreakers Helped Shape the Twentieth Century*, Washington, D.C., Brassey's, 2001, ISBN 1-57488-367-4.

- Oded Goldreich, Foundations of Cryptography, in two volumes, Cambridge University Press, 2001 and 2004.

- *Introduction to Modern Cryptography* by Jonathan Katz and Yehuda Lindell.

- *Alvin's Secret Code* by Clifford B. Hicks (children's novel that introduces some basic cryptography and cryptanalysis).

- Ibrahim A. Al-Kadi, "The Origins of Cryptology: the Arab Contributions," Cryptologia, vol. 16, no. 2 (April 1992), pp. 97–126.

- Christof Paar, Jan Pelzl, Understanding Cryptography, A Textbook for Students and Practitioners. Springer, 2009. (Slides, online cryptography lectures and other information are available on the companion web site.) Very accessible introduction to practical cryptography for non-mathematicians.

- *Introduction to Modern Cryptography* by Phillip Rogaway and Mihir Bellare, a mathematical introduction to theoretical cryptography including reduction-based security proofs. PDF download.

- Johann-Christoph Woltag, 'Coded Communications (Encryption)' in Rüdiger Wolfrum (ed) Max Planck Encyclopedia of Public International Law (Oxford University Press 2009). *"Max Planck Encyclopedia of Public International Law"., giving an overview of international law issues regarding cryptography.

- Jonathan Arbib & John Dwyer, Discrete Mathematics for Cryptography, 1st Edition ISBN 978-1-907934-01-8.

- Stallings, William (March 2013). *Cryptography and Network Security: Principles and Practice* (6th ed.). Prentice Hall. ISBN 978-0133354690.

16.8 External links

- The dictionary definition of cryptography at Wiktionary

- Media related to Cryptography at Wikimedia Commons

-

- Cryptography on *In Our Time* at the BBC. (listen now)

- Crypto Glossary and Dictionary of Technical Cryptography

- NSA's CryptoKids.

- Overview and Applications of Cryptology by the CrypTool Team; PDF; 3.8 MB—July 2008

- A Course in Cryptography by Raphael Pass & Abhi Shelat - offered at Cornell in the form of lecture notes.

- Cryptocorner.com by Chuck Easttom - A generalized resource on all aspects of cryptology.

- For more on the use of cryptographic elements in fiction, see: Dooley, John F., William and Marilyn Ingersoll Professor of Computer Science, Knox College (23 August 2012). "Cryptology in Fiction".

- The George Fabyan Collection at the Library of Congress has early editions of works of seventeenth-century English literature, publications relating to cryptography.

Variants of the Enigma machine, used by Germany's military and civil authorities from the late 1920s through World War II, implemented a complex electro-mechanical polyalphabetic cipher. Breaking and reading of the Enigma cipher at Poland's Cipher Bureau, for 7 years before the war, and subsequent decryption at Bletchley Park, was important to Allied victory.[8]

Poznań monument (center) to Polish cryptologists whose breaking of Germany's Enigma machine ciphers, beginning in 1932, altered the course of World War II

Chapter 17

Abstract algebra

This article is about the branch of mathematics. For the Swedish band, see Abstrakt Algebra.

"Modern algebra" redirects here. For van der Waerden's book, see Moderne Algebra.

In algebra, which is a broad division of mathematics, **abstract algebra** (occasionally called **modern algebra**) is the study of algebraic structures. Algebraic structures include groups, rings, fields, modules, vector spaces, lattices, and algebra over a field. The term *abstract algebra* was coined in the early 20th century to distinguish this area of study from the other parts of algebra.

Algebraic structures, with their associated homomorphisms, form mathematical categories. Category theory is a powerful formalism for analyzing and comparing different algebraic structures.

Universal algebra is a related subject that studies the nature and theories of various types of algebraic structures as a whole. For example, universal algebra studies the overall theory of groups, as distinguished from studying particular groups.

17.1   History

As in other parts of mathematics, concrete problems and examples have played important roles in the development of abstract algebra. Through the end of the nineteenth century, many -- perhaps most -- of these problems were in some way related to the theory of algebraic equations. Major themes include:

- Solving of systems of linear equations, which led to linear algebra

- Attempts to find formulae for solutions of general polynomial equations of higher degree that resulted in discovery of groups as abstract manifestations of symmetry

- Arithmetical investigations of quadratic and higher degree forms and diophantine equations, that directly produced the notions of a ring and ideal.

Numerous textbooks in abstract algebra start with axiomatic definitions of various algebraic structures and then proceed to establish their properties. This creates a false impression that in algebra axioms had come first and then served as a motivation and as a basis of further study. The true order of historical development was almost exactly the opposite. For example, the hypercomplex numbers of the nineteenth century had kinematic and physical motivations but challenged comprehension. Most theories that are now recognized as parts of algebra started as collections of disparate facts from various branches of mathematics, acquired a common theme that served as a core around which various results were grouped, and finally became unified on a basis of a common set of concepts. An archetypical example of this progressive synthesis can be seen in the history of group theory.

The permutations of Rubik's Cube have a group structure; the group is a fundamental concept within abstract algebra.

17.1.1   Early group theory

There were several threads in the early development of group theory, in modern language loosely corresponding to *number theory, theory of equations,* and *geometry.*

Leonhard Euler considered algebraic operations on numbers modulo an integer, modular arithmetic, in his generalization of Fermat's little theorem. These investigations were taken much further by Carl Friedrich Gauss, who considered the structure of multiplicative groups of residues mod n and established many properties of cyclic and more general abelian groups that arise in this way. In his investigations of composition of binary quadratic forms, Gauss explicitly stated the associative law for the composition of forms, but like Euler before him, he seems to have been more interested in concrete results than in general theory. In 1870, Leopold Kronecker gave a definition of an abelian group in the context of ideal class groups of a number field, generalizing Gauss's work; but it appears he did not tie his definition with previous work on groups, particularly permutation groups. In 1882, considering the same question, Heinrich M. Weber realized the connection and gave a similar definition that involved the cancellation property but omitted the existence of the inverse element, which was sufficient in his context (finite groups).

Permutations were studied by Joseph-Louis Lagrange in his 1770 paper *Réflexions sur la résolution algébrique des équations (Thoughts on the algebraic solution of equations)* devoted to solutions of algebraic equations, in which he introduced Lagrange resolvents. Lagrange's goal was to understand why equations of third and fourth degree admit formulae for solutions, and he identified as key objects permutations of the roots. An important novel step taken by Lagrange in this paper was the abstract view of the roots, i.e. as symbols and not as numbers. However, he did not consider composition of permutations. Serendipitously, the first edition of Edward Waring's *Meditationes Algebraicae (Meditations on Algebra)* appeared in the same year, with an expanded version published in 1782. Waring proved the main theorem on symmetric functions, and specially considered the relation between the roots of a quartic equation and its resolvent cubic. *Mémoire sur la résolution des équations (Memoire on the Solving of Equations)* of Alexandre Vandermonde (1771) developed the theory of symmetric functions from a slightly different angle, but like Lagrange, with the goal of understanding solvability of algebraic equations.

> Kronecker claimed in 1888 that the study of modern algebra began with this first paper of Vandermonde. Cauchy states quite clearly that Vandermonde had priority over Lagrange for this remarkable idea, which eventually led to the study of group theory.[1]

Paolo Ruffini was the first person to develop the theory of permutation groups, and like his predecessors, also in the context of solving algebraic equations. His goal was to establish the impossibility of an algebraic solution to a general algebraic equation of degree greater than four. En route to this goal he introduced the notion of the order of an element of a group, conjugacy, the cycle decomposition of elements of permutation groups and the notions of primitive and imprimitive and proved some important theorems relating these concepts, such as

> if G is a subgroup of S_n whose order is divisible by 5 then G contains an element of order 5.

Note, however, that he got by without formalizing the concept of a group, or even of a permutation group. The next step was taken by Évariste Galois in 1832, although his work remained unpublished until 1846, when he considered for the first time what is now called the *closure property* of a group of permutations, which he expressed as

> ... if in such a group one has the substitutions S and T then one has the substitution ST.

The theory of permutation groups received further far-reaching development in the hands of Augustin Cauchy and Camille Jordan, both through introduction of new concepts and, primarily, a great wealth of results about special classes of permutation groups and even some general theorems. Among other things, Jordan defined a notion of isomorphism, still in the context of permutation groups and, incidentally, it was he who put the term *group* in wide use.

The abstract notion of a group appeared for the first time in Arthur Cayley's papers in 1854. Cayley realized that a group need not be a permutation group (or even *finite*), and may instead consist of matrices, whose algebraic properties, such as multiplication and inverses, he systematically investigated in succeeding years. Much later Cayley would revisit the question whether abstract groups were more general than permutation groups, and establish that, in fact, any group is isomorphic to a group of permutations.

17.1.2   Modern algebra

The end of the 19th and the beginning of the 20th century saw a tremendous shift in the methodology of mathematics. Abstract algebra emerged around the start of the 20th century, under the name *modern algebra*. Its study was part of the drive for more intellectual rigor in mathematics. Initially, the assumptions in classical algebra, on which the whole of mathematics (and major parts of the natural sciences) depend, took the form of axiomatic systems. No longer satisfied with establishing properties of concrete objects, mathematicians started to turn their attention to general theory. Formal definitions of certain algebraic structures began to emerge in the 19th century. For example, results about various groups of permutations came to be seen as instances of general theorems that concern a general notion of an *abstract group*. Questions of structure and classification of various mathematical objects came to forefront.

These processes were occurring throughout all of mathematics, but became especially pronounced in algebra. Formal definition through primitive operations and axioms were proposed for many basic algebraic structures, such as groups, rings, and fields. Hence such things as group theory and ring theory took their places in pure mathematics. The algebraic investigations of general fields by Ernst Steinitz and of commutative and then general rings by David Hilbert, Emil Artin and Emmy Noether, building up on the work of Ernst Kummer, Leopold Kronecker and Richard Dedekind, who had considered ideals in commutative rings, and of Georg Frobenius and Issai Schur, concerning representation theory of groups, came to define abstract algebra. These developments of the last quarter of the 19th century and the first quarter of 20th century were systematically exposed in Bartel van der Waerden's *Moderne algebra*, the two-volume monograph published in 1930–1931 that forever changed for the mathematical world the meaning of the word *algebra* from *the theory of equations* to the *theory of algebraic structures*.

17.2   Basic concepts

Main article: Algebraic structures

By abstracting away various amounts of detail, mathematicians have created theories of various algebraic structures that apply to many objects. For instance, almost all systems studied are sets, to which the theorems of set theory apply. Those sets that have a certain binary operation defined on them form magmas, to which the concepts concerning magmas, as well those concerning sets, apply. We can add additional constraints on the algebraic structure, such as associativity (to form semigroups); identity, and inverses (to form groups); and other more complex structures. With additional structure, more theorems could be proved, but the generality is reduced. The "hierarchy" of algebraic objects (in terms of generality) creates a hierarchy of the corresponding theories: for instance, the theorems of group theory apply to rings (algebraic objects that have two binary operations with certain axioms) since a ring is a group over one of its operations. Mathematicians choose a balance between the amount of generality and the richness of the theory.

Examples of algebraic structures with a single binary operation are:

- Magmas
- Quasigroups
- Monoids
- Semigroups
- Groups

More complicated examples include:

- Rings
- Fields
- Modules
- Vector spaces
- Algebras over fields
- Associative algebras
- Lie algebras
- Lattices
- Boolean algebras

17.3 Applications

Because of its generality, abstract algebra is used in many fields of mathematics and science. For instance, algebraic topology uses algebraic objects to study topologies. The recently (As of 2006) proved Poincaré conjecture asserts that the fundamental group of a manifold, which encodes information about connectedness, can be used to determine whether a manifold is a sphere or not. Algebraic number theory studies various number rings that generalize the set of integers. Using tools of algebraic number theory, Andrew Wiles proved Fermat's Last Theorem.

In physics, groups are used to represent symmetry operations, and the usage of group theory could simplify differential equations. In gauge theory, the requirement of local symmetry can be used to deduce the equations describing a system. The groups that describe those symmetries are Lie groups, and the study of Lie groups and Lie algebras reveals much about the physical system; for instance, the number of force carriers in a theory is equal to dimension of the Lie algebra, and these bosons interact with the force they mediate if the Lie algebra is nonabelian.[2]

17.4 See also

Main article: Outline of abstract algebra

- Coding theory
- Publications in abstract algebra

17.5 References

[1] Vandermonde biography in Mac Tutor History of Mathematics Archive.

[2] Schumm, Bruce (2004), *Deep Down Things*, Baltimore: Johns Hopkins University Press, ISBN 0-8018-7971-X

17.6 Sources

- Allenby, R.B.J.T. (1991), *Rings, Fields and Groups*, Butterworth-Heinemann, ISBN 978-0-340-54440-2

- Artin, Michael (1991), *Algebra*, Prentice Hall, ISBN 978-0-89871-510-1

- Burris, Stanley N.; Sankappanavar, H. P. (1999) [1981], *A Course in Universal Algebra*

- Gilbert, Jimmie; Gilbert, Linda (2005), *Elements of Modern Algebra*, Thomson Brooks/Cole, ISBN 978-0-534-40264-8

- Lang, Serge (2002), *Algebra*, Graduate Texts in Mathematics 211 (Revised third ed.), New York: Springer-Verlag, ISBN 978-0-387-95385-4, MR 1878556

- Sethuraman, B. A. (1996), *Rings, Fields, Vector Spaces, and Group Theory: An Introduction to Abstract Algebra via Geometric Constructibility*, Berlin, New York: Springer-Verlag, ISBN 978-0-387-94848-5

- Whitehead, C. (2002), *Guide to Abstract Algebra* (2nd ed.), Houndmills: Palgrave, ISBN 978-0-333-79447-0

- W. Keith Nicholson (2012) *Introduction to Abstract Algebra*, 4th edition, John Wiley & Sons ISBN 978-1-118-13535-8 .

- John R. Durbin (1992) *Modern Algebra : an introduction*, John Wiley & Sons

17.7 External links

- John Beachy: *Abstract Algebra On Line*, Comprehensive list of definitions and theorems.

- Edwin Connell "Elements of Abstract and Linear Algebra ", Free online textbook.

- Fredrick M. Goodman: *Algebra: Abstract and Concrete*.

- Judson, Thomas W. (1997), *Abstract Algebra: Theory and Applications* An introductory undergraduate text in the spirit of texts by Gallian or Herstein, covering groups, rings, integral domains, fields and Galois theory. Free downloadable PDF with open-source GFDL license.

- Sethuraman, B.A.. (2015), *A Gentle Introduction to Abstract Algebra* A very gentle introduction that introduces rings and fields, vector spaces, and groups, with a focus on examples. Free downloadable PDF, under GNU-FDL license. Sized for both tablets and as a regular book. Contains many short video tutorials.

- Zeidler, A. Bernhard (2014), *Abstract Algebra* (PDF) A web-book on algebra and commutative algebra. Warning: work in progress! Free downloadable PDF under Open Publication License.

Chapter 18

Combinatorics

Not to be confused with combinatoriality.

Combinatorics is a branch of mathematics concerning the study of finite or countable discrete structures. Aspects of combinatorics include counting the structures of a given kind and size (enumerative combinatorics), deciding when certain criteria can be met, and constructing and analyzing objects meeting the criteria (as in combinatorial designs and matroid theory), finding "largest", "smallest", or "optimal" objects (extremal combinatorics and combinatorial optimization), and studying combinatorial structures arising in an algebraic context, or applying algebraic techniques to combinatorial problems (algebraic combinatorics).

Combinatorial problems arise in many areas of pure mathematics, notably in algebra, probability theory, topology, and geometry,[1] and combinatorics also has many applications in mathematical optimization, computer science, ergodic theory and statistical physics. Many combinatorial questions have historically been considered in isolation, giving an *ad hoc* solution to a problem arising in some mathematical context. In the later twentieth century, however, powerful and general theoretical methods were developed, making combinatorics into an independent branch of mathematics in its own right. One of the oldest and most accessible parts of combinatorics is graph theory, which also has numerous natural connections to other areas. Combinatorics is used frequently in computer science to obtain formulas and estimates in the analysis of algorithms.

A mathematician who studies combinatorics is called a **combinatorialist** or a **combinatorist**.

18.1 History

Main article: History of combinatorics

Basic combinatorial concepts and enumerative results appeared throughout the ancient world. In 6th century BCE, ancient Indian physician Sushruta asserts in Sushruta Samhita that 63 combinations can be made out of 6 different tastes, taken one at a time, two at a time, etc., thus computing all $2^6 - 1$ possibilities. Greek historian Plutarch discusses an argument between Chrysippus (3rd century BCE) and Hipparchus (2nd century BCE) of a rather delicate enumerative problem, which was later shown to be related to Schröder numbers.[2][3] In the *Ostomachion*, Archimedes (3rd century BCE) considers a tiling puzzle.

In the Middle Ages, combinatorics continued to be studied, largely outside of the European civilization. The Indian mathematician Mahāvira (c. 850) provided formulae for the number of permutations and combinations,[4][5] and these formulas may have been familiar to Indian mathematicians as early as the 6th century CE.[6] The philosopher and astronomer Rabbi Abraham ibn Ezra (c. 1140) established the symmetry of binomial coefficients, while a closed formula was obtained later by the talmudist and mathematician Levi ben Gerson (better known as Gersonides), in 1321.[7] The arithmetical triangle— a graphical diagram showing relationships among the binomial coefficients— was presented by mathematicians in treatises dating as far back as the 10th century, and would eventually become known as Pascal's triangle. Later, in Medieval England, campanology provided examples of what is now known as Hamiltonian cycles in certain Cayley graphs on permutations.[8]

During the Renaissance, together with the rest of mathematics and the sciences, combinatorics enjoyed a rebirth. Works of Pascal, Newton, Jacob Bernoulli and Euler became foundational in the emerging field. In modern times, the works of J. J. Sylvester (late 19th century) and Percy MacMahon (early 20th century) laid the foundation for enumerative and algebraic combinatorics. Graph theory also enjoyed an explosion of interest at the same time, especially in connection with the four color problem.

In the second half of 20th century, combinatorics enjoyed a rapid growth, which led to establishment of dozens of new journals and conferences in the subject.[9] In part, the growth was spurred by new connections and applications to other fields, ranging from algebra to probability, from functional analysis to number theory, etc. These connections shed the boundaries between combinatorics and parts of mathematics and theoretical computer science, but at the same time led to a partial fragmentation of the field.

18.2   Approaches and subfields of combinatorics

18.2.1   Enumerative combinatorics

Main article: Enumerative combinatorics

Enumerative combinatorics is the most classical area of combinatorics, and concentrates on counting the number of certain combinatorial objects. Although counting the number of elements in a set is a rather broad mathematical problem, many of the problems that arise in applications have a relatively simple combinatorial description. Fibonacci numbers is the basic example of a problem in enumerative combinatorics. The twelvefold way provides a unified framework for counting permutations, combinations and partitions.

18.2.2   Analytic combinatorics

Main article: Analytic combinatorics

Analytic combinatorics concerns the enumeration of combinatorial structures using tools from complex analysis and probability theory. In contrast with enumerative combinatorics, which uses explicit combinatorial formulae and generating functions to describe the results, analytic combinatorics aims at obtaining asymptotic formulae.

18.2.3   Partition theory

Main article: Partition theory

Partition theory studies various enumeration and asymptotic problems related to integer partitions, and is closely related to q-series, special functions and orthogonal polynomials. Originally a part of number theory and analysis, it is now considered a part of combinatorics or an independent field. It incorporates the bijective approach and various tools in analysis, analytic number theory, and has connections with statistical mechanics.

18.2.4   Graph theory

Main article: Graph theory

Graphs are basic objects in combinatorics. The questions range from counting (e.g., the number of graphs on n vertices with k edges) to structural (e.g., which graphs contain Hamiltonian cycles) to algebraic questions (e.g., given a graph G and two numbers x and y, does the Tutte polynomial $TG(x,y)$ have a combinatorial interpretation?). It should be noted that while there are very strong connections between graph theory and combinatorics, these two are sometimes thought of as separate subjects.[10]

18.2.5 Design theory

Main article: Combinatorial design

Design theory is a study of combinatorial designs, which are collections of subsets with certain intersection properties. Block designs are combinatorial designs of a special type. This area is one of the oldest parts of combinatorics, such as in Kirkman's schoolgirl problem proposed in 1850. The solution of the problem is a special case of a Steiner system, which systems play an important role in the classification of finite simple groups. The area has further connections to coding theory and geometric combinatorics.

18.2.6 Finite geometry

Main article: Finite geometry

Finite geometry is the study of geometric systems having only a finite number of points. Structures analogous to those found in continuous geometries (Euclidean plane, real projective space, etc.) but defined combinatorially are the main items studied. This area provides a rich source of examples for design theory. It should not be confused with discrete geometry (combinatorial geometry).

18.2.7 Order theory

Main article: Order theory

Order theory is the study of partially ordered sets, both finite and infinite. Various examples of partial orders appear in algebra, geometry, number theory and throughout combinatorics and graph theory. Notable classes and examples of partial orders include lattices and Boolean algebras.

18.2.8 Matroid theory

Main article: Matroid theory

Matroid theory abstracts part of geometry. It studies the properties of sets (usually, finite sets) of vectors in a vector space that do not depend on the particular coefficients in a linear dependence relation. Not only the structure but also enumerative properties belong to matroid theory. Matroid theory was introduced by Hassler Whitney and studied as a part of the order theory. It is now an independent field of study with a number of connections with other parts of combinatorics.

18.2.9 Extremal combinatorics

Main article: Extremal combinatorics

Extremal combinatorics studies extremal questions on set systems. The types of questions addressed in this case are about the largest possible graph which satisfies certain properties. For example, the largest triangle-free graph on $2n$ vertices is a complete bipartite graph $K_{n,n}$. Often it is too hard even to find the extremal answer $f(n)$ exactly and one can only give an asymptotic estimate.

Ramsey theory is another part of extremal combinatorics. It states that any sufficiently large configuration will contain some sort of order. It is an advanced generalization of the pigeonhole principle.

18.2.10 Probabilistic combinatorics

Main article: Probabilistic method

In probabilistic combinatorics, the questions are of the following type: what is the probability of a certain property for a random discrete object, such as a random graph? For instance, what is the average number of triangles in a random graph? Probabilistic methods are also used to determine the existence of combinatorial objects with certain prescribed properties (for which explicit examples might be difficult to find), simply by observing that the probability of randomly selecting an object with those properties is greater than 0. This approach (often referred to as *the* probabilistic method) proved highly effective in applications to extremal combinatorics and graph theory. A closely related area is the study of finite Markov chains, especially on combinatorial objects. Here again probabilistic tools are used to estimate the mixing time.

Often associated with Paul Erdős, who did the pioneer work on the subject, probabilistic combinatorics was traditionally viewed as a set of tools to study problems in other parts of combinatorics. However, with the growth of applications to analysis of algorithms in computer science, as well as classical probability, additive and probabilistic number theory, the area recently grew to become an independent field of combinatorics.

18.2.11 Algebraic combinatorics

Main article: Algebraic combinatorics

Algebraic combinatorics is an area of mathematics that employs methods of abstract algebra, notably group theory and representation theory, in various combinatorial contexts and, conversely, applies combinatorial techniques to problems in algebra. Algebraic combinatorics is continuously expanding its scope, in both topics and techniques, and can be seen as the area of mathematics where the interaction of combinatorial and algebraic methods is particularly strong and significant.

18.2.12 Combinatorics on words

Main article: Combinatorics on words

Combinatorics on words deals with formal languages. It arose independently within several branches of mathematics, including number theory, group theory and probability. It has applications to enumerative combinatorics, fractal analysis, theoretical computer science, automata theory and linguistics. While many applications are new, the classical Chomsky–Schützenberger hierarchy of classes of formal grammars is perhaps the best known result in the field.

18.2.13 Geometric combinatorics

Main article: Geometric combinatorics

Geometric combinatorics is related to convex and discrete geometry, in particular polyhedral combinatorics. It asks, for example, how many faces of each dimension can a convex polytope have. Metric properties of polytopes play an important role as well, e.g. the Cauchy theorem on rigidity of convex polytopes. Special polytopes are also considered, such as permutohedra, associahedra and Birkhoff polytopes. We should note that combinatorial geometry is an old fashioned name for discrete geometry.

18.2.14   Topological combinatorics

Main article: Topological combinatorics

Combinatorial analogs of concepts and methods in topology are used to study graph coloring, fair division, partitions, partially ordered sets, decision trees, necklace problems and discrete Morse theory. It should not be confused with combinatorial topology which is an older name for algebraic topology.

18.2.15   Arithmetic combinatorics

Main article: Arithmetic combinatorics

Arithmetic combinatorics arose out of the interplay between number theory, combinatorics, ergodic theory and harmonic analysis. It is about combinatorial estimates associated with arithmetic operations (addition, subtraction, multiplication, and division). *Additive combinatorics* refers to the special case when only the operations of addition and subtraction are involved. One important technique in arithmetic combinatorics is the ergodic theory of dynamical systems.

18.2.16   Infinitary combinatorics

Main article: Infinitary combinatorics

Infinitary combinatorics, or combinatorial set theory, is an extension of ideas in combinatorics to infinite sets. It is a part of set theory, an area of mathematical logic, but uses tools and ideas from both set theory and extremal combinatorics. Gian-Carlo Rota used the name *continuous combinatorics*[11] to describe probability and measure theory, since there are many analogies between *counting* and *measure*.

18.3   Related fields

18.3.1   Combinatorial optimization

Combinatorial optimization is the study of optimization on discrete and combinatorial objects. It started as a part of combinatorics and graph theory, but is now viewed as a branch of applied mathematics and computer science, related to operations research, algorithm theory and computational complexity theory.

18.3.2   Coding theory

Coding theory started as a part of design theory with early combinatorial constructions of error-correcting codes. The main idea of the subject is to design efficient and reliable methods of data transmission. It is now a large field of study, part of information theory.

18.3.3   Discrete and computational geometry

Discrete geometry (also called combinatorial geometry) also began a part of combinatorics, with early results on convex polytopes and kissing numbers. With the emergence of applications of discrete geometry to computational geometry, these two fields partially merged and became a separate field of study. There remain many connections with geometric and topological combinatorics, which themselves can be viewed as outgrowths of the early discrete geometry.

18.3.4 Combinatorics and dynamical systems

Combinatorial aspects of dynamical systems is another emerging field. Here dynamical systems can be defined on combinatorial objects. See for example graph dynamical system.

18.3.5 Combinatorics and physics

There are increasing interactions between combinatorics and physics, particularly statistical physics. Examples include an exact solution of the Ising model, and a connection between the Potts model on one hand, and the chromatic and Tutte polynomials on the other hand.

18.4 See also

- Combinatorial biology
- Combinatorial chemistry
- Combinatorial data analysis
- Combinatorial game theory
- Combinatorial group theory
- List of combinatorics topics
- Phylogenetics

18.5 Notes

[1] Björner and Stanley, p. 2

[2] Stanley, Richard P.; "Hipparchus, Plutarch, Schröder, and Hough", *American Mathematical Monthly* 104 (1997), no. 4, 344–350.

[3] Habsieger, Laurent; Kazarian, Maxim; and Lando, Sergei; "On the Second Number of Plutarch", *American Mathematical Monthly* 105 (1998), no. 5, 446.

[4] O'Connor, John J.; Robertson, Edmund F., "Combinatorics", *MacTutor History of Mathematics archive*, University of St Andrews.

[5] Puttaswamy, Tumkur K. (2000), "The Mathematical Accomplishments of Ancient Indian Mathematicians", in Selin, Helaine, *Mathematics Across Cultures: The History of Non-Western Mathematics*, Netherlands: Kluwer Academic Publishers, p. 417, ISBN 978-1-4020-0260-1

[6] Biggs, Norman L. (1979). "The Roots of Combinatorics". *Historia Mathematica* 6: 109–136.

[7] Maistrov, L. E. (1974), *Probability Theory: A Historical Sketch*, Academic Press, p. 35, ISBN 9781483218632. (Translation from 1967 Russian ed.)

[8] White, Arthur T.; "Ringing the Cosets", *American Mathematical Monthly*, 94 (1987), no. 8, 721–746; White, Arthur T.; "Fabian Stedman: The First Group Theorist?", *American Mathematical Monthly*, 103 (1996), no. 9, 771–778.

[9] See Journals in Combinatorics and Graph Theory

[10] Sanders, Daniel P.; *2-Digit MSC Comparison*

[11] *Continuous and profinite combinatorics*

18.6   References

- Björner, Anders; and Stanley, Richard P.; (2010); *A Combinatorial Miscellany*

- Bóna, Miklós; (2011); *A Walk Through Combinatorics (3rd Edition)*. ISBN 978-981-4335-23-2, ISBN 978-981-4460-00-2(pbk)

- Graham, Ronald L.; Groetschel, Martin; and Lovász, László; eds. (1996); *Handbook of Combinatorics*, Volumes 1 and 2. Amsterdam, NL, and Cambridge, MA: Elsevier (North-Holland) and MIT Press. ISBN 0-262-07169-X

- Lindner, Charles C.; and Rodger, Christopher A.; eds. (1997); *Design Theory*, CRC-Press; 1st. edition (October 31, 1997). ISBN 0-8493-3986-3.

- Riordan, John (1958); *An Introduction to Combinatorial Analysis*, New York, NY: Wiley & Sons (republished)

- Stanley, Richard P. (1997, 1999); *Enumerative Combinatorics*, Volumes 1 and 2, Cambridge University Press. ISBN 0-521-55309-1, ISBN 0-521-56069-1

- van Lint, Jacobus H.; and Wilson, Richard M.; (2001); *A Course in Combinatorics*, 2nd Edition, Cambridge University Press. ISBN 0-521-80340-3

18.7   External links

- Hazewinkel, Michiel, ed. (2001), "Combinatorial analysis", *Encyclopedia of Mathematics*, Springer, ISBN 978-1-55608-010-4

- Combinatorial Analysis – an article in Encyclopædia Britannica Eleventh Edition

- Combinatorics, a MathWorld article with many references.

- Combinatorics, from a *MathPages.com* portal.

- The Hyperbook of Combinatorics, a collection of math articles links.

- The Two Cultures of Mathematics by W. T. Gowers, article on problem solving vs theory building.

Plain Bob Minor

An example of change ringing (with six bells), one of the earliest nontrivial results in Graph Theory.

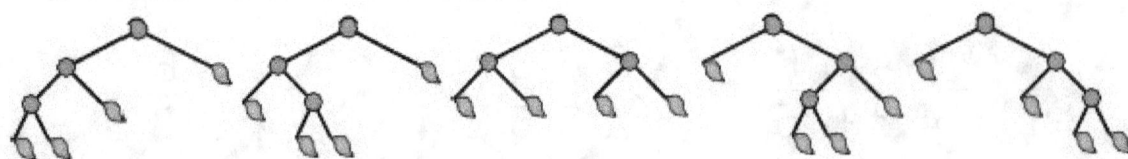

Five binary trees on three vertices, an example of Catalan numbers.

A plane partition.

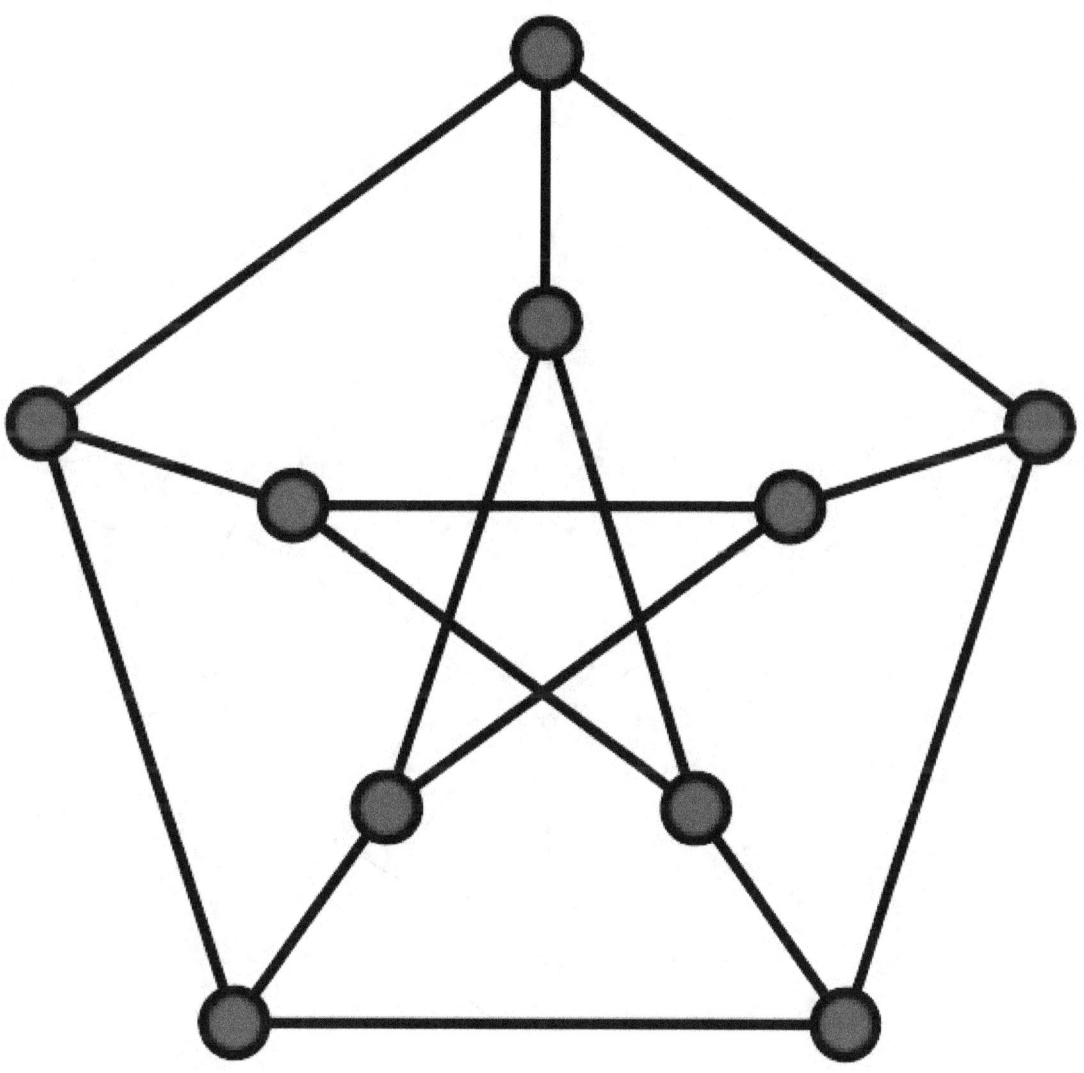

Petersen graph.

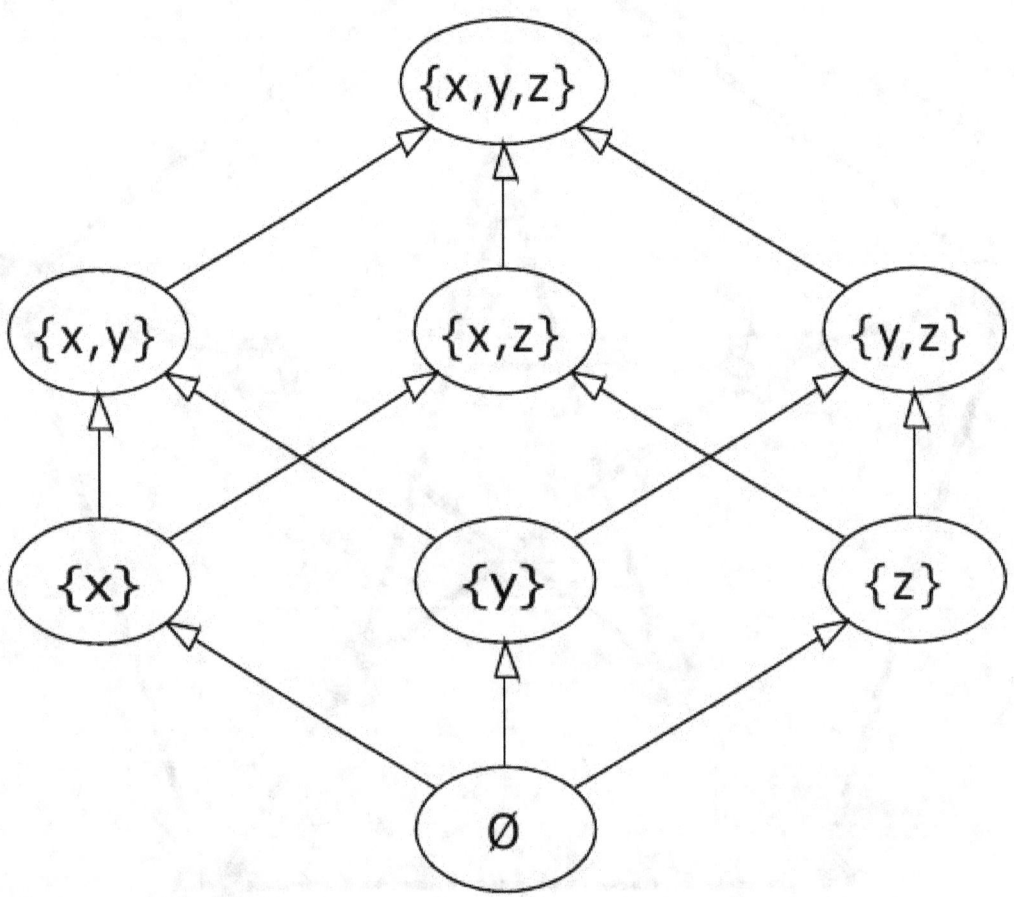

Hasse diagram of the powerset of {x,y,z} ordered by inclusion.

Self-avoiding walk in a square grid graph.

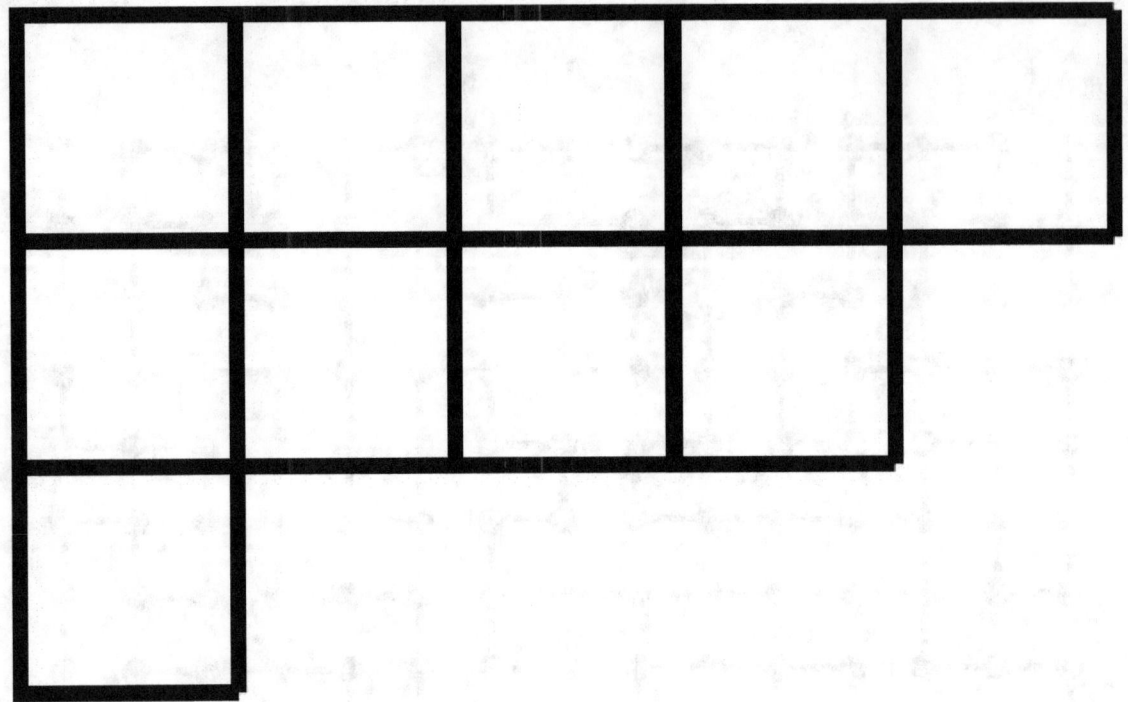

Young diagram of a partition (5,4,1).

0

Construction of a Thue–Morse infinite word.

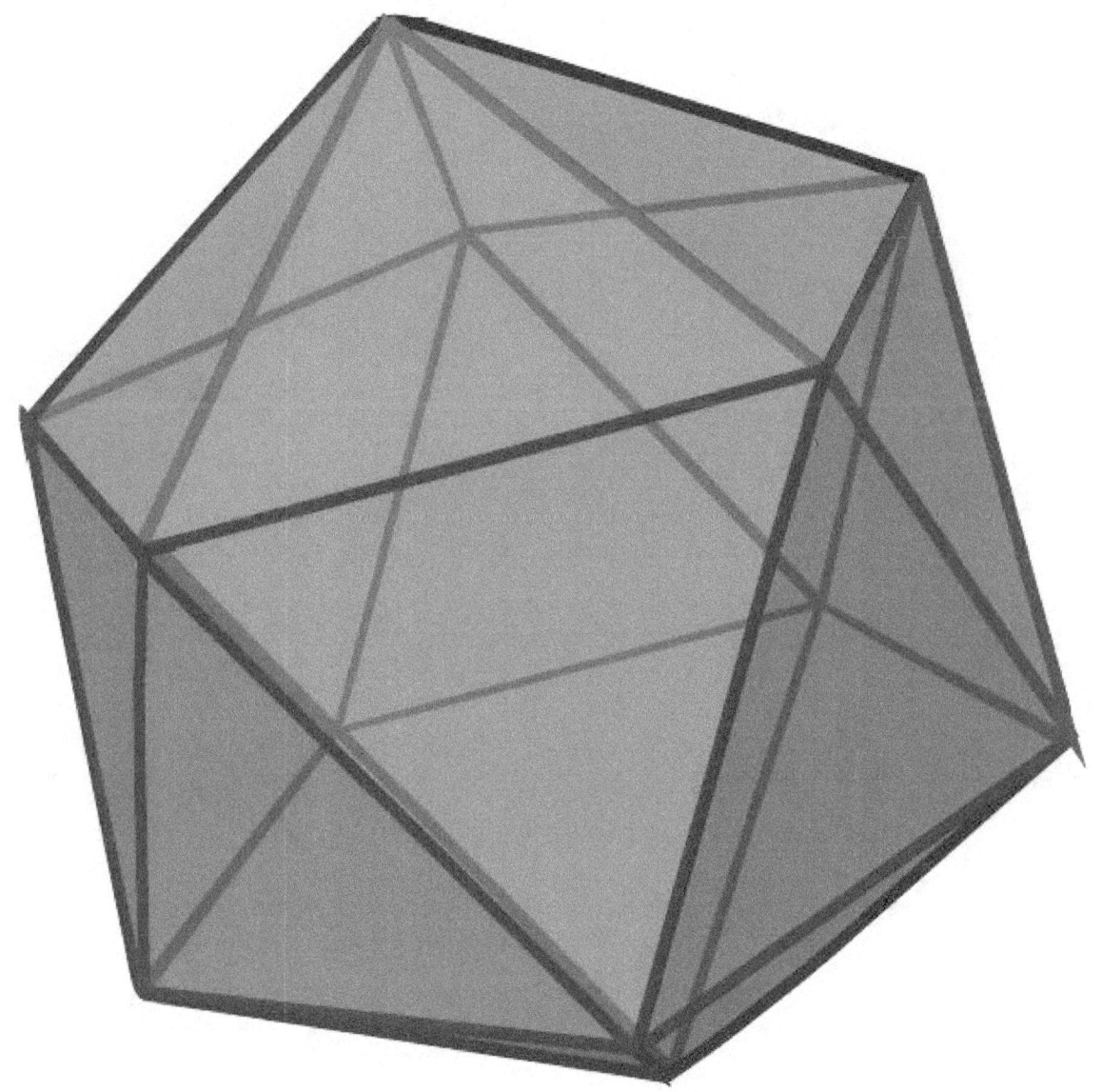

An icosahedron.

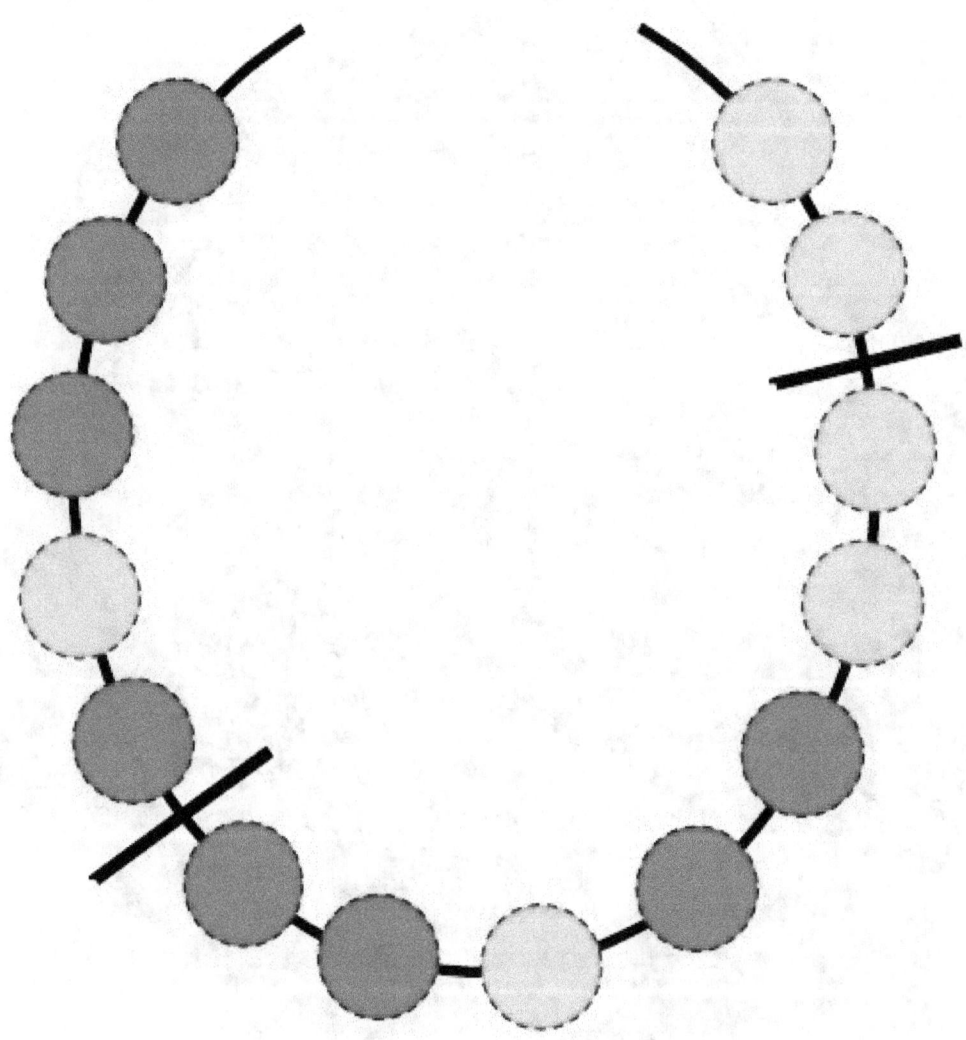

Splitting a necklace with two cuts.

Kissing spheres are connected to both coding theory and discrete geometry.

Chapter 19

Geometry

For other uses, see Geometry (disambiguation).

Geometry (from the Ancient Greek: γεωμετρία; *geo-* "earth", *-metron* "measurement") is a branch of mathematics

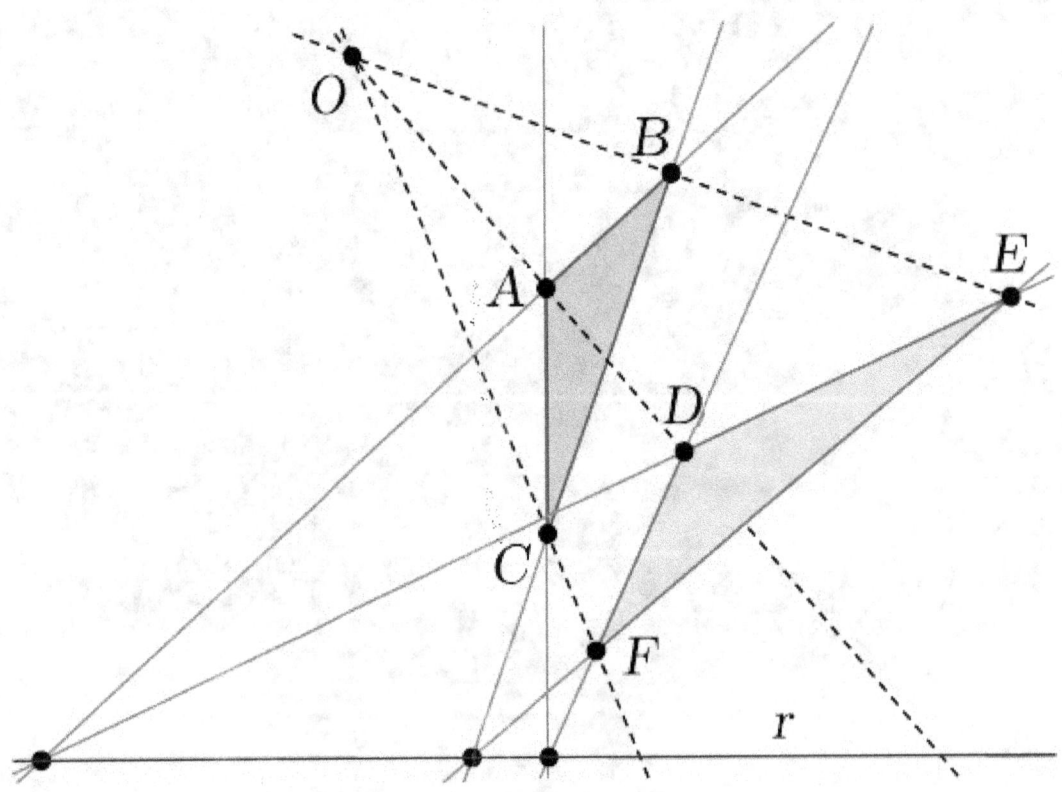

An illustration of Desargues' theorem, an important result in Euclidean and projective geometry

concerned with questions of shape, size, relative position of figures, and the properties of space. A mathematician who works in the field of geometry is called a geometer. Geometry arose independently in a number of early cultures as a body of practical knowledge concerning lengths, areas, and volumes, with elements of formal mathematical science emerging in the West as early as Thales (6th century BC). By the 3rd century BC, geometry was put into an axiomatic form by Euclid, whose treatment—Euclidean geometry—set a standard for many centuries to follow.[1] Archimedes developed ingenious techniques for calculating areas and volumes, in many ways anticipating modern integral calculus. The field of astronomy, especially as it relates to mapping the positions of stars and planets on the celestial sphere and describing the relationship between movements of celestial bodies, served as an important source of geometric problems during the next one and a half millennia. In the classical world, both geometry and astronomy were considered to be part of the Quadrivium, a subset of the seven liberal arts considered essential for a free citizen to master.

The introduction of coordinates by René Descartes and the concurrent developments of algebra marked a new stage for geometry, since geometric figures such as plane curves could now be represented analytically in the form of functions and equations. This played a key role in the emergence of infinitesimal calculus in the 17th century. Furthermore, the theory of perspective showed that there is more to geometry than just the metric properties of figures: perspective is the origin of projective geometry. The subject of geometry was further enriched by the study of the intrinsic structure of geometric objects that originated with Euler and Gauss and led to the creation of topology and differential geometry.

In Euclid's time, there was no clear distinction between physical and geometrical space. Since the 19th-century discovery of non-Euclidean geometry, the concept of space has undergone a radical transformation and raised the question of which geometrical space best fits physical space. With the rise of formal mathematics in the 20th century, 'space' (whether 'point', 'line', or 'plane') lost its intuitive contents, so today one has to distinguish between physical space, geometrical spaces (in which 'space', 'point' etc. still have their intuitive meanings) and abstract spaces. Contemporary geometry considers manifolds, spaces that are considerably more abstract than the familiar Euclidean space, which they only approximately resemble at small scales. These spaces may be endowed with additional structure which allow one to speak about length. Modern geometry has many ties to physics as is exemplified by the links between pseudo-Riemannian geometry and general relativity. One of the youngest physical theories, string theory, is also very geometric in flavour.

While the visual nature of geometry makes it initially more accessible than other mathematical areas such as algebra or number theory, geometric language is also used in contexts far removed from its traditional, Euclidean provenance (for example, in fractal geometry and algebraic geometry).[2]

19.1 Overview

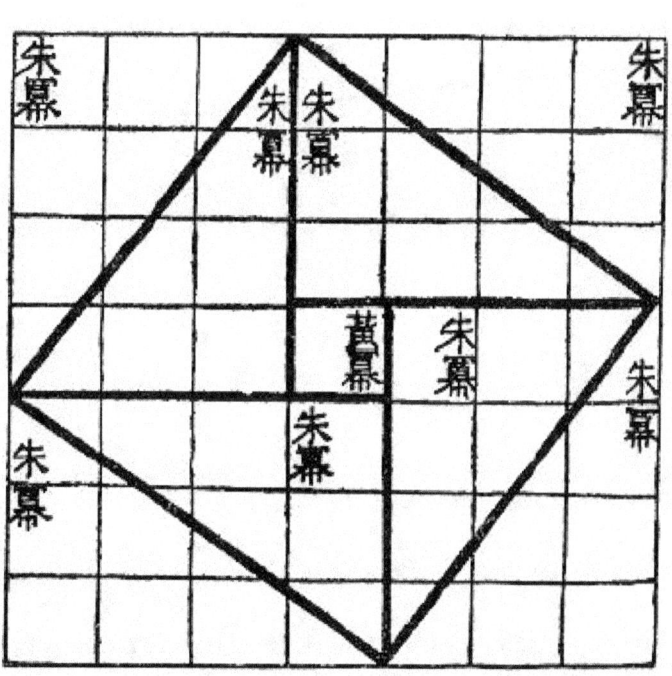

Visual checking of the Pythagorean theorem for the (3, 4, 5) triangle as in the Chou Pei Suan Ching 500–200 BC.

Because the recorded development of geometry spans more than two millennia, it is hardly surprising that perceptions of what constitutes geometry have evolved throughout the ages:

19.1.1   Practical geometry

Geometry originated as a practical science concerned with surveys, measurements, areas, and volumes. Among other highlights, notable accomplishments include formulas for lengths, areas and volumes, such as the Pythagorean theorem, circumference and area of a circle, area of a triangle, volume of a cylinder, sphere, and a pyramid. A method of computing certain inaccessible distances or heights based on similarity of geometric figures is attributed to Thales. The development of astronomy led to the emergence of trigonometry and spherical trigonometry, together with the attendant computational techniques.

19.1.2   Axiomatic geometry

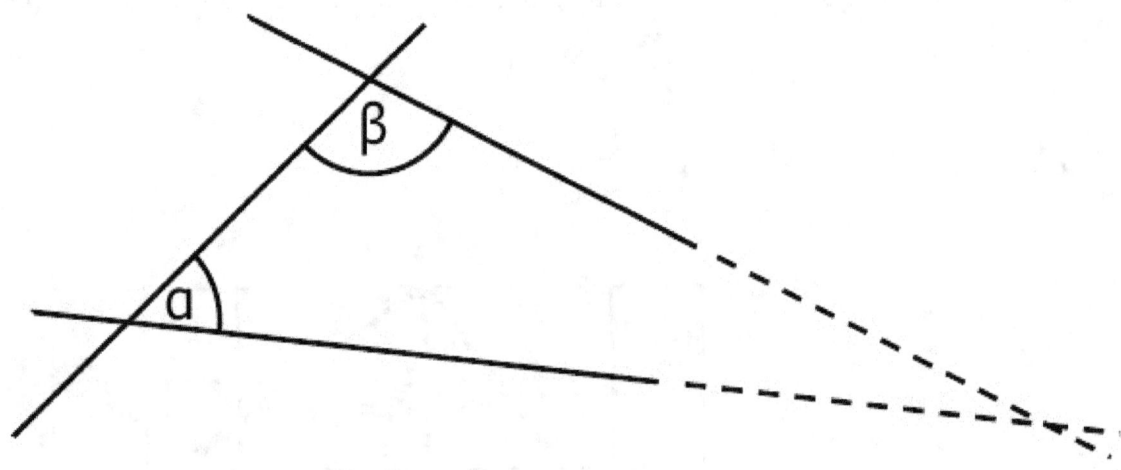

An illustration of Euclid's parallel postulate

See also: Euclidean geometry

Euclid took a more abstract approach in his Elements, one of the most influential books ever written. Euclid introduced certain axioms, or postulates, expressing primary or self-evident properties of points, lines, and planes. He proceeded to rigorously deduce other properties by mathematical reasoning. The characteristic feature of Euclid's approach to geometry was its rigor, and it has come to be known as *axiomatic* or *synthetic* geometry. At the start of the 19th century, the discovery of non-Euclidean geometries by Nikolai Ivanovich Lobachevsky (1792–1856), János Bolyai (1802–1860) and Carl Friedrich Gauss (1777–1855) and others led to a revival of interest in this discipline, and in the 20th century, David Hilbert (1862–1943) employed axiomatic reasoning in an attempt to provide a modern foundation of geometry.

Geometry lessons in the 20th century

19.1.3   Geometric constructions

Main article: Compass and straightedge constructions

Classical geometers paid special attention to constructing geometric objects that had been described in some other way. Classically, the only instruments allowed in geometric constructions are the compass and straightedge. Also, every construction had to be complete in a finite number of steps. However, some problems turned out to be difficult or impossible to solve by these means alone, and ingenious constructions using parabolas and other curves, as well as mechanical devices, were found.

19.1.4   Numbers in geometry

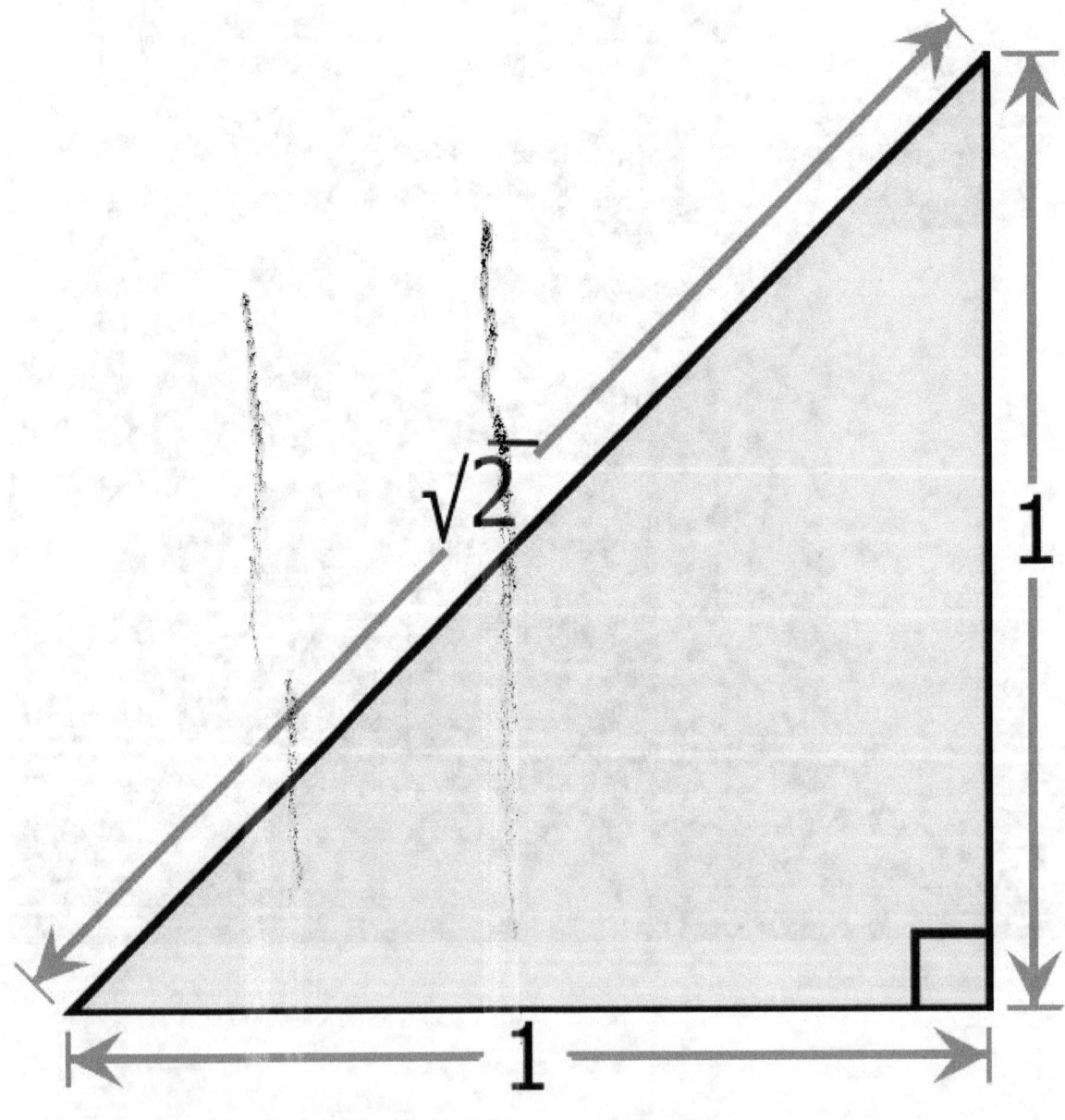

The Pythagoreans discovered that the sides of a triangle could have incommensurable lengths.

In ancient Greece the Pythagoreans considered the role of numbers in geometry. However, the discovery of incommensurable lengths, which contradicted their philosophical views, made them abandon abstract numbers in favor of concrete geometric quantities, such as length and area of figures. Numbers were reintroduced into geometry in the form of coordinates by Descartes, who realized that the study of geometric shapes can be facilitated by their algebraic representation, and for whom the Cartesian plane is named. Analytic geometry applies methods of algebra to geometric questions, typically by relating geometric curves to algebraic equations. These ideas played a key role in the development of calculus in the 17th century and led to the discovery of many new properties of plane curves. Modern algebraic geometry considers similar questions on a vastly more abstract level.

19.1.5 Geometry of position

Main articles: Projective geometry and Topology

Even in ancient times, geometers considered questions of relative position or spatial relationship of geometric figures and shapes. Some examples are given by inscribed and circumscribed circles of polygons, lines intersecting and tangent to conic sections, the Pappus and Menelaus configurations of points and lines. In the Middle Ages, new and more complicated questions of this type were considered: What is the maximum number of spheres simultaneously touching a given sphere of the same radius (kissing number problem)? What is the densest packing of spheres of equal size in space (Kepler conjecture)? Most of these questions involved 'rigid' geometrical shapes, such as lines or spheres. Projective, convex, and discrete geometry are three sub-disciplines within present day geometry that deal with these types of questions.

Leonhard Euler, in studying problems like the Seven Bridges of Königsberg, considered the most fundamental properties of geometric figures based solely on shape, independent of their metric properties. Euler called this new branch of geometry *geometria situs* (geometry of place), but it is now known as topology. Topology grew out of geometry, but turned into a large independent discipline. It does not differentiate between objects that can be continuously deformed into each other. The objects may nevertheless retain some geometry, as in the case of hyperbolic knots.

19.1.6 Geometry beyond Euclid

In the nearly two thousand years since Euclid, while the range of geometrical questions asked and answered inevitably expanded, the basic understanding of space remained essentially the same. Immanuel Kant argued that there is only one, *absolute*, geometry, which is known to be true *a priori* by an inner faculty of mind: Euclidean geometry was synthetic a priori.[3] This dominant view was overturned by the revolutionary discovery of non-Euclidean geometry in the works of Bolyai, Lobachevsky, and Gauss (who never published his theory). They demonstrated that ordinary Euclidean space is only one possibility for development of geometry. A broad vision of the subject of geometry was then expressed by Riemann in his 1867 inauguration lecture *Über die Hypothesen, welche der Geometrie zu Grunde liegen* (*On the hypotheses on which geometry is based*),[4] published only after his death. Riemann's new idea of space proved crucial in Einstein's general relativity theory, and Riemannian geometry, that considers very general spaces in which the notion of length is defined, is a mainstay of modern geometry.

19.1.7 Dimension

Where the traditional geometry allowed dimensions 1 (a line), 2 (a plane) and 3 (our ambient world conceived of as three-dimensional space), mathematicians have used higher dimensions for nearly two centuries. Dimension has gone through stages of being any natural number n, possibly infinite with the introduction of Hilbert space, and any positive real number in fractal geometry. Dimension theory is a technical area, initially within general topology, that discusses *definitions*; in common with most mathematical ideas, dimension is now defined rather than an intuition. Connected topological manifolds have a well-defined dimension; this is a theorem (invariance of domain) rather than anything *a priori*.

The issue of dimension still matters to geometry, in the absence of complete answers to classic questions. Dimensions 3 of space and 4 of space-time are special cases in geometric topology. Dimension 10 or 11 is a key number in string theory. Research may bring a satisfactory *geometric* reason for the significance of 10 and 11 dimensions.

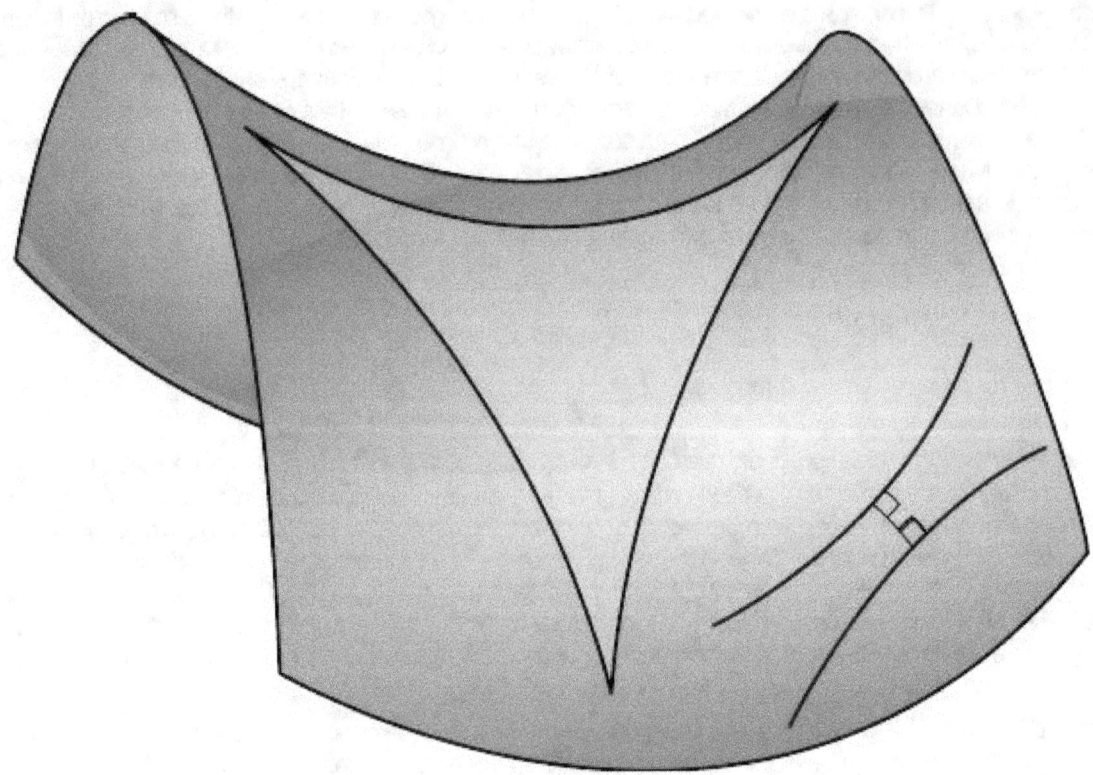

Differential geometry uses tools from calculus to study problems involving curvature.

19.1.8   Symmetry

The theme of symmetry in geometry is nearly as old as the science of geometry itself. Symmetric shapes such as the circle, regular polygons and platonic solids held deep significance for many ancient philosophers and were investigated in detail before the time of Euclid. Symmetric patterns occur in nature and were artistically rendered in a multitude of forms, including the graphics of M. C. Escher. Nonetheless, it was not until the second half of 19th century that the unifying role of symmetry in foundations of geometry was recognized. Felix Klein's Erlangen program proclaimed that, in a very precise sense, symmetry, expressed via the notion of a transformation group, determines what geometry *is*. Symmetry in classical Euclidean geometry is represented by congruences and rigid motions, whereas in projective geometry an analogous role is played by collineations, geometric transformations that take straight lines into straight lines. However it was in the new geometries of Bolyai and Lobachevsky, Riemann, Clifford and Klein, and Sophus Lie that Klein's idea to 'define a geometry via its symmetry group' proved most influential. Both discrete and continuous symmetries play prominent roles in geometry, the former in topology and geometric group theory, the latter in Lie theory and Riemannian geometry.

A different type of symmetry is the principle of duality in projective geometry (see Duality (projective geometry)) among other fields. This meta-phenomenon can roughly be described as follows: in any theorem, exchange *point* with *plane*, *join* with *meet*, *lies in* with *contains*, and you will get an equally true theorem. A similar and closely related form of duality exists between a vector space and its dual space.

19.2   History

Main article: History of geometry

The earliest recorded beginnings of geometry can be traced to ancient Mesopotamia and Egypt in the 2nd millennium BC.[5][6] Early geometry was a collection of empirically discovered principles concerning lengths, angles, areas, and

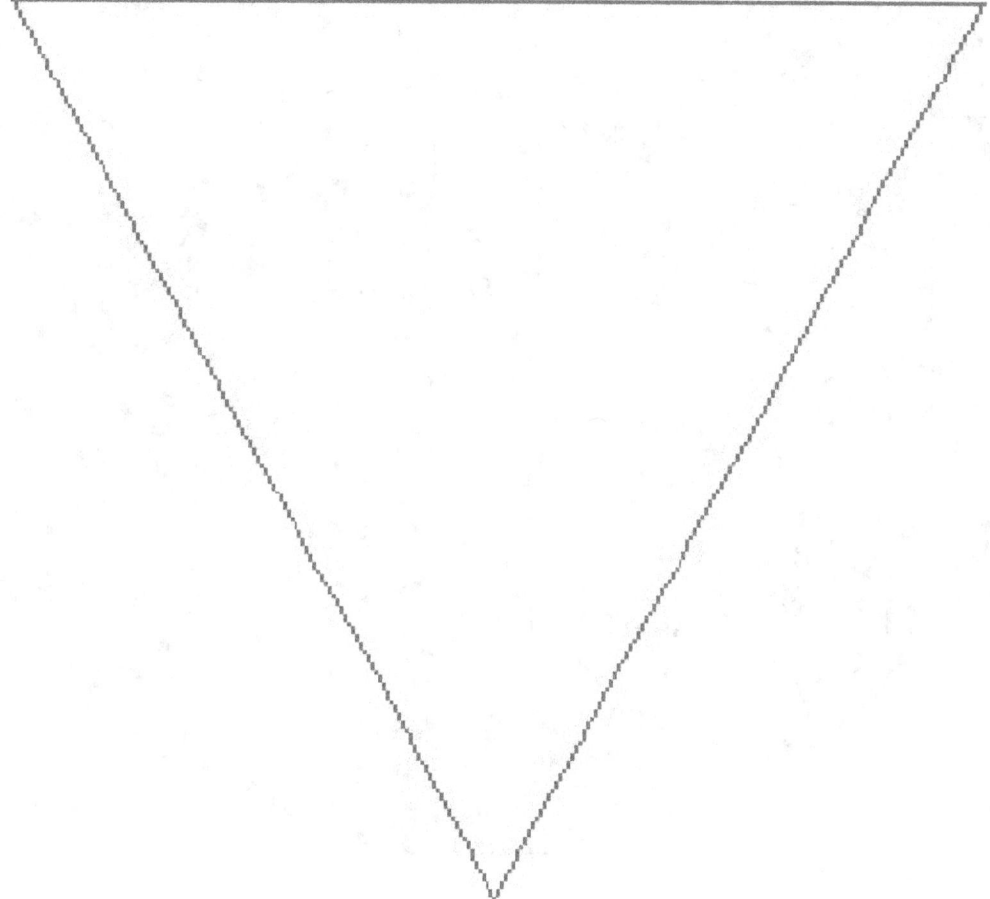

The Koch snowflake, with fractal dimension=log4/log3 and topological dimension=1

volumes, which were developed to meet some practical need in surveying, construction, astronomy, and various crafts. The earliest known texts on geometry are the Egyptian *Rhind Papyrus* (2000–1800 BC) and *Moscow Papyrus* (c. 1890 BC), the Babylonian clay tablets such as Plimpton 322 (1900 BC). For example, the Moscow Papyrus gives a formula for calculating the volume of a truncated pyramid, or frustum.[7] South of Egypt the ancient Nubians established a system of geometry including early versions of sun clocks.[8][9]

In the 7th century BC, the Greek mathematician Thales of Miletus used geometry to solve problems such as calculating the height of pyramids and the distance of ships from the shore. He is credited with the first use of deductive reasoning applied to geometry, by deriving four corollaries to Thales' Theorem.[10] Pythagoras established the Pythagorean School, which is credited with the first proof of the Pythagorean theorem,[11] though the statement of the theorem has a long history[12][13] Eudoxus (408–c. 355 BC) developed the method of exhaustion, which allowed the calculation of areas and volumes of curvilinear figures,[14] as well as a theory of ratios that avoided the problem of incommensurable magnitudes, which

A tiling of the hyperbolic plane

enabled subsequent geometers to make significant advances. Around 300 BC, geometry was revolutionized by Euclid, whose *Elements*, widely considered the most successful and influential textbook of all time,[15] introduced mathematical rigor through the axiomatic method and is the earliest example of the format still used in mathematics today, that of definition, axiom, theorem, and proof. Although most of the contents of the *Elements* were already known, Euclid arranged them into a single, coherent logical framework.[16] The *Elements* was known to all educated people in the West until the middle of the 20th century and its contents are still taught in geometry classes today.[17] Archimedes (c. 287–212 BC) of Syracuse used the method of exhaustion to calculate the area under the arc of a parabola with the summation of an infinite series, and gave remarkably accurate approximations of Pi.[18] He also studied the spiral bearing his name and obtained formulas for the volumes of surfaces of revolution.

Indian mathematicians also made many important contributions in geometry. The *Satapatha Brahmana* (ninth century BC) contains rules for ritual geometric constructions that are similar to the *Sulba Sutras*[19] According to (Hayashi 2005, p. 363), the *Śulba Sūtras* contain "the earliest extant verbal expression of the Pythagorean Theorem in the world, although it had already been known to the Old Babylonians. They contain lists of Pythagorean triples,[20] which are particular cases

A European and an Arab practicing geometry in the 15th century.

of Diophantine equations.[21] In the Bakhshali manuscript, there is a handful of geometric problems (including problems about volumes of irregular solids). The Bakhshali manuscript also "employs a decimal place value system with a dot for zero."[22] Aryabhata's *Aryabhatiya* (499) includes the computation of areas and volumes. Brahmagupta wrote his astronomical work *Brāhma Sphuṭa Siddhānta* in 628. Chapter 12, containing 66 Sanskrit verses, was divided into two sections: "basic operations" (including cube roots, fractions, ratio and proportion, and barter) and "practical mathematics" (including mixture, mathematical series, plane figures, stacking bricks, sawing of timber, and piling of grain).[23] In the latter section, he stated his famous theorem on the diagonals of a cyclic quadrilateral. Chapter 12 also included a formula for the area of a cyclic quadrilateral (a generalization of Heron's formula), as well as a complete description of rational triangles (*i.e.* triangles with rational sides and rational areas).[23]

In the Middle Ages, mathematics in medieval Islam contributed to the development of geometry, especially algebraic geometry[24] and geometric algebra.[25] Al-Mahani (b. 853) conceived the idea of reducing geometrical problems such as duplicating the cube to problems in algebra.[26] Thābit ibn Qurra (known as Thebit in Latin) (836–901) dealt with arithmetic operations applied to ratios of geometrical quantities, and contributed to the development of analytic geome-

Woman teaching geometry. *Illustration at the beginning of a medieval translation of Euclid's Elements, (c. 1310)*

try.[27] Omar Khayyám (1048–1131) found geometric solutions to cubic equations.[28] The theorems of Ibn al-Haytham (Alhazen), Omar Khayyam and Nasir al-Din al-Tusi on quadrilaterals, including the Lambert quadrilateral and Saccheri quadrilateral, were early results in hyperbolic geometry, and along with their alternative postulates, such as Playfair's axiom, these works had a considerable influence on the development of non-Euclidean geometry among later European geometers, including Witelo (c. 1230–c. 1314), Gersonides (1288–1344), Alfonso, John Wallis, and Giovanni Girolamo Saccheri.[29]

In the early 17th century, there were two important developments in geometry. The first was the creation of analytic geometry, or geometry with coordinates and equations, by René Descartes (1596–1650) and Pierre de Fermat (1601–1665). This was a necessary precursor to the development of calculus and a precise quantitative science of physics.

The second geometric development of this period was the systematic study of projective geometry by Girard Desargues (1591–1661). Projective geometry is a geometry without measurement or parallel lines, just the study of how points are related to each other.

Two developments in geometry in the 19th century changed the way it had been studied previously. These were the discovery of non-Euclidean geometries by Nikolai Ivanovich Lobachevsky, János Bolyai and Carl Friedrich Gauss and of the formulation of symmetry as the central consideration in the Erlangen Programme of Felix Klein (which generalized the Euclidean and non-Euclidean geometries). Two of the master geometers of the time were Bernhard Riemann (1826–1866), working primarily with tools from mathematical analysis, and introducing the Riemann surface, and Henri Poincaré, the founder of algebraic topology and the geometric theory of dynamical systems. As a consequence of these major changes in the conception of geometry, the concept of "space" became something rich and varied, and the natural background for theories as different as complex analysis and classical mechanics.

19.3 Contemporary geometry

19.3.1 Euclidean geometry

Euclidean geometry has become closely connected with computational geometry, computer graphics, convex geometry, incidence geometry, finite geometry, discrete geometry, and some areas of combinatorics. Attention was given to further work on Euclidean geometry and the Euclidean groups by crystallography and the work of H. S. M. Coxeter, and can be seen in theories of Coxeter groups and polytopes. Geometric group theory is an expanding area of the theory of more general discrete groups, drawing on geometric models and algebraic techniques.

19.3.2 Differential geometry

Differential geometry has been of increasing importance to mathematical physics due to Einstein's general relativity postulation that the universe is curved. Contemporary differential geometry is *intrinsic*, meaning that the spaces it considers are smooth manifolds whose geometric structure is governed by a Riemannian metric, which determines how distances are measured near each point, and not *a priori* parts of some ambient flat Euclidean space.

19.3.3 Topology and geometry

The field of topology, which saw massive development in the 20th century, is in a technical sense a type of transformation geometry, in which transformations are homeomorphisms. This has often been expressed in the form of the dictum 'topology is rubber-sheet geometry'. Contemporary geometric topology and differential topology, and particular subfields such as Morse theory, would be counted by most mathematicians as part of geometry. Algebraic topology and general topology have gone their own ways.

19.3.4 Algebraic geometry

The field of algebraic geometry is the modern incarnation of the Cartesian geometry of co-ordinates. From late 1950s through mid-1970s it had undergone major foundational development, largely due to work of Jean-Pierre Serre and Alexander Grothendieck. This led to the introduction of schemes and greater emphasis on topological methods, including various cohomology theories. One of seven Millennium Prize problems, the Hodge conjecture, is a question in algebraic geometry.

The study of low-dimensional algebraic varieties, algebraic curves, algebraic surfaces and algebraic varieties of dimension 3 ("algebraic threefolds"), has been far advanced. Gröbner basis theory and real algebraic geometry are among more applied subfields of modern algebraic geometry. Arithmetic geometry is an active field combining algebraic geometry and number theory. Other directions of research involve moduli spaces and complex geometry. Algebro-geometric methods are commonly applied in string and brane theory.

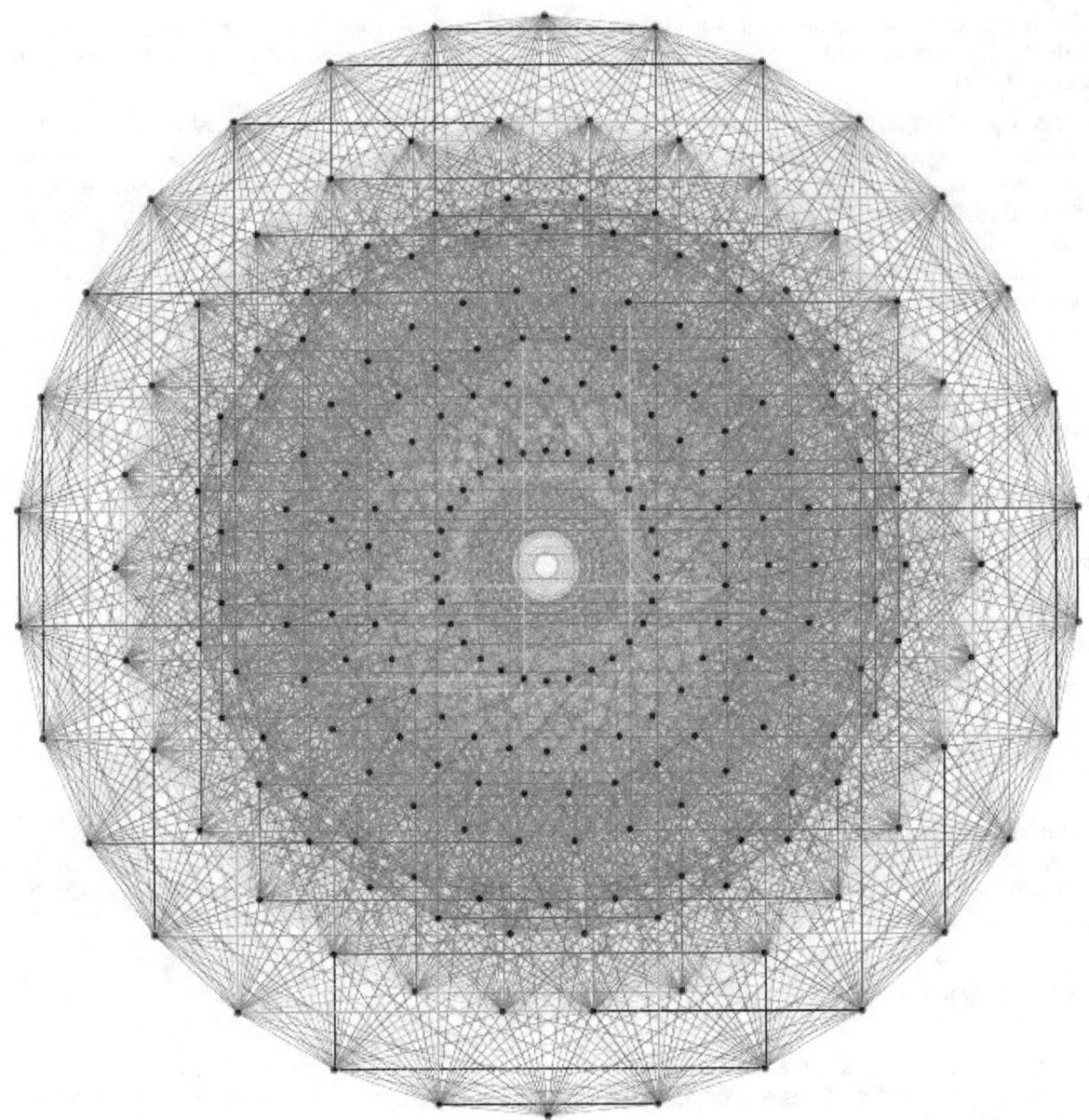

The 4_{21} polytope, orthogonally projected into the E_8 Lie group Coxeter plane

19.4   See also

19.4.1   Lists

- List of geometers
  - Category:Algebraic geometers
  - Category:Differential geometers
  - Category:Geometers
  - Category:Topologists
- List of formulas in elementary geometry

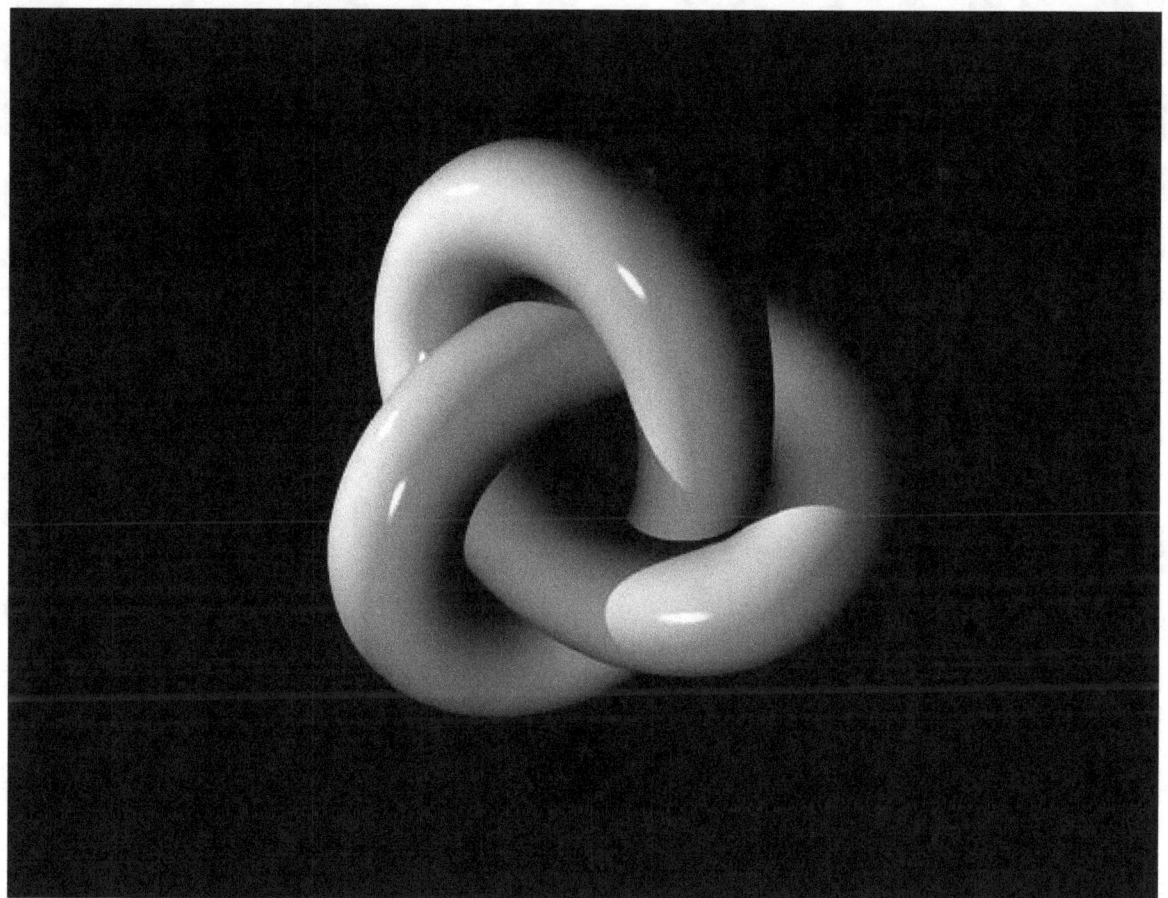

A thickening of the trefoil knot

- List of geometry topics

- List of important publications in geometry

- List of mathematics articles

19.4.2 Related topics

- Descriptive geometry

- *Flatland*, a book written by Edwin Abbott Abbott about two- and three-dimensional space, to understand the concept of four dimensions

- Interactive geometry software

19.4.3 Other fields

- Molecular geometry

Quintic Calabi–Yau threefold

19.5   Notes

[1] Martin J. Turner,Jonathan M. Blackledge,Patrick R. Andrews (1998). *Fractal geometry in digital imaging*. Academic Press. p. 1. ISBN 0-12-703970-8

[2] It is quite common in algebraic geometry to speak about *geometry of algebraic varieties over finite fields*, possibly singular. From a naive perspective, these objects are just finite sets of points, but by invoking powerful geometric imagery and using well developed geometric techniques, it is possible to find structure and establish properties that make them somewhat analogous to the ordinary spheres or cones.

[3] Kline (1972) "Mathematical thought from ancient to modern times", Oxford University Press, p. 1032. Kant did not reject the logical (analytic a priori) *possibility* of non-Euclidean geometry, see Jeremy Gray, "Ideas of Space Euclidean, Non-Euclidean, and Relativistic", Oxford, 1989; p. 85. Some have implied that, in light of this, Kant had in fact *predicted* the development of

non-Euclidean geometry, cf. Leonard Nelson, "Philosophy and Axiomatics," Socratic Method and Critical Philosophy, Dover, 1965, p. 164.

[4] "Ueber die Hypothesen, welche der Geometrie zu Grunde liegen.".

[5] J. Friberg, "Methods and traditions of Babylonian mathematics. Plimpton 322, Pythagorean triples, and the Babylonian triangle parameter equations", Historia Mathematica, 8, 1981, pp. 277—318.

[6] Neugebauer, Otto (1969) [1957]. The Exact Sciences in Antiquity (2 ed.). Dover Publications. ISBN 978-0-486-22332-2. Chap. IV "Egyptian Mathematics and Astronomy", pp. 71–96.

[7] (Boyer 1991, "Egypt" p. 19)

[8] The Journal of Egyptian Archaeology. Vol. 84, 1998 Gnomons at Meroë and Early Trigonometry. pg. 171

[9] Slayman, Andrew (May 27, 1998). "Neolithic Skywatchers". Archaeology Magazine Archive.

[10] (Boyer 1991, "Ionia and the Pythagoreans" p. 43)

[11] Eves, Howard, An Introduction to the History of Mathematics, Saunders, 1990, ISBN 0-03-029558-0.

[12] Kurt Von Fritz (1945). "The Discovery of Incommensurability by Hippasus of Metapontum". The Annals of Mathematics.

[13] James R. Choike (1980). "The Pentagram and the Discovery of an Irrational Number". The Two-Year College Mathematics Journal.

[14] (Boyer 1991, "The Age of Plato and Aristotle" p. 92)

[15] (Boyer 1991, "Euclid of Alexandria" p. 119)

[16] (Boyer 1991, "Euclid of Alexandria" p. 104)

[17] Howard Eves, An Introduction to the History of Mathematics, Saunders, 1990, ISBN 0-03-029558-0 p. 141: "No work, except The Bible, has been more widely used....".

[18] O'Connor, J.J. and Robertson, E.F. (February 1996). "A history of calculus". University of St Andrews. Retrieved 2007-08-07.

[19] (Staal 1999)

[20] Pythagorean triples are triples of integers (a, b, c) with the property: $a^2 + b^2 = c^2$. Thus, $3^2 + 4^2 = 5^2$, $8^2 + 15^2 = 17^2$, $12^2 + 35^2 = 37^2$ etc.

[21] (Cooke 2005, p. 198): "The arithmetic content of the Śulva Sūtras consists of rules for finding Pythagorean triples such as (3, 4, 5), (5, 12, 13), (8, 15, 17), and (12, 35, 37). It is not certain what practical use these arithmetic rules had. The best conjecture is that they were part of religious ritual. A Hindu home was required to have three fires burning at three different altars. The three altars were to be of different shapes, but all three were to have the same area. These conditions led to certain "Diophantine" problems, a particular case of which is the generation of Pythagorean triples, so as to make one square integer equal to the sum of two others."

[22] (Hayashi 2005, p. 371)

[23] (Hayashi 2003, pp. 121–122)

[24] R. Rashed (1994), The development of Arabic mathematics: between arithmetic and algebra, p. 35 London

[25] Boyer (1991). "The Arabic Hegemony". A History of Mathematics. pp. 241–242. Omar Khayyam (ca. 1050–1123), the "tent-maker," wrote an Algebra that went beyond that of al-Khwarizmi to include equations of third degree. Like his Arab predecessors, Omar Khayyam provided for quadratic equations both arithmetic and geometric solutions; for general cubic equations, he believed (mistakenly, as the 16th century later showed), arithmetic solutions were impossible; hence he gave only geometric solutions. The scheme of using intersecting conics to solve cubics had been used earlier by Menaechmus, Archimedes, and Alhazan, but Omar Khayyam took the praiseworthy step of generalizing the method to cover all third-degree equations (having positive roots). .. For equations of higher degree than three, Omar Khayyam evidently did not envision similar geometric methods, for space does not contain more than three dimensions, ...   One of the most fruitful contributions of Arabic eclecticism was the tendency to close the gap between numerical and geometric algebra. The decisive step in this direction came much later with Descartes, but Omar Khayyam was moving in this direction when he wrote, "Whoever thinks algebra is a trick in obtaining unknowns has thought it in vain. No attention should be paid to the fact that algebra and geometry are different in appearance. Algebras are geometric facts which are proved.""

[26] O'Connor, John J.; Robertson, Edmund F., "Al-Mahani", *MacTutor History of Mathematics archive*, University of St Andrews.

[27] O'Connor, John J.; Robertson, Edmund F., "Al-Sabi Thabit ibn Qurra al-Harrani", *MacTutor History of Mathematics archive*, University of St Andrews.

[28] O'Connor, John J.; Robertson, Edmund F., "Omar Khayyam", *MacTutor History of Mathematics archive*, University of St Andrews.

[29] Boris A. Rosenfeld and Adolf P. Youschkevitch (1996), "Geometry", in Roshdi Rashed, ed., *Encyclopedia of the History of Arabic Science*, Vol. 2, p. 447–494 [470], Routledge, London and New York:

> "Three scientists, Ibn al-Haytham, Khayyam, and al-Tusi, had made the most considerable contribution to this branch of geometry whose importance came to be completely recognized only in the 19th century. In essence, their propositions concerning the properties of quadrangles which they considered, assuming that some of the angles of these figures were acute of obtuse, embodied the first few theorems of the hyperbolic and the elliptic geometries. Their other proposals showed that various geometric statements were equivalent to the Euclidean postulate V. It is extremely important that these scholars established the mutual connection between this postulate and the sum of the angles of a triangle and a quadrangle. By their works on the theory of parallel lines Arab mathematicians directly influenced the relevant investigations of their European counterparts. The first European attempt to prove the postulate on parallel lines – made by Witelo, the Polish scientists of the 13th century, while revising Ibn al-Haytham's *Book of Optics* (*Kitab al-Manazir*) – was undoubtedly prompted by Arabic sources. The proofs put forward in the 14th century by the Jewish scholar Levi ben Gerson, who lived in southern France, and by the above-mentioned Alfonso from Spain directly border on Ibn al-Haytham's demonstration. Above, we have demonstrated that *Pseudo-Tusi's Exposition of Euclid* had stimulated both J. Wallis's and G. Saccheri's studies of the theory of parallel lines."

19.6   Sources

- Boyer, C. B. (1991) [1989]. *A History of Mathematics* (Second edition, revised by Uta C. Merzbach ed.). New York: Wiley. ISBN 0-471-54397-7.

- Nikolai I. Lobachevsky, *Pangeometry*, translator and editor: A. Papadopoulos, Heritage of European Mathematics Series, Vol. 4, European Mathematical Society, 2010.

19.7   Further reading

- Jay Kappraff, *A Participatory Approach to Modern Geometry*, 2014, World Scientific Publishing, ISBN 978-981-4556-70-5.

- Leonard Mlodinow, *Euclid's Window – The Story of Geometry from Parallel Lines to Hyperspace*, UK edn. Allen Lane, 1992.

19.8   External links

- A geometry course from Wikiversity
- *Unusual Geometry Problems*
- *The Math Forum* — Geometry
    - *The Math Forum* — K–12 Geometry
    - *The Math Forum* — College Geometry
    - *The Math Forum* — Advanced Geometry
- Nature Precedings — *Pegs and Ropes Geometry at Stonehenge*

- *The Mathematical Atlas* — Geometric Areas of Mathematics
- "4000 Years of Geometry", lecture by Robin Wilson given at Gresham College, 3 October 2007 (available for MP3 and MP4 download as well as a text file)
    - Finitism in Geometry at the Stanford Encyclopedia of Philosophy
- The Geometry Junkyard
- Interactive Geometry Applications (Java and Cabri 3D)
- Interactive geometry reference with hundreds of applets
- Dynamic Geometry Sketches (with some Student Explorations)
- Geometry classes at Khan Academy

Chapter 20

Convex geometry

In mathematics, **convex geometry** is the branch of geometry studying convex sets, mainly in Euclidean space. Convex sets occur naturally in many areas: computational geometry, convex analysis, discrete geometry, functional analysis, geometry of numbers, integral geometry, linear programming, probability theory, etc.

20.1 Classification

According to the Mathematics Subject Classification MSC2010,[1] the mathematical discipline *Convex and Discrete Geometry* includes three major branches:[2]

- general convexity
- polytopes and polyhedra
- discrete geometry

General convexity is further subdivided as follows:[3]

- axiomatic and generalized convexity
- convex sets without dimension restrictions
- convex sets in topological vector spaces
- convex sets in 2 dimensions (including convex curves)
- convex sets in 3 dimensions (including convex surfaces)
- convex sets in n dimensions (including convex hypersurfaces)
- finite-dimensional Banach spaces
- random convex sets and integral geometry
- asymptotic theory of convex bodies
- approximation by convex sets
- variants of convex sets (star-shaped, (m, n)-convex, etc.)
- Helly-type theorems and geometric transversal theory

- other problems of combinatorial convexity

- length, area, volume

- mixed volumes and related topics

- inequalities and extremum problems

- convex functions and convex programs

- spherical and hyperbolic convexity

The term *convex geometry* is also used in combinatorics as the name for one of the abstract models of convex sets, one that is equivalent to antimatroids.

20.2 Historical note

Convex geometry is a relatively young mathematical discipline. Although the first known contributions to convex geometry date back to antiquity and can be traced in the works of Euclid and Archimedes, it became an independent branch of mathematics at the turn of the 20th century, mainly due to the works of Hermann Brunn and Hermann Minkowski in dimensions two and three. A big part of their results was soon generalized to spaces of higher dimensions, and in 1934 T. Bonnesen and W. Fenchel gave a comprehensive survey of convex geometry in Euclidean space $\mathbf{R}^n$. Further development of convex geometry in the 20th century and its relations to numerous mathematical disciplines are summarized in the *Handbook of convex geometry* edited by P. M. Gruber and J. M. Wills.

20.3 See also

- List of convexity topics

20.4 Notes

[1] Website of Mathematics Subject Classification MSC2010

[2] Mathematics Subject Classification MSC2010, entry 52 "Convex and discrete geometry"

[3] Mathematics Subject Classification MSC2010, entry 52A "General convexity"

20.5 References

Expository articles on convex geometry

- K. Ball, *An elementary introduction to modern convex geometry*, in: Flavors of Geometry, pp. 1–58, Math. Sci. Res. Inst. Publ. Vol. 31, Cambridge Univ. Press, Cambridge, 1997, available online.

- M. Berger, *Convexity*, Amer. Math. Monthly, Vol. 97 (1990), 650—678.

- P. M. Gruber, *Aspects of convexity and its applications*, Exposition. Math., Vol. 2 (1984), 47—83.

- V. Klee, *What is a convex set?* Amer. Math. Monthly, Vol. 78 (1971), 616—631.

Books on convex geometry

- T. Bonnesen, W. Fenchel, *Theorie der konvexen Körper*, Julius Springer, Berlin, 1934. English translation: *Theory of convex bodies*, BCS Associates, Moscow, ID, 1987.

- R. J. Gardner, *Geometric tomography*, Cambridge University Press, New York, 1995. Second edition: 2006.

- P. M. Gruber, *Convex and discrete geometry*, Springer-Verlag, New York, 2007.

- P. M. Gruber, J. M. Wills (editors), *Handbook of convex geometry. Vol. A, B*, North-Holland, Amsterdam, 1993.

- G. Pisier, *The volume of convex bodies and Banach space geometry*, Cambridge University Press, Cambridge, 1989.

- R. Schneider, *Convex bodies: the Brunn-Minkowski theory*, Cambridge University Press, Cambridge, 1993.

- A. C. Thompson, *Minkowski geometry*, Cambridge University Press, Cambridge, 1996.

- A. Koldobsky, V. Yaskin, *The Interface between Convex Geometry and Harmonic Analysis*, American Mathematical Society, Providence, Rhode Island, 2008.

Articles on history of convex geometry

- W. Fenchel, *Convexity through the ages*, (Danish) Danish Mathematical Society (1929—1973), pp. 103–116, Dansk. Mat. Forening, Copenhagen, 1973. English translation: *Convexity through the ages*, in: P. M. Gruber, J. M. Wills (editors), Convexity and its Applications, pp. 120–130, Birkhauser Verlag, Basel, 1983.

- P. M. Gruber, *Zur Geschichte der Konvexgeometrie und der Geometrie der Zahlen*, in: G. Fischer, et al. (editors), Ein Jahrhundert Mathematik 1890—1990, pp. 421–455, Dokumente Gesch. Math., Vol. 6, F. Wieweg and Sohn, Braunschweig; Deutsche Mathematiker Vereinigung, Freiburg, 1990.

- P. M. Gruber, *History of convexity*, in: P. M. Gruber, J. M. Wills (editors), Handbook of convex geometry. Vol. A, pp. 1–15, North-Holland, Amsterdam, 1993.

Chapter 21

Discrete geometry

"Combinatorial geometry" redirects here. The term combinatorial geometry is also used in the theory of matroids to refer to a simple matroid, especially in older texts.

Discrete geometry and **combinatorial geometry** are branches of geometry that study combinatorial properties and constructive methods of discrete geometric objects. Most questions in discrete geometry involve finite or discrete sets of basic geometric objects, such as points, lines, planes, circles, spheres, polygons, and so forth. The subject focuses on the combinatorial properties of these objects, such as how they intersect one another, or how they may be arranged to cover a larger object.

Discrete geometry has large overlap with convex geometry and computational geometry, and is closely related to subjects such as finite geometry, combinatorial optimization, digital geometry, discrete differential geometry, geometric graph theory, toric geometry, and combinatorial topology.

21.1 History

Although polyhedra and tessellations had been studied for many years by people such as Kepler and Cauchy, modern discrete geometry has its origins in the late 19th century. Early topics studied were: the density of circle packings by Thue, projective configurations by Reye and Steinitz, the geometry of numbers by Minkowski, and map colourings by Tait, Heawood, and Hadwiger.

László Fejes Tóth, H.S.M. Coxeter and Paul Erdős, laid the foundations of *discrete geometry*. [1][2][3]

21.2 Topics in discrete geometry

21.2.1 Polyhedra and polytopes

Main articles: Polyhedron and Polytope

A **polytope** is a geometric object with flat sides, which exists in any general number of dimensions. A polygon is a polytope in two dimensions, a polyhedron in three dimensions, and so on in higher dimensions (such as a 4-polytope in four dimensions). Some theories further generalize the idea to include such objects as unbounded polytopes (apeirotopes and tessellations), and abstract polytopes.

The following are some of the aspects of polytopes studied in discrete geometry:

- Polyhedral combinatorics
- Lattice polytopes

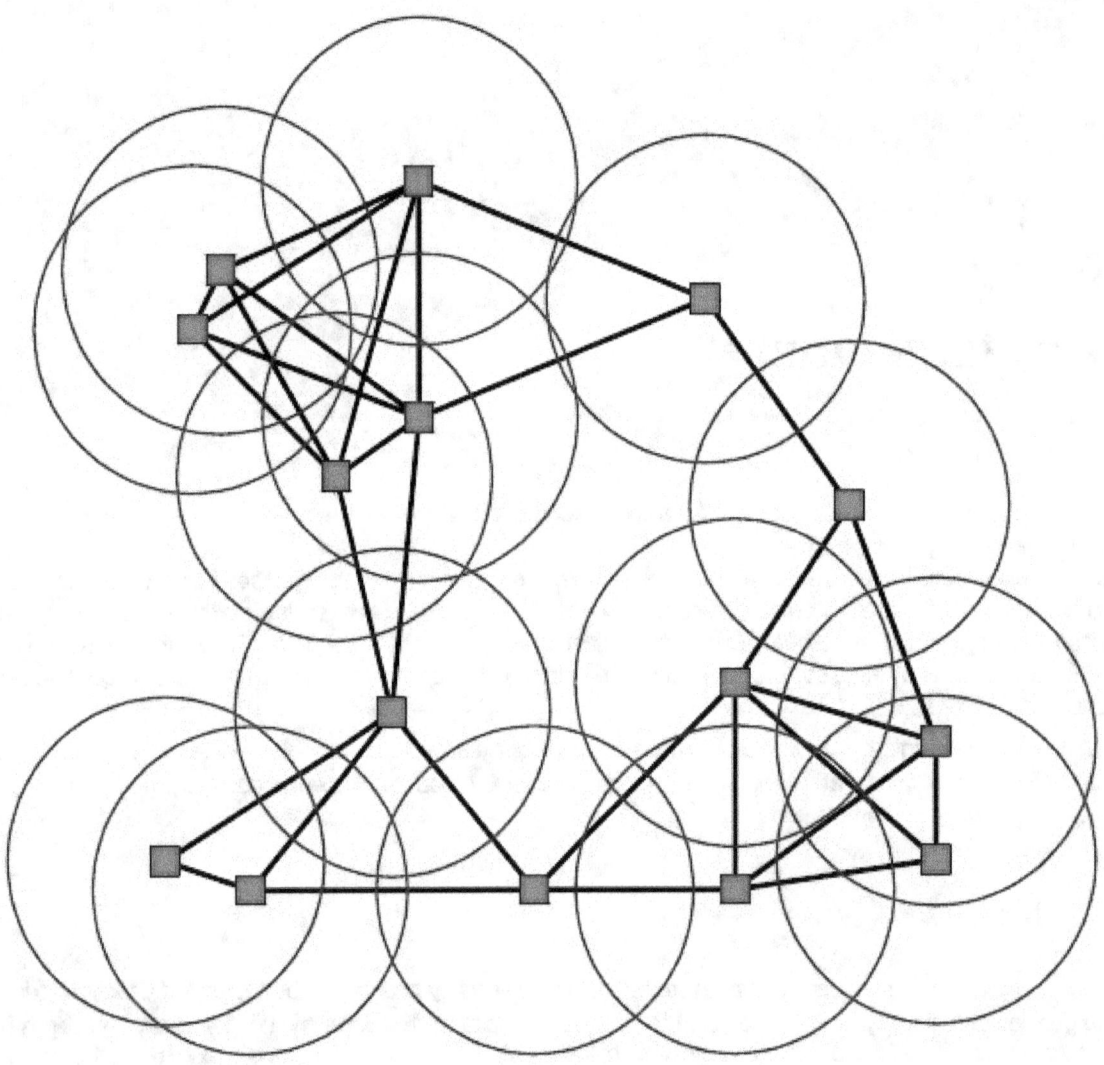

A collection of circles and the corresponding unit disk graph

- Ehrhart polynomials

- Pick's theorem

- Hirsch conjecture

21.2.2   Packings, coverings and tilings

Main articles: circle packing and tessellation

Packings, coverings, and tilings are all ways of arranging uniform objects (typically circles, spheres, or tiles) in a regular way on a surface or manifold.

A **sphere packing** is an arrangement of non-overlapping spheres within a containing space. The spheres considered are usually all of identical size, and the space is usually three-dimensional Euclidean space. However, sphere packing problems can be generalised to consider unequal spheres, n-dimensional Euclidean space (where the problem becomes circle packing in two dimensions, or hypersphere packing in higher dimensions) or to non-Euclidean spaces such as hyperbolic space.

A **tessellation** of a flat surface is the tiling of a plane using one or more geometric shapes, called tiles, with no overlaps and no gaps. In mathematics, tessellations can be generalized to higher dimensions.

Specific topics in this area include:

- Circle packings
- Sphere packings
- Kepler conjecture
- Quasicrystals
- Aperiodic tilings
- Periodic Graphs (Geometry)
- Finite subdivision rules

21.2.3   Structural rigidity and flexibility

Main article: Structural rigidity
Structural rigidity is a combinatorial theory for predicting the flexibility of ensembles formed by rigid bodies connected by flexible linkages or hinges.

Topics in this area include:

- Cauchy's theorem
- Flexible polyhedra

21.2.4   Incidence structures

Main article: Incidence structure
Incidence structures generalize planes (such as affine, projective, and Möbius planes) as can be seen from their axiomatic definitions. Incidence structures also generalize the higher-dimensional analogs and the finite structures are sometimes called finite geometries.

Formally, an **incidence structure** is a triple

$$C = (P, L, I).$$

where P is a set of "points", L is a set of "lines" and $I \subseteq P \times L$ is the incidence relation. The elements of I are called **flags**. If

$$(p, l) \in I,$$

we say that point p "lies on" line l.

Topics in this area include:

- Configurations

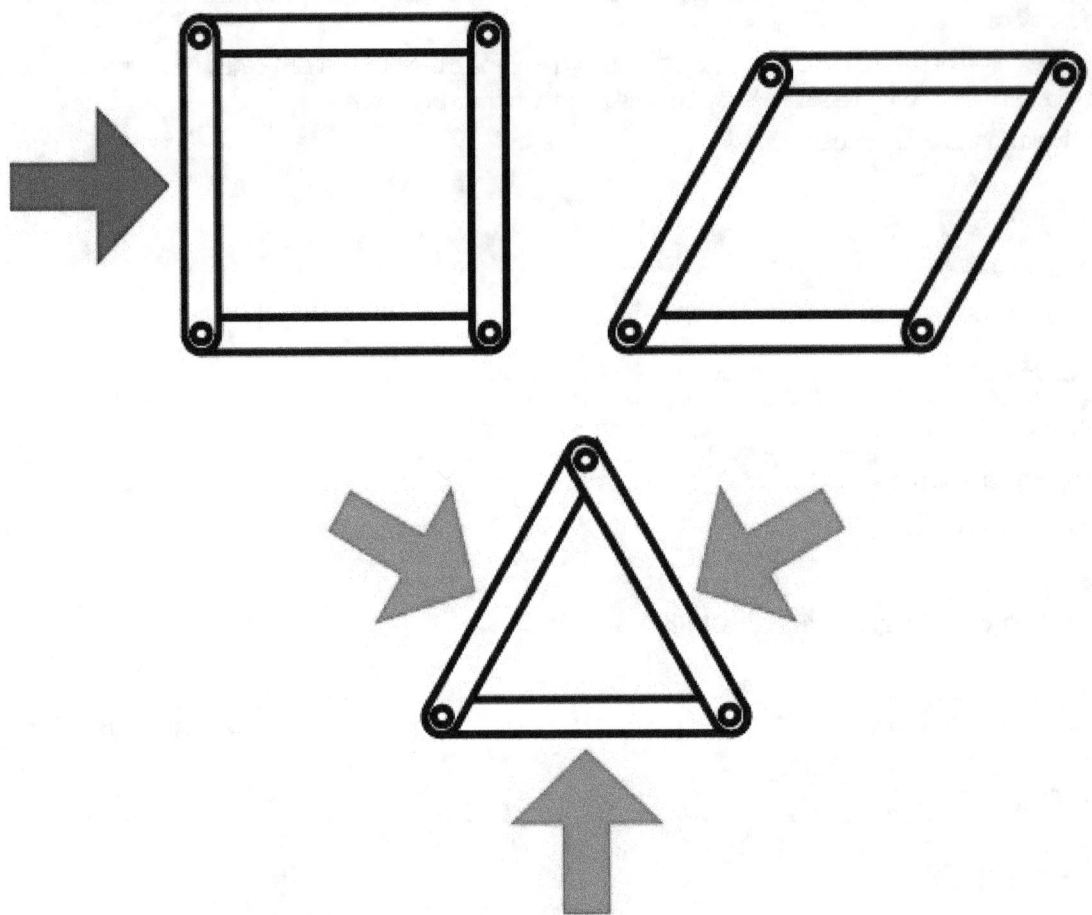

Graphs are drawn as rods connected by rotating hinges. The cycle graph C_4 drawn as a square can be tilted over by the blue force into a parallelogram, so it is a flexible graph. K_3, drawn as a triangle, cannot be altered by any force that is applied to it, so it is a rigid graph.

- Line arrangements

- Hyperplane arrangements

- Buildings

21.2.5   Oriented matroids

Main article: Oriented matroid

An **oriented matroid** is a mathematical structure that abstracts the properties of directed graphs and of arrangements of vectors in a vector space over an ordered field (particularly for partially ordered vector spaces).[4] In comparison, an ordinary (i.e., non-oriented) matroid abstracts the dependence properties that are common both to graphs, which are not necessarily *directed,* and to arrangements of vectors over fields, which are not necessarily *ordered.*[5] [6]

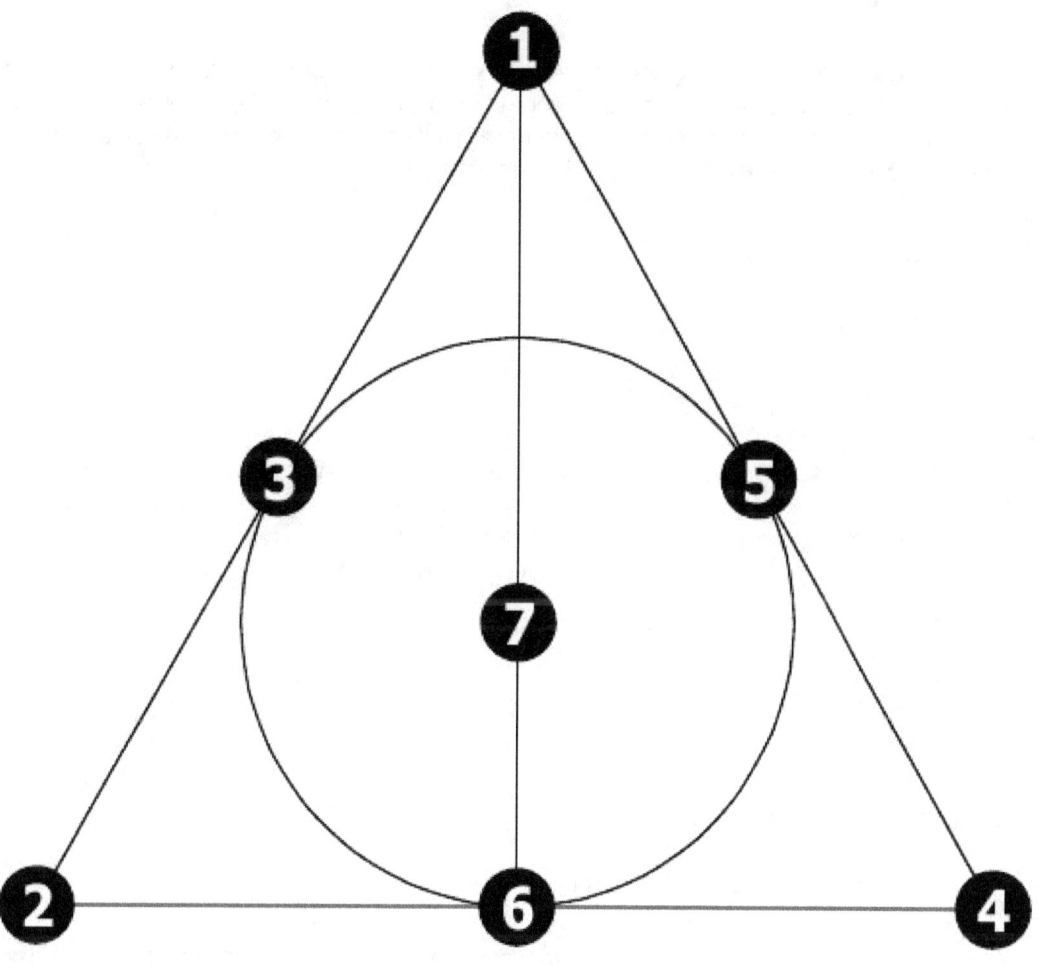

Seven points are elements of seven lines in the Fano plane, an example of an incidence structure.

21.2.6   Geometric graph theory

Main article: Geometric graph theory

A **geometric graph** is a graph in which the vertices or edges are associated with geometric objects. Examples include Euclidean graphs, the 1-skeleton of a polyhedron or polytope, intersection graphs, and visibility graphs.

Topics in this area include:

- Graph drawing
- Polyhedral graphs
- Voronoi diagrams and Delaunay triangulations

21.2.7   Simplicial complexes

Main article: Simplicial complex

A **simplicial complex** is a topological space of a certain kind, constructed by "gluing together" points, line segments, triangles, and their n-dimensional counterparts (see illustration). Simplicial complexes should not be confused with the more abstract notion of a simplicial set appearing in modern simplicial homotopy theory. The purely combinatorial counterpart to a simplicial complex is an abstract simplicial complex.

21.2.8   Topological combinatorics

Main article: Topological combinatorics

The discipline of combinatorial topology used combinatorial concepts in topology and in the early 20th century this turned into the field of algebraic topology.

In 1978 the situation was reversed – methods from algebraic topology were used to solve a problem in combinatorics – when László Lovász proved the Kneser conjecture, thus beginning the new study of **topological combinatorics**. Lovász's proof used the Borsuk-Ulam theorem and this theorem retains a prominent role in this new field. This theorem has many equivalent versions and analogs and has been used in the study of fair division problems.

Topics in this are include:

- Sperner's lemma
- Regular maps

21.2.9   Lattices and discrete groups

Main articles: Lattice (discrete group) and discrete group

A **discrete group** is a group G equipped with the discrete topology. With this topology, G becomes a topological group. A **discrete subgroup** of a topological group G is a subgroup H whose relative topology is the discrete one. For example, the integers, Z, form a discrete subgroup of the reals, R (with the standard metric topology), but the rational numbers, Q, do not.

A **lattice** in a locally compact topological group is a discrete subgroup with the property that the quotient space has finite invariant measure. In the special case of subgroups of R^n, this amounts to the usual geometric notion of a lattice, and both the algebraic structure of lattices and the geometry of the totality of all lattices are relatively well understood. Deep results of Borel, Harish-Chandra, Mostow, Tamagawa, M. S. Raghunathan, Margulis, Zimmer obtained from the 1950s through the 1970s provided examples and generalized much of the theory to the setting of nilpotent Lie groups and semisimple algebraic groups over a local field. In the 1990s, Bass and Lubotzky initiated the study of *tree lattices*, which remains an active research area.

Topics in this area include:

- Reflection groups
- Triangle groups

21.2.10   Digital geometry

Main article: Digital geometry

Digital geometry deals with discrete sets (usually discrete point sets) considered to be digitized models or images of objects of the 2D or 3D Euclidean space.

Simply put, digitizing is replacing an object by a discrete set of its points. The images we see on the TV screen, the raster display of a computer, or in newspapers are in fact digital images.

Its main application areas are computer graphics and image analysis.

21.2.11   Discrete differential geometry

Main article: Discrete differential geometry

Discrete differential geometry is the study of discrete counterparts of notions in differential geometry. Instead of smooth curves and surfaces, there are polygons, meshes, and simplicial complexes. It is used in the study of computer graphics and topological combinatorics.

Topics in this area include:

- Discrete Laplace operator
- Discrete exterior calculus
- Discrete Morse theory
- Topological combinatorics
- Spectral shape analysis
- Abstract differential geometry
- Analysis on fractals

21.3   See also

- *Discrete and Computational Geometry*
- Discrete mathematics
- Paul Erdős

21.4   Notes

[1] Pach, János et al. (2008), *Intuitive Geometry, in Memoriam László Fejes Tóth*, Alfréd Rényi Institute of Mathematics

[2] Katona, G. O. H. (2005), "Laszlo Fejes Toth – Obituary", *Studia Scientiarum Mathematicarum Hungarica* 42 (2): 113

[3] Bárány, Imre (2010), "Discrete and convex geometry", in Horváth, János, *A Panorama of Hungarian Mathematics in the Twentieth Century, I*, New York: Springer, pp. 431–441, ISBN 9783540307211

[4] Rockafellar 1969. Björner et alia, Chapters 1-3. Bokowski, Chapter 1. Ziegler, Chapter 7.

[5] Björner et alia, Chapters 1-3. Bokowski, Chapters 1-4.

[6] Because matroids and oriented matroids are abstractions of other mathematical abstractions, nearly all the relevant books are written for mathematical scientists rather than for the general public. For learning about oriented matroids, a good preparation is to study the textbook on linear optimization by Nering and Tucker, which is infused with oriented-matroid ideas, and then to proceed to Ziegler's lectures on polytopes.

21.5   References

- Bezdek, András, (2003). *Discrete geometry: in honor of W. Kuperberg's 60th birthday.* New York, N.Y: Marcel Dekker. ISBN 0-8247-0968-3.

- Bezdek, Károly (2010). *Classical Topics in Discrete Geometry.* New York, N.Y: Springer. ISBN 978-1-4419-0599-4.

- Bezdek, Károly (2013). *Lectures on Sphere Arrangements - the Discrete Geometric Side.* New York, N.Y: Springer. ISBN 978-1-4614-8117-1.

- Bezdek, Károly; Deza, Antoine; Ye, Yinyu (2013). *Discrete Geometry and Optimization.* New York, N.Y: Springer. ISBN 978-3-319-00200-2.

- Brass, Peter; Moser, William; Pach, János (2005). *Research problems in discrete geometry.* Berlin: Springer. ISBN 0-387-23815-8.

- Pach, János; Agarwal, Pankaj K. (1995). *Combinatorial geometry.* New York: Wiley-Interscience. ISBN 0-471-58890-3.

- Goodman, Jacob E. and O'Rourke, Joseph (2004). *Handbook of Discrete and Computational Geometry, Second Edition.* Boca Raton: Chapman & Hall/CRC. ISBN 1-58488-301-4.

- Gruber, Peter M. (2007). *Convex and Discrete Geometry.* Berlin: Springer. ISBN 3-540-71132-5.

- Matoušek, Jiří (2002). *Lectures on discrete geometry.* Berlin: Springer. ISBN 0-387-95374-4.

- Vladimir Boltyanski, Horst Martini, Petru S. Soltan, (1997). *Excursions into Combinatorial Geometry.* Springer. ISBN 3-540-61341-2.

Chapter 22

Differential geometry

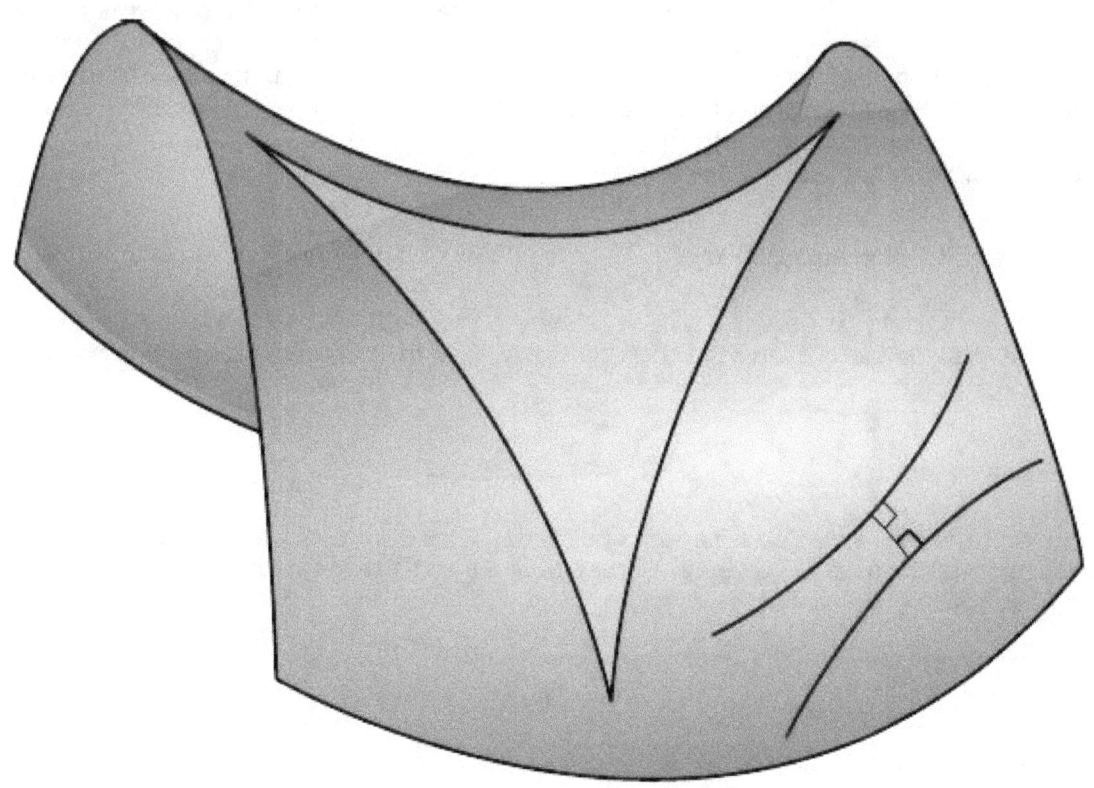

A triangle immersed in a saddle-shape plane (a hyperbolic paraboloid), as well as two diverging ultraparallel lines.

Differential geometry is a mathematical discipline that uses the techniques of differential calculus, integral calculus, linear algebra and multilinear algebra to study problems in geometry. The theory of plane and space curves and surfaces in the three-dimensional Euclidean space formed the basis for development of differential geometry during the 18th century and the 19th century.

Since the late 19th century, differential geometry has grown into a field concerned more generally with the geometric structures on differentiable manifolds. Differential geometry is closely related to differential topology and the geometric aspects of the theory of differential equations. The differential geometry of surfaces captures many of the key ideas and techniques characteristic of this field.

22.1   Branches of differential geometry

22.1.1   Riemannian geometry

Main article: Riemannian geometry

Riemannian geometry studies Riemannian manifolds, smooth manifolds with a *Riemannian metric*. This is a concept of distance expressed by means of a smooth positive definite symmetric bilinear form defined on the tangent space at each point. Riemannian geometry generalizes Euclidean geometry to spaces that are not necessarily flat, although they still resemble the Euclidean space at each point infinitesimally, i.e. in the first order of approximation. Various concepts based on length, such as the arc length of curves, area of plane regions, and volume of solids all possess natural analogues in Riemannian geometry. The notion of a directional derivative of a function from multivariable calculus is extended in Riemannian geometry to the notion of a covariant derivative of a tensor. Many concepts and techniques of analysis and differential equations have been generalized to the setting of Riemannian manifolds.

A distance-preserving diffeomorphism between Riemannian manifolds is called an isometry. This notion can also be defined *locally*, i.e. for small neighborhoods of points. Any two regular curves are locally isometric. However, the Theorema Egregium of Carl Friedrich Gauss showed that already for surfaces, the existence of a local isometry imposes strong compatibility conditions on their metrics: the Gaussian curvatures at the corresponding points must be the same. In higher dimensions, the Riemann curvature tensor is an important pointwise invariant associated to a Riemannian manifold that measures how close it is to being flat. An important class of Riemannian manifolds is the Riemannian symmetric spaces, whose curvature is not necessarily constant. These are the closest analogues to the "ordinary" plane and space considered in Euclidean and non-Euclidean geometry.

22.1.2   Pseudo-Riemannian geometry

Pseudo-Riemannian geometry generalizes Riemannian geometry to the case in which the metric tensor need not be positive-definite. A special case of this is a Lorentzian manifold, which is the mathematical basis of Einstein's general relativity theory of gravity.

22.1.3   Finsler geometry

Finsler geometry has the *Finsler manifold* as the main object of study. This is a differential manifold with a Finsler metric, i.e. a Banach norm defined on each tangent space. A Finsler metric is a much more general structure than a Riemannian metric. A Finsler structure on a manifold M is a function $F : TM \rightarrow [0,\infty)$ such that:

1. $F(x, my) = |m|F(x,y)$ for all x, y in TM,

2. F is infinitely differentiable in $TM - \{0\}$,

3. The vertical Hessian of F^2 is positive definite.

22.1.4   Symplectic geometry

Main article: Symplectic geometry

Symplectic geometry is the study of symplectic manifolds. An **almost symplectic manifold** is a differentiable manifold equipped with a smoothly varying non-degenerate skew-symmetric bilinear form on each tangent space, i.e., a nondegenerate 2-form ω, called the *symplectic form*. A symplectic manifold is an almost symplectic manifold for which the symplectic form ω is closed: $d\omega = 0$.

A diffeomorphism between two symplectic manifolds which preserves the symplectic form is called a symplectomorphism. Non-degenerate skew-symmetric bilinear forms can only exist on even-dimensional vector spaces, so symplectic manifolds necessarily have even dimension. In dimension 2, a symplectic manifold is just a surface endowed with an area form and a symplectomorphism is an area-preserving diffeomorphism. The phase space of a mechanical system is a symplectic manifold and they made an implicit appearance already in the work of Joseph Louis Lagrange on analytical mechanics and later in Carl Gustav Jacobi's and William Rowan Hamilton's formulations of classical mechanics.

By contrast with Riemannian geometry, where the curvature provides a local invariant of Riemannian manifolds, Darboux's theorem states that all symplectic manifolds are locally isomorphic. The only invariants of a symplectic manifold are global in nature and topological aspects play a prominent role in symplectic geometry. The first result in symplectic topology is probably the Poincaré-Birkhoff theorem, conjectured by Henri Poincaré and then proved by G.D. Birkhoff in 1912. It claims that if an area preserving map of an annulus twists each boundary component in opposite directions, then the map has at least two fixed points.[1]

22.1.5   Contact geometry

Main article: Contact geometry

Contact geometry deals with certain manifolds of odd dimension. It is close to symplectic geometry and like the latter, it originated in questions of classical mechanics. A *contact structure* on a $(2n + 1)$ - dimensional manifold M is given by a smooth hyperplane field H in the tangent bundle that is as far as possible from being associated with the level sets of a differentiable function on M (the technical term is "completely nonintegrable tangent hyperplane distribution"). Near each point p, a hyperplane distribution is determined by a nowhere vanishing 1-form a, which is unique up to multiplication by a nowhere vanishing function:

$$H_p = \ker a_p \subset T_p M.$$

A local 1-form on M is a *contact form* if the restriction of its exterior derivative to H is a non-degenerate two-form and thus induces a symplectic structure on H_p at each point. If the distribution H can be defined by a global one-form a then this form is contact if and only if the top-dimensional form

$$a \wedge (da)^n$$

is a volume form on **M**, i.e. does not vanish anywhere. A contact analogue of the Darboux theorem holds: all contact structures on an odd-dimensional manifold are locally isomorphic and can be brought to a certain local normal form by a suitable choice of the coordinate system.

22.1.6   Complex and Kähler geometry

Complex differential geometry is the study of complex manifolds. An almost complex manifold is a *real* manifold M, endowed with a tensor of type $(1, 1)$, i.e. a vector bundle endomorphism (called an *almost complex structure*)

$$J : TM \to TM, \text{ such that } J^2 = -1.$$

It follows from this definition that an almost complex manifold is even-dimensional.

An almost complex manifold is called *complex* if $N_J = 0$, where N_J is a tensor of type $(2, 1)$ related to J, called the Nijenhuis tensor (or sometimes the *torsion*). An almost complex manifold is complex if and only if it admits a holomorphic coordinate atlas. An *almost Hermitian structure* is given by an almost complex structure J, along with a Riemannian metric g, satisfying the compatibility condition

$$g(JX, JY) = g(X, Y)$$

An almost Hermitian structure defines naturally a differential two-form

$$\omega_{J,g}(X, Y) := g(JX, Y)$$

The following two conditions are equivalent:

1. $N_J = 0$ and $d\omega = 0$
2. $\nabla J = 0$

where ∇ is the Levi-Civita connection of g. In this case, (J, g) is called a *Kähler structure*, and a *Kähler manifold* is a manifold endowed with a Kähler structure. In particular, a Kähler manifold is both a complex and a symplectic manifold. A large class of Kähler manifolds (the class of Hodge manifolds) is given by all the smooth complex projective varieties.

22.1.7   CR geometry

CR geometry is the study of the intrinsic geometry of boundaries of domains in complex manifolds.

22.1.8   Differential topology

Differential topology is the study of (global) geometric invariants without a metric or symplectic form. It starts from the natural operations such as Lie derivative of natural vector bundles and de Rham differential of forms. Beside Lie algebroids, also Courant algebroids start playing a more important role.

22.1.9   Lie groups

A Lie group is a group in the category of smooth manifolds. Beside the algebraic properties this enjoys also differential geometric properties. The most obvious construction is that of a Lie algebra which is the tangent space at the unit endowed with the Lie bracket between left-invariant vector fields. Beside the structure theory there is also the wide field of representation theory.

22.2   Bundles and connections

The apparatus of vector bundles, principal bundles, and connections on bundles plays an extraordinarily important role in modern differential geometry. A smooth manifold always carries a natural vector bundle, the tangent bundle. Loosely speaking, this structure by itself is sufficient only for developing analysis on the manifold, while doing geometry requires, in addition, some way to relate the tangent spaces at different points, i.e. a notion of parallel transport. An important example is provided by affine connections. For a surface in $\mathbf{R}^3$, tangent planes at different points can be identified using a natural path-wise parallelism induced by the ambient Euclidean space, which has a well-known standard definition of metric and parallelism. In Riemannian geometry, the Levi-Civita connection serves a similar purpose. (The Levi-Civita connection defines path-wise parallelism in terms of a given arbitrary Riemannian metric on a manifold.) More generally, differential geometers consider spaces with a vector bundle and an arbitrary affine connection which is not defined in terms of a metric. In physics, the manifold may be the space-time continuum and the bundles and connections are related to various physical fields.

22.3   Intrinsic versus extrinsic

From the beginning and through the middle of the 18th century, differential geometry was studied from the *extrinsic* point of view: curves and surfaces were considered as lying in a Euclidean space of higher dimension (for example a

surface in an ambient space of three dimensions). The simplest results are those in the differential geometry of curves and differential geometry of surfaces. Starting with the work of Riemann, the *intrinsic* point of view was developed, in which one cannot speak of moving "outside" the geometric object because it is considered to be given in a free-standing way. The fundamental result here is Gauss's theorema egregium, to the effect that Gaussian curvature is an intrinsic invariant.

The intrinsic point of view is more flexible. For example, it is useful in relativity where space-time cannot naturally be taken as extrinsic (what would be "outside" of it?). However, there is a price to pay in technical complexity: the intrinsic definitions of curvature and connections become much less visually intuitive.

These two points of view can be reconciled, i.e. the extrinsic geometry can be considered as a structure additional to the intrinsic one. (See the Nash embedding theorem.) In the formalism of geometric calculus both extrinsic and intrinsic geometry of a manifold can be characterized by a single bivector-valued one-form called the shape operator.[2]

22.4   Applications

Below are some examples of how differential geometry is applied to other fields of science and mathematics.

- In physics, four uses will be mentioned:
  - Differential geometry is the language in which Einstein's general theory of relativity is expressed. According to the theory, the universe is a smooth manifold equipped with a pseudo-Riemannian metric, which describes the curvature of space-time. Understanding this curvature is essential for the positioning of satellites into orbit around the earth. Differential geometry is also indispensable in the study of gravitational lensing and black holes.
  - Differential forms are used in the study of electromagnetism.
  - Differential geometry has applications to both Lagrangian mechanics and Hamiltonian mechanics. Symplectic manifolds in particular can be used to study Hamiltonian systems.
  - Riemannian geometry and contact geometry have been used to construct the formalism of geometrothermodynamics which has found applications in classical equilibrium thermodynamics.
- In economics, differential geometry has applications to the field of econometrics.[3]
- Geometric modeling (including computer graphics) and computer-aided geometric design draw on ideas from differential geometry.
- In engineering, differential geometry can be applied to solve problems in digital signal processing.[4]
- In control theory, differential geometry can be used to analyze nonlinear controllers, particularly geometric control [5]
- In probability, statistics, and information theory, one can interpret various structures as Riemannian manifolds, which yields the field of information geometry, particularly via the Fisher information metric.
- In structural geology, differential geometry is used to analyze and describe geologic structures.
- In computer vision, differential geometry is used to analyze shapes.[6]
- In image processing, differential geometry is used to process and analyse data on non-flat surfaces.[7]
- Grigori Perelman's proof of the Poincaré conjecture using the techniques of Ricci flows demonstrated the power of the differential-geometric approach to questions in topology and it highlighted the important role played by its analytic methods.
- In wireless communications, Grassmannian manifolds are used for beamforming techniques in multiple antenna systems.[8]

22.5   See also

- Abstract differential geometry

- Affine differential geometry

- Analysis on fractals

- Basic introduction to the mathematics of curved spacetime

- Discrete differential geometry

- Glossary of differential geometry and topology

- Integral geometry

- List of differential geometry topics

- Important publications in differential geometry

- Important publications in differential topology

- Noncommutative geometry

- Projective differential geometry

- Synthetic differential geometry

22.6   References

[1] It is easy to show that the area preserving condition (or the twisting condition) cannot be removed. Note that if one tries to extend such a theorem to higher dimensions, one would probably guess that a volume preserving map of a certain type must have fixed points. This is false in dimensions greater than 3.

[2] David Hestenes "The Shape of Differential Geometry in Geometric Calculus" http://geocalc.clas.asu.edu/pdf/Shape%20in%20GC-2012.pdf there is also a pdf available of a scientific talk on the subject http://staff.science.uva.nl/~{}leo/agacse2010/talks_world/Hestenes.pdf

[3] Paul Marriott and Mark Salmon (editors), "Applications of Differential Geometry to Econometrics", Cambridge University Press; 1 edition (September 18, 2000).

[4] Jonathan H. Manton, "On the role of differential geometry in signal processing" .

[5] Francesco Bullo and Andrew Lewis, "Geometric Control of Simple Mechanical Systems." Springer-Verlag, 2001.

[6] Mario Micheli, "The Differential Geometry of Landmark Shape Manifolds: Metrics, Geodesics, and Curvature", http://www.math.ucla.edu/~{}micheli/PUBLICATIONS/micheli_phd.pdf

[7] Anand A. Joshi, "Geometric methods for image processing and signal analysis",

[8] David J. Love and Robert W. Heath, Jr. "Grassmannian Beamforming for Multiple-Input Multiple-Output Wireless Systems," IEEE Transactions on Information Theory, Vol. 49, No. 10, October 2003

22.7   Further reading

- Wolfgang Kühnel (2002). *Differential Geometry: Curves - Surfaces - Manifolds* (2nd ed.). ISBN 0-8218-3988-8.

- Theodore Frankel (2004). *The geometry of physics: an introduction* (2nd ed.). ISBN 0-521-53927-7.

- Spivak, Michael (1999). *A Comprehensive Introduction to Differential Geometry (5 Volumes)* (3rd ed.).

- do Carmo, Manfredo (1976). *Differential Geometry of Curves and Surfaces.* ISBN 0-13-212589-7. Classical geometric approach to differential geometry without tensor analysis.

- Kreyszig, Erwin (1991). *Differential Geometry.* ISBN 0-486-66721-9. Good classical geometric approach to differential geometry with tensor machinery.

- do Carmo, Manfredo Perdigao (1994). *Riemannian Geometry.*

- McCleary, John (1994). *Geometry from a Differentiable Viewpoint.*

- Bloch, Ethan D. (1996). *A First Course in Geometric Topology and Differential Geometry.*

- Gray, Alfred (1998). *Modern Differential Geometry of Curves and Surfaces with Mathematica* (2nd ed.).

- Burke, William L. (1985). *Applied Differential Geometry.*

- ter Haar Romeny, Bart M. (2003). *Front-End Vision and Multi-Scale Image Analysis.* ISBN 1-4020-1507-0.

22.8   External links

- Hazewinkel, Michiel, ed. (2001), "Diff erential geometry", *Encyclopedia of Mathematics,* Springer, ISBN 978-1-55608-010-4

- B. Conrad. Differential Geometry handouts, Stanford University

- Michael Murray's online differential geometry course, 1996

- A Modern Course on Curves and Surface, Richard S Palais, 2003

- Richard Palais's 3DXM Surfaces Gallery

- Balázs Csikós's Notes on Differential Geometry

- N. J. Hicks, Notes on Differential Geometry, Van Nostrand.

- MIT OpenCourseWare: Differential Geometry, Fall 2008

Chapter 23

Algebraic geometry

Not to be confused with Geometric algebra, an application of Clifford algebra to geometry.

For the book by Robin Hartshorne, see Algebraic Geometry (book).

Algebraic geometry is a branch of mathematics, classically studying zeros of multivariate polynomials. Modern algebraic geometry is based on the use of abstract algebraic techniques, mainly from commutative algebra, for solving geometrical problems about these sets of zeros.

The fundamental objects of study in algebraic geometry are algebraic varieties, which are geometric manifestations of solutions of systems of polynomial equations. Examples of the most studied classes of algebraic varieties are: plane algebraic curves, which include lines, circles, parabolas, ellipses, hyperbolas, cubic curves like elliptic curves and quartic curves like lemniscates, and Cassini ovals. A point of the plane belongs to an algebraic curve if its coordinates satisfy a given polynomial equation. Basic questions involve the study of the points of special interest like the singular points, the inflection points and the points at infinity. More advanced questions involve the topology of the curve and relations between the curves given by different equations.

Algebraic geometry occupies a central place in modern mathematics and has multiple conceptual connections with such diverse fields as complex analysis, topology and number theory. Initially a study of systems of polynomial equations in several variables, the subject of algebraic geometry starts where equation solving leaves off, and it becomes even more important to understand the intrinsic properties of the totality of solutions of a system of equations, than to find a specific solution; this leads into some of the deepest areas in all of mathematics, both conceptually and in terms of technique.

In the 20th century, algebraic geometry has split into several subareas.

- The main stream of algebraic geometry is devoted to the study of the complex points of the algebraic varieties and more generally to the points with coordinates in an algebraically closed field.

- The study of the points of an algebraic variety with coordinates in the field of the rational numbers or in a number field became arithmetic geometry (or more classically Diophantine geometry), a subfield of algebraic number theory.

- The study of the real points of an algebraic variety is the subject of real algebraic geometry.

- A large part of singularity theory is devoted to the singularities of algebraic varieties.

- With the rise of the computers, a computational algebraic geometry area has emerged, which lies at the intersection of algebraic geometry and computer algebra. It consists essentially in developing algorithms and software for studying and finding the properties of explicitly given algebraic varieties.

Much of the development of the main stream of algebraic geometry in the 20th century occurred within an abstract algebraic framework, with increasing emphasis being placed on "intrinsic" properties of algebraic varieties not dependent on any particular way of embedding the variety in an ambient coordinate space; this parallels developments in topology, differential and complex geometry. One key achievement of this abstract algebraic geometry is Grothendieck's scheme

This Togliatti surface is an algebraic surface of degree five. The picture represents a portion of its real locus.

theory which allows one to use sheaf theory to study algebraic varieties in a way which is very similar to its use in the study of differential and analytic manifolds. This is obtained by extending the notion of point: In classical algebraic geometry, a point of an affine variety may be identified, through Hilbert's Nullstellensatz, with a maximal ideal of the coordinate ring, while the points of the corresponding affine scheme are all prime ideals of this ring. This means that a point of such a scheme may be either a usual point or a subvariety. This approach also enables a unification of the language and the tools of classical algebraic geometry, mainly concerned with complex points, and of algebraic number theory. Wiles's proof of the longstanding conjecture called Fermat's last theorem is an example of the power of this approach.

23.1 Basic notions

Further information: Algebraic variety

23.1.1   Zeros of simultaneous polynomials

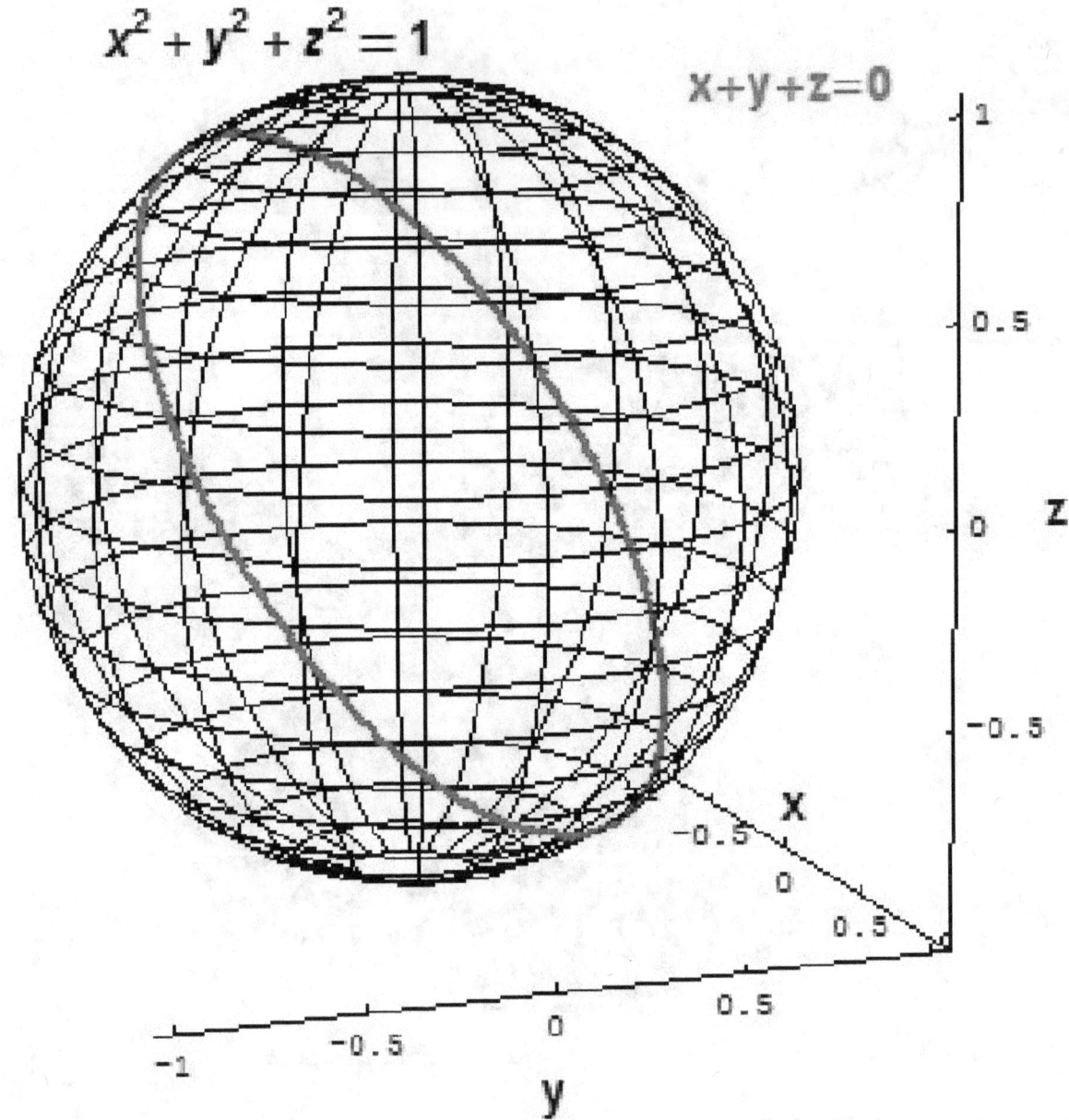

Sphere and slanted circle

In classical algebraic geometry, the main objects of interest are the vanishing sets of collections of polynomials, meaning the set of all points that simultaneously satisfy one or more polynomial equations. For instance, the two-dimensional sphere in three-dimensional Euclidean space $\mathbf{R}^3$ could be defined as the set of all points (x,y,z) with

$$x^2 + y^2 + z^2 - 1 = 0.$$

A "slanted" circle in $\mathbf{R}^3$ can be defined as the set of all points (x,y,z) which satisfy the two polynomial equations

$$x^2 + y^2 + z^2 - 1 = 0,$$

$x + y + z = 0.$

23.1.2 Affine varieties

Main article: Affine variety

First we start with a field k. In classical algebraic geometry, this field was always the complex numbers $\mathbb{C}$, but many of the same results are true if we assume only that k is algebraically closed. We consider the affine space of dimension n over k, denoted $\mathbf{A}^n(k)$ (or more simply $\mathbf{A}^n$, when k is clear from the context). When one fixes a coordinates system, one may identify $\mathbf{A}^n(k)$ with k^n. The purpose of not working with k^n is to emphasize that one "forgets" the vector space structure that k^n carries.

A function $f : \mathbf{A}^n \to \mathbf{A}^1$ is said to be *polynomial* (or *regular*) if it can be written as a polynomial, that is, if there is a polynomial p in $k[x_1,...,x_n]$ such that $f(M) = p(t_1,...,t_n)$ for every point M with coordinates $(t_1,...,t_n)$ in $\mathbf{A}^n$. The property of a function to be polynomial (or regular) does not depend on the choice of a coordinate system in $\mathbf{A}^n$.

When a coordinate system is chosen, the regular functions on the affine n-space may be identified with the ring of polynomial functions in n variables over k. Therefore the set of the regular functions on $\mathbf{A}^n$ is a ring, which is denoted $k[\mathbf{A}^n]$.

We say that a polynomial *vanishes* at a point if evaluating it at that point gives zero. Let S be a set of polynomials in $k[\mathbf{A}^n]$. The *vanishing set of S* (or *vanishing locus* or *zero set*) is the set $V(S)$ of all points in $\mathbf{A}^n$ where every polynomial in S vanishes. In other words,

$$V(S) = \{(t_1,\ldots,t_n) / p \in S, p(t_1,\ldots,t_n) = 0\}.$$

A subset of $\mathbf{A}^n$ which is $V(S)$, for some S, is called an *algebraic set*. The V stands for *variety* (a specific type of algebraic set to be defined below).

Given a subset U of $\mathbf{A}^n$, can one recover the set of polynomials which generate it? If U is *any* subset of $\mathbf{A}^n$, define $I(U)$ to be the set of all polynomials whose vanishing set contains U. The I stands for ideal: if two polynomials f and g both vanish on U, then $f+g$ vanishes on U, and if h is any polynomial, then hf vanishes on U, so $I(U)$ is always an ideal of the polynomial ring $k[\mathbf{A}^n]$.

Two natural questions to ask are:

- Given a subset U of $\mathbf{A}^n$, when is $U = V(I(U))$?
- Given a set S of polynomials, when is $S = I(V(S))$?

The answer to the first question is provided by introducing the Zariski topology, a topology on $\mathbf{A}^n$ whose closed sets are the algebraic sets, and which directly reflects the algebraic structure of $k[\mathbf{A}^n]$. Then $U = V(I(U))$ if and only if U is an algebraic set or equivalently a Zariski-closed set. The answer to the second question is given by Hilbert's Nullstellensatz. In one of its forms, it says that $I(V(S))$ is the radical of the ideal generated by S. In more abstract language, there is a Galois connection, giving rise to two closure operators; they can be identified, and naturally play a basic role in the theory; the example is elaborated at Galois connection.

For various reasons we may not always want to work with the entire ideal corresponding to an algebraic set U. Hilbert's basis theorem implies that ideals in $k[\mathbf{A}^n]$ are always finitely generated.

An algebraic set is called *irreducible* if it cannot be written as the union of two smaller algebraic sets. Any algebraic set is a finite union of irreducible algebraic sets and this decomposition is unique. Thus its elements are called the *irreducible components* of the algebraic set. An irreducible algebraic set is also called a *variety*. It turns out that an algebraic set is a variety if and only if it may be defined as the vanishing set of a prime ideal of the polynomial ring.

Some authors do not make a clear distinction between algebraic sets and varieties and use *irreducible variety* to make the distinction when needed.

23.1.3   Regular functions

Main article: Regular function

Just as continuous functions are the natural maps on topological spaces and smooth functions are the natural maps on differentiable manifolds, there is a natural class of functions on an algebraic set, called *regular functions* or *polynomial functions*. A regular function on an algebraic set V contained in A^n is the restriction to V of a regular function on A^n. For an algebraic set defined on the field of the complex numbers, the regular functions are smooth and even analytic.

It may seem unnaturally restrictive to require that a regular function always extend to the ambient space, but it is very similar to the situation in a normal topological space, where the Tietze extension theorem guarantees that a continuous function on a closed subset always extends to the ambient topological space.

Just as with the regular functions on affine space, the regular functions on V form a ring, which we denote by $k[V]$. This ring is called the *coordinate ring of V*.

Since regular functions on V come from regular functions on A^n, there is a relationship between the coordinate rings. Specifically, if a regular function on V is the restriction of two functions f and g in $k[A^n]$, then $f - g$ is a polynomial function which is null on V and thus belongs to $I(V)$. Thus $k[V]$ may be identified with $k[A^n]/I(V)$.

23.1.4   Morphism of affine varieties

Using regular functions from an affine variety to A^1, we can define regular maps from one affine variety to another. First we will define a regular map from a variety into affine space: Let V be a variety contained in A^n. Choose m regular functions on V, and call them $f_1, ..., f_m$. We define a *regular map f* from V to A^m by letting $f = (f_1, ..., f_m)$. In other words, each f_i determines one coordinate of the range of f.

If V' is a variety contained in A^m, we say that f is a *regular map* from V to V' if the range of f is contained in V'.

The definition of the regular maps apply also to algebraic sets. The regular maps are also called *morphisms*, as they make the collection of all affine algebraic sets into a category, where the objects are the affine algebraic sets and the morphisms are the regular maps. The affine varieties is a subcategory of the category of the algebraic sets.

Given a regular map g from V to V' and a regular function f of $k[V']$, then $f \circ g \in k[V]$. The map $f \to f \circ g$ is a ring homomorphism from $k[V']$ to $k[V]$. Conversely, every ring homomorphism from $k[V']$ to $k[V]$ defines a regular map from V to V'. This defines an equivalence of categories between the category of algebraic sets and the opposite category of the finitely generated reduced k-algebras. This equivalence is one of the starting points of scheme theory.

23.1.5   Rational function and birational equivalence

Main article: Rational mapping

Contrarily to the preceding ones, this section concerns only varieties and not algebraic sets. On the other hand the definitions extend naturally to projective varieties (next section), as an affine variety and its projective completion have the same field of functions.

If V is an affine variety, its coordinate ring is an integral domain and has thus a field of fractions which is denoted $k(V)$ and called the *field of the rational functions* on V or, shortly, the *function field* of V. Its elements are the restrictions to V of the rational functions over the affine space containing V. The domain of a rational function f is not V but the complement of the subvariety (a hypersurface) where the denominator of f vanishes.

Like for regular maps, one may define a *rational map* from a variety V to a variety V'. Like for the regular maps, the rational maps from V to V' may be identified to the field homomorphisms from $k(V')$ to $k(V)$.

Two affine varieties are *birationally equivalent* if there are two rational functions between them which are inverse one to the other in the regions where both are defined. Equivalently, they are birationally equivalent if their function fields are isomorphic.

An affine variety is a *rational variety* if it is birationally equivalent to an affine space. This means that the variety admits a rational parameterization. For example, the circle of equation $x^2 + y^2 - 1 = 0$ is a rational curve, as it has the parameterization

$$x = \frac{2t}{1+t^2}$$

$$\frac{}{1+t^2}'$$

which may also be viewed as a rational map from the line to the circle.

The problem of resolution of singularities is to know if every algebraic variety is birationally equivalent to a variety whose projective completion is nonsingular (see also smooth completion). It has been positively solved in characteristic 0 by Heisuke Hironaka in 1964 and is yet unsolved in finite characteristic.

23.1.6   Projective variety

Main article: Algebraic geometry of projective spaces
Just as the formulas for the roots of 2nd, 3rd and 4th degree polynomials suggest extending real numbers to the more

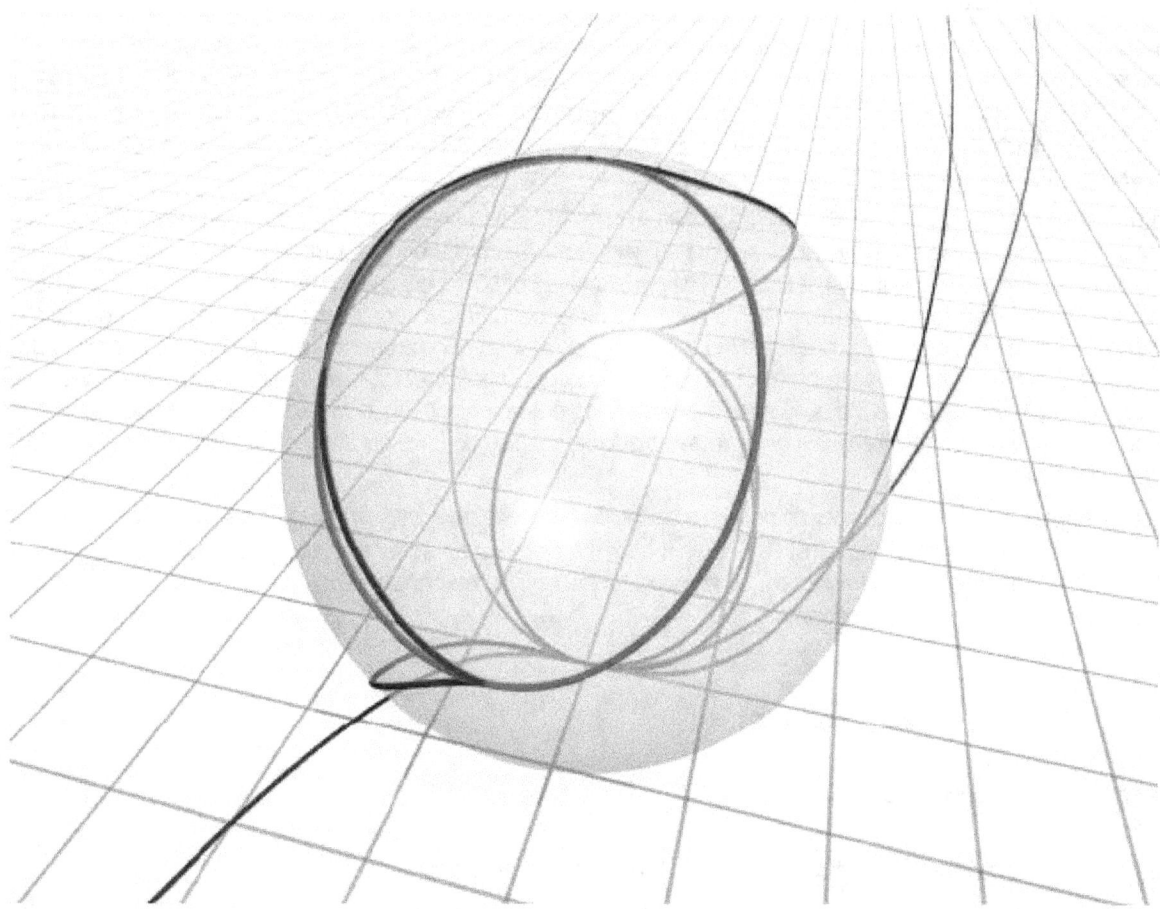

parabola ($y = x^2$, red) and cubic ($y = x^3$, blue) in projective space

algebraically complete setting of the complex numbers, many properties of algebraic varieties suggest extending affine space to a more geometrically complete projective space. Whereas the complex numbers are obtained by adding the number i, a root of the polynomial $x^2 + 1$, projective space is obtained by adding in appropriate points "at infinity", points where parallel lines may meet.

To see how this might come about, consider the variety $V(y - x^2)$. If we draw it, we get a parabola. As x goes to positive infinity, the slope of the line from the origin to the point (x, x^2) also goes to positive infinity. As x goes to negative infinity, the slope of the same line goes to negative infinity.

Compare this to the variety $V(y - x^3)$. This is a cubic curve. As x goes to positive infinity, the slope of the line from the origin to the point (x, x^3) goes to positive infinity just as before. But unlike before, as x goes to negative infinity, the slope of the same line goes to positive infinity as well; the exact opposite of the parabola. So the behavior "at infinity" of $V(y - x^3)$ is different from the behavior "at infinity" of $V(y - x^2)$.

The consideration of the *projective completion* of the two curves, which is their prolongation "at infinity" in the projective plane, allows to quantify this difference: the point at infinity of the parabola is a regular point, whose tangent is the line at infinity, while the point at infinity of the cubic curve is a cusp. Also, both curves are rational, as they are parameterized by x, and Riemann-Roch theorem implies that the cubic curve must have a singularity, which must be at infinity, as all its points in the affine space are regular.

Thus many of the properties of algebraic varieties, including birational equivalence and all the topological properties, depends on the behavior "at infinity" and so it is natural to study the varieties in projective space. Furthermore, the introduction of projective techniques made many theorems in algebraic geometry simpler and sharper: For example, Bézout's theorem on the number of intersection points between two varieties can be stated in its sharpest form only in projective space. For these reasons, projective space plays a fundamental role in algebraic geometry.

Nowadays, the *projective space* P^n of dimension n is usually defined as the set of the lines passing through a point, considered as the origin, in the affine space of dimension $n+1$, or equivalently to the set of the vector lines in a vector space of dimension $n+1$. When a coordinate system has been chosen in the space of dimension $n+1$, all the points of a line have the same set of coordinates, up to the multiplication by an element of k. This defines the homogeneous coordinates of a point of P^n as a sequence of $n+1$ elements of the base field k, defined up to the multiplication by a nonzero element of k (the same for the whole sequence).

Given a polynomial in $n+1$ variables, it vanishes at all the point of a line passing through the origin if and only if it is homogeneous. In this case, one says that the polynomial *vanishes* at the corresponding point of P^n. This allows to define a *projective algebraic set* in P^n as the set $V(f_1, ..., f_k)$ where a finite set of homogeneous polynomials $\{f_1, ..., f_k\}$ vanishes. Like for affine algebraic sets, there is a bijection between the projective algebraic sets and the reduced homogeneous ideals which define them. The *projective varieties* are the projective algebraic sets whose defining ideal is prime. In other words, a projective variety is a projective algebraic set, whose homogeneous coordinate ring is an integral domain, the *projective coordinates ring* being defined as the quotient of the graded ring or the polynomials in $n+1$ variables by the homogeneous (reduced) ideal defining the variety. Every projective algebraic set may be uniquely decomposed into a finite union of projective varieties.

The only regular functions which may be defined properly on a projective variety are the constant functions. Thus this notion is not used in projective situations. On the other hand the *field of the rational functions* or *function field* is a useful notion, which, similarly as in the affine case, is defined as the set of the quotients of two homogeneous elements of the same degree in the homogeneous coordinate ring.

23.2   Real algebraic geometry

Main article: Real algebraic geometry

The real algebraic geometry is the study of the real points of the algebraic geometry.

The fact that the field of the reals number is an ordered field may not be occulted in such a study. For example, the curve of equation $x^2 + y^2 - a = 0$ is a circle if $a > 0$, but does not have any real point if $a < 0$. It follows that real algebraic geometry is not only the study of the real algebraic varieties, but has been generalized to the study of the *semi-algebraic sets*, which are the solutions of systems of polynomial equations and polynomial inequalities. For example, a branch of the hyperbola of equation $xy - 1 = 0$ is not an algebraic variety, but is a semi-algebraic set defined by $xy - 1 = 0$ and

$x > 0$ or by $xy - 1 = 0$ and $x + y > 0$.

One of the challenging problems of real algebraic geometry is the unsolved Hilbert's sixteenth problem: Decide which respective positions are possible for the ovals of a nonsingular plane curve of degree 8.

23.3 Computational algebraic geometry

One may date the origin of computational algebraic geometry to meeting EUROSAM'79 (International Symposium on Symbolic and Algebraic Manipulation) held at Marseille, France in June 1979. At this meeting,

- Dennis S. Arnon showed that George E. Collins's Cylindrical algebraic decomposition (CAD) allows the computation of the topology of semi-algebraic sets,

- Bruno Buchberger presented the Gröbner bases and his algorithm to compute them,

- Daniel Lazard presented a new algorithm for solving systems of homogeneous polynomial equations with a computational complexity which is essentially polynomial in the expected number of solutions and thus simply exponential in the number of the unknowns. This algorithm is strongly related with Macaulay's multivariate resultant.

Since then, most results in this area are related to one or several of these items either by using or improving one of these algorithms, or by finding algorithms whose complexity is simply exponential in the number of the variables.

23.3.1 Gröbner basis

Main article: Gröbner basis

A Gröbner basis is a system of generators of a polynomial ideal whose computation allows the deduction of many properties of the affine algebraic variety defined by the ideal.

Given an ideal I defining an algebraic set V:

- V is empty (over an algebraically closed extension of the basis field), if and only if the Gröbner basis for any monomial ordering is reduced to $\{1\}$.

- By mean of the Hilbert series one may compute the dimension and the degree of V from any Gröbner basis of I for a monomial ordering refining the total degree.

- If the dimension of V is 0, one may compute the points (finite in number) of V from any Gröbner basis of I (see systems of polynomial equations).

- A Gröbner basis computation allows to remove from V all irreducible components which are contained in a given hyper surface.

- A Gröbner basis computation allows to compute the Zariski closure of the image of V by the projection on the k first coordinates, and the subset of the image where the projection is not proper.

- More generally Gröbner basis computations allows to compute the Zariski closure of the image and the critical points of a rational function of V into another affine variety.

Gröbner basis computations do not allow to compute directly the primary decomposition of I nor the prime ideals defining the irreducible components of V, but most algorithms for this involve Gröbner basis computation. The algorithms which are not based on Gröbner bases use regular chains but may need Gröbner bases in some exceptional situations.

Gröbner base are deemed to be difficult to compute. In fact they may contain, in the worst case, polynomials whose degree is doubly exponential in the number of variables and a number of polynomials which is also doubly exponential.

However, this is only a worst case complexity, and the complexity bound of Lazard's algorithm of 1979 may frequently apply. Faugère's F4 and F5 algorithms realize this complexity, as F5 algorithm may be viewed as an improvement of Lazard's 1979 algorithm. It follows that the best implementations allow to compute almost routinely with algebraic sets of degree more than 100. This means that, presently, the difficulty of computing a Gröbner basis is strongly related to the intrinsic difficulty of the problem.

23.3.2 Cylindrical Algebraic Decomposition (CAD)

CAD is an algorithm which had been introduced in 1973 by G. Collins to implement with an acceptable complexity the Tarski–Seidenberg theorem on quantifier elimination over the real numbers.

This theorem concerns the formulas of the first-order logic whose atomic formulas are polynomial equalities or inequalities between polynomials with real coefficients. These formulas are thus the formulas which may be constructed from the atomic formulas by the logical operators *and* ($\land$), *or* ($\lor$), *not* ($\neg$), *for all* ($\forall$) and *exists* ($\exists$). Tarski's theorem asserts that, from such a formula, one may compute an equivalent formula without quantifier ($\forall$, $\exists$).

The complexity of CAD is doubly exponential in the number of variables. This means that CAD allow, in theory, to solve every problem of real algebraic geometry which may be expressed by such a formula, that is almost every problem concerning explicitly given varieties and semi-algebraic sets.

While Gröbner basis computation has doubly exponential complexity only in rare cases, CAD has almost always this high complexity. This implies that, unless if most polynomials appearing in the input are linear, it may not solve problems with more than four variables.

Since 1973, most of the research on this subject is devoted either to improve CAD or to find alternate algorithms in special cases of general interest.

As an example of the state of art, there are efficient algorithms to find at least a point in every connected component of a semi-algebraic set, and thus to test if a semi-algebraic set is empty. On the other hand CAD is yet, in practice, the best algorithm to count the number of connected components.

23.3.3 Asymptotic complexity vs. practical efficiency

The basic general algorithms of computational geometry have a double exponential worst case complexity. More precisely, if d is the maximal degree of the input polynomials and n the number of variables, their complexity is at most $d^{2^{cn}}$ for some constant c, and, for some inputs, the complexity is at least $d^{2^{c'n}}$ for another constant c'.

During the last 20 years of 20th century, various algorithms have been introduced to solve specific subproblems with a better complexity. Most of these algorithms have a complexity $d^{O(n)^2}$.

Among these algorithms which solve a sub problem of the problems solved by Gröbner bases, one may cite *testing if an affine variety is empty* and *solving nonhomogeneous polynomial systems which have a finite number of solutions*. Such algorithms are rarely implemented because, on most entries Faugère's F4 and F5 algorithms have a better practical efficiency and probably a similar or better complexity (*probably* because the evaluation of the complexity of Gröbner basis algorithms on a particular class of entries is a difficult task which has be done only in few special cases).

The main algorithms of real algebraic geometry which solve a problem solved by CAD are related to the topology of semi-algebraic sets. One may cite *counting the number of connected components, testing if two points are in the same components* or *computing a Whitney stratification of a real algebraic set*. They have a complexity of $d^{O(n)^2}$, but the constant involved by O notation is so high that using them to solve any nontrivial problem effectively solved by CAD, is impossible even if one could use all the existing computing power in the world. Therefore these algorithms have never been implemented and this is an active research area to search for algorithms with have together a good asymptotic complexity and a good practical efficiency.

23.4   Abstract modern viewpoint

The modern approaches to algebraic geometry redefine and effectively extend the range of basic objects in various levels of generality to schemes, formal schemes, ind-schemes, algebraic spaces, algebraic stacks and so on. The need for this arises already from the useful ideas within theory of varieties, e.g. the formal functions of Zariski can be accommodated by introducing nilpotent elements in structure rings; considering spaces of loops and arcs, constructing quotients by group actions and developing formal grounds for natural intersection theory and deformation theory lead to some of the further extensions.

Most remarkably, in late 1950s, algebraic varieties were subsumed into Alexander Grothendieck's concept of a scheme. Their local objects are affine schemes or prime spectra which are locally ringed spaces which form a category which is antiequivalent to the category of commutative unital rings, extending the duality between the category of affine algebraic varieties over a field k, and the category of finitely generated reduced k-algebras. The gluing is along Zariski topology; one can glue within the category of locally ringed spaces, but also, using the Yoneda embedding, within the more abstract category of presheaves of sets over the category of affine schemes. The Zariski topology in the set theoretic sense is then replaced by a Grothendieck topology. Grothendieck introduced Grothendieck topologies having in mind more exotic but geometrically finer and more sensitive examples than the crude Zariski topology, namely the étale topology, and the two flat Grothendieck topologies: fppf and fpqc; nowadays some other examples became prominent including Nisnevich topology. Sheaves can be furthermore generalized to stacks in the sense of Grothendieck, usually with some additional representability conditions leading to Artin stacks and, even finer, Deligne-Mumford stacks, both often called algebraic stacks.

Sometimes other algebraic sites replace the category of affine schemes. For example, Nikolai Durov has introduced commutative algebraic monads as a generalization of local objects in a generalized algebraic geometry. Versions of a tropical geometry, of an absolute geometry over a field of one element and an algebraic analogue of Arakelov's geometry were realized in this setup.

Another formal generalization is possible to Universal algebraic geometry in which every variety of algebras has its own algebraic geometry. The term *variety of algebras* should not be confused with *algebraic variety*.

The language of schemes, stacks and generalizations has proved to be a valuable way of dealing with geometric concepts and became cornerstones of modern algebraic geometry.

Algebraic stacks can be further generalized and for many practical questions like deformation theory and intersection theory, this is often the most natural approach. One can extend the Grothendieck site of affine schemes to a higher categorical site of derived affine schemes, by replacing the commutative rings with an infinity category of differential graded commutative algebras, or of simplicial commutative rings or a similar category with an appropriate variant of a Grothendieck topology. One can also replace presheaves of sets by presheaves of simplicial sets (or of infinity groupoids). Then, in presence of an appropriate homotopic machinery one can develop a notion of derived stack as such a presheaf on the infinity category of derived affine schemes, which is satisfying certain infinite categorical version of a sheaf axiom (and to be algebraic, inductively a sequence of representability conditions). Quillen model categories, Segal categories and quasicategories are some of the most often used tools to formalize this yielding the *derived algebraic geometry*, introduced by the school of Carlos Simpson, including Andre Hirschowitz, Bertrand Toën, Gabrielle Vezzosi, Michel Vaquié and others; and developed further by Jacob Lurie, Bertrand Toën, and Gabrielle Vezzosi. Another (noncommutative) version of derived algebraic geometry, using A-infinity categories has been developed from early 1990-s by Maxim Kontsevich and followers.

23.5   History

23.5.1   Prehistory: before the 16th century

Some of the roots of algebraic geometry date back to the work of the Hellenistic Greeks from the 5th century BC. The Delian problem, for instance, was to construct a length x so that the cube of side x contained the same volume as the rectangular box $a^2 b$ for given sides a and b. Menaechmus (circa 350 BC) considered the problem geometrically by intersecting the pair of plane conics $ay = x^2$ and $xy = ab$.[1] The later work, in the 3rd century BC, of Archimedes and Apollonius studied more systematically problems on conic sections,[2] and also involved the use of coordinates.[1] The Arab mathematicians were able to solve by purely algebraic means certain cubic equations, and then to interpret the results geometrically. This was done, for instance, by Ibn al-Haytham in the 10th century AD.[3] Subsequently, Persian mathematician Omar Khayyám (born 1048 A.D.) discovered the general method of solving cubic equations by intersecting a parabola with a circle.[4] Each of these early developments in algebraic geometry dealt with questions of finding and describing the intersections of algebraic curves.

23.5.2   Renaissance

Such techniques of applying geometrical constructions to algebraic problems were also adopted by a number of Renaissance mathematicians such as Gerolamo Cardano and Niccolò Fontana "Tartaglia" on their studies of the cubic equation. The geometrical approach to construction problems, rather than the algebraic one, was favored by most 16th and 17th century mathematicians, notably Blaise Pascal who argued against the use of algebraic and analytical methods in geometry.[5] The French mathematicians Franciscus Vieta and later René Descartes and Pierre de Fermat revolutionized the conventional way of thinking about construction problems through the introduction of coordinate geometry. They were interested primarily in the properties of *algebraic curves*, such as those defined by Diophantine equations (in the case of Fermat), and the algebraic reformulation of the classical Greek works on conics and cubics (in the case of Descartes).

During the same period, Blaise Pascal and Gérard Desargues approached geometry from a different perspective, developing the synthetic notions of projective geometry. Pascal and Desargues also studied curves, but from the purely geometrical point of view: the analog of the Greek *ruler and compass construction*. Ultimately, the analytic geometry of Descartes and Fermat won out, for it supplied the 18th century mathematicians with concrete quantitative tools needed to study physical problems using the new calculus of Newton and Leibniz. However, by the end of the 18th century, most of the algebraic character of coordinate geometry was subsumed by the *calculus of infinitesimals* of Lagrange and Euler.

23.5.3   19th and early 20th century

It took the simultaneous 19th century developments of non-Euclidean geometry and Abelian integrals in order to bring the old algebraic ideas back into the geometrical fold. The first of these new developments was seized up by Edmond Laguerre and Arthur Cayley, who attempted to ascertain the generalized metric properties of projective space. Cayley introduced the idea of *homogeneous polynomial forms*, and more specifically quadratic forms, on projective space. Subsequently, Felix Klein studied projective geometry (along with other types of geometry) from the viewpoint that the geometry on a space is encoded in a certain class of transformations on the space. By the end of the 19th century, projective geometers were studying more general kinds of transformations on figures in projective space. Rather than the projective linear transformations which were normally regarded as giving the fundamental Kleinian geometry on projective space, they concerned themselves also with the higher degree birational transformations. This weaker notion of congruence would later lead members of the 20th century Italian school of algebraic geometry to classify algebraic surfaces up to birational isomorphism.

The second early 19th century development, that of Abelian integrals, would lead Bernhard Riemann to the development of Riemann surfaces.

In the same period began the algebraization of the algebraic geometry through commutative algebra. The prominent results in this direction are Hilbert's basis theorem and Hilbert's Nullstellensatz, which are the basis of the connexion between algebraic geometry and commutative algebra, and Macaulay's multivariate resultant, which is the basis of elimination theory. Probably because of the size of the computation which is implied by multivariate resultants, elimination theory was forgotten during the middle of the 20th century until it was renewed by singularity theory and computational algebraic geometry.[6]

23.5.4   20th century

B. L. van der Waerden, Oscar Zariski and André Weil developed a foundation for algebraic geometry based on contemporary commutative algebra, including valuation theory and the theory of ideals. One of the goals was to give a rigorous framework for proving the results of Italian school of algebraic geometry. In particular, this school used systematically the notion of generic point without any precise definition, which was first given by these authors during the 1930s.

In the 1950s and 1960s Jean-Pierre Serre and Alexander Grothendieck recast the foundations making use of sheaf theory. Later, from about 1960, and largely lead by Grothendieck, the idea of schemes was worked out, in conjunction with a very refined apparatus of homological techniques. After a decade of rapid development the field stabilized in the 1970s, and new applications were made, both to number theory and to more classical geometric questions on algebraic varieties, singularities and moduli.

An important class of varieties, not easily understood directly from their defining equations, are the abelian varieties, which are the projective varieties whose points form an abelian group. The prototypical examples are the elliptic curves, which have a rich theory. They were instrumental in the proof of Fermat's last theorem and are also used in elliptic curve cryptography.

In parallel with the abstract trend of the algebraic geometry, which is concerned with general statements about varieties, methods for effective computation with concretely-given varieties have also been developed, which lead to the new area of computational algebraic geometry. One of the founding methods of this area is the theory of Gröbner bases, introduced by Bruno Buchberger in 1965. Another founding method, more specially devoted to real algebraic geometry, is the cylindrical algebraic decomposition, introduced by George E. Collins in 1973.

23.6 Analytic geometry

An **analytic variety** is defined locally as the set of common solutions of several equations involving analytic functions. It is analogous to the included concept of real or complex algebraic variety. Any complex manifold is an analytic variety. Since analytic varieties may have singular points, not all analytic varieties are manifolds.

Modern analytic geometry is essentially equivalent to real and complex algebraic geometry, as has been shown by Jean-Pierre Serre in his paper *GAGA*, the name of which is French for *Algebraic geometry and analytic geometry*. Nevertheless, the two fields remain distinct, as the methods of proof are quite different and algebraic geometry includes also geometry in finite characteristic.

23.7 Applications

Algebraic geometry now finds applications in statistics,[7] control theory,[8][9] robotics,[10] error-correcting codes,[11] phylogenetics[12] and geometric modelling.[13] There are also connections to string theory,[14] game theory,[15] graph matchings,[16] solitons[17] and integer programming.[18]

23.8 See also

- Algebraic statistics
- Differential geometry
- Geometric algebra
- Glossary of classical algebraic geometry
- Intersection theory
- Important publications in algebraic geometry
- List of algebraic surfaces
- Noncommutative algebraic geometry

- Differential algebraic geometry

- Real algebraic geometry

23.9   Notes

[1] Dieudonné, Jean (1972). "The historical development of algebraic geometry". *The American Mathematical Monthly* 79 (8): 827–866. doi:10.2307/2317664. JSTOR 2317664.

[2] Kline, M. (1972) *Mathematical Thought from Ancient to Modern Times* (Volume 1). Oxford University Press. pp. 108, 90.

[3] Kline, M. (1972) *Mathematical Thought from Ancient to Modern Times* (Volume 1). Oxford University Press. p. 193.

[4] Kline, M. (1972) *Mathematical Thought from Ancient to Modern Times* (Volume 1). Oxford University Press. pp. 193–195.

[5] Kline, M. (1972) *Mathematical Thought from Ancient to Modern Times* (Volume 1). Oxford University Press. p. 279.

[6] A witness of this oblivion is the fact that Van der Waerden removed the chapter on elimination theory from the third edition (and all the subsequent ones) of his treatise *Moderne algebra* (in German).

[7] Drton, Mathias; Sturmfels, Bernd; Sullivant, Seth (2009). *Lectures on Algebraic Statistics.* Springer. ISBN 978-3-7643-8904-8.

[8] Falb, Peter (1990). *Methods of Algebraic Geometry in Control Theory Part II Multivariable Linear Systems and Projective Algebraic Geometry.* Springer. ISBN 978-0-8176-4113-9.

[9] Allen Tannenbaum (1982), Invariance and Systems Theory: Algebraic and Geometric Aspects, Lecture Notes in Mathematics, volume 845, Springer-Verlag, ISBN 9783540105657

[10] Selig, J.M. (2005). *Geometric Fundamentals of Robotics.* Springer. ISBN 978-0-387-20874-9.

[11] Tsfasman, Michael A.; Vlăduţ, Serge G.; Nogin, Dmitry (1990). *Algebraic Geometric Codes Basic Notions.* American Mathematical Soc. ISBN 978-0-8218-7520-9.

[12] Barry A. Cipra (2007), Algebraic Geometers See Ideal Approach to Biology, SIAM News, Volume 40, Number 6

[13] Jüttler, Bert; Piene, Ragni (2007). *Geometric Modeling and Algebraic Geometry.* Springer. ISBN 978-3-540-72185-7.

[14] Cox, David A.; Katz, Sheldon (1999). *Mirror Symmetry and Algebraic Geometry.* American Mathematical Soc. ISBN 978-0-8218-2127-5.

[15] Blume, L. E.; Zame, W. R. (1994). "The algebraic geometry of perfect and sequential equilibrium" (PDF). *Econometrica* 62 (4): 783–794. JSTOR 2951732.

[16] Kenyon, Richard; Okounkov, Andrei; Sheffield, Scott (2003). "Dimers and Amoebae". arXiv:math-ph/0311005 [math-ph].

[17] Fordy, Allan P. (1990). *Soliton Theory A Survey of Results.* Manchester University Press. ISBN 978-0-7190-1491-8.

[18] Cox, David A.; Sturmfels, Bernd. Manocha, Dinesh N., ed. *Applications of Computational Algebraic Geometry.* American Mathematical Soc. ISBN 978-0-8218-6758-7.

23.10   Further reading

Some classic textbooks that predate schemes

- van der Waerden, B. L. (1945). *Einfuehrung in die algebraische Geometrie.* Dover.

- Hodge, W. V. D.; Pedoe, Daniel (1994). *Methods of Algebraic Geometry Volume 1.* Cambridge University Press. ISBN 0-521-46900-7. Zbl 0796.14001.

- Hodge, W. V. D.; Pedoe, Daniel (1994). *Methods of Algebraic Geometry Volume 2.* Cambridge University Press. ISBN 0-521-46901-5. Zbl 0796.14002.

- Hodge, W. V. D.; Pedoe, Daniel (1994). *Methods of Algebraic Geometry Volume 3*. Cambridge University Press. ISBN 0-521-46775-6. Zbl 0796.14003.

Modern textbooks that do not use the language of schemes

- Garrity, Thomas et al. (2013). *Algebraic Geometry A Problem Solving Approach*. American Mathematical Society. ISBN 0-821-89396-3.

- Griffiths, Phillip; Harris, Joe (1994). *Principles of Algebraic Geometry*. Wiley-Interscience. ISBN 0-471-05059-8. Zbl 0836.14001.

- Harris, Joe (1995). *Algebraic Geometry A First Course*. Springer-Verlag. ISBN 0-387-97716-3. Zbl 0779.14001.

- Mumford, David (1995). *Algebraic Geometry I Complex Projective Varieties* (2nd ed.). Springer-Verlag. ISBN 3-540-58657-1. Zbl 0821.14001.

- Reid, Miles (1988). *Undergraduate Algebraic Geometry*. Cambridge University Press. ISBN 0-521-35662-8. Zbl 0701.14001.

- Shafarevich, Igor (1995). *Basic Algebraic Geometry I Varieties in Projective Space* (2nd ed.). Springer-Verlag. ISBN 0-387-54812-2. Zbl 0797.14001.

Textbooks in computational algebraic geometry

- Cox, David A.; Little, John; O'Shea, Donal (1997). *Ideals, Varieties, and Algorithms* (2nd ed.). Springer-Verlag. ISBN 0-387-94680-2. Zbl 0861.13012.

- Basu, Saugata; Pollack, Richard; Roy, Marie-Françoise (2006). *Algorithms in real algebraic geometry*. Springer-Verlag.

- González-Vega, Laureano; Recio, Tómas (1996). *Algorithms in algebraic geometry and applications*. Birkhaüser.

- Elkadi, Mohamed; Mourrain, Bernard; Piene, Ragni, eds. (2006). *Algebraic geometry and geometric modeling*. Springer-Verlag.

- Dickenstein, Alicia; Schreyer, Frank-Olaf; Sommese, Andrew J., eds. (2008). *Algorithms in Algebraic Geometry*. The IMA Volumes in Mathematics and its Applications 146. Springer. ISBN 9780387751559. LCCN 2007938208.

- Cox, David A.; Little, John B.; O'Shea, Donal (1998). *Using algebraic geometry*. Springer-Verlag.

- Caviness, Bob F.; Johnson, Jeremy R. (1998). *Quantifier elimination and cylindrical algebraic decomposition*. Springer-Verlag.

Textbooks and references for schemes

- Eisenbud, David; Harris, Joe (1998). *The Geometry of Schemes*. Springer-Verlag. ISBN 0-387-98637-5. Zbl 0960.14002.

- Grothendieck, Alexander (1960). *Éléments de géométrie algébrique*. Publications Mathématiques de l'IHÉS. Zbl 0118.36206.

- Grothendieck, Alexander; Dieudonné, Jean Alexandre (1971). *Éléments de géométrie algébrique* 1 (2nd ed.). Springer-Verlag. ISBN 3-540-05113-9. Zbl 0203.23301.

- Hartshorne, Robin (1977). *Algebraic Geometry*. Springer-Verlag. ISBN 0-387-90244-9. Zbl 0367.14001.

- Mumford, David (1999). *The Red Book of Varieties and Schemes Includes the Michigan Lectures on Curves and Their Jacobians* (2nd ed.). Springer-Verlag. ISBN 3-540-63293-X. Zbl 0945.14001.

- Shafarevich, Igor (1995). *Basic Algebraic Geometry II Schemes and complex manifolds* (2nd ed.). Springer-Verlag. ISBN 3-540-57554-5. Zbl 0797.14002.

23.11   External links

- *Foundations of Algebraic Geometry* by Ravi Vakil, 764 pp.

- *Algebraic geometry* entry on PlanetMath

- English translation of the van der Waerden textbook

- The History of Algebraic Geometry (1.425 Gigabyte MOV file), a 1972 talk by Jean Dieudonné at the Department of Mathematics of the University of Wisconsin-Milwaukee

- The Stacks Project, an open source textbook and reference work on algebraic stacks and algebraic geometry

Chapter 24

Topology

Not to be confused with topography.
This article is about the branch of mathematics. For other uses, see Topology (disambiguation).
 In mathematics, **topology** (from the Greek τόπος, "place", and λόγος, "study"), the study of topological spaces, is an area

Möbius strips, which have only one surface and one edge, are a kind of object studied in topology.

of mathematics concerned with the properties of space that are preserved under continuous deformations, such as stretching and bending, but not tearing or gluing. Important topological properties include connectedness and compactness.

Topology developed as a field of study out of geometry and set theory, through analysis of such concepts as space, dimension, and transformation. Such ideas go back to Gottfried Leibniz, who in the 17th century envisioned the *geometria situs* (Greek-Latin for "geometry of place") and *analysis situs* (Greek-Latin for "picking apart of place"). Leonhard Euler's Seven Bridges of Königsberg Problem and Polyhedron Formula are arguably the field's first theorems. The term *topology* was introduced by Johann Benedict Listing in the 19th century, although it was not until the first decades of the 20th century that the idea of a topological space was developed. By the middle of the 20th century, topology had become a major branch of mathematics.

Topology has many subfields:

- **General topology** establishes the foundational aspects of topology and investigates properties of topological spaces and investigates concepts inherent to topological spaces. It includes point-set topology, which is the foundational topology used in all other branches (including topics like compactness and connectedness).

- **Algebraic topology** tries to measure degrees of connectivity using algebraic constructs such as homology and homotopy groups.

- **Differential topology** is the field dealing with differentiable functions on differentiable manifolds. It is closely related to differential geometry and together they make up the geometric theory of differentiable manifolds.

- **Geometric topology** primarily studies manifolds and their embeddings (placements) in other manifolds. A particularly active area is **low dimensional topology**, which studies manifolds of four or fewer dimensions. This includes **knot theory**, the study of mathematical knots.

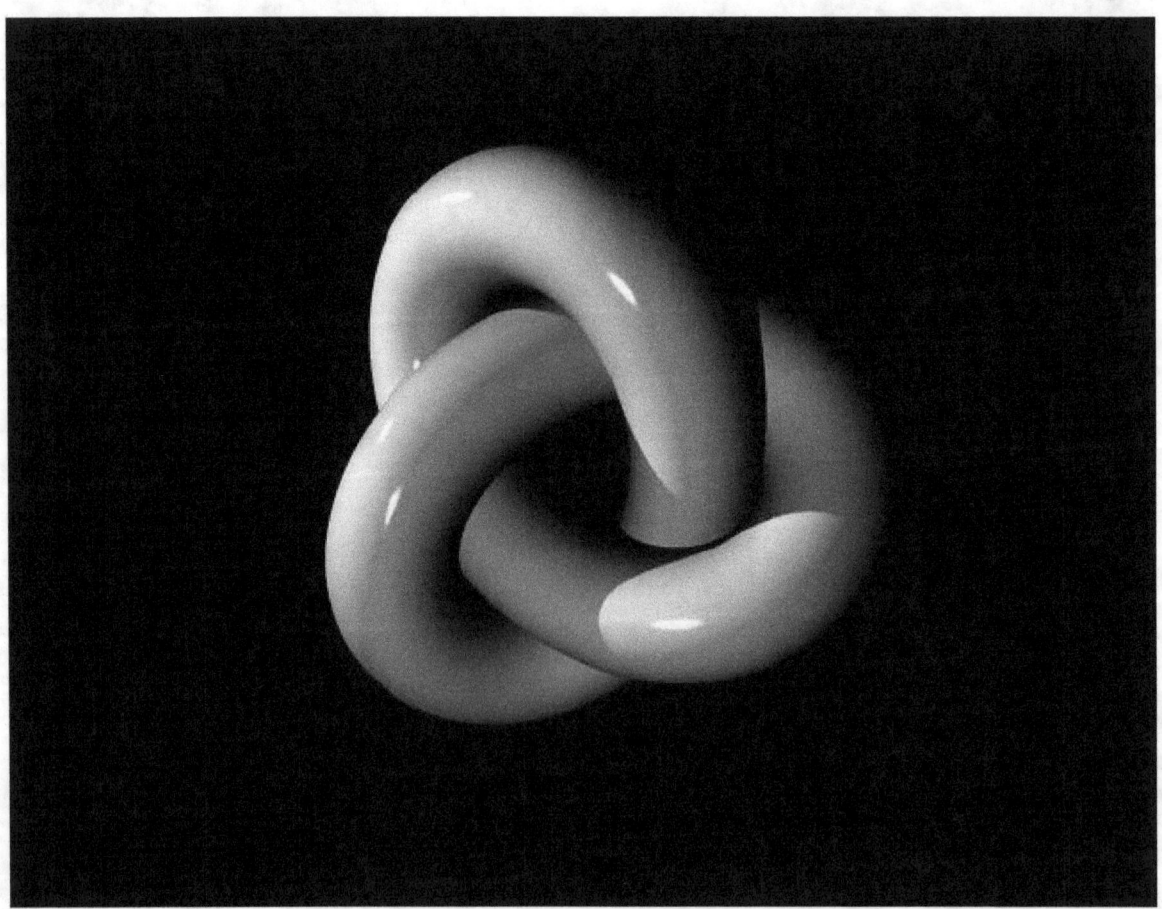

A three-dimensional depiction of a thickened trefoil knot, the simplest non-trivial knot

See also: topology glossary for definitions of some of the terms used in topology, and topological space for a more technical treatment of the subject.

The Seven Bridges of Königsberg was a problem solved by Euler.

24.1   History

Topology began with the investigation of certain questions in geometry. Leonhard Euler's 1736 paper on the Seven Bridges of Königsberg[1] is regarded as one of the first academic treatises in modern topology.

The term "Topologie" was introduced in German in 1847 by Johann Benedict Listing in *Vorstudien zur Topologie*,[2] who had used the word for ten years in correspondence before its first appearance in print. The English form topology was first used in 1883 in Listing's obituary in the journal *Nature*[3] to distinguish "qualitative geometry from the ordinary geometry in which quantitative relations chiefly are treated". The term **topologist** in the sense of a specialist in topology was used in 1905 in the magazine *Spectator*. However, none of these uses corresponds exactly to the modern definition of topology.

Modern topology depends strongly on the ideas of set theory, developed by Georg Cantor in the later part of the 19th century. In addition to establishing the basic ideas of set theory, Cantor considered point sets in Euclidean space as part of his study of Fourier series.

Henri Poincaré published *Analysis Situs* in 1895,[4] introducing the concepts of homotopy and homology, which are now considered part of algebraic topology.

Unifying the work on function spaces of Georg Cantor, Vito Volterra, Cesare Arzelà, Jacques Hadamard, Giulio Ascoli and others, Maurice Fréchet introduced the metric space in 1906.[5] A metric space is now considered a special case of a general topological space. In 1914, Felix Hausdorff coined the term "topological space" and gave the definition for what is now called a Hausdorff space.[6] Currently, a topological space is a slight generalization of Hausdorff spaces, given in 1922 by Kazimierz Kuratowski.

For further developments, see point-set topology and algebraic topology.

24.2   Introduction

Topology can be formally defined as "the study of qualitative properties of certain objects (called topological spaces) that are invariant under a certain kind of transformation (called a continuous map), especially those properties that are invariant under a certain kind of transformation (called homeomorphism)."

Topology is also used to refer to a structure imposed upon a set X, a structure that essentially 'characterizes' the set X as a topological space by taking proper care of properties such as convergence, connectedness and continuity, upon transformation.

Topological spaces show up naturally in almost every branch of mathematics. This has made topology one of the great unifying ideas of mathematics.

The motivating insight behind topology is that some geometric problems depend not on the exact shape of the objects involved, but rather on the way they are put together. For example, the square and the circle have many properties in common: they are both one dimensional objects (from a topological point of view) and both separate the plane into two parts, the part inside and the part outside.

One of the first papers in topology was the demonstration, by Leonhard Euler, that it was impossible to find a route through the town of Königsberg (now Kaliningrad) that would cross each of its seven bridges exactly once. This result did not depend on the lengths of the bridges, nor on their distance from one another, but only on connectivity properties: which bridges are connected to which islands or riverbanks. This problem in introductory mathematics called *Seven Bridges of Königsberg* led to the branch of mathematics known as graph theory.

Similarly, the hairy ball theorem of algebraic topology says that "one cannot comb the hair flat on a hairy ball without creating a cowlick." This fact is immediately convincing to most people, even though they might not recognize the more formal statement of the theorem, that there is no nonvanishing continuous tangent vector field on the sphere. As with the *Bridges of Königsberg*, the result does not depend on the shape of the sphere; it applies to any kind of smooth blob, as long as it has no holes.

To deal with these problems that do not rely on the exact shape of the objects, one must be clear about just what properties these problems *do* rely on. From this need arises the notion of homeomorphism. The impossibility of crossing each bridge just once applies to any arrangement of bridges homeomorphic to those in Königsberg, and the hairy ball theorem applies to any space homeomorphic to a sphere.

Intuitively, two spaces are homeomorphic if one can be deformed into the other without cutting or gluing. A traditional joke is that a topologist cannot distinguish a coffee mug from a doughnut, since a sufficiently pliable doughnut could be reshaped to a coffee cup by creating a dimple and progressively enlarging it, while shrinking the hole into a handle.

Homeomorphism can be considered the most basic *topological equivalence*. Another is homotopy equivalence. This is harder to describe without getting technical, but the essential notion is that two objects are homotopy equivalent if they both result from "squishing" some larger object.

An introductory exercise is to classify the uppercase letters of the English alphabet according to homeomorphism and homotopy equivalence. The result depends partially on the font used. The figures use the sans-serif Myriad font. Homotopy equivalence is a rougher relationship than homeomorphism; a homotopy equivalence class can contain several homeomorphism classes. The simple case of homotopy equivalence described above can be used here to show two letters are homotopy equivalent. For example, O fits inside P and the tail of the P can be squished to the "hole" part.

Homeomorphism classes are:

- no holes,
- no holes three tails,
- no holes four tails,
- one hole no tail,

A continuous deformation (a type of homeomorphism) of a mug into a doughnut (torus) and back

- one hole one tail,

- one hole two tails,

- two holes no tail, and

- a bar with four tails (the "bar" on the *K* is almost too short to see).

Homotopy classes are larger, because the tails can be squished down to a point. They are:

- one hole,

- two holes, and

- no holes.

To be sure that the letters are classified correctly, we need to show that two letters in the same class are equivalent and two letters in different classes are not equivalent. In the case of homeomorphism, this can be done by selecting points and showing their removal disconnects the letters differently. For example, X and Y are not homeomorphic because removing the center point of the X leaves four pieces; whatever point in Y corresponds to this point, its removal can leave at most three pieces. The case of homotopy equivalence is harder and requires a more elaborate argument showing an algebraic invariant, such as the fundamental group, is different on the supposedly differing classes.

Letter topology has practical relevance in stencil typography. For instance, Braggadocio font stencils are made of one connected piece of material.

24.3 Concepts

24.3.1 Topologies on Sets

Main article: Topological space

The term **topology** also refers to a specific mathematical idea which is central to the area of mathematics called topology. Informally, a topology is used to tell how elements of a set are related spatially to each other. The same set can have different topologies. For instance, the real line, the complex plane, and the Cantor set can be thought of as the same set with different topologies.

Formally, let X be a set and let τ be a family of subsets of X. Then τ is called a *topology on X* if:

1. Both the empty set and X are elements of τ

2. Any union of elements of τ is an element of τ

3. Any intersection of finitely many elements of τ is an element of τ

If τ is a topology on X, then the pair (X, τ) is called a *topological space*. The notation $X\tau$ may be used to denote a set X endowed with the particular topology τ.

The members of τ are called *open sets* in X. A subset of X is said to be closed if its complement is in τ (i.e., its complement is open). A subset of X may be open, closed, both (clopen set), or neither. The empty set and X itself are always both closed and open. An open set containing a point x is called a 'neighborhood' of x.

A set with a topology is called a topological space.

24.3.2 Continuous functions and homeomorphisms

Main articles: Continuous function and homeomorphism

A function or map from one topological space to another is called *continuous* if the inverse image of any open set is open. If the function maps the real numbers to the real numbers (both spaces with the Standard Topology), then this definition of continuous is equivalent to the definition of continuous in calculus. If a continuous function is one-to-one and onto, and if the inverse of the function is also continuous, then the function is called a homeomorphism and the domain of the function is said to be homeomorphic to the range. Another way of saying this is that the function has a natural extension to the topology. If two spaces are homeomorphic, they have identical topological properties, and are considered topologically the same. The cube and the sphere are homeomorphic, as are the coffee cup and the doughnut. But the circle is not homeomorphic to the doughnut.

24.3.3 Manifolds

Main article: Manifold

While topological spaces can be extremely varied and exotic, many areas of topology focus on the more familiar class of spaces known as manifolds. A **manifold** is a topological space that resembles Euclidean space near each point. More precisely, each point of an n-dimensional manifold has a neighbourhood that is homeomorphic to the Euclidean space of dimension n. Lines and circles, but not figure eights, are one-dimensional manifolds. Two-dimensional manifolds are also called surfaces. Examples include the plane, the sphere, and the torus, which can all be realized in three dimensions, but also the Klein bottle and real projective plane which cannot.

24.4 Topics

24.4.1 General topology

Main article: General topology

General topology is the branch of topology dealing with the basic set-theoretic definitions and constructions used in topology.[7][8] It is the foundation of most other branches of topology, including differential topology, geometric topology, and algebraic topology. Another name for general topology is **point-set topology**.

The fundamental concepts in point-set topology are *continuity*, *compactness*, and *connectedness*. Intuitively, continuous functions take nearby points to nearby points; compact sets are those which can be covered by finitely many sets of arbitrarily small size; and connected sets are sets which cannot be divided into two pieces which are far apart. The words 'nearby', 'arbitrarily small', and 'far apart' can all be made precise by using open sets. If we change the definition of 'open set', we change what continuous functions, compact sets, and connected sets are. Each choice of definition for 'open set' is called a *topology*. A set with a topology is called a *topological space*.

Metric spaces are an important class of topological spaces where distances can be assigned a number called a *metric*. Having a metric simplifies many proofs, and many of the most common topological spaces are metric spaces.

24.4.2 Algebraic topology

Main article: Algebraic topology

Algebraic topology is a branch of mathematics that uses tools from abstract algebra to study topological spaces.[9] The basic goal is to find algebraic invariants that classify topological spaces up to homeomorphism, though usually most classify up to homotopy equivalence.

The most important of these invariants are homotopy groups, homology, and cohomology.

Although algebraic topology primarily uses algebra to study topological problems, using topology to solve algebraic problems is sometimes also possible. Algebraic topology, for example, allows for a convenient proof that any subgroup of a free group is again a free group.

24.4.3 Differential topology

Main article: Differential topology

Differential topology is the field dealing with differentiable functions on differentiable manifolds.[10] It is closely related to differential geometry and together they make up the geometric theory of differentiable manifolds.

More specifically, differential topology considers the properties and structures that require only a smooth structure on a manifold to be defined. Smooth manifolds are 'softer' than manifolds with extra geometric structures, which can act as obstructions to certain types of equivalences and deformations that exist in differential topology. For instance, volume and Riemannian curvature are invariants that can distinguish different geometric structures on the same smooth manifold—that is, one can smoothly "flatten out" certain manifolds, but it might require distorting the space and affecting the curvature or volume.

24.4.4   Geometric topology

Main article: Geometric topology

Geometric topology is a branch of topology that primarily focuses on low-dimensional manifolds (i.e. dimensions 2,3 and 4) and their interaction with geometry, but it also includes some higher-dimensional topology.[11] [12] Some examples of topics in geometric topology are orientability, handle decompositions, local flatness, and the planar and higher-dimensional Schönflies theorem.

In high-dimensional topology, characteristic classes are a basic invariant, and surgery theory is a key theory.

Low-dimensional topology is strongly geometric, as reflected in the uniformization theorem in 2 dimensions – every surface admits a constant curvature metric; geometrically, it has one of 3 possible geometries: positive curvature/spherical, zero curvature/flat, negative curvature/hyperbolic – and the geometrization conjecture (now theorem) in 3 dimensions – every 3-manifold can be cut into pieces, each of which has one of 8 possible geometries.

2-dimensional topology can be studied as complex geometry in one variable (Riemann surfaces are complex curves) – by the uniformization theorem every conformal class of metrics is equivalent to a unique complex one, and 4-dimensional topology can be studied from the point of view of complex geometry in two variables (complex surfaces), though not every 4-manifold admits a complex structure.

24.4.5   Generalizations

Occasionally, one needs to use the tools of topology but a "set of points" is not available. In pointless topology one considers instead the lattice of open sets as the basic notion of the theory,[13] while Grothendieck topologies are structures defined on arbitrary categories that allow the definition of sheaves on those categories, and with that the definition of general cohomology theories.[14]

24.5   Applications

24.5.1   Biology

Knot theory, a branch of topology, is used in biology to study the effects of certain enzymes on DNA. These enzymes cut, twist, and reconnect the DNA, causing knotting with observable effects such as slower electrophoresis.[15] Topology is also used in evolutionary biology to represent the relationship between phenotype and genotype.[16] Phenotypic forms which appear quite different can be separated by only a few mutations depending on how genetic changes map to phenotypic changes during development.

24.5.2   Computer science

Topological data analysis uses techniques from algebraic topology to determine the large scale structure of a set (for instance, determining if a cloud of points is spherical or toroidal). The main method used by topological data analysis is:

1. Replace a set of data points with a family of simplicial complexes, indexed by a proximity parameter.

2. Analyse these topological complexes via algebraic topology — specifically, via the theory of persistent homology.[17]

3. Encode the persistent homology of a data set in the form of a parameterized version of a Betti number which is called a barcode.[17]

24.5.3 Physics

In physics, topology is used in several areas such as quantum field theory and cosmology.

A **topological quantum field theory** (or **topological field theory** or **TQFT**) is a quantum field theory which computes topological invariants.

Although TQFTs were invented by physicists, they are also of mathematical interest, being related to, among other things, knot theory and the theory of four-manifolds in algebraic topology, and to the theory of moduli spaces in algebraic geometry. Donaldson, Jones, Witten, and Kontsevich have all won Fields Medals for work related to topological field theory.

In cosmology, topology can be used to describe the overall shape of the universe.[18] This area is known as spacetime topology.

24.5.4 Robotics

The various possible positions of a robot can be described by a manifold called configuration space.[19] In the area of motion planning, one finds paths between two points in configuration space. These paths represent a motion of the robot's joints and other parts into the desired location and pose.

24.6 See also

- Equivariant topology
- General topology
- List of algebraic topology topics
- List of examples in general topology
- List of general topology topics
- List of geometric topology topics
- List of topology topics
- Publications in topology
- Topology glossary

24.7 References

[1] Euler, Leonhard, Solutio problematis ad geometriam situs pertinentis

[2] Listing, Johann Benedict, "Vorstudien zur Topologie", Vandenhoeck und Ruprecht, Göttingen, p. 67, 1848

[3] Tait, Peter Guthrie, "Johann Benedict Listing (obituary)", Nature *27*, 1 February 1883, pp. 316–317

[4] Poincaré, Henri, "Analysis situs", Journal de l'École Polytechnique ser 2, 1 (1895) pp. 1–123

[5] Fréchet, Maurice, "Sur quelques points du calcul fonctionnel", PhD dissertation, 1906

[6] Hausdorff, Felix, "Grundzüge der Mengenlehre", Leipzig: Veit. In (Hausdorff Werke, II (2002), 91–576)

[7] Munkres, James R. Topology. Vol. 2. Upper Saddle River: Prentice Hall, 2000.

[8] Adams, Colin Conrad, and Robert David Franzosa. Introduction to topology: pure and applied. Pearson Prentice Hall, 2008.

[9] Allen Hatcher, *Algebraic topology.* (2002) Cambridge University Press, xii+544 pp. ISBN 0-521-79160-X and ISBN 0-521-79540-0.

[10] Lee, John M. (2006). *Introduction to Smooth Manifolds.* Springer-Verlag. ISBN 978-0-387-95448-6.

[11] Budney, Ryan (2011). "What is geometric topology?". *mathoverflow.net.* Retrieved 29 December 2013.

[12] R.B. Sher and R.J. Daverman (2002), *Handbook of Geometric Topology,* North-Holland. ISBN 0-444-82432-4

[13] Johnstone, Peter T., 1983, "The point of pointless topology," *Bulletin of the American Mathematical Society 8(1):* 41-53.

[14] Artin, Michael (1962). *Grothendieck topologies.* Cambridge, MA: Harvard University, Dept. of Mathematics. Zbl 0208.48701.

[15] Adams, Colin (2004). *The Knot Book: An Elementary Introduction to the Mathematical Theory of Knots.* American Mathematical Society. ISBN 0-8218-3678-1

[16] Barble M R Stadler et al. "The Topology of the Possible: Formal Spaces Underlying Patterns of Evolutionary Change". *Journal of Theoretical Biology* 213: 241–274. doi:10.1006/jtbi.2001.2423.

[17] Gunnar Carlsson (April 2009). "Topology and data" (PDF). *BULLETIN (New Series) OF THE AMERICAN MATHEMATICAL SOCIETY* 46 (2): 255–308. doi:10.1090/S0273-0979-09-01249-X.

[18] *The Shape of Space: How to Visualize Surfaces and Three-dimensional Manifolds* 2nd ed (Marcel Dekker, 1985, ISBN 0-8247-7437-X)

[19] John J. Craig, **Introduction to Robotics: Mechanics and Control,** 3rd Ed. Prentice-Hall, 2004

24.8   Further reading

- Ryszard Engelking, *General Topology,* Heldermann Verlag, Sigma Series in Pure Mathematics, December 1989, ISBN 3-88538-006-4.

- Bourbaki; *Elements of Mathematics: General Topology,* Addison–Wesley (1966).

- Breitenberger, E. (2006). "Johann Benedict Listing". In James, I. M. *History of Topology.* North Holland. ISBN 978-0-444-82375-5.

- Kelley, John L. (1975). *General Topology.* Springer-Verlag. ISBN 0-387-90125-6.

- Brown, Ronald (2006). *Topology and Groupoids.* Booksurge. ISBN 1-4196-2722-8. (Provides a well motivated, geometric account of general topology, and shows the use of groupoids in discussing van Kampen's theorem, covering spaces, and orbit spaces.)

- Wacław Sierpiński, *General Topology,* Dover Publications, 2000, ISBN 0-486-41148-6

- Pickover, Clifford A. (2006). *The Möbius Strip: Dr. August Möbius's Marvelous Band in Mathematics, Games, Literature, Art, Technology, and Cosmology.* Thunder's Mouth Press. ISBN 1-56025-826-8. (Provides a popular introduction to topology and geometry)

- Gemignani, Michael C. (1990) [1967], *Elementary Topology* (2nd ed.), Dover Publications Inc., ISBN 0-486-66522-4

24.9 External links

- Hazewinkel, Michiel, ed. (2001), "Topology, general", *Encyclopedia of Mathematics*, Springer, ISBN 978-1-55608-010-4

- Elementary Topology: A First Course Viro, Ivanov, Netsvetaev, Kharlamov.

- Topology at DMOZ

- The Topological Zoo at The Geometry Center.

- Topology Atlas

- Topology Course Lecture Notes Aisling McCluskey and Brian McMaster, Topology Atlas.

- Topology Glossary

- Moscow 1935: Topology moving towards America, a historical essay by Hassler Whitney.

Chapter 25

General topology

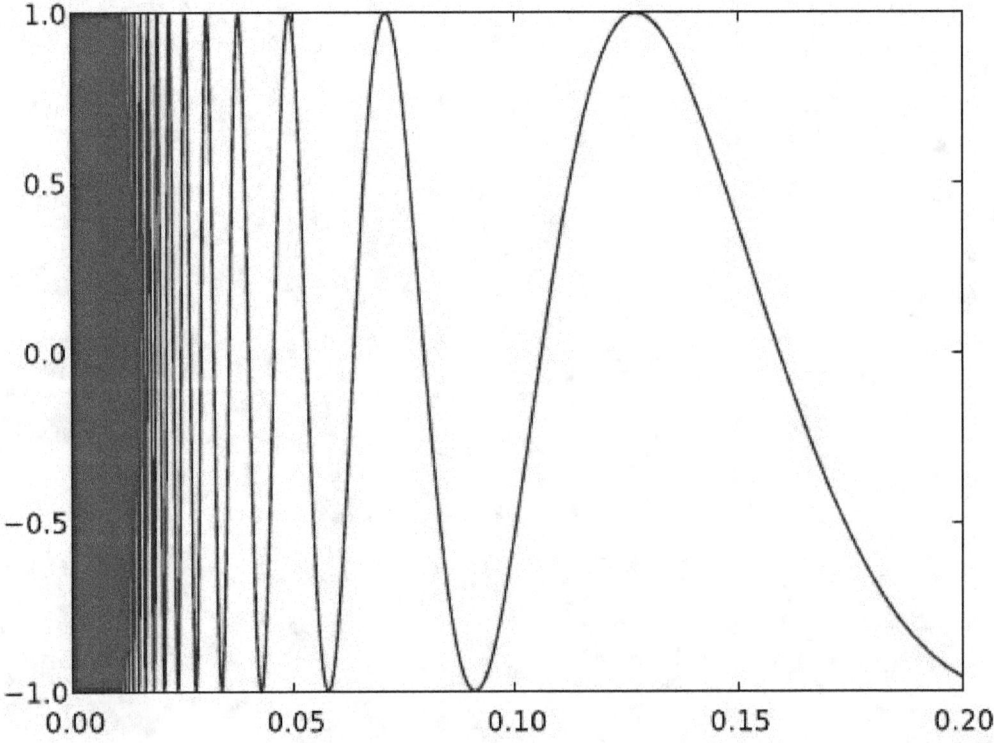

The Topologist's sine curve, a useful example in point-set topology. It is connected but not path-connected.

In mathematics, **general topology** is the branch of topology that deals with the basic set-theoretic definitions and constructions used in topology. It is the foundation of most other branches of topology, including differential topology, geometric topology, and algebraic topology. Another name for general topology is **point-set topology**.

The fundamental concepts in point-set topology are *continuity*, *compactness*, and *connectedness*:

- Continuous functions, intuitively, take nearby points to nearby points.
- Compact sets are those that can be covered by finitely many sets of arbitrarily small size.

- Connected sets are sets that cannot be divided into two pieces that are far apart.

The words 'nearby', 'arbitrarily small', and 'far apart' can all be made precise by using open sets, as described below. If we change the definition of 'open set', we change what continuous functions, compact sets, and connected sets are. Each choice of definition for 'open set' is called a *topology*. A set with a topology is called a *topological space*.

Metric spaces are an important class of topological spaces where distances can be assigned a number called a *metric*. Having a metric simplifies many proofs, and many of the most common topological spaces are metric spaces.

25.1 History

General topology grew out of a number of areas, most importantly the following:

- the detailed study of subsets of the real line (once known as the *topology of point sets*; this usage is now obsolete)

- the introduction of the manifold concept

- the study of metric spaces, especially normed linear spaces, in the early days of functional analysis.

General topology assumed its present form around 1940. It captures, one might say, almost everything in the intuition of continuity, in a technically adequate form that can be applied in any area of mathematics.

25.2 A topology on a set

Main article: Topological space

Let X be a set and let τ be a family of subsets of X. Then τ is called a *topology on X* if:[1][2]

1. Both the empty set and X are elements of τ

2. Any union of elements of τ is an element of τ

3. Any intersection of finitely many elements of τ is an element of τ

If τ is a topology on X, then the pair (X, τ) is called a *topological space*. The notation $X\tau$ may be used to denote a set X endowed with the particular topology τ.

The members of τ are called *open sets* in X. A subset of X is said to be closed if its complement is in τ (i.e., its complement is open). A subset of X may be open, closed, both (clopen set), or neither. The empty set and X itself are always both closed and open.

25.2.1 Basis for a topology

Main article: Basis (topology)

A **base** (or **basis**) B for a topological space X with topology T is a collection of open sets in T such that every open set in T can be written as a union of elements of B.[3][4] We say that the base *generates* the topology T. Bases are useful because many properties of topologies can be reduced to statements about a base that generates that topology—and because many topologies are most easily defined in terms of a base that generates them.

25.2.2   Subspace and quotient

Every subset of a topological space can be given the subspace topology in which the open sets are the intersections of the open sets of the larger space with the subset. For any indexed family of topological spaces, the product can be given the product topology, which is generated by the inverse images of open sets of the factors under the projection mappings. For example, in finite products, a basis for the product topology consists of all products of open sets. For infinite products, there is the additional requirement that in a basic open set, all but finitely many of its projections are the entire space.

A quotient space is defined as follows: if X is a topological space and Y is a set, and if $f : X \to Y$ is a surjective function, then the quotient topology on Y is the collection of subsets of Y that have open inverse images under f. In other words, the quotient topology is the finest topology on Y for which f is continuous. A common example of a quotient topology is when an equivalence relation is defined on the topological space X. The map f is then the natural projection onto the set of equivalence classes.

25.2.3   Examples of topological spaces

A given set may have many different topologies. If a set is given a different topology, it is viewed as a different topological space. Any set can be given the discrete topology in which every subset is open. The only convergent sequences or nets in this topology are those that are eventually constant. Also, any set can be given the trivial topology (also called the indiscrete topology), in which only the empty set and the whole space are open. Every sequence and net in this topology converges to every point of the space. This example shows that in general topological spaces, limits of sequences need not be unique. However, often topological spaces must be Hausdorff spaces where limit points are unique.

There are many ways to define a topology on $\mathbf{R}$, the set of real numbers. The standard topology on $\mathbf{R}$ is generated by the open intervals. The set of all open intervals forms a base or basis for the topology, meaning that every open set is a union of some collection of sets from the base. In particular, this means that a set is open if there exists an open interval of non zero radius about every point in the set. More generally, the Euclidean spaces $\mathbf{R}^n$ can be given a topology. In the usual topology on $\mathbf{R}^n$ the basic open sets are the open balls. Similarly, $\mathbf{C}$, the set of complex numbers, and $\mathbf{C}^n$ have a standard topology in which the basic open sets are open balls.

Every metric space can be given a metric topology, in which the basic open sets are open balls defined by the metric. This is the standard topology on any normed vector space. On a finite-dimensional vector space this topology is the same for all norms.

Many sets of linear operators in functional analysis are endowed with topologies that are defined by specifying when a particular sequence of functions converges to the zero function.

Any local field has a topology native to it, and this can be extended to vector spaces over that field.

Every manifold has a natural topology since it is locally Euclidean. Similarly, every simplex and every simplicial complex inherits a natural topology from $\mathbf{R}^n$.

The Zariski topology is defined algebraically on the spectrum of a ring or an algebraic variety. On $\mathbf{R}^n$ or $\mathbf{C}^n$, the closed sets of the Zariski topology are the solution sets of systems of polynomial equations.

A linear graph has a natural topology that generalises many of the geometric aspects of graphs with vertices and edges.

The Sierpiński space is the simplest non-discrete topological space. It has important relations to the theory of computation and semantics.

There exist numerous topologies on any given finite set. Such spaces are called finite topological spaces. Finite spaces are sometimes used to provide examples or counterexamples to conjectures about topological spaces in general.

Any set can be given the cofinite topology in which the open sets are the empty set and the sets whose complement is finite. This is the smallest T_1 topology on any infinite set.

Any set can be given the cocountable topology, in which a set is defined as open if it is either empty or its complement is countable. When the set is uncountable, this topology serves as a counterexample in many situations.

The real line can also be given the lower limit topology. Here, the basic open sets are the half open intervals $[a, b)$. This topology on $\mathbf{R}$ is strictly finer than the Euclidean topology defined above; a sequence converges to a point in this topology if and only if it converges from above in the Euclidean topology. This example shows that a set may have many distinct topologies defined on it.

If Γ is an ordinal number, then the set Γ = [0, Γ) may be endowed with the order topology generated by the intervals (*a*, *b*), [0, *b*) and (*a*, Γ) where *a* and *b* are elements of Γ.

25.3 Continuous functions

Main article: Continuous function

Continuity is expressed in terms of neighborhoods: *f* is continuous at some point *x* ∈ *X* if and only if for any neighborhood *V* of *f*(*x*), there is a neighborhood *U* of *x* such that *f*(*U*) ⊆ *V*. Intuitively, continuity means no matter how "small" *V* becomes, there is always a *U* containing *x* that maps inside *V* and whose image under *f* contains *f*(*x*). This is equivalent to the condition that the preimages of the open (closed) sets in *Y* are open (closed) in *X*. In metric spaces, this definition is equivalent to the ε–δ-definition that is often used in analysis.

An extreme example: if a set *X* is given the discrete topology, all functions

$$f : X \to T$$

to any topological space *T* are continuous. On the other hand, if *X* is equipped with the indiscrete topology and the space *T* set is at least T₀, then the only continuous functions are the constant functions. Conversely, any function whose range is indiscrete is continuous.

25.3.1 Alternative definitions

Several equivalent definitions for a topological structure exist and thus there are several equivalent ways to define a continuous function.

Neighborhood definition

Definitions based on preimages are often difficult to use directly. The following criterion expresses continuity in terms of neighborhoods: *f* is continuous at some point *x* ∈ *X* if and only if for any neighborhood *V* of *f*(*x*), there is a neighborhood *U* of *x* such that *f*(*U*) ⊆ *V*. Intuitively, continuity means no matter how "small" *V* becomes, there is always a *U* containing *x* that maps inside *V*.

If *X* and *Y* are metric spaces, it is equivalent to consider the neighborhood system of open balls centered at *x* and *f*(*x*) instead of all neighborhoods. This gives back the above δ-ε definition of continuity in the context of metric spaces. However, in general topological spaces, there is no notion of nearness or distance.

Note, however, that if the target space is Hausdorff, it is still true that *f* is continuous at *a* if and only if the limit of *f* as *x* approaches *a* is *f*(*a*). At an isolated point, every function is continuous.

Sequences and nets

In several contexts, the topology of a space is conveniently specified in terms of limit points. In many instances, this is accomplished by specifying when a point is the limit of a sequence, but for some spaces that are too large in some sense, one specifies also when a point is the limit of more general sets of points indexed by a directed set, known as nets.[5] A function is continuous only if it takes limits of sequences to limits of sequences. In the former case, preservation of limits is also sufficient; in the latter, a function may preserve all limits of sequences yet still fail to be continuous, and preservation of nets is a necessary and sufficient condition.

In detail, a function $f: X \to Y$ is **sequentially continuous** if whenever a sequence (x_n) in X converges to a limit x, the sequence $(f(x_n))$ converges to $f(x)$.[6] Thus sequentially continuous functions "preserve sequential limits". Every continuous function is sequentially continuous. If X is a first-countable space and countable choice holds, then the converse also holds: any function preserving sequential limits is continuous. In particular, if X is a metric space, sequential continuity and continuity are equivalent. For non first-countable spaces, sequential continuity might be strictly weaker than continuity. (The spaces for which the two properties are equivalent are called sequential spaces.) This motivates the consideration of nets instead of sequences in general topological spaces. Continuous functions preserve limits of nets, and in fact this property characterizes continuous functions.

Closure operator definition

Instead of specifying the open subsets of a topological space, the topology can also be determined by a closure operator (denoted cl), which assigns to any subset $A \subseteq X$ its closure, or an interior operator (denoted int), which assigns to any subset A of X its interior. In these terms, a function

$$f: (X, \text{cl}) \to (X', \text{cl}')$$

between topological spaces is continuous in the sense above if and only if for all subsets A of X

$$f(\text{cl}(A)) \subseteq \text{cl}'(f(A)).$$

That is to say, given any element x of X that is in the closure of any subset A, $f(x)$ belongs to the closure of $f(A)$. This is equivalent to the requirement that for all subsets A' of X'

$$f^{-1}(\text{cl}'(A')) \supseteq \text{cl}(f^{-1}(A')).$$

Moreover,

$$f: (X, \text{int}) \to (X', \text{int}')$$

is continuous if and only if

$$f^{-1}(\text{int}'(A)) \subseteq \text{int}(f^{-1}(A))$$

for any subset A of X.

25.3.2   Properties

If $f: X \to Y$ and $g: Y \to Z$ are continuous, then so is the composition $g \circ f: X \to Z$. If $f: X \to Y$ is continuous and

- X is compact, then $f(X)$ is compact.
- X is connected, then $f(X)$ is connected.
- X is path-connected, then $f(X)$ is path-connected.
- X is Lindelöf, then $f(X)$ is Lindelöf.
- X is separable, then $f(X)$ is separable.

The possible topologies on a fixed set X are partially ordered: a topology τ_1 is said to be coarser than another topology τ_2 (notation: $\tau_1 \subseteq \tau_2$) if every open subset with respect to τ_1 is also open with respect to τ_2. Then, the identity map

$$\mathrm{idX}: (X, \tau_2) \rightarrow (X, \tau_1)$$

is continuous if and only if $\tau_1 \subseteq \tau_2$ (see also comparison of topologies). More generally, a continuous function

$$(X, \tau_X) \rightarrow (Y, \tau_Y)$$

stays continuous if the topology τY is replaced by a coarser topology and/or τX is replaced by a finer topology.

25.3.3   Homeomorphisms

Symmetric to the concept of a continuous map is an open map, for which *images* of open sets are open. In fact, if an open map f has an inverse function, that inverse is continuous, and if a continuous map g has an inverse, that inverse is open. Given a bijective function f between two topological spaces, the inverse function f^{-1} need not be continuous. A bijective continuous function with continuous inverse function is called a *homeomorphism*.

If a continuous bijection has as its domain a compact space and its codomain is Hausdorff, then it is a homeomorphism.

25.3.4   Defining topologies via continuous functions

Given a function

$$f: X \rightarrow S,$$

where X is a topological space and S is a set (without a specified topology), the final topology on S is defined by letting the open sets of S be those subsets A of S for which $f^{-1}(A)$ is open in X. If S has an existing topology, f is continuous with respect to this topology if and only if the existing topology is coarser than the final topology on S. Thus the final topology can be characterized as the finest topology on S that makes f continuous. If f is surjective, this topology is canonically identified with the quotient topology under the equivalence relation defined by f.

Dually, for a function f from a set S to a topological space, the initial topology on S has as open subsets A of S those subsets for which $f(A)$ is open in X. If S has an existing topology, f is continuous with respect to this topology if and only if the existing topology is finer than the initial topology on S. Thus the initial topology can be characterized as the coarsest topology on S that makes f continuous. If f is injective, this topology is canonically identified with the subspace topology of S, viewed as a subset of X.

More generally, given a set S, specifying the set of continuous functions

$$S \rightarrow X$$

into all topological spaces X defines a topology. Dually, a similar idea can be applied to maps

$$X \rightarrow S.$$

This is an instance of a universal property.

25.4   Compact sets

Main article: Compact (mathematics)

Formally, a topological space X is called *compact* if each of its open covers has a finite subcover. Otherwise it is called *non-compact*. Explicitly, this means that for every arbitrary collection

$$\{U_\alpha\}_{\alpha \in A}$$

of open subsets of X such that

$$X = \bigcup_{\alpha \in A} U_\alpha,$$

there is a finite subset J of A such that

$$X = \bigcup_{i \in J} U_i.$$

Some branches of mathematics such as algebraic geometry, typically influenced by the French school of Bourbaki, use the term *quasi-compact* for the general notion, and reserve the term *compact* for topological spaces that are both Hausdorff and *quasi-compact*. A compact set is sometimes referred to as a *compactum*, plural *compacta*.

Every closed interval in $\mathbb{R}$ of finite length is compact. More is true: In $\mathbb{R}^n$, a set is compact if and only if it is closed and bounded. (See Heine–Borel theorem).

Every continuous image of a compact space is compact.

A compact subset of a Hausdorff space is closed.

Every continuous bijection from a compact space to a Hausdorff space is necessarily a homeomorphism.

Every sequence of points in a compact metric space has a convergent subsequence.

Every compact finite-dimensional manifold can be embedded in some Euclidean space $\mathbb{R}^n$.

25.5   Connected sets

Main article: connected space

A topological space X is said to be **disconnected** if it is the union of two disjoint nonempty open sets. Otherwise, X is said to be **connected**. A subset of a topological space is said to be connected if it is connected under its subspace topology. Some authors exclude the empty set (with its unique topology) as a connected space, but this article does not follow that practice.

For a topological space X the following conditions are equivalent:

1. X is connected.

2. X cannot be divided into two disjoint nonempty closed sets.

3. The only subsets of X that are both open and closed (clopen sets) are X and the empty set.

4. The only subsets of X with empty boundary are X and the empty set.

5. X cannot be written as the union of two nonempty separated sets.

6. The only continuous functions from X to $\{0,1\}$, the two-point space endowed with the discrete topology, are constant.

Every interval in **R** is connected.

The continuous image of a connected space is connected.

25.5.1   Connected components

The maximal connected subsets (ordered by inclusion) of a nonempty topological space are called the **connected components** of the space. The components of any topological space X form a partition of X: they are disjoint, nonempty, and their union is the whole space. Every component is a closed subset of the original space. It follows that, in the case where their number is finite, each component is also an open subset. However, if their number is infinite, this might not be the case; for instance, the connected components of the set of the rational numbers are the one-point sets, which are not open.

Let Γ_x be the connected component of x in a topological space X, and Γ_x be the intersection of all open-closed sets containing x (called quasi-component of x.) Then $\Gamma_x \subset \Gamma_x$ where the equality holds if X is compact Hausdorff or locally connected.

25.5.2   Disconnected spaces

A space in which all components are one-point sets is called totally disconnected. Related to this property, a space X is called **totally separated** if, for any two distinct elements x and y of X, there exist disjoint open neighborhoods U of x and V of y such that X is the union of U and V. Clearly any totally separated space is totally disconnected, but the converse does not hold. For example take two copies of the rational numbers **Q**, and identify them at every point except zero. The resulting space, with the quotient topology, is totally disconnected. However, by considering the two copies of zero, one sees that the space is not totally separated. In fact, it is not even Hausdorff, and the condition of being totally separated is strictly stronger than the condition of being Hausdorff.

25.5.3   Path-connected sets

A **path** from a point x to a point y in a topological space X is a continuous function f from the unit interval $[0,1]$ to X with $f(0) = x$ and $f(1) = y$. A **path-component** of X is an equivalence class of X under the equivalence relation, which makes x equivalent to y if there is a path from x to y. The space X is said to be **path-connected** (or **pathwise connected** or **0-connected**) if there is at most one path-component, i.e. if there is a path joining any two points in X. Again, many authors exclude the empty space.

Every path-connected space is connected. The converse is not always true: examples of connected spaces that are not path-connected include the extended long line L^* and the *topologist's sine curve*.

However, subsets of the real line **R** are connected if and only if they are path-connected; these subsets are the intervals of **R**. Also, open subsets of $\mathbf{R}^n$ or $\mathbf{C}^n$ are connected if and only if they are path-connected. Additionally, connectedness and path-connectedness are the same for finite topological spaces.

25.6   Products of spaces

Main article: Product topology

Given X such that

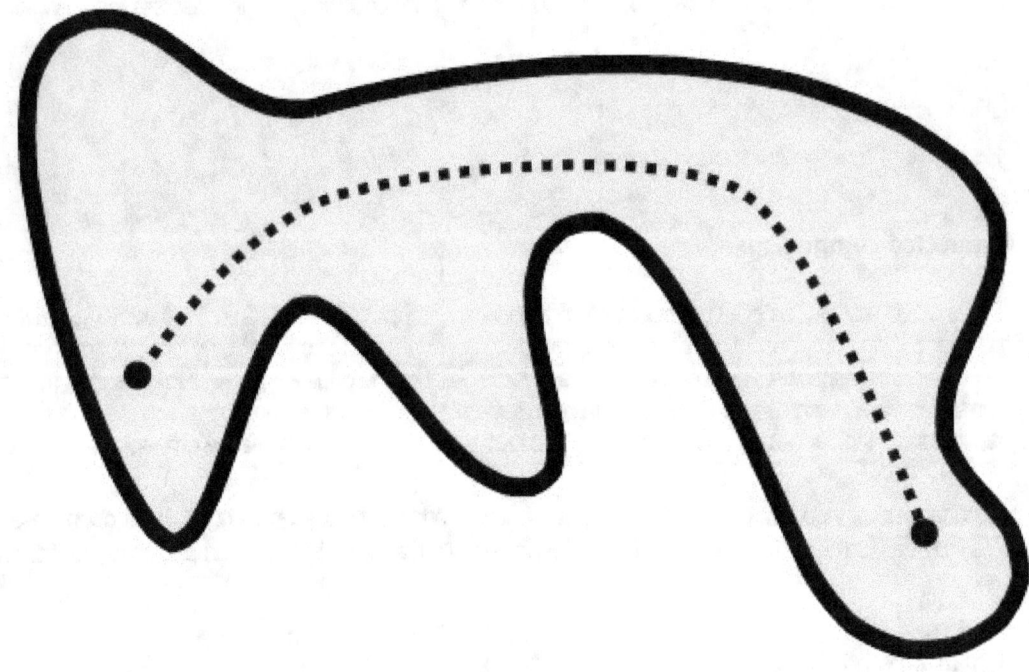

This subspace of $\mathbf{R}^2$ is path-connected, because a path can be drawn between any two points in the space.

$$X := \prod_{i \in I} X_i,$$

is the Cartesian product of the topological spaces X_i, indexed by $i \in I$, and the **canonical projections** $p_i : X \to X_i$, the **product topology** on X is defined as the coarsest topology (i.e. the topology with the fewest open sets) for which all the projections p_i are continuous. The product topology is sometimes called the **Tychonoff topology**.

The open sets in the product topology are unions (finite or infinite) of sets of the form $\prod_{i \in I} U_i$, where each U_i is open in X_i and $U_i \neq X_i$ only finitely many times. In particular, for a finite product (in particular, for the product of two topological spaces), the products of base elements of the X_i gives a basis for the product $\prod_{i \in I} X_i$.

The product topology on X is the topology generated by sets of the form $p_i^{-1}(U)$, where i is in I and U is an open subset of X_i. In other words, the sets $\{p_i^{-1}(U)\}$ form a subbase for the topology on X. A subset of X is open if and only if it is a (possibly infinite) union of intersections of finitely many sets of the form $p_i^{-1}(U)$. The $p_i^{-1}(U)$ are sometimes called open cylinders, and their intersections are cylinder sets.

In general, the product of the topologies of each X_i forms a basis for what is called the box topology on X. In general, the box topology is finer than the product topology, but for finite products they coincide.

Related to compactness is Tychonoff's theorem: the (arbitrary) product of compact spaces is compact.

25.7   Separation axioms

Main article: Separation axiom

Many of these names have alternative meanings in some of mathematical literature, as explained on History of the separation axioms; for example, the meanings of "normal" and "T_4" are sometimes interchanged, similarly "regular" and "T_3", etc. Many of the concepts also have several names; however, the one listed first is always least likely to be ambiguous.

Most of these axioms have alternative definitions with the same meaning; the definitions given here fall into a consistent pattern that relates the various notions of separation defined in the previous section. Other possible definitions can be found in the individual articles.

In all of the following definitions, X is again a topological space.

- X is T_0, or *Kolmogorov*, if any two distinct points in X are topologically distinguishable. (It is a common theme among the separation axioms to have one version of an axiom that requires T_0 and one version that doesn't.)

- X is T_1, or *accessible* or *Fréchet*, if any two distinct points in X are separated. Thus, X is T_1 if and only if it is both T_0 and R_0. (Though you may say such things as T_1 *space, Fréchet topology,* and *Suppose that the topological space* X *is Fréchet*, avoid saying *Fréchet space* in this context, since there is another entirely different notion of Fréchet space in functional analysis.)

- X is *Hausdorff*, or T_2 or *separated*, if any two distinct points in X are separated by neighbourhoods. Thus, X is Hausdorff if and only if it is both T_0 and R_1. A Hausdorff space must also be T_1.

- X is $T2\frac{1}{2}$, or *Urysohn*, if any two distinct points in X are separated by closed neighbourhoods. A $T_2\frac{1}{2}$ space must also be Hausdorff.

- X is *regular*, or T_3, if it is T_0 and if given any point x and closed set F in X such that x does not belong to F, they are separated by neighbourhoods. (In fact, in a regular space, any such x and F is also separated by closed neighbourhoods.)

- X is *Tychonoff*, or $T3\frac{1}{2}$, *completely* T_3, or *completely regular*, if it is T_0 and if f, given any point x and closed set F in X such that x does not belong to F, they are separated by a continuous function.

- X is *normal*, or T_4, if it is Hausdorff and if any two disjoint closed subsets of X are separated by neighbourhoods. (In fact, a space is normal if and only if any two disjoint closed sets can be separated by a continuous function; this is Urysohn's lemma.)

- X is *completely normal*, or T_5 or *completely* T_4, if it is T_1. and if any two separated sets are separated by neighbourhoods. A completely normal space must also be normal.

- X is *perfectly normal*, or T_6 or *perfectly* T_4, if it is T_1 and if any two disjoint closed sets are precisely separated by a continuous function. A perfectly normal Hausdorff space must also be completely normal Hausdorff.

The Tietze extension theorem: In a normal space, every continuous real-valued function defined on a closed subspace can be extended to a continuous map defined on the whole space.

25.8   Countability axioms

Main article: axiom of countability

An **axiom of countability** is a property of certain mathematical objects (usually in a category) that requires the existence of a countable set with certain properties, while without it such sets might not exist.

Important countability axioms for topological spaces:

- sequential space: a set is open if every sequence convergent to a point in the set is eventually in the set

- first-countable space: every point has a countable neighbourhood basis (local base)

- second-countable space: the topology has a countable base

- separable space: there exists a countable dense subspace

- Lindelöf space: every open cover has a countable subcover

- σ-compact space: there exists a countable cover by compact spaces

Relations:

- Every first countable space is sequential.

- Every second-countable space is first-countable, separable, and Lindelöf.

- Every σ-compact space is Lindelöf.

- A metric space is first-countable.

- For metric spaces second-countability, separability, and the Lindelöf property are all equivalent.

25.9   Metric spaces

Main article: Metric space

A **metric space**[7] is an ordered pair (M, d) where M is a set and d is a metric on M, i.e., a function

$$d: M \times M \to \mathbb{R}$$

such that for any $x, y, z \in M$, the following holds:

1. $d(x, y) \geq 0$ (*non-negative*),

2. $d(x, y) = 0$ iff $x = y$ (*identity of indiscernibles*),

3. $d(x, y) = d(y, x)$ (*symmetry*) and

4. $d(x, z) \leq d(x, y) + d(y, z)$ (*triangle inequality*) .

The function d is also called *distance function* or simply *distance*. Often, d is omitted and one just writes M for a metric space if it is clear from the context what metric is used.

Every metric space is paracompact and Hausdorff, and thus normal.

The metrization theorems provide necessary and sufficient conditions for a topology to come from a metric.

25.10   Baire category theory

Main article: Baire category theorem

The Baire category theorem says: If X is a complete metric space or a locally compact Hausdorff space, then the interior of every union of countably many nowhere dense sets is empty.[8]

Any open subspace of a Baire space is itself a Baire space.

25.11 Main areas of research

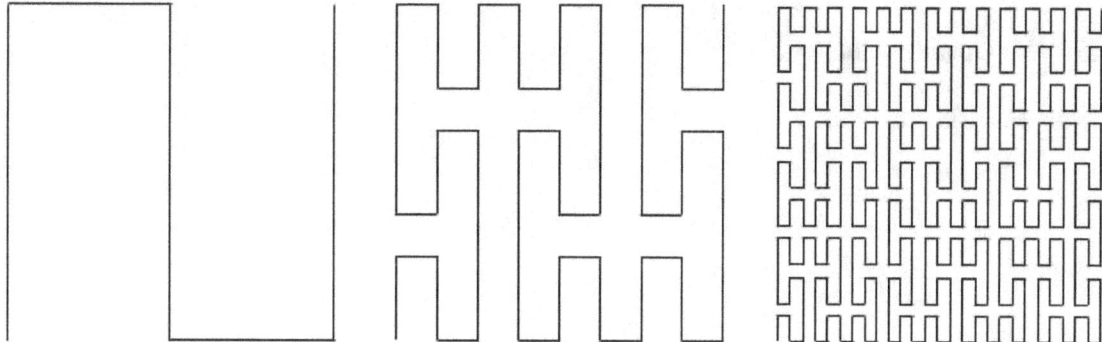

*Three iterations of a Peano curve construction, whose limit is a space-filling curve. The Peano curve is studied in continuum theory, a branch of **general topology**.*

25.11.1 Continuum theory

Main article: Continuum theory

A **continuum** (pl *continua*) is a nonempty compact connected metric space, or less frequently, a compact connected Hausdorff space. **Continuum theory** is the branch of topology devoted to the study of continua.

25.11.2 Pointless topology

Main article: Pointless topology

Pointless topology (also called **point-free** or **pointfree topology**) is an approach to topology that avoids mentioning points. The name 'pointless topology' is due to John von Neumann.[9] The ideas of pointless topology are closely related to mereotopologies, in which regions (sets) are treated as foundational without explicit reference to underlying point sets.

25.11.3 Dimension theory

Main article: Dimension theory

Dimension theory is a branch of general topology dealing with dimensional invariants of topological spaces.

25.11.4 Topological algebras

Main article: Topological algebra

A **topological algebra** A over a topological field K is a topological vector space together with a continuous multiplication

$$\cdot : A \times A \longrightarrow A$$

$$(a, b) \longrightarrow a \cdot b$$

that makes it an algebra over **K**. A unital associative topological algebra is a topological ring.

The term was coined by David van Dantzig; it appears in the title of his doctoral dissertation (1931).

25.11.5   Metrizability theory

Main article: Metrization theorem

In topology and related areas of mathematics, a **metrizable space** is a topological space that is homeomorphic to a metric space. That is, a topological space (X, τ) is said to be metrizable if there is a metric

$$d: X \times X \to [0, \infty)$$

such that the topology induced by d is τ. **Metrization theorems** are theorems that give sufficient conditions for a topological space to be metrizable.

25.11.6   Set-theoretic topology

Main article: Set-theoretic topology

Set-theoretic topology is a subject that combines set theory and general topology. It focuses on topological questions that are independent of Zermelo–Fraenkel set theory(ZFC). A famous problem is the normal Moore space question, a question in general topology that was the subject of intense research. The answer to the normal Moore space question was eventually proved to be independent of ZFC.

25.12   See also

- List of examples in general topology
- Glossary of general topology for detailed definitions
- List of general topology topics for related articles
- Category of topological spaces

25.13   References

[1] Munkres, James R. Topology. Vol. 2. Upper Saddle River: Prentice Hall, 2000.

[2] Adams, Colin Conrad, and Robert David Franzosa. Introduction to topology: pure and applied. Pearson Prentice Hall, 2008.

[3] Merrifield, Richard E.; Simmons, Howard E. (1989). *Topological Methods in Chemistry*. New York: John Wiley & Sons. p. 16. ISBN 0-471-83817-9. Retrieved 27 July 2012. **Definition.** A collection B of subsets of a topological space (X, T) is called a *basis* for T if every open set can be expressed as a union of members of B.

[4] Armstrong, M. A. (1983). *Basic Topology*. Springer. p. 30. ISBN 0-387-90839-0. Retrieved 13 June 2013. Suppose we have a topology on a set X, and a collection β of open sets such that every open set is a union of members of β. Then β is called a *base* for the topology...

[5] Moore, E. H.; Smith, H. L. (1922). "A General Theory of Limits". *American Journal of Mathematics* 44 (2): 102–121. doi:10.2307/2370388. JSTOR 2370388

[6] Heine, E.. "Die Elemente der Functionenlehre.." *Journal für die reine und angewandte Mathematik* 74 (1872): 172-188. <http://eudml.org/doc/148175>.

[7] Maurice Fréchet introduced metric spaces in his work *Sur quelques points du calcul fonctionnel*, Rendic. Circ. Mat. Palermo 22 (1906) 1–74.

[8] R. Baire. Sur les fonctions de variables réelles. Ann. di Mat., 3:1–123, 1899.

[9] Garrett Birkhoff, *VON NEUMANN AND LATTICE THEORY*, *John Von Neumann 1903-1957*, J. C. Oxtoley, B. J. Pettis, American Mathematical Soc., 1958, page 50-5

25.14   Further reading

Some standard books on general topology include:

- Bourbaki, Topologie Générale (General Topology), ISBN 0-387-19374-X.

- John L. Kelley (1955) *General Topology*, link from Internet Archive, originally published by David Van Nostrand Company.

- Stephen Willard, General Topology, ISBN 0-486-43479-6.

- James Munkres, Topology, ISBN 0-13-181629-2.

- George F. Simmons, Introduction to Topology and Modern Analysis, ISBN 1-575-24238-9.

- Paul L. Shick, Topology: Point-Set and Geometric, ISBN 0-470-09605-5.

- Ryszard Engelking, General Topology, ISBN 3-88538-006-4.

- Steen, Lynn Arthur; Seebach, J. Arthur Jr. (1995) [1978], *Counterexamples in Topology* (Dover reprint of 1978 ed.), Berlin, New York: Springer-Verlag, ISBN 978-0-486-68735-3, MR 507446

- O.Ya. Viro, O.A. Ivanov, V.M. Kharlamov and N.Yu. Netsvetaev, Elementary Topology: Textbook in Problems, ISBN 978-0-8218-4506-6.

The arXiv subject code is math.GN.

Chapter 26

Algebraic topology

For the topology of pointwise convergence, see Algebraic topology (object).

Algebraic topology is a branch of mathematics that uses tools from abstract algebra to study topological spaces. The

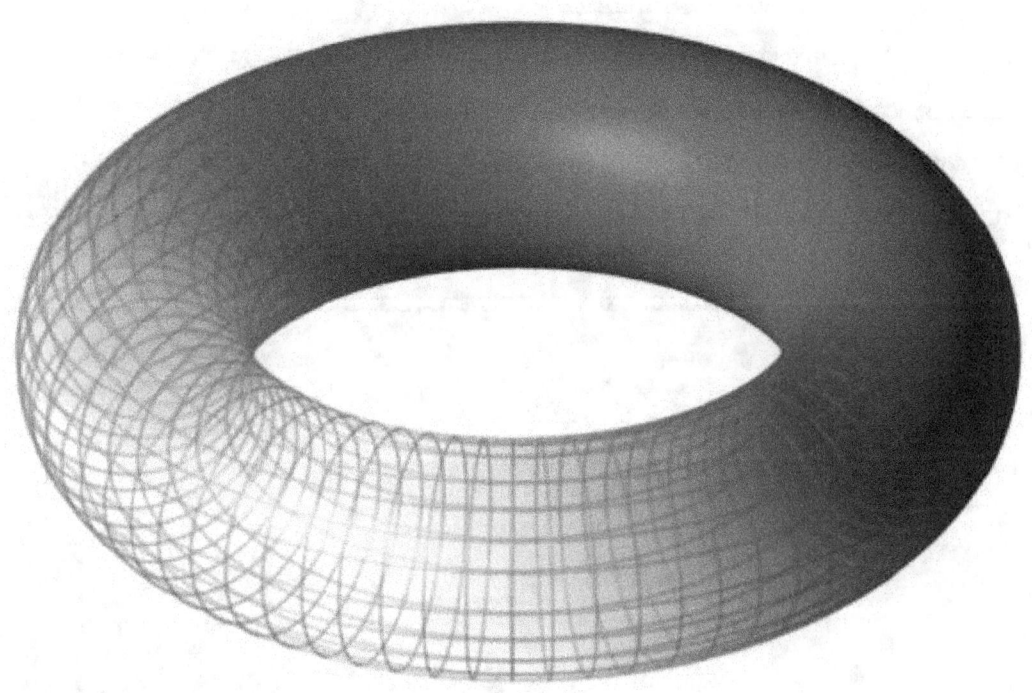

A torus, one of the most frequently studied objects in algebraic topology

basic goal is to find algebraic invariants that classify topological spaces up to homeomorphism, though usually most classify up to homotopy equivalence.

Although algebraic topology primarily uses algebra to study topological problems, using topology to solve algebraic problems is sometimes also possible. Algebraic topology, for example, allows for a convenient proof that any subgroup of a free group is again a free group.

26.1 Main branches of algebraic topology

Below are some of the main areas studied in algebraic topology:

26.1.1 Homotopy groups

Main article: Homotopy group

In mathematics, homotopy groups are used in algebraic topology to classify topological spaces. The first and simplest homotopy group is the fundamental group, which records information about loops in a space. Intuitively, homotopy groups record information about the basic shape, or holes, of a topological space.

26.1.2 Homology

Main article: Homology

In algebraic topology and abstract algebra, homology (in part from Greek ὁμός *homos* "identical") is a certain general procedure to associate a sequence of abelian groups or modules with a given mathematical object such as a topological space or a group.[1]

26.1.3 Cohomology

Main article: Cohomology

In homology theory and algebraic topology, cohomology is a general term for a sequence of abelian groups defined from a co-chain complex. That is, cohomology is defined as the abstract study of cochains, cocycles, and coboundaries. Cohomology can be viewed as a method of assigning algebraic invariants to a topological space that has a more refined algebraic structure than does homology. Cohomology arises from the algebraic dualization of the construction of homology. In less abstract language, cochains in the fundamental sense should assign 'quantities' to the *chains* of homology theory.

26.1.4 Manifolds

Main article: Manifold

A manifold is a topological space that near each point resembles Euclidean space. More precisely, each point of an n-dimensional manifold has a neighbourhood that is diffeomorphic to the Euclidean space of dimension n. Lines and circles, but not figure eights, are one-dimensional manifolds. Two-dimensional manifolds are also called surfaces. Examples include the plane, the sphere, and the torus, which can all be realized in three dimensions, but also the Klein bottle and real projective plane which cannot be realized in three dimensions, but can be realized in four dimensions.

26.1.5 Knot theory

Main article: Knot theory

Knot theory is the study of mathematical knots. While inspired by knots that appear in daily life in shoelaces and rope, a mathematician's knot differs in that the ends are joined together so that it cannot be undone. In precise mathematical language, a knot is an embedding of a circle in 3-dimensional Euclidean space, $\mathbf{R}^3$. Two mathematical knots are equivalent if one can be transformed into the other via a deformation of $\mathbf{R}^3$ upon itself (known as an ambient isotopy); these transformations correspond to manipulations of a knotted string that do not involve cutting the string or passing the string through itself.

26.1.6   Complexes

Main articles: Simplicial complex and CW complex

A **simplicial complex** is a topological space of a certain kind, constructed by "gluing together" points, line segments, triangles, and their *n*-dimensional counterparts (see illustration). Simplicial complexes should not be confused with the more abstract notion of a simplicial set appearing in modern simplicial homotopy theory. The purely combinatorial counterpart to a simplicial complex is an abstract simplicial complex.

A **CW complex** is a type of topological space introduced by J. H. C. Whitehead to meet the needs of homotopy theory. This class of spaces is broader and has some better categorical properties than simplicial complexes, but still retains a combinatorial nature that allows for computation (often with a much smaller complex).

26.2   Method of algebraic invariants

An older name for the subject was combinatorial topology, implying an emphasis on how a space X was constructed from simpler ones[2] (the modern standard tool for such construction is the CW-complex). In the 1920s and 1930s, there was growing emphasis on investigating topological spaces by finding correspondences from them to algebraic groups, which led to the change of name to algebraic topology.[3] The combinatorial topology name is still sometimes used to emphasize an algorithmic approach based on decomposition of spaces.[4]

In the algebraic approach, one finds a correspondence between spaces and groups that respects the relation of homeomorphism (or more general homotopy) of spaces. This allows one to recast statements about topological spaces into statements about groups, which have a great deal of manageable structure, often making these statement easier to prove. Two major ways in which this can be done are through fundamental groups, or more generally homotopy theory, and through homology and cohomology groups. The fundamental groups give us basic information about the structure of a topological space, but they are often nonabelian and can be difficult to work with. The fundamental group of a (finite) simplicial complex does have a finite presentation.

Homology and cohomology groups, on the other hand, are abelian and in many important cases finitely generated. Finitely generated abelian groups are completely classified and are particularly easy to work with.

26.3   Setting in category theory

In general, all constructions of algebraic topology are functorial; the notions of category, functor and natural transformation originated here. Fundamental groups and homology and cohomology groups are not only *invariants* of the underlying topological space, in the sense that two topological spaces which are homeomorphic have the same associated groups, but their associated morphisms also correspond — a continuous mapping of spaces induces a group homomorphism on the associated groups, and these homomorphisms can be used to show non-existence (or, much more deeply, existence) of mappings.

One of the first mathematicians to work with different types of cohomology was Georges de Rham. One can use the differential structure of smooth manifolds via de Rham cohomology, or Čech or sheaf cohomology to investigate the solvability of differential equations defined on the manifold in question. De Rham showed that all of these approaches were interrelated and that, for a closed, oriented manifold, the Betti numbers derived through simplicial homology were the same Betti numbers as those derived through de Rham cohomology. This was extended in the 1950s, when Eilenberg and Steenrod generalized this approach. They defined homology and cohomology as functors equipped with natural transformations subject to certain axioms (e.g., a weak equivalence of spaces passes to an isomorphism of homology groups), verified that all existing (co)homology theories satisfied these axioms, and then proved that such an axiomatization uniquely characterized the theory.

26.4   Applications of algebraic topology

Classic applications of algebraic topology include:

- The Brouwer fixed point theorem: every continuous map from the unit n-disk to itself has a fixed point.

- The free rank of the nth homology group of a simplicial complex is the n-th Betti number, which allows one to calculate the Euler-Poincaré characteristic.

- One can use the differential structure of smooth manifolds via de Rham cohomology, or Čech or sheaf cohomology to investigate the solvability of differential equations defined on the manifold in question.

- A manifold is orientable when the top-dimensional integral homology group is the integers, and is non-orientable when it is 0.

- The n-sphere admits a nowhere-vanishing continuous unit vector field if and only if n is odd. (For $n = 2$, this is sometimes called the "hairy ball theorem".)

- The Borsuk–Ulam theorem: any continuous map from the n-sphere to Euclidean n-space identifies at least one pair of antipodal points.

- Any subgroup of a free group is free. This result is quite interesting, because the statement is purely algebraic yet the simplest proof is topological. Namely, any free group G may be realized as the fundamental group of a graph X. The main theorem on covering spaces tells us that every subgroup H of G is the fundamental group of some covering space Y of X; but every such Y is again a graph. Therefore its fundamental group H is free. On the other hand this type of application is also handled more simply by the use of covering morphisms of groupoids, and that technique has yielded subgroup theorems not yet proved by methods of algebraic topology. (See the book by Higgins listed under groupoids.)

- Topological combinatorics

26.5   Notable algebraic topologists

- Frank Adams

- Enrico Betti

- Armand Borel

- Karol Borsuk

- Luitzen Egbertus Jan Brouwer

- William Browder

- Ronald Brown (mathematician)

- Henri Cartan

- Charles Ehresmann

- Samuel Eilenberg

- Hans Freudenthal

- Peter Freyd

- Pierre Gabriel

- Alexander Grothendieck

- Friedrich Hirzebruch

- Heinz Hopf

- Michael J. Hopkins

- Witold Hurewicz

- Egbert van Kampen

- Daniel Kan

- Hermann Künneth

- Solomon Lefschetz

- Jean Leray

- Saunders Mac Lane

- Mark Mahowald

- J. Peter May

- Barry Mazur

- John Milnor

- John Coleman Moore

- Jack Morava

- Emmy Noether

- Sergei Novikov

- Grigori Perelman

- Lev Pontryagin

- Nicolae Popescu

- Mikhail Postnikov

- Daniel Quillen

- Jean-Pierre Serre

- Stephen Smale

- Edwin Spanier

- Norman Steenrod

- Dennis Sullivan

- René Thom

- Hiroshi Toda

- Leopold Vietoris
- Hassler Whitney
- J. H. C. Whitehead
- Allen Hatcher

26.6 Important theorems in algebraic topology

- Borsuk–Ulam theorem
- Brouwer fixed point theorem
- Cellular approximation theorem
- Eilenberg–Zilber theorem
- Freudenthal suspension theorem
- Hurewicz theorem
- Künneth theorem
- Poincaré duality theorem
- Universal coefficient theorem
- Van Kampen's theorem
- Generalized van Kampen's theorems
- Whitehead's theorem

26.7 See also

26.8 Notes

[1] Fraleigh (1976, p. 163)

[2]Fréchet, Maurice; Fan, Ky (2012),*Invitation to Combinatorial Topology,* Courier Dover Publications, p. 101,ISBN97804861478.

[3] Henle, Michael (1994), *A Combinatorial Introduction to Topology,* Courier Dover Publications, p. 221, ISBN 9780486679662.

[4] Spreer, Jonathan (2011), *Blowups, slicings and permutation groups in combinatorial topology,* Logos Verlag Berlin GmbH, p. 23, ISBN 9783832529833.

26.9 References

- Dylan G. L. Allegretti, *Simplicial Sets and van Kampen's Theorem (Discusses generalized versions of van Kampen's theorem applied to topological spaces and simplicial sets).*
- Bredon, Glen E. (1993), *Topology and Geometry,* Graduate Texts in Mathematics 139, Springer, ISBN 0-387-97926-3, retrieved 2008-04-01.

- Ronald Brown, *Higher dimensional group theory* (2007) *(Gives a broad view of higher-dimensional van Kampen theorems involving multiple groupoids).*

- R. Brown and A. Razak, *A van Kampen theorem for unions of non-connected spaces,* Archiv. Math. 42 (1984) 85–88. "Gives a general theorem on the fundamental groupoid with a set of base points of a space which is the union of open sets."

- R. Brown, K. Hardie, H. Kamps, T. Porter: The homotopy double groupoid of a Hausdorff space., *Theory Appl. Categories,* 10:71--93 (2002).

- R. Brown and P.J. Higgins, *On the connection between the second relative homotopy groups of some related spaces,* Proc. London Math. Soc. (3) 36 (1978) 193–212. "The first 2-dimensional version of van Kampen's theorem."

- R. Brown, P.J. Higgins, and R. Sivera. *Non-Abelian Algebraic Topology: filtered spaces, crossed complexes, cubical higher homotopy groupoids;* European Mathematical Society Tracts in Mathematics Vol. 15, 2011, This provides a homotopy theoretic approach to basic algebraic topology, without needing a basis in singular homology, or the method of simplicial approximation. It contains a lot of material on crossed modules.

- Fraleigh, John B. (1976), *A First Course In Abstract Algebra* (2nd ed.), Reading: Addison-Wesley, ISBN 0-201-01984-1

- Greenberg, Marvin J. and John R. Harper. (1981), *Algebraic Topology: A First Course, Revised edition,* Mathematics Lecture Note Series, Westview/Perseus, ISBN 9780805335576. A functorial, algebraic approach originally by Greenberg with geometric flavoring added by Harper.

- Hatcher, Allen (2002), *Algebraic Topology,* Cambridge: Cambridge University Press, ISBN 0-521-79540-0. A modern, geometrically flavoured introduction to algebraic topology.

- P. J. Higgins, *Categories and groupoids* (1971) Van Nostrand-Reinhold.

- Maunder, C. R. F. (1970), *Algebraic Topology,* London: Van Nostrand Reinhold, ISBN 0-486-69131-4.

- tom Dieck, T., Algebraic topology. EMS Textbooks in Mathematics. European Mathematical Society (EMS), Zürich (2008).

- E. R. van Kampen. *On the connection between the fundamental groups of some related spaces.* American Journal of Mathematics, vol. 55 (1933), pp. 261–267.

- Van Kampen's theorem at PlanetMath.org.

- Van Kampen's theorem result at PlanetMath.org.

26.10   Further reading

- Allen Hatcher, *Algebraic topology.* (2002) Cambridge University Press, xii+544 pp. ISBN 0-521-79160-X and ISBN 0-521-79540-0.

- Hazewinkel, Michiel, ed. (2001), "Algebraic topology", *Encyclopedia of Mathematics,* Springer, ISBN 978-1-55608-010-4

- May JP (1999). *A Concise Course in Algebraic Topology* (PDF). U. Chicago Press. Retrieved 2008-09-27. Section 2.7 provides a category-theoretic presentation of the theorem as a colimit in the category of groupoids.

- Ronald Brown, *Topology and groupoids* (2006) Booksurge LLC ISBN 1-4196-2722-8.

Chapter 27

Manifold

For other uses, see Manifold (disambiguation).

In mathematics, a **manifold** is a topological space that resembles Euclidean space near each point. More precisely, each point of an n-dimensional manifold has a neighbourhood that is homeomorphic to the Euclidean space of dimension n. Lines and circles, but not figure eights, are one-dimensional manifolds. Two-dimensional manifolds are also called surfaces. Examples include the plane, the sphere, and the torus, which can all be embedded in three dimensional real space, but also the Klein bottle and real projective plane which cannot.

Although a manifold resembles Euclidean space near each point, globally it may not. For example, the surface of the sphere is not a Euclidean space, but in a region it can be charted by means of map projections of the region into the Euclidean plane (in the context of manifolds they are called *charts*). When a region appears in two neighbouring charts, the two representations do not coincide exactly and a transformation is needed to pass from one to the other, called a *transition map*.

The concept of a manifold is central to many parts of geometry and modern mathematical physics because it allows more complicated structures to be described and understood in terms of the relatively well-understood properties of Euclidean space. Manifolds naturally arise as solution sets of systems of equations and as graphs of functions. Manifolds may have additional features. One important class of manifolds is the class of differentiable manifolds. This differentiable structure allows calculus to be done on manifolds. A Riemannian metric on a manifold allows distances and angles to be measured. Symplectic manifolds serve as the phase spaces in the Hamiltonian formalism of classical mechanics, while four-dimensional Lorentzian manifolds model spacetime in general relativity.

27.1 Motivational examples

27.1.1 Circle

Main article: Circle

After a line, the circle is the simplest example of a topological manifold. Topology ignores bending, so a small piece of a circle is treated exactly the same as a small piece of a line. Consider, for instance, the top part of the unit circle, $x^2 + y^2 = 1$, where the y-coordinate is positive (indicated by the yellow circular arc in *Figure 1*). Any point of this arc can be uniquely described by its x-coordinate. So, projection onto the first coordinate is a continuous, and invertible, mapping from the upper arc to the open interval $(-1,1)$:

$$\chi_{top}(x, y) = x.$$

Such functions along with the open regions they map are called *charts*. Similarly, there are charts for the bottom (red), left (blue), and right (green) parts of the circle:

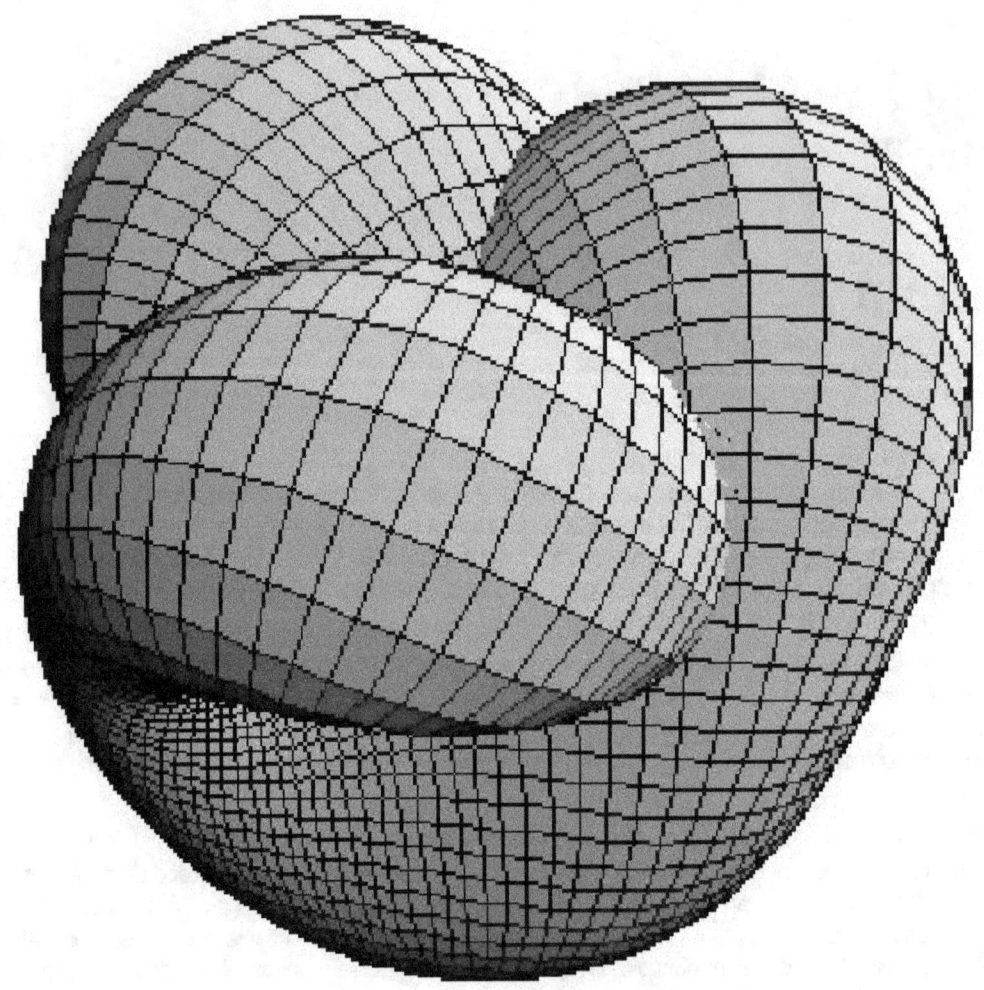

The real projective plane is a two-dimensional manifold that cannot be realized in three dimensions without self-intersection, shown here as Boy's surface.

$\chi_{bottom}(x, y) = x$

$\chi_{left}(x, y) = y$

$\chi_{right}(x, y) = y.$

Together, these parts cover the whole circle and the four charts form an atlas for the circle.

The top and right charts overlap: their intersection lies in the quarter of the circle where both the x- and the y-coordinates are positive. The two charts χ_{to} and χ_{ri} each map this part into the interval $(0, 1)$. Thus a function T from $(0, 1)$ to itself can be constructed, which first uses the inverse of the top chart to reach the circle and then follows the right chart back to the interval. Let a be any number in $(0, 1)$, then:

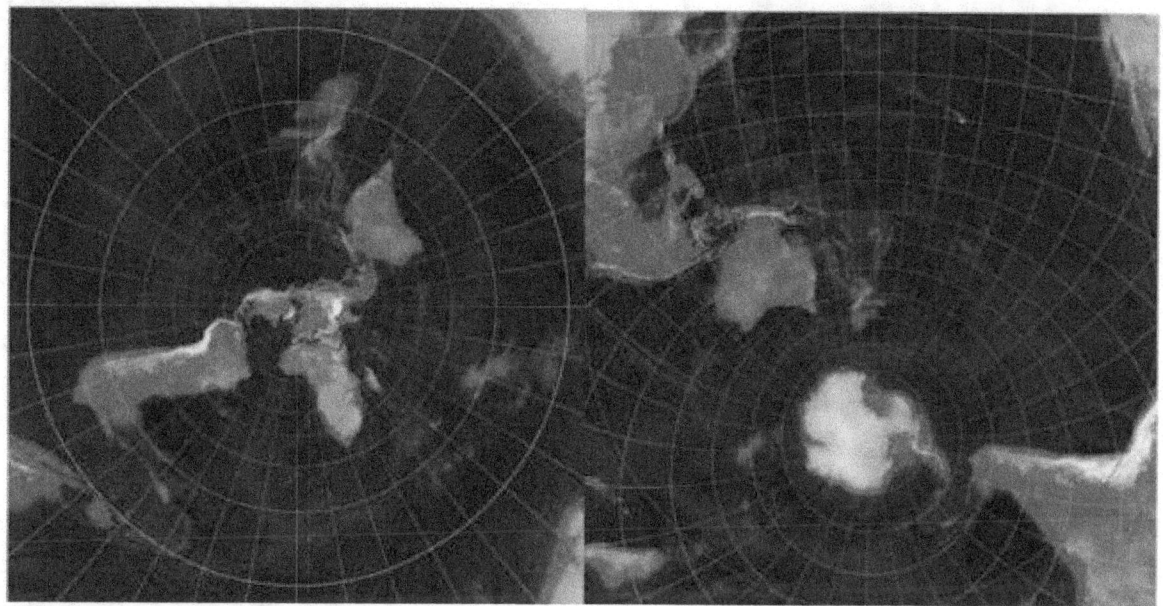

The surface of the Earth requires (at least) two charts to include every point. Here the globe is decomposed into charts around the North and South Poles.

$$= \chi_{\text{right}} \left(a, \quad \begin{pmatrix} \text{top} \\ \sqrt{} \end{pmatrix} \right)$$
$$= \sqrt{1 - a^2}$$

Such a function is called a *transition map*.

The top, bottom, left, and right charts show that the circle is a manifold, but they do not form the only possible atlas. Charts need not be geometric projections, and the number of charts is a matter of some choice. Consider the charts

$$\chi_{\text{minus}}(x, y) = s = \frac{y}{1 + x}$$

and

$$\chi_{\text{plus}}(x, y) = t = \frac{y}{1 - x}$$

Here s is the slope of the line through the point at coordinates (x,y) and the fixed pivot point $(-1, 0)$; t follows similarly, but with pivot point $(+1, 0)$. The inverse mapping from s to (x, y) is given by

$$x = \frac{1 - s^2}{1 + s^2}$$
$$y = \frac{2s}{1 + s^2}$$

It can easily be confirmed that $x^2 + y^2 = 1$ for all values of the slope s. These two charts provide a second atlas for the circle, with

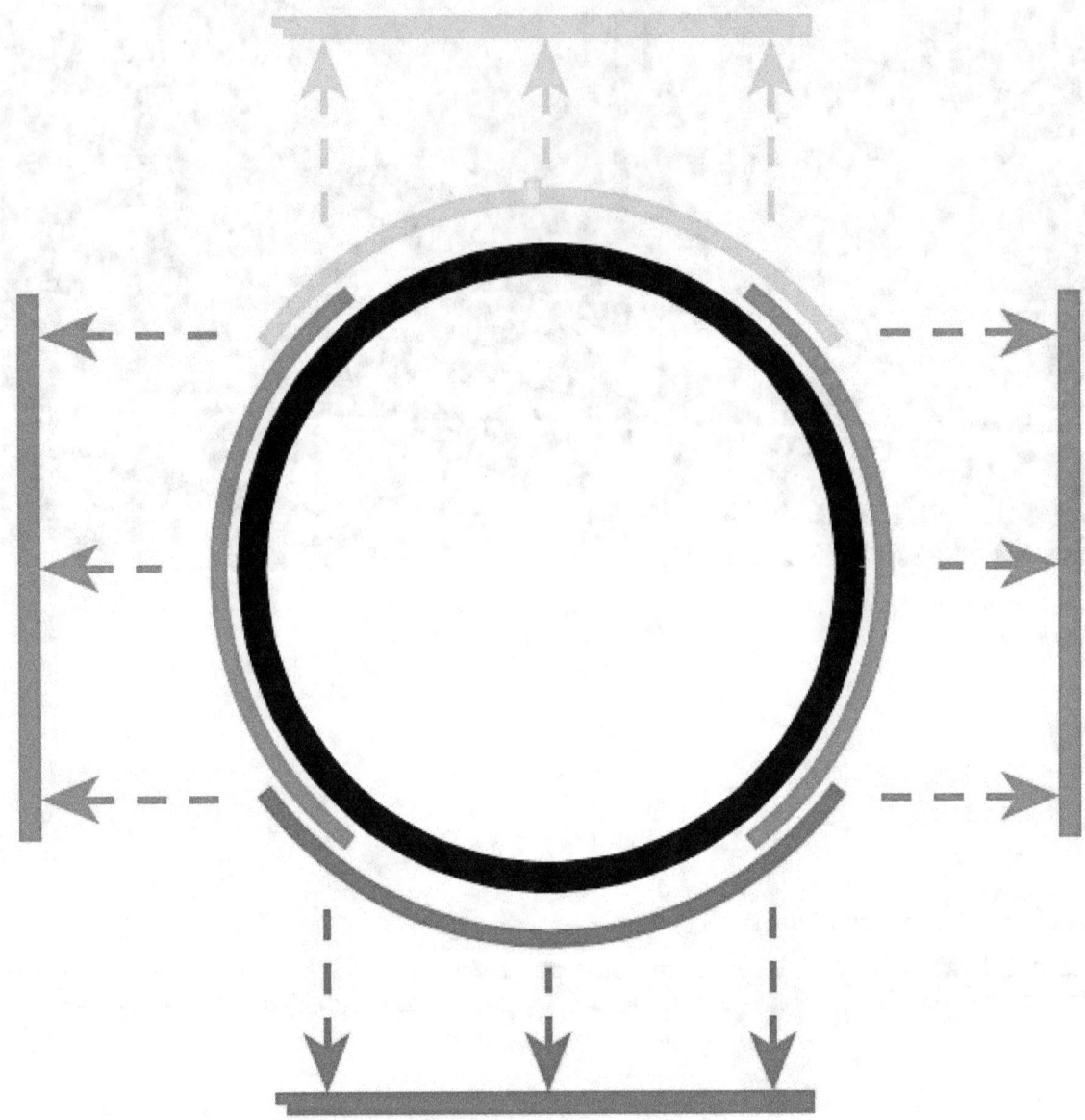

Figure 1: The four charts each map part of the circle to an open interval, and together cover the whole circle.

$$t = \frac{1}{s}$$

Each chart omits a single point, either $(-1, 0)$ for s or $(+1, 0)$ for t so neither chart alone is sufficient to cover the whole circle. It can be proved that it is not possible to cover the full circle with a single chart. For example, although it is possible to construct a circle from a single line interval by overlapping and "gluing" the ends, this does not produce a chart; a portion of the circle will be mapped to both ends at once, losing invertibility.

27.1.2   Enriched circle

Viewed using calculus, the circle transition function T is simply a function between open intervals, which gives a meaning to the statement that T is differentiable. The transition map T, and all the others, are differentiable on $(0, 1)$; therefore,

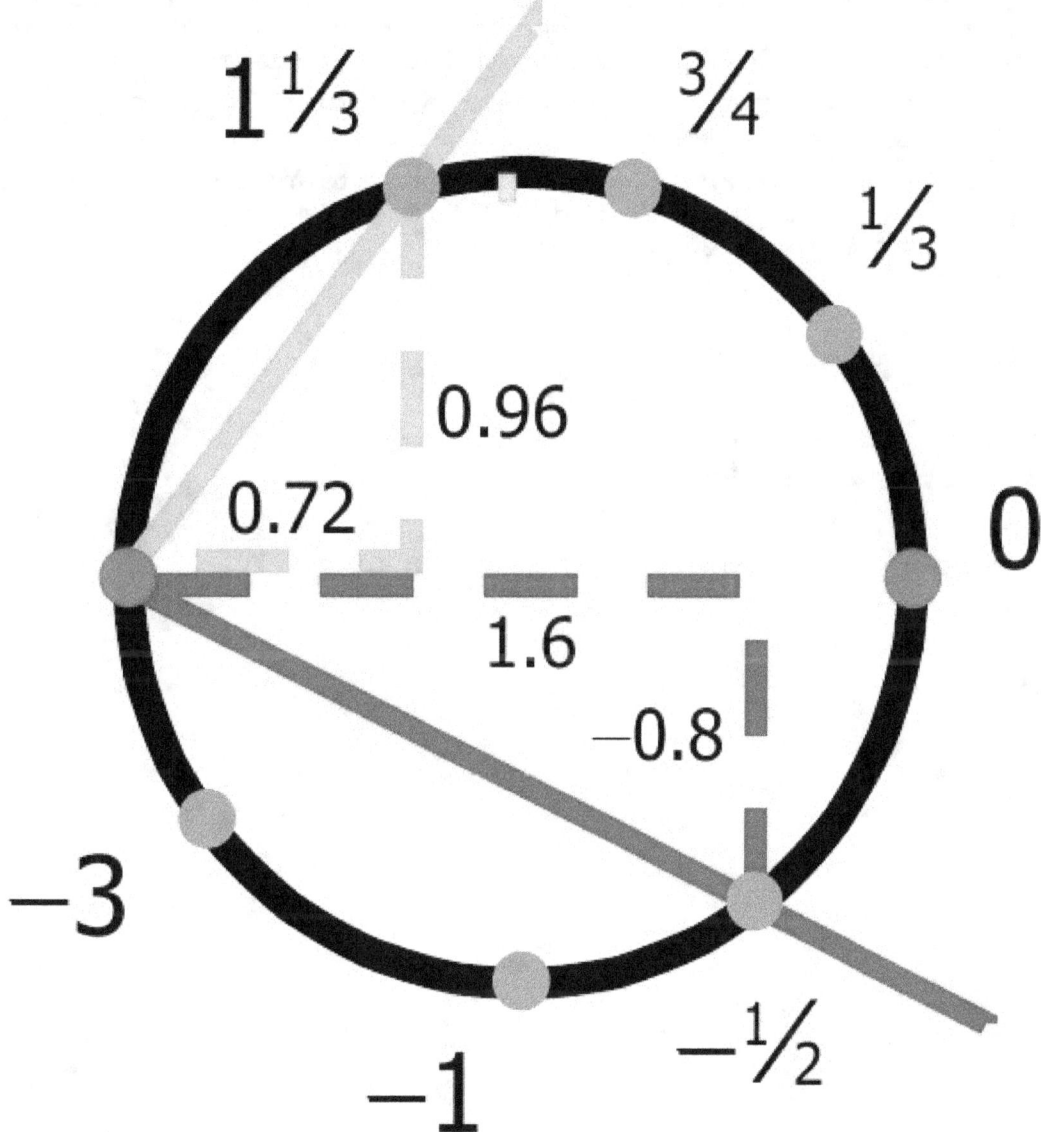

Figure 2: A circle manifold chart based on slope, covering all but one point of the circle.

with this atlas the circle is a *differentiable manifold*. It is also *smooth* and *analytic* because the transition functions have these properties as well.

Other circle properties allow it to meet the requirements of more specialized types of manifold. For example, the circle has a notion of distance between two points, the arc-length between the points; hence it is a *Riemannian manifold*.

27.1.3   Sphere

The sphere is an example of a manifold of dimension 2. The unit sphere of implicit equation

$$x^2 + y^2 + z^2 - 1 = 0$$

may be covered by an atlas of six charts: the plane $z = 0$ divides the sphere into two half spheres ($z > 0$ and $z < 0$), which may both be mapped on the disc $x^2 + y^2 < 1$ by the projection on the xy plane of coordinates. This provides two charts; the four other charts are provided by a similar construction with the two other coordinate planes.

As for the circle, one may define one chart that covers the whole sphere excluding one point. Thus two charts are sufficient, but the sphere cannot be covered by a single chart.

This example is historically significant, as it has motivated the terminology; it became apparent that the whole surface of the Earth cannot have a plane representation consisting of a single map (also called "chart", see nautical chart), and therefore one needs atlases for covering the whole Earth surface.

27.1.4   Other curves

Four manifolds from algebraic curves:
circles,   parabola,   hyperbola,   cubic.

Manifolds need not be connected (all in "one piece"); an example is a pair of separate circles.

Manifolds need not be closed; thus a line segment without its end points is a manifold. And they are never countable, unless the dimension of the manifold is 0. Putting these freedoms together, other examples of manifolds are a parabola, a hyperbola (two open, infinite pieces), and the locus of points on a cubic curve $y^2 = x^3 - x$ (a closed loop piece and an open, infinite piece).

However, excluded are examples like two touching circles that share a point to form a figure-8; at the shared point a satisfactory chart cannot be created. Even with the bending allowed by topology, the vicinity of the shared point looks like a "+", not a line. A "+" is not homeomorphic to a closed interval (line segment), since deleting the center point from the "+" gives a space with four components (i.e. pieces), whereas deleting a point from a closed interval gives a space with at most two pieces; topological operations always preserve the number of pieces.

27.2 History

For more details on this topic, see History of manifolds and varieties.

The study of manifolds combines many important areas of mathematics: it generalizes concepts such as curves and surfaces as well as ideas from linear algebra and topology.

27.2.1 Early development

Before the modern concept of a manifold there were several important results.

Non-Euclidean geometry considers spaces where Euclid's parallel postulate fails. Saccheri first studied them in 1733. Lobachevsky, Bolyai, and Riemann developed them 100 years later. Their research uncovered two types of spaces whose geometric structures differ from that of classical Euclidean space; these gave rise to hyperbolic geometry and elliptic geometry. In the modern theory of manifolds, these notions correspond to Riemannian manifolds with constant negative and positive curvature, respectively.

Carl Friedrich Gauss may have been the first to consider abstract spaces as mathematical objects in their own right. His theorema egregium gives a method for computing the curvature of a surface without considering the ambient space in which the surface lies. Such a surface would, in modern terminology, be called a manifold; and in modern terms, the theorem proved that the curvature of the surface is an intrinsic property. Manifold theory has come to focus exclusively on these intrinsic properties (or invariants), while largely ignoring the extrinsic properties of the ambient space.

Another, more topological example of an intrinsic property of a manifold is its Euler characteristic. Leonhard Euler showed that for a convex polytope in the three-dimensional Euclidean space with V vertices (or corners), E edges, and F faces,

$$V - E + F = 2.$$

The same formula will hold if we project the vertices and edges of the polytope onto a sphere, creating a topological map with V vertices, E edges, and F faces, and in fact, will remain true for any spherical map, even if it does not arise from any convex polytope.[1] Thus 2 is a topological invariant of the sphere, called its **Euler characteristic**. On the other hand, a torus can be sliced open by its 'parallel' and 'meridian' circles, creating a map with $V = 1$ vertex, $E = 2$ edges, and $F = 1$ face. Thus the Euler characteristic of the torus is $1 - 2 + 1 = 0$. The Euler characteristic of other surfaces is a useful topological invariant, which can be extended to higher dimensions using Betti numbers. In the mid nineteenth century, the Gauss–Bonnet theorem linked the Euler characteristic to the Gaussian curvature.

27.2.2 Synthesis

Investigations of Niels Henrik Abel and Carl Gustav Jacobi on inversion of elliptic integrals in the first half of 19th century led them to consider special types of complex manifolds, now known as Jacobians. Bernhard Riemann further contributed to their theory, clarifying the geometric meaning of the process of analytic continuation of functions of complex variables.

Another important source of manifolds in 19th century mathematics was analytical mechanics, as developed by Siméon Poisson, Jacobi, and William Rowan Hamilton. The possible states of a mechanical system are thought to be points of an abstract space, phase space in Lagrangian and Hamiltonian formalisms of classical mechanics. This space is, in fact, a high-dimensional manifold, whose dimension corresponds to the degrees of freedom of the system and where the points are specified by their generalized coordinates. For an unconstrained movement of free particles the manifold is equivalent to the Euclidean space, but various conservation laws constrain it to more complicated formations, e.g. Liouville tori. The theory of a rotating solid body, developed in the 18th century by Leonhard Euler and Joseph-Louis Lagrange, gives another example where the manifold is nontrivial. Geometrical and topological aspects of classical mechanics were emphasized by Henri Poincaré, one of the founders of topology.

Riemann was the first one to do extensive work generalizing the idea of a surface to higher dimensions. The name *manifold* comes from Riemann's original German term, *Mannigfaltigkeit*, which William Kingdon Clifford translated as "manifoldness". In his Göttingen inaugural lecture, Riemann described the set of all possible values of a variable with certain constraints as a *Mannigfaltigkeit*, because the variable can have *many* values. He distinguishes between *stetige Mannigfaltigkeit* and *diskrete Mannigfaltigkeit* (*continuous manifoldness* and *discontinuous manifoldness*), depending on whether the value changes continuously or not. As continuous examples, Riemann refers to not only colors and the locations of objects in space, but also the possible shapes of a spatial figure. Using induction, Riemann constructs an *n-fach ausgedehnte Mannigfaltigkeit* (*n times extended manifoldness* or *n-dimensional manifoldness*) as a continuous stack of (n−1) dimensional manifoldnesses. Riemann's intuitive notion of a *Mannigfaltigkeit* evolved into what is today formalized as a manifold. Riemannian manifolds and Riemann surfaces are named after Riemann.

27.2.3 Poincaré's definition

In his very influential paper, Analysis Situs,[2] Henri Poincaré gave a definition of a (differentiable) manifold (*variété*) which served as a precursor to the modern concept of a manifold.[3]

In the first section of Analysis Situs, Poincaré defines a manifold as the level set of a continuously differentiable function between Euclidean spaces that satisfies the nondegeneracy hypothesis of the implicit function theorem. In the third section, he begins by remarking that the graph of a continuously differentiable function is a manifold in the latter sense. He then proposes a new, more general, definition of manifold based on a 'chain of manifolds' (*une chaîne des variétés*).

Poincaré's notion of a 'chain of manifolds' is a precursor to the modern notion of atlas. In particular, he considers two manifolds defined respectively as graphs of functions $\theta(y)$ and $\theta'(y')$. If these manifolds overlap (*a une partie commune*), then he requires that the coordinates y depend continuously differentiably on the coordinates y' and vice versa ('...*les y sont fonctions analytiques des y' et inversement*'). In this way he introduces a precursor to the notion of a chart and of a transition map. Note that it is implicit in Analysis Situs that a manifold obtained as a 'chain' is a subset of Euclidean space.

For example, the unit circle in the plane can be thought of as the graph of the function $y = \sqrt{1 - x^2}$ or else the function $y = -\sqrt{1 - x^2}$ in a neighborhood of every point except the points (1,0) and (−1,0); and in a neighborhood of those

can be represented by a graph in the neighborhood of every point is because the left hand side of its defining equation $x^2 + y^2 - 1 = 0$ has nonzero gradient at every point of the circle. By the implicit function theorem, every submanifold of Euclidean space is locally the graph of a function.

Hermann Weyl gave an intrinsic definition for differentiable manifolds in his lecture course on Riemann surfaces in 1911–1912, opening the road to the general concept of a topological space that followed shortly. During the 1930s Hassler Whitney and others clarified the foundational aspects of the subject, and thus intuitions dating back to the latter half of the 19th century became precise, and developed through differential geometry and Lie group theory. Notably, the Whitney embedding theorem[4] showed that the intrinsic definition in terms of charts was equivalent to Poincaré's definition in terms of subsets of Euclidean space.

27.2.4   Topology of manifolds: highlights

Two-dimensional manifolds, also known as a 2D *surfaces* embedded in our common 3D space, were considered by Riemann under the guise of Riemann surfaces, and rigorously classified in the beginning of the 20th century by Poul Heegaard and Max Dehn. Henri Poincaré pioneered the study of three-dimensional manifolds and raised a fundamental question about them, today known as the Poincaré conjecture. After nearly a century of effort by many mathematicians, starting with Poincaré himself, a consensus among experts (as of 2006) is that Grigori Perelman has proved the Poincaré conjecture (see the Solution of the Poincaré conjecture). William Thurston's geometrization program, formulated in the 1970s, provided a far-reaching extension of the Poincaré conjecture to the general three-dimensional manifolds. Four-dimensional manifolds were brought to the forefront of mathematical research in the 1980s by Michael Freedman and in a different setting, by Simon Donaldson, who was motivated by the then recent progress in theoretical physics (Yang–Mills theory), where they serve as a substitute for ordinary 'flat' spacetime. Andrey Markov Jr. showed in 1960 that no algorithm exists for classifying four-dimensional manifolds. Important work on higher-dimensional manifolds, including analogues of the Poincaré conjecture, had been done earlier by René Thom, John Milnor, Stephen Smale and Sergei Novikov. One of the most pervasive and flexible techniques underlying much work on the topology of manifolds is Morse theory.

27.3   Mathematical definition

For more details on this topic, see Categories of manifolds.

Informally, a manifold is a space that is "modeled on" Euclidean space.

There are many different kinds of manifolds and generalizations. In geometry and topology, all manifolds are topological manifolds, possibly with additional structure, most often a differentiable structure. In terms of constructing manifolds via patching, a manifold has an additional structure if the transition maps between different patches satisfy axioms beyond just continuity. For instance, differentiable manifolds have homeomorphisms on overlapping neighborhoods diffeomorphic with each other, so that the manifold has a well-defined set of functions which are differentiable in each neighborhood, and so differentiable on the manifold as a whole.

Formally, a **topological manifold**[5] is a second countable Hausdorff space that is locally homeomorphic to Euclidean space.

Second countable and *Hausdorff* are point-set conditions; *second countable* excludes spaces which are in some sense 'too large' such as the long line, while *Hausdorff* excludes spaces such as "the line with two origins" (these generalizations of manifolds are discussed in non-Hausdorff manifolds).

Locally homeomorphic to Euclidean space means[6] that every point has a neighborhood homeomorphic to an open Euclidean n-ball,

$$B^n = \{(x_1, x_2, \ldots , x_n) \in R^n \mid x_1 + x_2 + \cdots + x_n^2 < 1\}.$$

Generally manifolds are taken to have a fixed dimension (the space must be locally homeomorphic to a fixed n-ball), and such a space is called an ***n*-manifold**; however, some authors admit manifolds where different points can have different dimensions.[7] If a manifold has a fixed dimension, it is called a **pure manifold**. For example, the sphere has a constant dimension of 2 and is therefore a pure manifold whereas the disjoint union of a sphere and a line in three-dimensional space is *not* a pure manifold. Since dimension is a local invariant (i.e. the map sending each point to the dimension of its neighbourhood over which a chart is defined, is locally constant), each connected component has a fixed dimension.

Scheme-theoretically, a manifold is a locally ringed space, whose structure sheaf is locally isomorphic to the sheaf of continuous (or differentiable, or complex-analytic, etc.) functions on Euclidean space. This definition is mostly used when discussing analytic manifolds in algebraic geometry.

27.3.1   Broad definition

Main article: Banach manifold

The broadest common definition of manifold is a topological space locally homeomorphic to a topological vector space over the reals. This omits the point-set axioms, allowing higher cardinalities and non-Hausdorff manifolds; and it omits finite dimension, allowing structures such as Hilbert manifolds to be modeled on Hilbert spaces, Banach manifolds to be modeled on Banach spaces, and Fréchet manifolds to be modeled on Fréchet spaces. Usually one relaxes one or the other condition: manifolds with the point-set axioms are studied in general topology, while infinite-dimensional manifolds are studied in functional analysis.

27.4   Charts, atlases, and transition maps

Main article: Atlas (topology)
See also: Differentiable manifold

The spherical Earth is navigated using flat maps or charts, collected in an atlas. Similarly, a differentiable manifold can be described using mathematical maps, called *coordinate charts*, collected in a mathematical *atlas*. It is not generally possible to describe a manifold with just one chart, because the global structure of the manifold is different from the simple structure of the charts. For example, no single flat map can represent the entire Earth without separation of adjacent features across the map's boundaries or duplication of coverage. When a manifold is constructed from multiple overlapping charts, the regions where they overlap carry information essential to understanding the global structure.

27.4.1   Charts

A **coordinate map**, a **coordinate chart**, or simply a **chart**, of a manifold is an invertible map between a subset of the manifold and a simple space such that both the map and its inverse preserve the desired structure.[8] For a topological manifold, the simple space is some Euclidean space $\mathbf{R}^n$ and interest focuses on the topological structure. This structure is preserved by homeomorphisms, invertible maps that are continuous in both directions.

In the case of a differentiable manifold, a set of **charts** called an **atlas** allows us to do calculus on manifolds. Polar coordinates, for example, form a chart for the plane $\mathbf{R}^2$ minus the positive x-axis and the origin. Another example of a chart is the map χ_0 mentioned in the section above, a chart for the circle.

27.4.2   Atlases

Main article: Atlas (topology)

The description of most manifolds requires more than one chart (a single chart is adequate for only the simplest manifolds). A specific collection of charts which covers a manifold is called an **atlas**. An atlas is not unique as all manifolds can be covered multiple ways using different combinations of charts. Two atlases are said to be C^k-equivalent if their union is also a C^k atlas.

The atlas containing all possible charts consistent with a given atlas is called the **maximal atlas** (i.e. an equivalence class containing that given atlas (under the already defined equivalence relation given in the previous paragraph)). Unlike an ordinary atlas, the maximal atlas of a given manifold is unique. Though it is useful for definitions, it is an abstract object and not used directly (e.g. in calculations).

27.4.3 Transition maps

Charts in an atlas may overlap and a single point of a manifold may be represented in several charts. If two charts overlap, parts of them represent the same region of the manifold, just as a map of Europe and a map of Asia may both contain Moscow. Given two overlapping charts, a transition function can be defined which goes from an open ball in R^n to the manifold and then back to another (or perhaps the same) open ball in R^n. The resultant map, like the map T in the circle example above, is called a **change of coordinates**, a **coordinate transformation**, a **transition function**, or a **transition map**.

27.4.4 Additional structure

An atlas can also be used to define additional structure on the manifold. The structure is first defined on each chart separately. If all the transition maps are compatible with this structure, the structure transfers to the manifold.

This is the standard way differentiable manifolds are defined. If the transition functions of an atlas for a topological manifold preserve the natural differential structure of R^n (that is, if they are diffeomorphisms), the differential structure transfers to the manifold and turns it into a differentiable manifold. Complex manifolds are introduced in an analogous way by requiring that the transition functions of an atlas are holomorphic functions. For symplectic manifolds, the transition functions must be symplectomorphisms.

The structure on the manifold depends on the atlas, but sometimes different atlases can be said to give rise to the same structure. Such atlases are called **compatible**.

These notions are made precise in general through the use of pseudogroups.

27.5 Manifold with boundary

See also: Topological manifold § Manifolds with boundary

A **manifold with boundary** is a manifold with an edge. For example a sheet of paper is a 2-manifold with a 1-dimensional boundary. The boundary of an n-manifold with boundary is an $(n - 1)$-manifold. A disk (circle plus interior) is a 2-manifold with boundary. Its boundary is a circle, a 1-manifold. A square with interior is also a 2-manifold with boundary. A ball (sphere plus interior) is a 3-manifold with boundary. Its boundary is a sphere, a 2-manifold. (See also Boundary (topology)).

In technical language, a manifold with boundary is a space containing both interior points and boundary points. Every interior point has a neighborhood homeomorphic to the open n-ball $\{(x_1, x_2, ..., x_n) \mid \Sigma\, x_i^2 < 1\}$. Every boundary point has a neighborhood homeomorphic to the "half" n-ball $\{(x_1, x_2, ..., x_n) \mid \Sigma\, x_i^2 < 1 \text{ and } x_1 \geq 0\}$. The homeomorphism must send each boundary point to a point with $x_1 = 0$.

27.5.1 Boundary and interior

Let M be a manifold with boundary. The **interior** of M, denoted Int M, is the set of points in M which have neighborhoods homeomorphic to an open subset of R^n. The **boundary** of M, denoted ∂M, is the complement of Int M in M. The boundary points can be characterized as those points which land on the boundary hyperplane ($x_n = 0$) of R^n_+ under some coordinate chart.

If M is a manifold with boundary of dimension n, then Int M is a manifold (without boundary) of dimension n and ∂M is a manifold (without boundary) of dimension $n - 1$.

27.6   Construction

A single manifold can be constructed in different ways, each stressing a different aspect of the manifold, thereby leading to a slightly different viewpoint.

27.6.1   Charts

Perhaps the simplest way to construct a manifold is the one used in the example above of the circle. First, a subset of R^2 is identified, and then an atlas covering this subset is constructed. The concept of *manifold* grew historically from constructions like this. Here is another example, applying this method to the construction of a sphere:

Sphere with charts

A sphere can be treated in almost the same way as the circle. In mathematics a sphere is just the surface (not the solid interior), which can be defined as a subset of R^3:

$$S = \{(x, y, z) \in R^3 | x^2 + y^2 + z^2 = 1\}.$$

The sphere is two-dimensional, so each chart will map part of the sphere to an open subset of R^2. Consider the northern hemisphere, which is the part with positive z coordinate (coloured red in the picture on the right). The function χ defined by

$$\chi(x, y, z) = (x, y),$$

maps the northern hemisphere to the open unit disc by projecting it on the (x, y) plane. A similar chart exists for the southern hemisphere. Together with two charts projecting on the (x, z) plane and two charts projecting on the (y, z) plane, an atlas of six charts is obtained which covers the entire sphere.

This can be easily generalized to higher-dimensional spheres.

27.6.2   Patchwork

A manifold can be constructed by gluing together pieces in a consistent manner, making them into overlapping charts. This construction is possible for any manifold and hence it is often used as a characterisation, especially for differentiable and Riemannian manifolds. It focuses on an atlas, as the patches naturally provide charts, and since there is no exterior space involved it leads to an intrinsic view of the manifold.

The manifold is constructed by specifying an atlas, which is itself defined by transition maps. A point of the manifold is therefore an equivalence class of points which are mapped to each other by transition maps. Charts map equivalence classes to points of a single patch. There are usually strong demands on the consistency of the transition maps. For topological manifolds they are required to be homeomorphisms; if they are also diffeomorphisms, the resulting manifold is a differentiable manifold.

This can be illustrated with the transition map $t = 1/s$ from the second half of the circle example. Start with two copies of the line. Use the coordinate s for the first copy, and t for the second copy. Now, glue both copies together by identifying the point t on the second copy with the point $s = 1/t$ on the first copy (the points $t = 0$ and $s = 0$ are not identified with any point on the first and second copy, respectively). This gives a circle.

Intrinsic and extrinsic view

The first construction and this construction are very similar, but they represent rather different points of view. In the first construction, the manifold is seen as embedded in some Euclidean space. This is the *extrinsic view*. When a manifold is viewed in this way, it is easy to use intuition from Euclidean spaces to define additional structure. For example, in a Euclidean space it is always clear whether a vector at some point is tangential or normal to some surface through that point.

The patchwork construction does not use any embedding, but simply views the manifold as a topological space by itself. This abstract point of view is called the *intrinsic view*. It can make it harder to imagine what a tangent vector might be, and there is no intrinsic notion of a normal bundle, but instead there is an intrinsic stable normal bundle.

n-Sphere as a patchwork

The *n*-sphere S^n is a generalisation of the idea of a circle (1-sphere) and sphere (2-sphere) to higher dimensions. An *n*-sphere S^n can be constructed by gluing together two copies of $\mathbf{R}^n$. The transition map between them is defined as

$$\mathbf{R}^n \mid \{0\} \to \mathbf{R}^n \mid \{0\} : x \to x/\mid x \mid^2.$$

This function is its own inverse and thus can be used in both directions. As the transition map is a smooth function, this atlas defines a smooth manifold. In the case $n = 1$, the example simplifies to the circle example given earlier.

27.6.3   Identifying points of a manifold

Main articles: Orbifold and Group action

It is possible to define different points of a manifold to be same. This can be visualized as gluing these points together in a single point, forming a quotient space. There is, however, no reason to expect such quotient spaces to be manifolds. Among the possible quotient spaces that are not necessarily manifolds, orbifolds and CW complexes are considered to be relatively well-behaved. An example of a quotient space of a manifold that is also a manifold is the real projective space identified as a quotient space of the corresponding sphere.

One method of identifying points (gluing them together) is through a right (or left) action of a group, which acts on the manifold. Two points are identified if one is moved onto the other by some group element. If M is the manifold and G is the group, the resulting quotient space is denoted by M / G (or $G \setminus M$).

Manifolds which can be constructed by identifying points include tori and real projective spaces (starting with a plane and a sphere, respectively).

27.6.4   Gluing along boundaries

Main article: Quotient space (topology)

Two manifolds with boundaries can be glued together along a boundary. If this is done the right way, the result is also a manifold. Similarly, two boundaries of a single manifold can be glued together.

Formally, the gluing is defined by a bijection between the two boundaries. Two points are identified when they are mapped onto each other. For a topological manifold this bijection should be a homeomorphism, otherwise the result will not be a topological manifold. Similarly for a differentiable manifold it has to be a diffeomorphism. For other manifolds other structures should be preserved.

A finite cylinder may be constructed as a manifold by starting with a strip $[0, 1] \times [0, 1]$ and gluing a pair of opposite edges on the boundary by a suitable diffeomorphism. A projective plane may be obtained by gluing a sphere with a hole in it to a Möbius strip along their respective circular boundaries.

27.6.5   Cartesian products

The Cartesian product of manifolds is also a manifold.

The dimension of the product manifold is the sum of the dimensions of its factors. Its topology is the product topology, and a Cartesian product of charts is a chart for the product manifold. Thus, an atlas for the product manifold can be constructed using atlases for its factors. If these atlases define a differential structure on the factors, the corresponding atlas defines a differential structure on the product manifold. The same is true for any other structure defined on the factors. If one of the factors has a boundary, the product manifold also has a boundary. Cartesian products may be used to construct tori and finite cylinders, for example, as $S^1 \times S^1$ and $S^1 \times [0, 1]$, respectively.

27.7   Manifolds with additional structure

Main article: Categories of manifolds

27.7.1   Topological manifolds

Main article: topological manifold

The simplest kind of manifold to define is the topological manifold, which looks locally like some "ordinary" Euclidean space R^n. Formally, a topological manifold is a topological space locally homeomorphic to a Euclidean space. This means that every point has a neighbourhood for which there exists a homeomorphism (a bijective continuous function whose inverse is also continuous) mapping that neighbourhood to R^n. These homeomorphisms are the charts of the manifold.

It is to be noted that a *topological* manifold looks locally like a Euclidean space in a rather weak manner: while for each individual chart it is possible to distinguish differentiable functions or measure distances and angles, merely by virtue of being a topological manifold a space does not have any *particular* and *consistent* choice of such concepts. In order to discuss such properties for a manifold, one needs to specify further structure and consider differentiable manifolds and Riemannian manifolds discussed below. In particular, the same underlying topological manifold can have several mutually incompatible classes of differentiable functions and an infinite number of ways to specify distances and angles.

Usually additional technical assumptions on the topological space are made to exclude pathological cases. It is customary to require that the space be Hausdorff and second countable.

The *dimension* of the manifold at a certain point is the dimension of the Euclidean space that the charts at that point map to (number n in the definition). All points in a connected manifold have the same dimension. Some authors require that all charts of a topological manifold map to Euclidean spaces of same dimension. In that case every topological manifold has a topological invariant, its dimension. Other authors allow disjoint unions of topological manifolds with differing dimensions to be called manifolds.

27.7.2   Differentiable manifolds

Main article: Differentiable manifold

For most applications a special kind of topological manifold, namely a **differentiable manifold**, is used. If the local charts on a manifold are compatible in a certain sense, one can define directions, tangent spaces, and differentiable functions on that manifold. In particular it is possible to use calculus on a differentiable manifold. Each point of an n-dimensional differentiable manifold has a tangent space. This is an n-dimensional Euclidean space consisting of the tangent vectors of the curves through the point.

Two important classes of differentiable manifolds are **smooth** and **analytic manifolds**. For smooth manifolds the transition maps are smooth, that is infinitely differentiable. Analytic manifolds are smooth manifolds with the additional condition that the transition maps are analytic (they can be expressed as power series). The sphere can be given analytic structure, as can most familiar curves and surfaces.

There are also topological manifolds, i.e., locally Euclidean spaces, which possess no differentiable structures at all.[9]

A rectifiable set generalizes the idea of a piecewise smooth or rectifiable curve to higher dimensions; however, rectifiable sets are not in general manifolds.

27.7.3 Riemannian manifolds

Main article: Riemannian manifold

To measure distances and angles on manifolds, the manifold must be Riemannian. A 'Riemannian manifold' is a differentiable manifold in which each tangent space is equipped with an inner product ⟨·,·⟩ in a manner which varies smoothly from point to point. Given two tangent vectors u and v, the inner product ⟨u,v⟩ gives a real number. The dot (or scalar) product is a typical example of an inner product. This allows one to define various notions such as length, angles, areas (or volumes), curvature, gradients of functions and divergence of vector fields.

All differentiable manifolds (of constant dimension) can be given the structure of a Riemannian manifold. The Euclidean space itself carries a natural structure of Riemannian manifold (the tangent spaces are naturally identified with the Euclidean space itself and carry the standard scalar product of the space). Many familiar curves and surfaces, including for example all n-spheres, are specified as subspaces of a Euclidean space and inherit a metric from their embedding in it.

27.7.4 Finsler manifolds

Main article: Finsler manifold

A **Finsler manifold** allows the definition of distance but does not require the concept of angle; it is an analytic manifold in which each tangent space is equipped with a norm, $\|\cdot\|$, in a manner which varies smoothly from point to point. This norm can be extended to a metric, defining the length of a curve; but it cannot in general be used to define an inner product.

Any Riemannian manifold is a Finsler manifold.

27.7.5 Lie groups

Main article: Lie group

Lie groups, named after Sophus Lie, are differentiable manifolds that carry also the structure of a group which is such that the group operations are defined by smooth maps.

A Euclidean vector space with the group operation of vector addition is an example of a non-compact Lie group. A simple example of a compact Lie group is the circle: the group operation is simply rotation. This group, known as U(1), can be also characterised as the group of complex numbers of modulus 1 with multiplication as the group operation. Other examples of Lie groups include special groups of matrices, which are all subgroups of the general linear group, the group of n by n matrices with non-zero determinant. If the matrix entries are real numbers, this will be an n^2-dimensional disconnected manifold. The orthogonal groups, the symmetry groups of the sphere and hyperspheres, are $n(n-1)/2$ dimensional manifolds, where $n-1$ is the dimension of the sphere. Further examples can be found in the table of Lie groups.

27.7.6 Other types of manifolds

Main articles: Complex manifold and Symplectic manifold

- A 'complex manifold' is a manifold modeled on C^n with holomorphic transition functions on chart overlaps. These manifolds are the basic objects of study in complex geometry. A one-complex-dimensional manifold is called a Riemann surface. Note that an n-dimensional complex manifold has dimension $2n$ as a real differentiable manifold.

- A 'CR manifold' is a manifold modeled on boundaries of domains in C^n.

- 'Infinite dimensional manifolds': to allow for infinite dimensions, one may consider Banach manifolds which are locally homeomorphic to Banach spaces. Similarly, Fréchet manifolds are locally homeomorphic to Fréchet spaces.

- A 'symplectic manifold' is a kind of manifold which is used to represent the phase spaces in classical mechanics. They are endowed with a 2-form that defines the Poisson bracket. A closely related type of manifold is a contact manifold.

- A 'combinatorial manifold' is a kind of manifold which is discretization of a manifold. It usually means a piecewise linear manifold made by simplicial complexes.

- A 'digital manifold' is a special kind of combinatorial manifold which is defined in digital space. See digital topology

27.8 Classification and invariants

For more details on this topic, see Classification of manifolds.

Different notions of manifolds have different notions of classification and invariant; in this section we focus on smooth closed manifolds.

The classification of smooth closed manifolds is well-understood *in principle*, except in dimension 4: in low dimensions (2 and 3) it is geometric, via the uniformization theorem and the solution of the Poincaré conjecture, and in high dimension (5 and above) it is algebraic, via surgery theory. This is a classification in principle: the general question of whether two smooth manifolds are diffeomorphic is not computable in general. Further, specific computations remain difficult, and there are many open questions.

Orientable surfaces can be visualized, and their diffeomorphism classes enumerated, by genus. Given two orientable surfaces, one can determine if they are diffeomorphic by computing their respective genera and comparing: they are diffeomorphic if and only if the genera are equal, so the genus forms a complete set of invariants.

This is much harder in higher dimensions: higher-dimensional manifolds cannot be directly visualized (though visual intuition is useful in understanding them), nor can their diffeomorphism classes be enumerated, nor can one in general determine if two different descriptions of a higher-dimensional manifold refer to the same object.

However, one can determine if two manifolds are *different* if there is some intrinsic characteristic that differentiates them. Such criteria are commonly referred to as invariants, because, while they may be defined in terms of some presentation (such as the genus in terms of a triangulation), they are the same relative to all possible descriptions of a particular manifold: they are *invariant* under different descriptions.

Naively, one could hope to develop an arsenal of invariant criteria that would definitively classify all manifolds up to isomorphism. Unfortunately, it is known that for manifolds of dimension 4 and higher, no program exists that can decide whether two manifolds are diffeomorphic.

Smooth manifolds have a rich set of invariants, coming from point-set topology, classic algebraic topology, and geometric topology. The most familiar invariants, which are visible for surfaces, are orientability (a normal invariant, also detected by homology) and genus (a homological invariant).

Smooth closed manifolds have no local invariants (other than dimension), though geometric manifolds have local invariants, notably the curvature of a Riemannian manifold and the torsion of a manifold equipped with an affine connection. This distinction between local invariants and no local invariants is a common way to distinguish between geometry and topology. All invariants of a smooth closed manifold are thus global.

Algebraic topology is a source of a number of important global invariant properties. Some key criteria include the *simply connected* property and orientability (see below). Indeed several branches of mathematics, such as homology and homotopy theory, and the theory of characteristic classes were founded in order to study invariant properties of manifolds.

27.9 Examples of surfaces

27.9.1 Orientability

Main article: Orientable manifold

In dimensions two and higher, a simple but important invariant criterion is the question of whether a manifold admits a meaningful orientation. Consider a topological manifold with charts mapping to R^n. Given an ordered basis for R^n, a chart causes its piece of the manifold to itself acquire a sense of ordering, which in 3-dimensions can be viewed as either right-handed or left-handed. Overlapping charts are not required to agree in their sense of ordering, which gives manifolds an important freedom. For some manifolds, like the sphere, charts can be chosen so that overlapping regions agree on their "handedness"; these are *orientable* manifolds. For others, this is impossible. The latter possibility is easy to overlook, because any closed surface embedded (without self-intersection) in three-dimensional space is orientable.

Some illustrative examples of non-orientable manifolds include: (1) the Möbius strip, which is a manifold with boundary, (2) the Klein bottle, which must intersect itself in its 3-space representation, and (3) the real projective plane, which arises naturally in geometry.

Möbius strip

Main article: Möbius strip

Begin with an infinite circular cylinder standing vertically, a manifold without boundary. Slice across it high and low to produce two circular boundaries, and the cylindrical strip between them. This is an orientable manifold with boundary, upon which "surgery" will be performed. Slice the strip open, so that it could unroll to become a rectangle, but keep a grasp on the cut ends. Twist one end 180°, making the inner surface face out, and glue the ends back together seamlessly. This results in a strip with a permanent half-twist: the Möbius strip. Its boundary is no longer a pair of circles, but (topologically) a single circle; and what was once its "inside" has merged with its "outside", so that it now has only a *single* side.

Klein bottle

Main article: Klein bottle

Take two Möbius strips; each has a single loop as a boundary. Straighten out those loops into circles, and let the strips distort into cross-caps. Gluing the circles together will produce a new, closed manifold without boundary, the Klein bottle. Closing the surface does nothing to improve the lack of orientability, it merely removes the boundary. Thus, the Klein bottle is a closed surface with no distinction between inside and outside. Note that in three-dimensional space, a Klein bottle's surface must pass through itself. Building a Klein bottle which is not self-intersecting requires four or more dimensions of space.

Real projective plane

Main article: Real projective space

Begin with a sphere centered on the origin. Every line through the origin pierces the sphere in two opposite points called *antipodes*. Although there is no way to do so physically, it is possible (by considering a quotient space) to mathematically merge each antipode pair into a single point. The closed surface so produced is the real projective plane, yet another non-orientable surface. It has a number of equivalent descriptions and constructions, but this route explains its name: all the points on any given line through the origin project to the same "point" on this "plane".

27.9.2   Genus and the Euler characteristic

For two dimensional manifolds a key invariant property is the genus, or the "number of handles" present in a surface. A torus is a sphere with one handle, a double torus is a sphere with two handles, and so on. Indeed it is possible to fully characterize compact, two-dimensional manifolds on the basis of genus and orientability. In higher-dimensional manifolds genus is replaced by the notion of Euler characteristic, and more generally Betti numbers and homology and cohomology.

27.10   Maps of manifolds

Main article: Maps of manifolds

Just as there are various types of manifolds, there are various types of maps of manifolds. In addition to continuous functions and smooth functions generally, there are maps with special properties. In geometric topology a basic type are embeddings, of which knot theory is a central example, and generalizations such as immersions, submersions, covering spaces, and ramified covering spaces. Basic results include the Whitney embedding theorem and Whitney immersion theorem.

In Riemannian geometry, one may ask for maps to preserve the Riemannian metric, leading to notions of isometric embeddings, isometric immersions, and Riemannian submersions; a basic result is the Nash embedding theorem.

27.10.1   Scalar-valued functions

A basic example of maps between manifolds are scalar-valued functions on a manifold,

$$f: M \to \mathbb{R} \text{ or } f: M \to \mathbb{C},$$

sometimes called regular functions or functionals, by analogy with algebraic geometry or linear algebra. These are of interest both in their own right, and to study the underlying manifold.

In geometric topology, most commonly studied are Morse functions, which yield handlebody decompositions, while in mathematical analysis, one often studies solution to partial differential equations, an important example of which is harmonic analysis, where one studies harmonic functions: the kernel of the Laplace operator. This leads to such functions as the spherical harmonics, and to heat kernel methods of studying manifolds, such as hearing the shape of a drum and some proofs of the Atiyah–Singer index theorem.

27.11   Generalizations of manifolds

- **Orbifolds:** An orbifold is a generalization of manifold allowing for certain kinds of "singularities" in the topology. Roughly speaking, it is a space which locally looks like the quotients of some simple space (*e.g.* Euclidean space) by the actions of various finite groups. The singularities correspond to fixed points of the group actions, and the actions must be compatible in a certain sense.

- **Algebraic varieties and schemes:** Non-singular algebraic varieties over the real or complex numbers are manifolds. One generalizes this first by allowing singularities, secondly by allowing different fields, and thirdly by emulating the patching construction of manifolds: just as a manifold is glued together from open subsets of Euclidean space, an algebraic variety is glued together from affine algebraic varieties, which are zero sets of polynomials over algebraically closed fields. Schemes are likewise glued together from affine schemes, which are a generalization of algebraic varieties. Both are related to manifolds, but are constructed algebraically using sheaves instead of atlases.

  Because of singular points, a variety is in general not a manifold, though linguistically the French *variété*, German *Mannigfaltigkeit* and English *manifold* are largely synonymous. In French an algebraic variety is called *une variété algébrique* (an *algebraic variety*), while a smooth manifold is called *une variété différentielle* (a *differential variety*).

- **Stratified space:** A "stratified space" is a space that can be divided into pieces ("strata"), with each stratum a manifold, with the strata fitting together in prescribed ways (formally, a filtration by closed subsets). There are various technical definitions, notably a Whitney stratified space (see Whitney conditions) for smooth manifolds and a topologically stratified space for topological manifolds. Basic examples include manifold with boundary (top dimensional manifold and codimension 1 boundary) and manifold with corners (top dimensional manifold, codimension 1 boundary, codimension 2 corners). Whitney stratified spaces are a broad class of spaces, including algebraic varieties, analytic varieties, semialgebraic sets, and subanalytic sets.

- **CW-complexes:** A CW complex is a topological space formed by gluing disks of different dimensionality together. In general the resulting space is singular, and hence not a manifold. However, they are of central interest in algebraic topology, especially in homotopy theory, as they are easy to compute with and singularities are not a concern.

- **Homology manifolds:** A homology manifold is a space that behaves like a manifold from the point of view of homology theory. These are not all manifolds, but (in high dimension) can be analyzed by surgery theory similarly to manifolds, and failure to be a manifold is a local obstruction, as in surgery theory.[10]

- **Differential spaces:** Let M be a nonempty set. Suppose that some family of real functions on M was chosen. Denote it by $C \varepsilon R^M$. It is an algebra with respect to the pointwise addition and multiplication. Let M be equipped with the topology induced by C. Suppose also that the following conditions hold. First: for every $H \varepsilon C^\infty(R^i)$, where $i \varepsilon N$, and arbitrary $f_1, \ldots, f_n \varepsilon C$, the composition $H \circ (f_1, \ldots, f_n) \varepsilon C$. Second: every function, which in every point of M locally coincides with some function from C, also belongs to C. A pair (M, C) for which the above conditions hold, is called a Sikorski differential space.[11] [12]

27.12   See also

- Affine geodesic: paths on manifolds

- Directional statistics: statistics on manifolds

- List of manifolds

- Mathematics of general relativity

- Submanifold

27.12.1   By dimension

- Curve (1-manifold)

- Surface (2-manifold)

- 3-manifold

- 4-manifold

- 5-manifold

- Banach manifold

- Fréchet manifold

- Manifolds of mappings

27.13   Notes

[1] The notion of a map can formalized as a cell decomposition.

[2] Poincaré, H.: Analysis Situs. (French) *Journal de l'Ecole Polytechnique*, Serié 11 Gauthier-Villars (1895).

[3] Arnol'd, V. I.: On the teaching of mathematics.(Russian) Uspekhi Mat. Nauk 53 (1998), no. 1(319), 229–234; translation in Russian Math. Surveys 53 (1998), no. 1, 229–236

[4] Whitney H., *Differentiable manifolds*, Ann. of Math. (2), 37 (1936), 645–680.

[5] In the narrow sense of requiring point-set axioms and finite dimension.

[6] Formally, locally homeomorphic means that each point m in the manifold M has a neighborhood homeomorphic to a *neighborhood* in Euclidean space, not to the unit ball specifically. However, given such a homeomorphism, the pre-image of an ϵ-ball gives a homeomorphism between the unit ball and a smaller neighborhood of m, so this is no loss of generality. For topological or differentiable manifolds, one can also ask that every point have a neighborhood homeomorphic to all of Euclidean space (as this is diffeomorphic to the unit ball), but this cannot be done for complex manifolds, as the complex unit ball is not holomorphic to complex space.

[7] E.g. see Riaza, Ricardo (2008), *Differential-Algebraic Systems: Analytical Aspects and Circuit Applications*, World Scientific, p. 110, ISBN 9789812791818; Gunning, R. C. (1990), *Introduction to Holomorphic Functions of Several Variables, Volume 2*, CRC Press, p. 73, ISBN 9780534133092.

[8] Shigeyuki Morita, Teruko Nagase, Katsumi Nomizu (2001). *Geometry of Differential Forms*. American Mathematical Society Bookstore. p. 12. ISBN 0-8218-1045-6.

[9] Kervaire M., *A Manifold which does not admit any differentiable structure*, Comment. Math. Helv., 35 (1961), 1–14.

[10] J. Bryant, S. Ferry, W. Mio, and S. Weinberger, *Topology of homology manifolds*, Annals of Maths. 143, 435–467 (1996)

[11] R. Sikorski, *Abstract covariant derivative*, Coll. Math. 18, 251-272 (1967)

[12] K. Drachal, *Introduction to d–spaces theory*, Math. Aeterna 3, 753-770 (2013)

27.14 References

- Freedman, Michael H., and Quinn, Frank (1990) *Topology of 4-Manifolds.* Princeton University Press. ISBN 0-691-08577-3.

- Guillemin, Victor and Pollack, Alan (1974) *Differential Topology.* Prentice-Hall. ISBN 0-13-212605-2. Inspired by Milnor and commonly used in undergraduate courses.

- Hempel, John (1976) *3-Manifolds.* Princeton University Press. ISBN 0-8218-3695-1.

- Hirsch, Morris, (1997) *Differential Topology.* Springer Verlag. ISBN 0-387-90148-5. The most complete account, with historical insights and excellent, but difficult, problems. The standard reference for those wishing to have a deep understanding of the subject.

- Kirby, Robion C. and Siebenmann, Laurence C. (1977) *Foundational Essays on Topological Manifolds. Smoothings, and Triangulations.* Princeton University Press. ISBN 0-691-08190-5. A detailed study of the category of topological manifolds.

- Lee, John M. (2000) *Introduction to Topological Manifolds.* Springer-Verlag. ISBN 0-387-98759-2.

- Lee, John M. (2002), *Introduction to Smooth Manifolds,* Springer, ISBN 978-0-387-95448-6

- Lee, John M. (2003) *Introduction to Smooth Manifolds.* Springer-Verlag. ISBN 0-387-95495-3.

- Massey, William S. (1977) *Algebraic Topology: An Introduction.* Springer-Verlag. ISBN 0-387-90271-6.

- Milnor, John (1997) *Topology from the Differentiable Viewpoint.* Princeton University Press. ISBN 0-691-04833-9.

- Munkres, James R. (2000) *Topology.* Prentice Hall. ISBN 0-13-181629-2.

- Neuwirth, L. P., ed. (1975) *Knots, Groups, and 3-Manifolds. Papers Dedicated to the Memory of R. H. Fox.* Princeton University Press. ISBN 978-0-691-08170-0.

- Riemann, Bernhard, *Gesammelte mathematische Werke und wissenschaftlicher Nachlass,* Sändig Reprint. ISBN 3-253-03059-8.

  - *Grundlagen für eine allgemeine Theorie der Functionen einer veränderlichen complexen Grösse.* The 1851 doctoral thesis in which "manifold" (*Mannigfaltigkeit*) first appears.

  - *Ueber die Hypothesen, welche der Geometrie zu Grunde liegen.* The 1854 Göttingen inaugural lecture (*Habilitationsschrift*).

- Spivak, Michael (1965) *Calculus on Manifolds: A Modern Approach to Classical Theorems of Advanced Calculus.* HarperCollins Publishers. ISBN 0-8053-9021-9. The standard graduate text.

27.15 External links

- Hazewinkel, Michiel, ed. (2001), "Manifold", *Encyclopedia of Mathematics,* Springer, ISBN 978-1-55608-010-4

- Dimensions-math.org (A film explaining and visualizing manifolds up to fourth dimension.)

- The manifold atlas project of the Max Planck Institute for Mathematics in Bonn

The chart maps the part of the sphere with positive z coordinate to a disc.

A finite cylinder is a manifold with boundary.

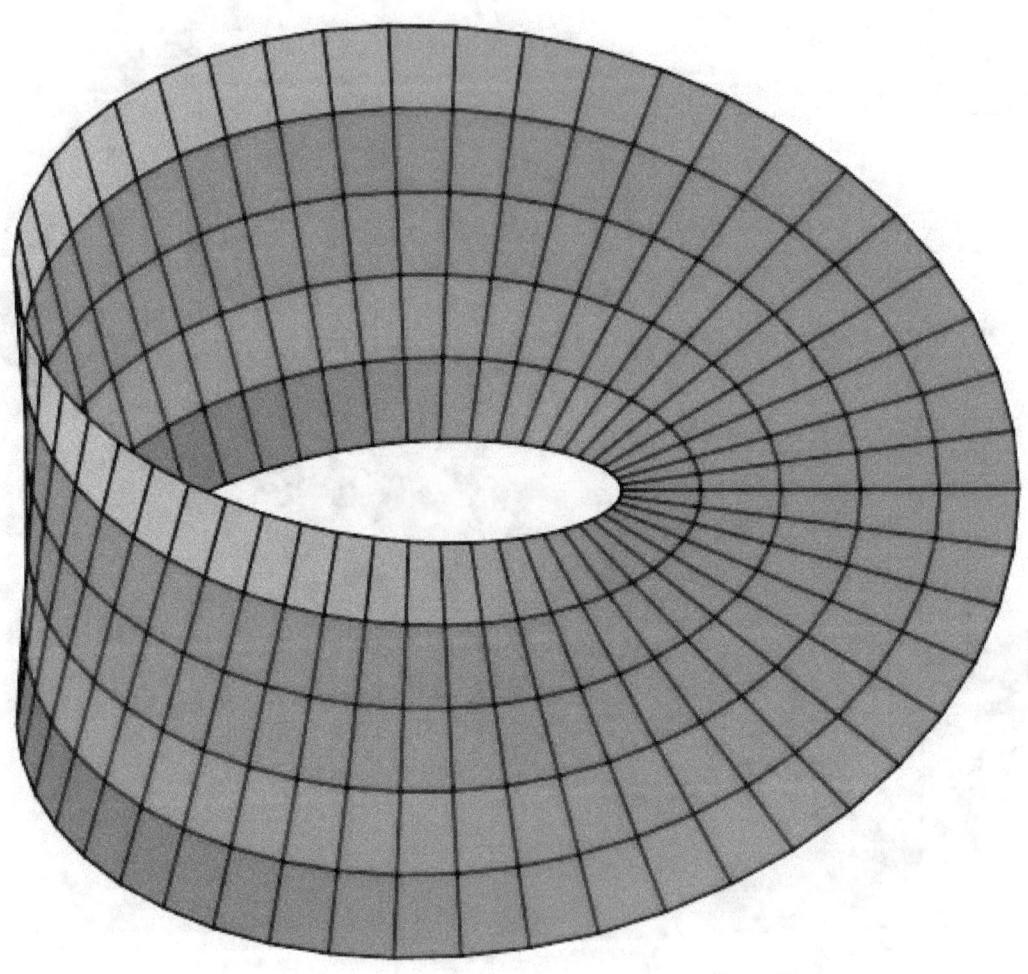

Möbius strip

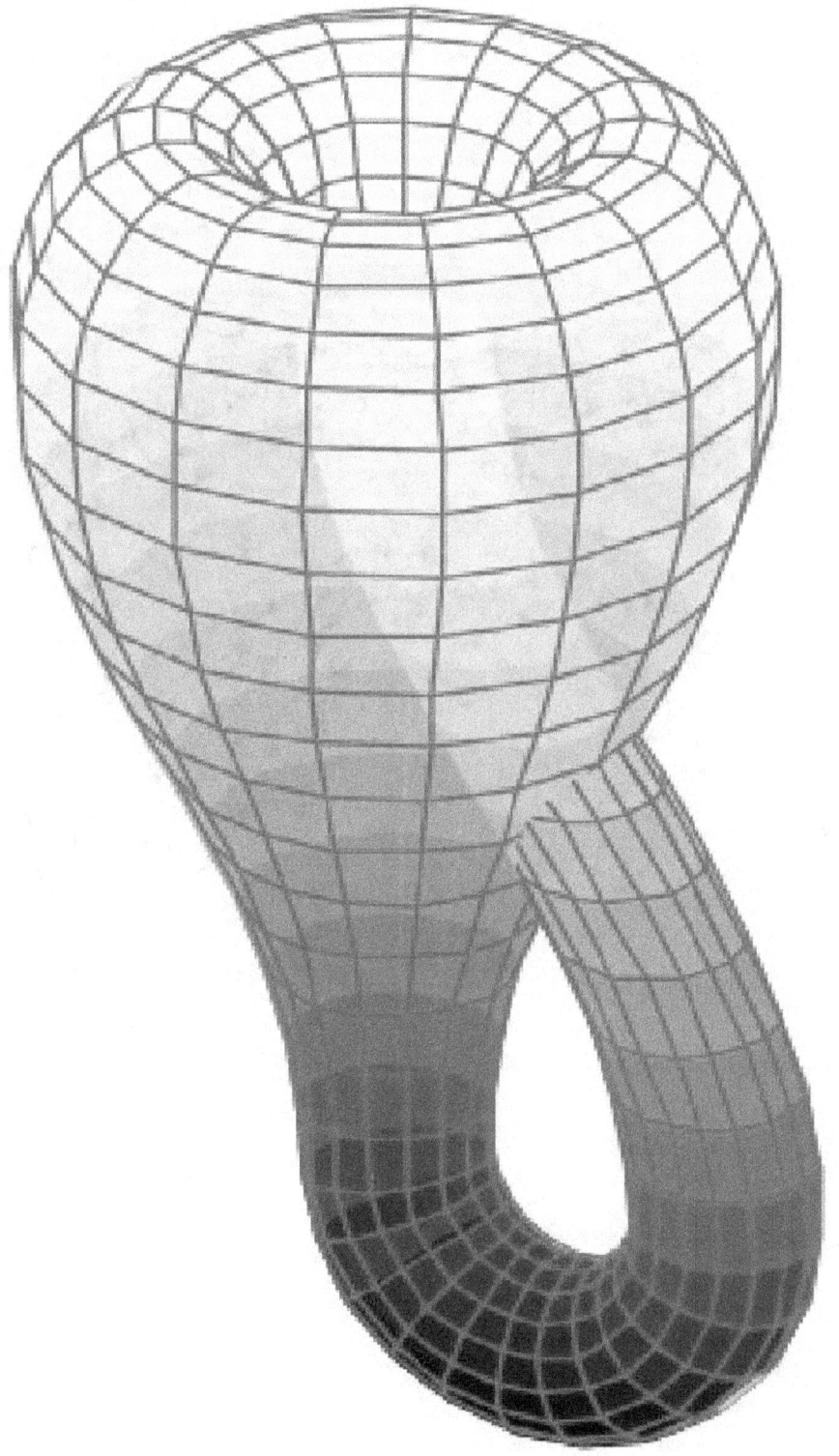

The Klein bottle immersed in three-dimensional space

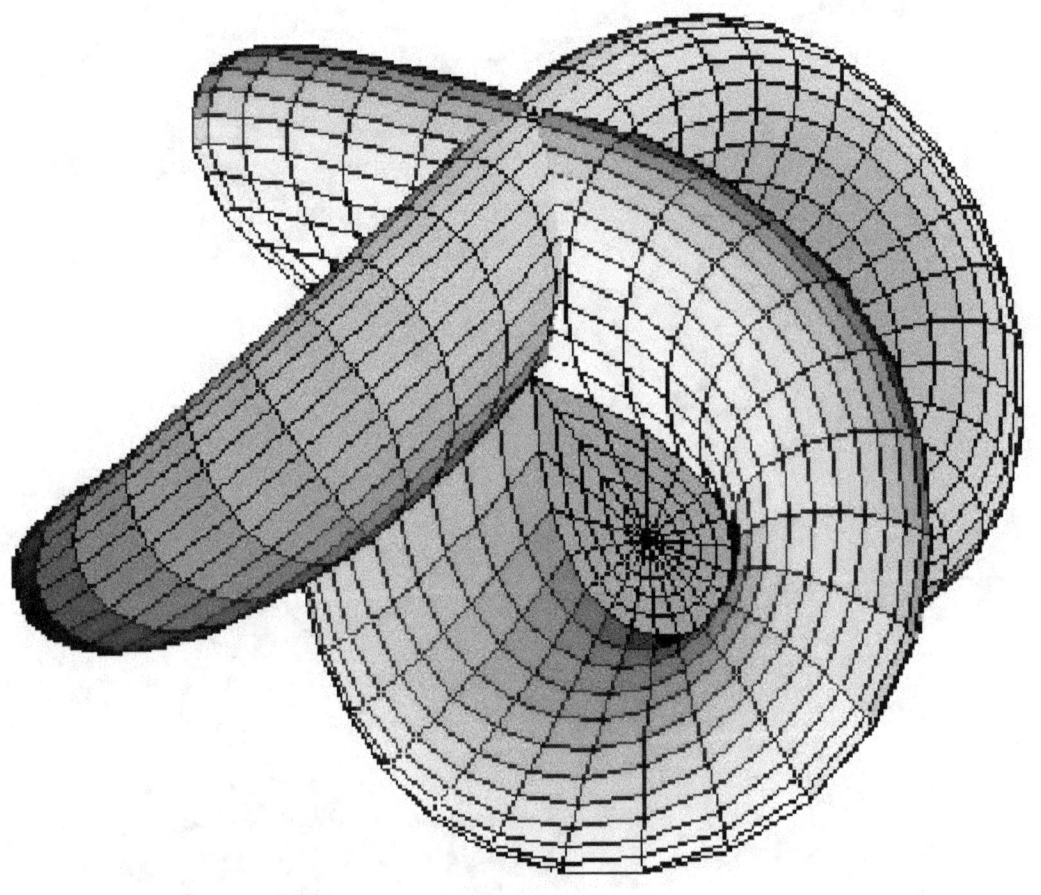

A Morin surface, an immersion used in sphere eversion

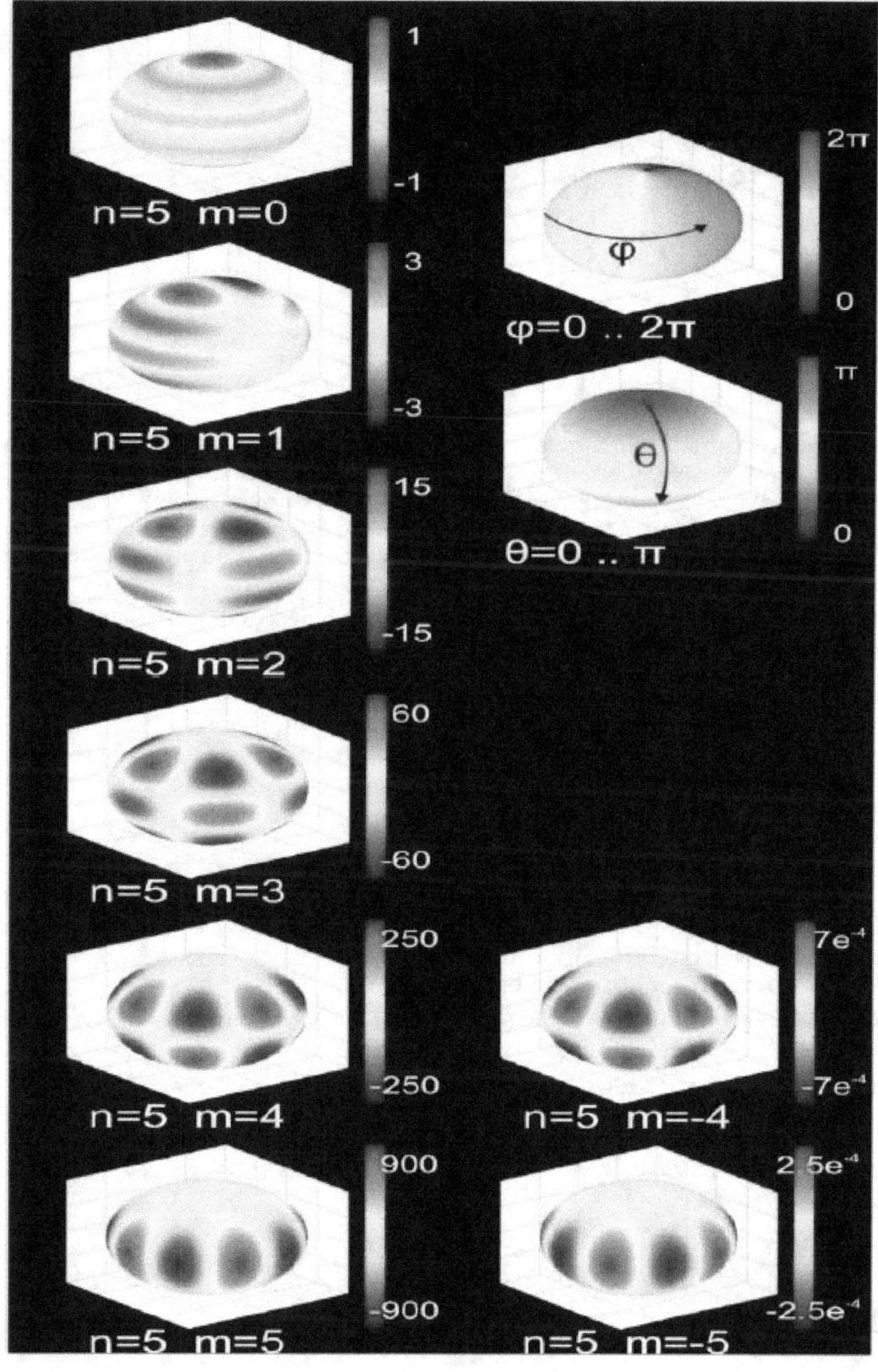

Chapter 28

Probability theory

Probability theory is the branch of mathematics concerned with probability, the analysis of random phenomena.[1] The central objects of probability theory are random variables, stochastic processes, and events: mathematical abstractions of non-deterministic events or measured quantities that may either be single occurrences or evolve over time in an apparently random fashion. If an individual coin toss or the roll of dice is considered to be a random event, then if repeated many times the sequence of random events will exhibit certain patterns, which can be studied and predicted. Two representative mathematical results describing such patterns are the law of large numbers and the central limit theorem.

As a mathematical foundation for statistics, probability theory is essential to many human activities that involve quantitative analysis of large sets of data. Methods of probability theory also apply to descriptions of complex systems given only partial knowledge of their state, as in statistical mechanics. A great discovery of twentieth century physics was the probabilistic nature of physical phenomena at atomic scales, described in quantum mechanics.

28.1   History

The mathematical theory of probability has its roots in attempts to analyze games of chance by Gerolamo Cardano in the sixteenth century, and by Pierre de Fermat and Blaise Pascal in the seventeenth century (for example the "problem of points"). Christiaan Huygens published a book on the subject in 1657[2] and in the 19th century a big work was done by Laplace in what can be considered today as the classic interpretation.[3]

Initially, probability theory mainly considered **discrete** events, and its methods were mainly combinatorial. Eventually, analytical considerations compelled the incorporation of **continuous** variables into the theory.

This culminated in modern probability theory, on foundations laid by Andrey Nikolaevich Kolmogorov. Kolmogorov combined the notion of sample space, introduced by Richard von Mises, and **measure theory** and presented his axiom system for probability theory in 1933. Fairly quickly this became the mostly undisputed axiomatic basis for modern probability theory but alternatives exist, in particular the adoption of finite rather than countable additivity by Bruno de Finetti.[4]

28.2   Treatment

Most introductions to probability theory treat discrete probability distributions and continuous probability distributions separately. The more mathematically advanced measure theory based treatment of probability covers both the discrete, the continuous, any mix of these two and more.

28.2.1 Motivation

Consider an experiment that can produce a number of outcomes. The set of all outcomes is called the *sample space* of the experiment. The *power set* of the sample space is formed by considering all different collections of possible results. For example, rolling an honest die produces one of six possible results. One collection of possible results corresponds to getting an odd number. Thus, the subset {1,3,5} is an element of the power set of the sample space of die rolls. These collections are called *events*. In this case, {1,3,5} is the event that the die falls on some odd number. If the results that actually occur fall in a given event, that event is said to have occurred.

Probability is a way of assigning every "event" a value between zero and one, with the requirement that the event made up of all possible results (in our example, the event {1,2,3,4,5,6}) be assigned a value of one. To qualify as a probability distribution, the assignment of values must satisfy the requirement that if you look at a collection of mutually exclusive events (events that contain no common results, e.g., the events {1,6}, {3}, and {2,4} are all mutually exclusive), the probability that one of the events will occur is given by the sum of the probabilities of the individual events.[5]

The probability that any one of the events {1,6}, {3}, or {2,4} will occur is 5/6. This is the same as saying that the probability of event {1,2,3,4,6} is 5/6. This event encompasses the possibility of any number except five being rolled. The mutually exclusive event {5} has a probability of 1/6, and the event {1,2,3,4,5,6} has a probability of 1, that is, absolute certainty.

28.2.2 Discrete probability distributions

Main article: Discrete probability distribution
Discrete probability theory deals with events that occur in countable sample spaces.

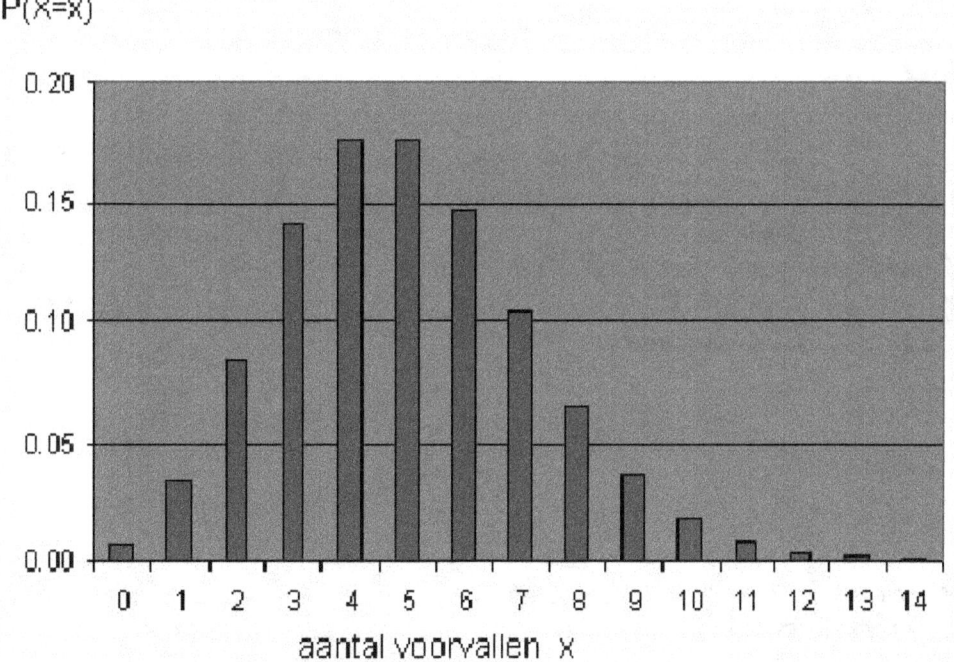

The Poisson distribution, a discrete probability distribution.

Examples: Throwing dice, experiments with decks of cards, random walk, and tossing coins

Classical definition: Initially the probability of an event to occur was defined as number of cases favorable for the event, over the number of total outcomes possible in an equiprobable sample space: see Classical definition of probability.

For example, if the event is "occurrence of an even number when a die is rolled", the probability is given by $\frac{3}{6} = \frac{1}{2}$ 3 faces out of the 6 have even numbers and each face has the same probability of appearing.

Modern definition: The modern definition starts with a finite or countable set called the **sample space**, which relates to the set of all *possible outcomes* in classical sense, denoted by Ω . It is then assumed that for each element $x \in \Omega$, an intrinsic "probability" value $f(x)$ is attached, which satisfies the following properties:

1. $f(x) \in [0, 1]$ for all $x \in \Omega$;
2. $\sum_{x \in \Omega} f(x) = 1$.

That is, the probability function $f(x)$ lies between zero and one for every value of x in the sample space Ω, and the sum of $f(x)$ over all values x in the sample space Ω is equal to 1. An **event** is defined as any subset E of the sample space Ω . The **probability** of the event E is defined as

$$P(E) = \sum_{x \in E} f(x).$$

So, the probability of the entire sample space is 1, and the probability of the null event is 0.

The function $f(x)$ mapping a point in the sample space to the "probability" value is called a **probability mass function** abbreviated as **pmf**. The modern definition does not try to answer how probability mass functions are obtained; instead it builds a theory that assumes their existence.

28.2.3   Continuous probability distributions

Main article: Continuous probability distribution
 Continuous probability theory deals with events that occur in a continuous sample space.

Classical definition: The classical definition breaks down when confronted with the continuous case. See Bertrand's paradox.

Modern definition: If the outcome space of a random variable X is the set of real numbers ($\mathbb{R}$) or a subset thereof, then a function called the **cumulative distribution function** (or **cdf**) F exists, defined by $F(x) = P(X \leq x)$. That is, $F(x)$ returns the probability that X will be less than or equal to x.

The cdf necessarily satisfies the following properties.

1. F is a monotonically non-decreasing, right-continuous function;
2. $\lim_{x \to -\infty} F(x) = 0$;
3. $\lim_{x \to \infty} F(x) = 1$.

If F is absolutely continuous, i.e., its derivative exists and integrating the derivative gives us the cdf back again, then the random variable X is said to have a **probability density function** or **pdf** or simply density $f(x) = \frac{dF(x)}{dx}$.

For a set $E \subseteq \mathbb{R}$, the probability of the random variable X being in E is

$$P(X \in E) = \int_{x \in E} dF(x).$$

In case the probability density function exists, this can be written as

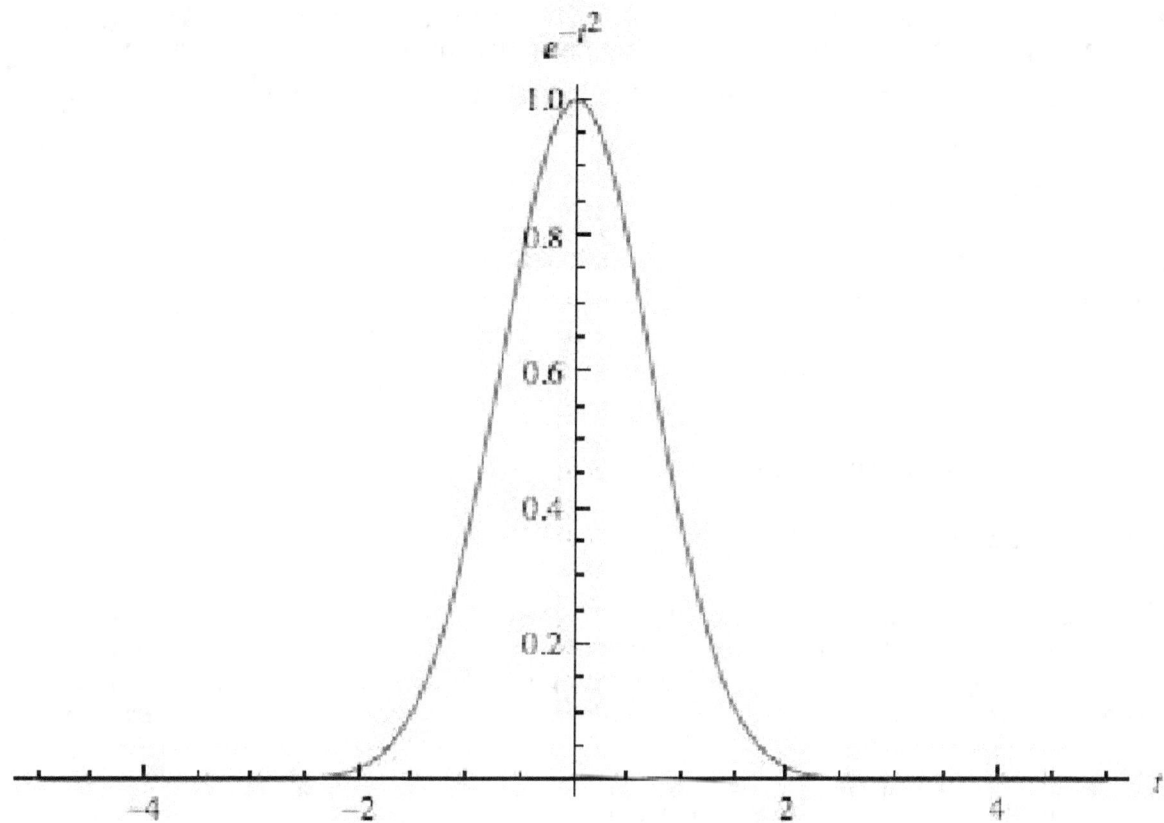

The normal distribution, a continuous probability distribution.

$$P(X_\epsilon\ E) = \int_{x_\epsilon\ E} f(x)\ dx.$$

Whereas the *pdf* exists only for continuous random variables, the *cdf* exists for all random variables (including discrete random variables) that take values in $\mathbb{R}$.

These concepts can be generalized for multidimensional cases on $\mathbb{R}^n$ and other continuous sample spaces.

28.2.4   Measure-theoretic probability theory

The *raison d'être* of the measure-theoretic treatment of probability is that it unifies the discrete and the continuous cases, and makes the difference a question of which measure is used. Furthermore, it covers distributions that are neither discrete nor continuous nor mixtures of the two.

An example of such distributions could be a mix of discrete and continuous distributions—for example, a random variable that is 0 with probability 1/2, and takes a random value from a normal distribution with probability 1/2. It can still be studied to some extent by considering it to have a pdf of $(\delta[x] + \varphi(x))/2$, where $\delta[x]$ is the Dirac delta function.

Other distributions may not even be a mix, for example, the Cantor distribution has no positive probability for any single point, neither does it have a density. The modern approach to probability theory solves these problems using measure theory to define the probability space:

Given any set Ω , (also called **sample space**) and a σ-algebra F on it, a measure P defined on F is called a **probability measure** if $P(\Omega) = 1$.

If F is the Borel σ-algebra on the set of real numbers, then there is a unique probability measure on F for any cdf, and vice versa. The measure corresponding to a cdf is said to be **induced** by the cdf. This measure coincides with the pmf for discrete variables and pdf for continuous variables, making the measure-theoretic approach free of fallacies.

The *probability* of a set E in the σ-algebra F is defined as

$$P(E) = \int_{\omega \in E} \mu_F(d\omega)$$

where the integration is with respect to the measure μ_F induced by F.

Along with providing better understanding and unification of discrete and continuous probabilities, measure-theoretic treatment also allows us to work on probabilities outside R^n, as in the theory of stochastic processes. For example to study Brownian motion, probability is defined on a space of functions.

28.3   Classical probability distributions

Main article: Probability distributions

Certain random variables occur very often in probability theory because they well describe many natural or physical processes. Their distributions therefore have gained *special importance* in probability theory. Some fundamental *discrete distributions* are the discrete uniform, Bernoulli, binomial, negative binomial, Poisson and geometric distributions. Important *continuous distributions* include the continuous uniform, normal, exponential, gamma and beta distributions.

28.4   Convergence of random variables

Main article: Convergence of random variables

In probability theory, there are several notions of convergence for random variables. They are listed below in the order of strength, i.e., any subsequent notion of convergence in the list implies convergence according to all of the preceding notions.

  Weak convergence: A sequence of random variables $X_1, X_2, \ldots$, converges weakly to the random variable X if their respective cumulative *distribution functions* $F_1, F_2, \ldots$ converge to the cumulative distribution function F of X, wherever F is continuous. Weak convergence is also called **convergence in distribution.**

  Most common shorthand notation: $X_n \xrightarrow{D} X$.

  Convergence in probability: The sequence of random variables $X_1, X_2, \ldots$ is said to converge towards the random variable X in probability if $\lim_{n \to \infty} P(|X_n - X| \geq \varepsilon) = 0$ for every $\varepsilon > 0$.

  Most common shorthand notation: $X_n \xrightarrow{P} X$.

  Strong convergence: The sequence of random variables $X_1, X_2, \ldots$ is said to converge towards the random variable X **strongly** if $P(\lim_{n \to \infty} X_n = X) = 1$. Strong convergence is also known as **almost sure convergence.**

  Most common shorthand notation: $X_n \xrightarrow{a.s.} X$.

As the names indicate, weak convergence is weaker than strong convergence. In fact, strong convergence implies convergence in probability, and convergence in probability implies weak convergence. The reverse statements are not always true.

28.4.1 Law of large numbers

Main article: Law of large numbers

Common intuition suggests that if a fair coin is tossed many times, then *roughly* half of the time it will turn up *heads*, and the other half it will turn up *tails*. Furthermore, the more often the coin is tossed, the more likely it should be that the ratio of the number of *heads* to the number of *tails* will approach unity. Modern probability provides a formal version of this intuitive idea, known as the **law of large numbers**. This law is remarkable because it is not assumed in the foundations of probability theory, but instead emerges from these foundations as a theorem. Since it links theoretically derived probabilities to their actual frequency of occurrence in the real world, the law of large numbers is considered as a pillar in the history of statistical theory and has had widespread influence.[6]

The **law of large numbers** (LLN) states that the sample average

$$\overline{X_n} = \frac{1}{n}\sum_{k=1}^{n} X_k$$

of a sequence of independent and identically distributed random variables X_k converges towards their common expectation μ, provided that the expectation of $|X_k|$ is finite.

It is in the different forms of convergence of random variables that separates the *weak* and the *strong* law of large numbers

law: Weak $\overline{X}_n \xrightarrow{P} \mu$ for $n \to \infty$
law: Strong $\overline{X}_n \xrightarrow{a.s.} \mu$ for $n \to \infty$.

It follows from the LLN that if an event of probability p is observed repeatedly during independent experiments, the ratio of the observed frequency of that event to the total number of repetitions converges towards p.

For example, if $Y_1, Y_2, \ldots$ are independent Bernoulli random variables taking values 1 with probability p and 0 with probability $1-p$, then $E(Y_i) = p$ for all i, so that $\overline{Y}_n$ converges to p almost surely.

28.4.2 Central limit theorem

Main article: Central limit theorem

"The central limit theorem (CLT) is one of the great results of mathematics." (Chapter 18 in[7]) It explains the ubiquitous occurrence of the normal distribution in nature.

The theorem states that the average of many independent and identically distributed random variables with finite variance tends towards a normal distribution *irrespective* of the distribution followed by the original random variables. Formally, let $X_1, X_2, \ldots$ be independent random variables with mean μ and variance $\sigma^2 > 0$. Then the sequence of random variables

$$Z_n = \frac{\Sigma_n}{\sigma\sqrt{n}}$$

converges in distribution to a standard normal random variable.

Notice that for some classes of random variables the classic central limit theorem works rather fast (see Berry–Esseen theorem), for example the distributions with finite first, second and third moment from the exponential family, on the other hand for some random variables of the heavy tail and fat tail variety, it works very slow or may not work at all: in such cases one may use the Generalized Central Limit Theorem (GCLT).

28.5   See also

28.6   Notes

[1] "Probability theory, Encyclopaedia Britannica". Britannica.com. Retrieved 2012-02-12.

[2] Grinstead, Charles Miller; James Laurie Snell. "Introduction". *Introduction to Probability*. pp. vii.

[3] Hájek, Alan. "Interpretations of Probability". Retrieved 2012-06-20.

[4] ""The origins and legacy of Kolmogorov's Grundbegriff e", by Glenn Shafer and Vladimir Vovk" (PDF). Retrieved 2012-02-12.

[5] Ross, Sheldon. *A First course in Probability*, 8th Edition. Page 26–27.

[6] "Leithner & Co Pty Ltd - Value Investing, Risk and Risk Management - Part I". Leithner.com.au. 2000-09-15. Retrieved 2012-02-12.

[7] David Williams, "Probability with martingales", Cambridge 1991/2008

28.7   References

- Pierre Simon de Laplace (1812). *Analytical Theory of Probability*.

  The first major treatise blending calculus with probability theory, originally in French: *Théorie Analytique des Probabilités*.

- A. Kolmogoroff (1933). *Grundbegriffe der Wahrscheinlichkeitsrechnung*. doi:10.1007/978-3-642-49888-6. ISBN 978-3-642-49888-6.

  An English translation by Nathan Morrison appeared under the title *Foundations of the Theory of Probability* (Chelsea, New York) in 1950, with a second edition in 1956.

- Patrick Billingsley (1979). *Probability and Measure*. New York, Toronto, London: John Wiley and Sons.

- Olav Kallenberg; *Foundations of Modern Probability*, 2nd ed. Springer Series in Statistics. (2002). 650 pp. ISBN 0-387-95313-2

- Henk Tijms (2004). *Understanding Probability*. Cambridge Univ. Press.

  A lively introduction to probability theory for the beginner.

- Olav Kallenberg; *Probabilistic Symmetries and Invariance Principles*. Springer -Verlag, New York (2005). 510 pp. ISBN 0-387-25115-4

- Gut, Allan (2005). *Probability: A Graduate Course*. Springer-Verlag. ISBN 0-387-22833-0.

28.8   External links

- Animation on YouTube on the probability space of dice.

Chapter 29

Statistics

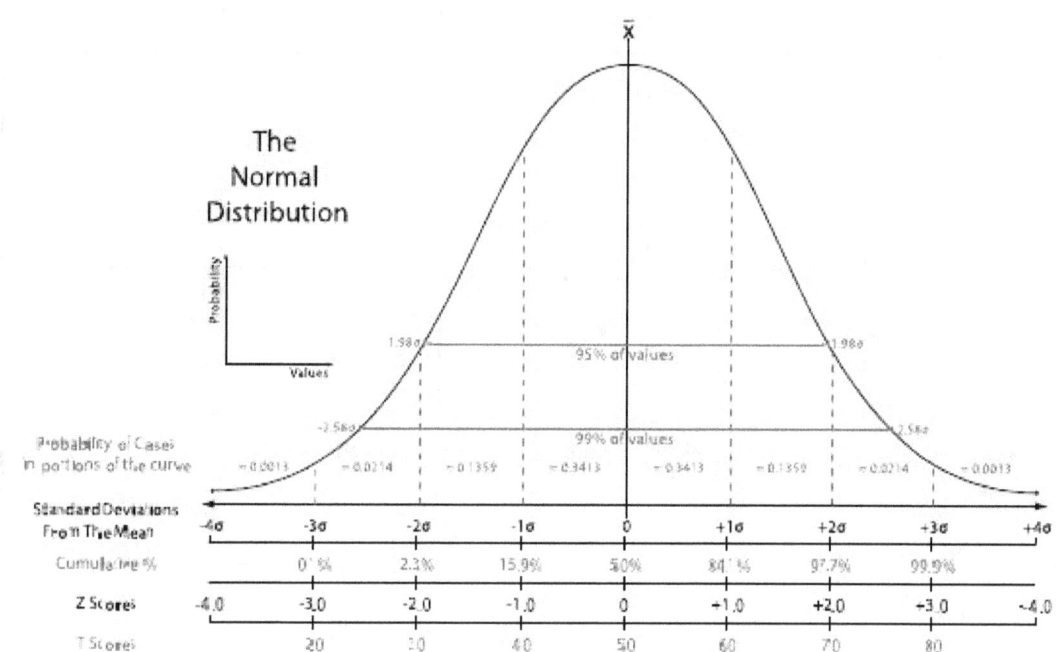

The
Normal
Distribution

Probability of Cases
in portions of the curve

Standard Deviations
From The Mean

Cumulative %

Z Scores

T Scores

More probability density is found as one gets closer to the expected (mean) value in a normal distribution. Statistics used in standardized testing assessment are shown. The scales include standard deviations, cumulative percentages, percentile equivalents, Z-scores, T-scores, standard nines, and percentages in standard nines.

Statistics is the study of the collection, analysis, interpretation, presentation, and organization of data.[1] In applying statistics to, e.g., a scientific, industrial, or societal problem, it is conventional to begin with a statistical population or a statistical model process to be studied. Populations can be diverse topics such as "all persons living in a country" or "every atom composing a crystal". Statistics deals with all aspects of data including the planning of data collection in terms of the design of surveys and experiments.[1]

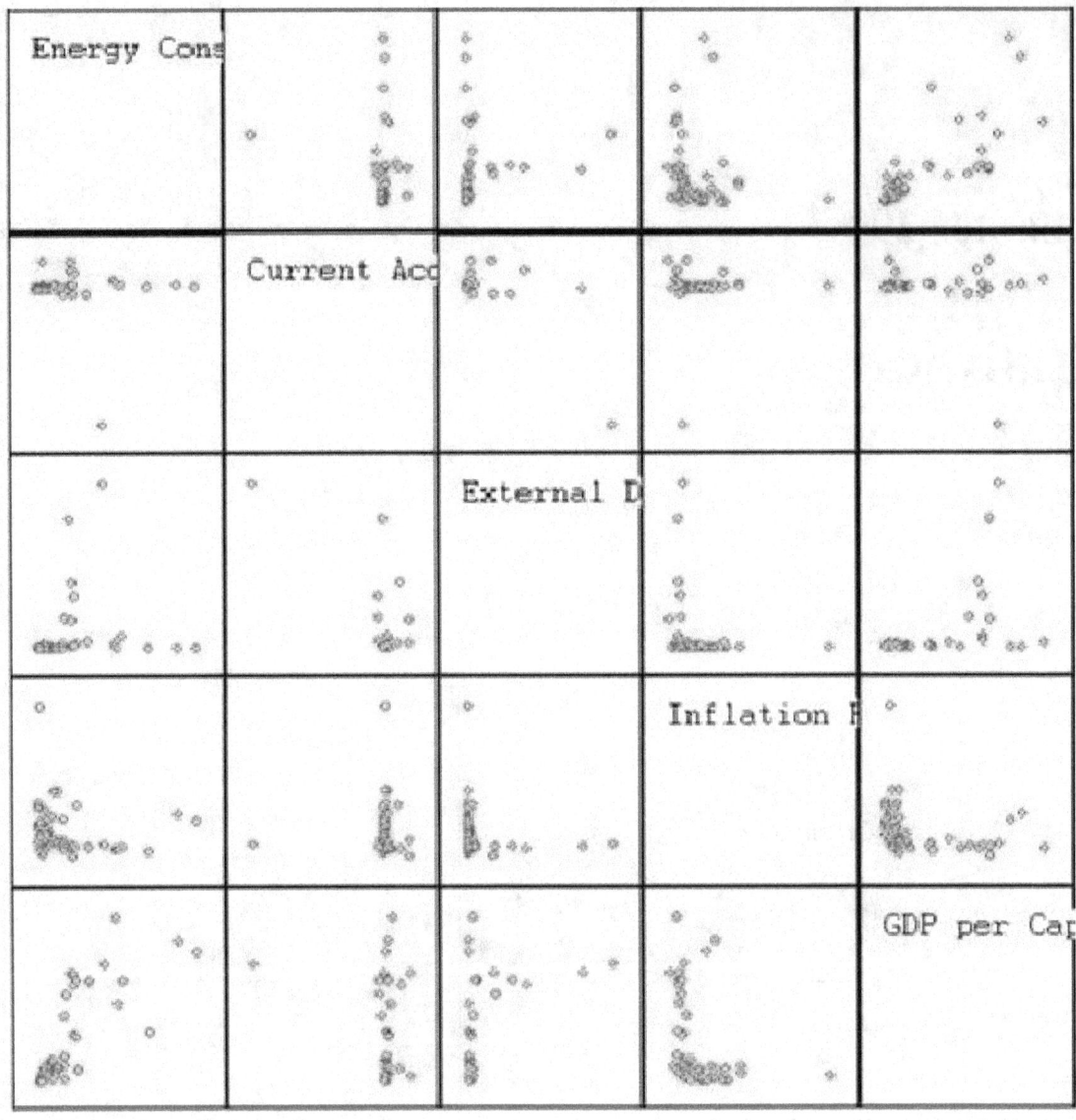

Scatter plots are used in descriptive statistics to show the observed relationships between different variables.

When census data cannot be collected, statisticians collect data by developing specific experiment designs and survey samples. Representative sampling assures that inferences and conclusions can safely extend from the sample to the population as a whole. An experimental study involves taking measurements of the system under study, manipulating the system, and then taking additional measurements using the same procedure to determine if the manipulation has modified the values of the measurements. In contrast, an observational study does not involve experimental manipulation.

Two main statistical methodologies are used in data analysis: descriptive statistics, which summarizes data from a sample using indexes such as the mean or standard deviation, and inferential statistics, which draws conclusions from data that are subject to random variation (e.g., observational errors, sampling variation).[2] Descriptive statistics are most often concerned with two sets of properties of a *distribution* (sample or population): *central tendency* (or *location*) seeks to characterize the distribution's central or typical value, while *dispersion* (or *variability*) characterizes the extent to which members of the distribution depart from its center and each other. Inferences on mathematical statistics are made under the framework of probability theory, which deals with the analysis of random phenomena.

Standard statistical procedure commonly involve the development of a null hypothesis, a general statement or default position that there is no relationship between two quantities. Rejecting or disproving the null hypothesis is a common task in the modern practice of science, and gives a precise sense in which a claim is capable of being proven false. What statisticians call an alternative hypothesis is simply a hypothesis that contradicts the null hypothesis. Working from a null hypothesis, two basic forms of error are recognized: Type I errors (null hypothesis is falsely rejected giving a "false positive") and Type II errors (null hypothesis fails to be rejected and an actual difference between populations is missed giving a "false negative"). Multiple problems have come to be associated with this framework: ranging from obtaining a sufficient sample size to specifying an adequate null hypothesis.

Measurement processes that generate statistical data are also subject to error. Many of these errors are classified as random (noise) or systematic (bias), but other important types of errors (e.g., blunder, such as when an analyst reports incorrect units) can also be important. The presence of missing data and/or censoring may result in biased estimates and specific techniques have been developed to address these problems.

Statistics can be said to have begun in ancient civilization, going back at least to the 5th century BC, but it was not until the 18th century that it started to draw more heavily from calculus and probability theory. Statistics continues to be an area of active research, for example on the problem of how to analyze Big data.

29.1 Scope

Statistics is a mathematical body of science that pertains to the collection, analysis, interpretation or explanation, and presentation of data,[3] or as a branch of mathematics.[4] Some consider statistics to be a distinct mathematical science rather than a branch of mathematics.[5][6]

29.1.1 Mathematical statistics

Main article: Mathematical statistics

Mathematical statistics is the application of mathematics to statistics, which was originally conceived as the science of the state — the collection and analysis of facts about a country: its economy, land, military, population, and so forth. Mathematical techniques used for this include mathematical analysis, linear algebra, stochastic analysis, differential equations, and measure-theoretic probability theory.[7][8]

29.2 Overview

In applying statistics to e.g. a scientific, industrial, or societal problem, it is necessary to begin with a population or process to be studied. Populations can be diverse topics such as "all persons living in a country" or "every atom composing a crystal".

Ideally, statisticians compile data about the entire population (an operation called census). This may be organized by governmental statistical institutes. Descriptive statistics can be used to summarize the population data. Numerical descriptors include mean and standard deviation for continuous data types (like income), while frequency and percentage are more useful in terms of describing categorical data (like race).

When a census is not feasible, a chosen subset of the population called a sample is studied. Once a sample that is representative of the population is determined, data is collected for the sample members in an observational or experimental setting. Again, descriptive statistics can be used to summarize the sample data. However, the drawing of the sample has been subject to an element of randomness, hence the established numerical descriptors from the sample are also due to uncertainty. To still draw meaningful conclusions about the entire population, inferential statistics is needed. It uses patterns in the sample data to draw inferences about the population represented, accounting for randomness. These inferences may take the form of: answering yes/no questions about the data (hypothesis testing), estimating numerical characteristics of the data (estimation), describing associations within the data (correlation) and modeling relationships within the data (for example, using regression analysis). Inference can extend to forecasting, prediction and estimation of unobserved values either in or associated with the population being studied; it can include extrapolation and interpolation of time series or spatial data, and can also include data mining.

29.3   Data collection

29.3.1   Sampling

When full census data cannot be collected, statisticians collect sample data by developing specific experiment designs and survey samples. Statistics itself also provides tools for prediction and forecasting the use of data through statistical models. To use a sample as a guide to an entire population, it is important that it truly represents the overall population. Representative sampling assures that inferences and conclusions can safely extend from the sample to the population as a whole. A major problem lies in determining the extent that the sample chosen is actually representative. Statistics offers methods to estimate and correct for any bias within the sample and data collection procedures. There are also methods of experimental design for experiments that can lessen these issues at the outset of a study, strengthening its capability to discern truths about the population.

Sampling theory is part of the mathematical discipline of probability theory. Probability is used in mathematical statistics to study the sampling distributions of sample statistics and, more generally, the properties of statistical procedures. The use of any statistical method is valid when the system or population under consideration satisfies the assumptions of the method. The difference in point of view between classic probability theory and sampling theory is, roughly, that probability theory starts from the given parameters of a total population to deduce probabilities that pertain to samples. Statistical inference, however, moves in the opposite direction—inductively inferring from samples to the parameters of a larger or total population.

29.3.2   Experimental and observational studies

A common goal for a statistical research project is to investigate causality, and in particular to draw a conclusion on the effect of changes in the values of predictors or independent variables on dependent variables. There are two major types of causal statistical studies: experimental studies and observational studies. In both types of studies, the effect of differences of an independent variable (or variables) on the behavior of the dependent variable are observed. The difference between the two types lies in how the study is actually conducted. Each can be very effective. An experimental study involves taking measurements of the system under study, manipulating the system, and then taking additional measurements using the same procedure to determine if the manipulation has modified the values of the measurements. In contrast, an observational study does not involve experimental manipulation. Instead, data are gathered and correlations between predictors and response are investigated. While the tools of data analysis work best on data from randomized studies, they are also applied to other kinds of data – like natural experiments and observational studies[9] – for which a statistician would use a modified, more structured estimation method (e.g., Difference in differences estimation and instrumental variables, among many others) that produce consistent estimators.

Experiments

The basic steps of a statistical experiment are:

1. Planning the research, including finding the number of replicates of the study, using the following information: preliminary estimates regarding the size of treatment effects, alternative hypotheses, and the estimated experimental variability. Consideration of the selection of experimental subjects and the ethics of research is necessary. Statisticians recommend that experiments compare (at least) one new treatment with a standard treatment or control, to allow an unbiased estimate of the difference in treatment effects.

2. Design of experiments, using blocking to reduce the influence of confounding variables, and randomized assignment of treatments to subjects to allow unbiased estimates of treatment effects and experimental error. At this stage, the

experimenters and statisticians write the *experimental protocol* that will guide the performance of the experiment and which specifies the *primary analysis* of the experimental data.

3. Performing the experiment following the experimental protocol and analyzing the data following the experimental protocol.

4. Further examining the data set in secondary analyses, to suggest new hypotheses for future study.

5. Documenting and presenting the results of the study.

Experiments on human behavior have special concerns. The famous Hawthorne study examined changes to the working environment at the Hawthorne plant of the Western Electric Company. The researchers were interested in determining whether increased illumination would increase the productivity of the assembly line workers. The researchers first measured the productivity in the plant, then modified the illumination in an area of the plant and checked if the changes in illumination affected productivity. It turned out that productivity indeed improved (under the experimental conditions). However, the study is heavily criticized today for errors in experimental procedures, specifically for the lack of a control group and blindness. The Hawthorne effect refers to finding that an outcome (in this case, worker productivity) changed due to observation itself. Those in the Hawthorne study became more productive not because the lighting was changed but because they were being observed.[10]

Observational study

An example of an observational study is one that explores the association between smoking and lung cancer. This type of study typically uses a survey to collect observations about the area of interest and then performs statistical analysis. In this case, the researchers would collect observations of both smokers and non-smokers, perhaps through a case-control study, and then look for the number of cases of lung cancer in each group.

29.4   Types of data

Main articles: Statistical data type and Levels of measurement

Various attempts have been made to produce a taxonomy of levels of measurement. The psychophysicist Stanley Smith Stevens defined nominal, ordinal, interval, and ratio scales. Nominal measurements do not have meaningful rank order among values, and permit any one-to-one transformation. Ordinal measurements have imprecise differences between consecutive values, but have a meaningful order to those values, and permit any order-preserving transformation. Interval measurements have meaningful distances between measurements defined, but the zero value is arbitrary (as in the case with longitude and temperature measurements in Celsius or Fahrenheit), and permit any linear transformation. Ratio measurements have both a meaningful zero value and the distances between different measurements defined, and permit any rescaling transformation.

Because variables conforming only to nominal or ordinal measurements cannot be reasonably measured numerically, sometimes they are grouped together as categorical variables, whereas ratio and interval measurements are grouped together as quantitative variables, which can be either discrete or continuous, due to their numerical nature. Such distinctions can often be loosely correlated with data type in computer science, in that dichotomous categorical variables may be represented with the Boolean data type, polytomous categorical variables with arbitrarily assigned integers in the integral data type, and continuous variables with the real data type involving floating point computation. But the mapping of computer science data types to statistical data types depends on which categorization of the latter is being implemented.

Other categorizations have been proposed. For example, Mosteller and Tukey (1977)[11] distinguished grades, ranks, counted fractions, counts, amounts, and balances. Nelder (1990)[12] described continuous counts, continuous ratios, count ratios, and categorical modes of data. See also Chrisman (1998),[13] van den Berg (1991).[14]

The issue of whether or not it is appropriate to apply different kinds of statistical methods to data obtained from different kinds of measurement procedures is complicated by issues concerning the transformation of variables and the precise interpretation of research questions. "The relationship between the data and what they describe merely reflects the fact that certain kinds of statistical statements may have truth values which are not invariant under some transformations. Whether or not a transformation is sensible to contemplate depends on the question one is trying to answer" (Hand, 2004, p. 82).[15]

29.5   Terminology and theory of inferential statistics

29.5.1   Statistics, estimators and pivotal quantities

Consider an independent identically distributed (IID) random variables with a given probability distribution: standard statistical inference and estimation theory defines a random sample as the random vector given by the column vector of these IID variables.[16] The population being examined is described by a probability distribution that may have unknown parameters.

A statistic is a random variable that is a function of the random sample, but *not a function of unknown parameters*. The probability distribution of the statistic, though, may have unknown parameters.

Consider now a function of the unknown parameter: an estimator is a statistic used to estimate such function. Commonly used estimators include sample mean, unbiased sample variance and sample covariance.

A random variable that is a function of the random sample and of the unknown parameter, but whose probability distribution *does not depend on the unknown parameter* is called a pivotal quantity or pivot. Widely used pivots include the z-score, the chi square statistic and Student's t-value.

Between two estimators of a given parameter, the one with lower mean squared error is said to be more efficient. Furthermore, an estimator is said to be unbiased if its expected value is equal to the true value of the unknown parameter being estimated, and asymptotically unbiased if its expected value converges at the limit to the true value of such parameter.

Other desirable properties for estimators include: UMVUE estimators that have the lowest variance for all possible values of the parameter to be estimated (this is usually an easier property to verify than efficiency) and consistent estimators which converges in probability to the true value of such parameter.

This still leaves the question of how to obtain estimators in a given situation and carry the computation, several methods have been proposed: the method of moments, the maximum likelihood method, the least squares method and the more recent method of estimating equations.

29.5.2   Null hypothesis and alternative hypothesis

Interpretation of statistical information can often involve the development of a null hypothesis in that the assumption is that whatever is proposed as a cause has no effect on the variable being measured.

The best illustration for a novice is the predicament encountered by a jury trial. The null hypothesis, H_0, asserts that the defendant is innocent, whereas the alternative hypothesis, H_1, asserts that the defendant is guilty. The indictment comes because of suspicion of the guilt. The H_0 (status quo) stands in opposition to H_1 and is maintained unless H_1 is supported by evidence "beyond a reasonable doubt". However, "failure to reject H_0" in this case does not imply innocence, but merely that the evidence was insufficient to convict. So the jury does not necessarily *accept* H_0 but *fails to reject* H_0. While one can not "prove" a null hypothesis, one can test how close it is to being true with a power test, which tests for type II errors.

What statisticians call an alternative hypothesis is simply an hypothesis that contradicts the null hypothesis.

29.5.3   Error

Working from a null hypothesis two basic forms of error are recognized:

- Type I errors where the null hypothesis is falsely rejected giving a "false positive".

- Type II errors where the null hypothesis fails to be rejected and an actual difference between populations is missed giving a "false negative".

Standard deviation refers to the extent to which individual observations in a sample differ from a central value, such as the sample or population mean, while Standard error refers to an estimate of difference between sample mean and population mean.

A statistical error is the amount by which an observation differs from its expected value, a residual is the amount an observation differs from the value the estimator of the expected value assumes on a given sample (also called prediction).

Mean squared error is used for obtaining efficient estimators, a widely used class of estimators. Root mean square error is simply the square root of mean squared error.

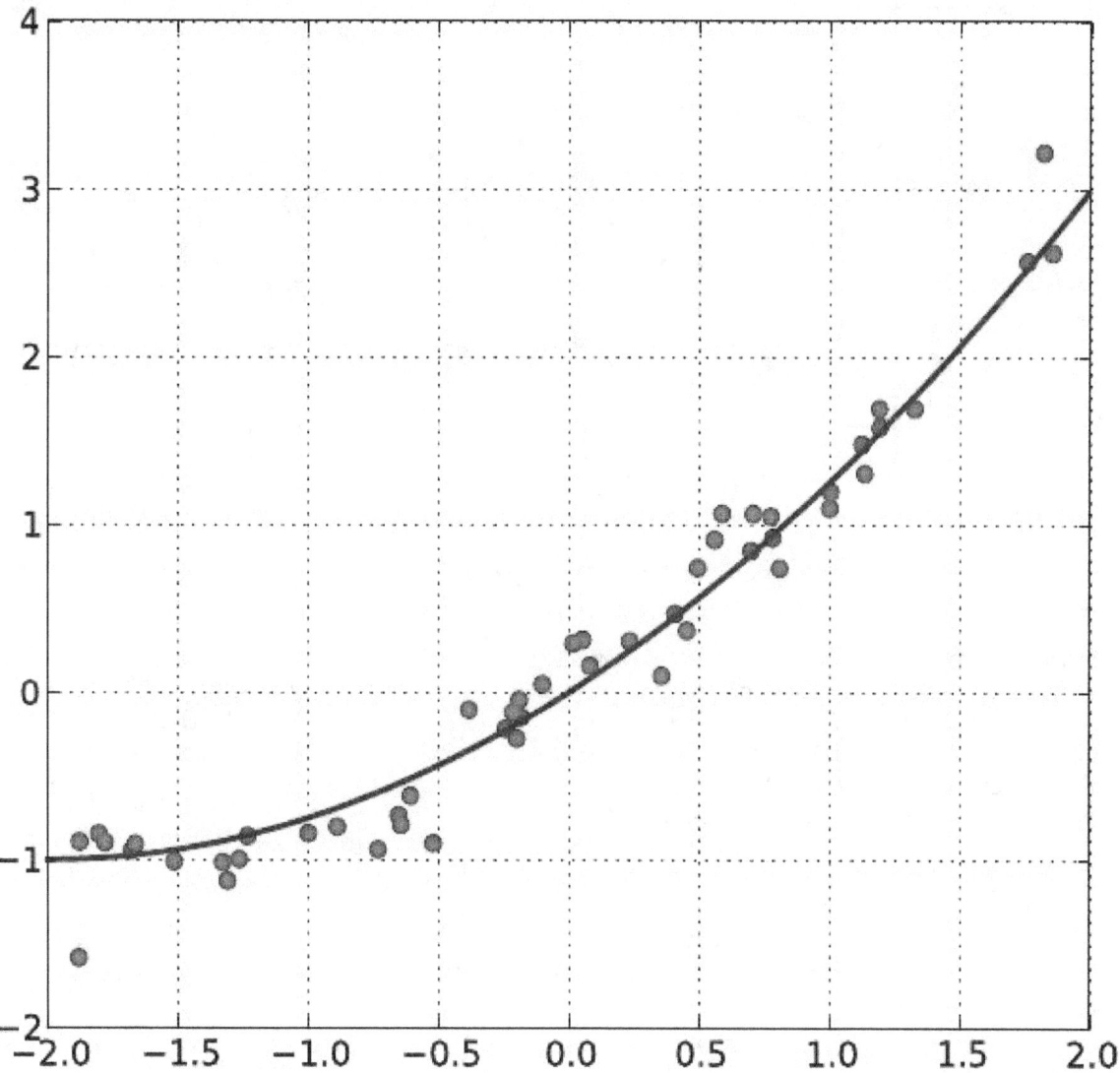

A least squares fit: in red the points to be fitted, in blue the fitted line.

Many statistical methods seek to minimize the residual sum of squares, and these are called "methods of least squares" in contrast to Least absolute deviations. The later gives equal weight to small and big errors, while the former gives more weight to large errors. Residual sum of squares is also differentiable, which provides a handy property for doing regression. Least squares applied to linear regression is called ordinary least squares method and least squares applied to nonlinear regression is called non-linear least squares. Also in a linear regression model the non deterministic part of the model is called error term, disturbance or more simply noise. Both linear regression and non-linear regression are addressed in polynomial least squares, which also describes the variance in a prediction of the dependent variable (y axis) as a function of the independent variable (x axis) and the deviations (errors, noise, disturbances) from the estimated (fitted) curve.

Measurement processes that generate statistical data are also subject to error. Many of these errors are classified as random (noise) or systematic (bias), but other types of errors (e.g., blunder, such as when an analyst reports incorrect units) can also be important. The presence of missing data and/or censoring may result in biased estimates and specific techniques have been developed to address these problems.[17]

29.5.4 Interval estimation

Main article: Interval estimation

Most studies only sample part of a population, so results don't fully represent the whole population. Any estimates

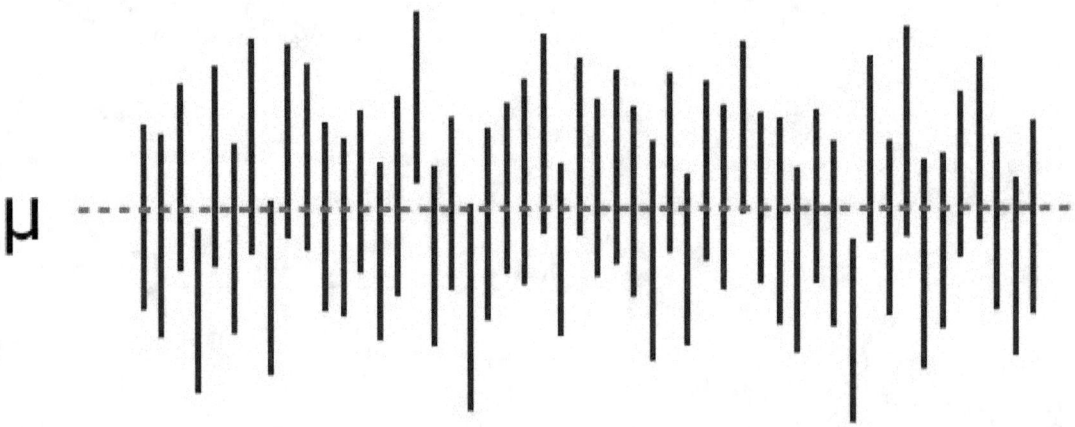

Confidence intervals: the red line is true value for the mean in this example, the blue lines are random confidence intervals for 100 realizations.

obtained from the sample only approximate the population value. Confidence intervals allow statisticians to express how closely the sample estimate matches the true value in the whole population. Often they are expressed as 95% confidence intervals. Formally, a 95% confidence interval for a value is a range where, if the sampling and analysis were repeated under the same conditions (yielding a different dataset), the interval would include the true (population) value in 95% of all possible cases. This does *not* imply that the probability that the true value is in the confidence interval is 95%. From the frequentist perspective, such a claim does not even make sense, as the true value is not a random variable. Either the true value is or is not within the given interval. However, it is true that, before any data are sampled and given a plan for how to construct the confidence interval, the probability is 95% that the yet-to-be-calculated interval will cover the true value: at this point, the limits of the interval are yet-to-be-observed random variables. One approach that does yield an interval that can be interpreted as having a given probability of containing the true value is to use a credible interval from Bayesian statistics: this approach depends on a different way of interpreting what is meant by "probability", that is as a Bayesian probability.

In principle confidence intervals can be symmetrical or asymmetrical. An interval can be asymmetrical because it works as lower or upper bound for a parameter (left-sided interval or right sided interval), but it can also be asymmetrical because the two sided interval is built violating symmetry around the estimate. Sometimes the bounds for a confidence interval are reached asymptotically and these are used to approximate the true bounds.

29.5.5 Significance

Main article: Statistical significance

Statistics rarely give a simple Yes/No type answer to the question under analysis. Interpretation often comes down to the level of statistical significance applied to the numbers and often refers to the probability of a value accurately rejecting the null hypothesis (sometimes referred to as the p-value).

Important:

Pr (observation | hypothesis) ≠ Pr (hypothesis | observation)

The probability of observing a result given that some hypothesis is true is *not equivalent* to the probability that a hypothesis is true given that some result has been observed.

Using the p-value as a "score" is committing an egregious logical error: **the transposed conditional fallacy.**

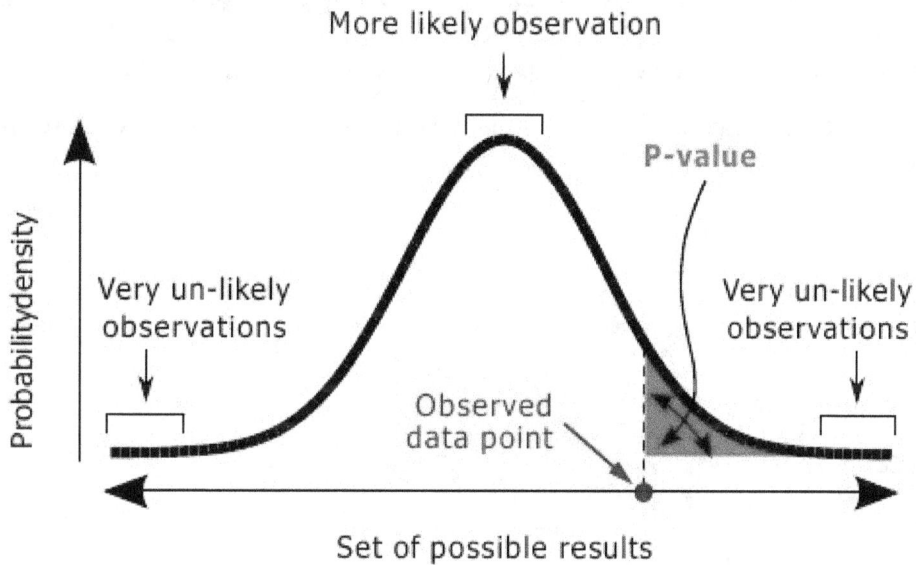

A **p-value** (shaded green area) is the probability of an observed (or more extreme) result assuming that the null hypothesis is true.

In this graph the black line is probability distribution for the test statistic, the critical region is the set of values to the right of the observed data point (observed value of the test statistic) and the p-value is represented by the green area.

The standard approach[16] is to test a null hypothesis against an alternative hypothesis. A critical region is the set of values of the estimator that leads to refuting the null hypothesis. The probability of type I error is therefore the probability that the estimator belongs to the critical region given that null hypothesis is true (statistical significance) and the probability of type II error is the probability that the estimator doesn't belong to the critical region given that the alternative hypothesis is true. The statistical power of a test is the probability that it correctly rejects the null hypothesis when the null hypothesis is false.

Referring to statistical significance does not necessarily mean that the overall result is significant in real world terms. For example, in a large study of a drug it may be shown that the drug has a statistically significant but very small beneficial effect, such that the drug is unlikely to help the patient noticeably.

While in principle the acceptable level of statistical significance may be subject to debate, the p-value is the smallest significance level that allows the test to reject the null hypothesis. This is logically equivalent to saying that the p-value is the probability, assuming the null hypothesis is true, of observing a result at least as extreme as the test statistic. Therefore the smaller the p-value, the lower the probability of committing type I error.

Some problems are usually associated with this framework (See criticism of hypothesis testing):

- A difference that is highly statistically significant can still be of no practical significance, but it is possible to properly formulate tests to account for this. One response involves going beyond reporting only the significance level to include the p-value when reporting whether a hypothesis is rejected or accepted. The p-value, however, does not indicate the size or importance of the observed effect and can also seem to exaggerate the importance of minor differences in large studies. A better and increasingly common approach is to report confidence intervals. Although these are produced from the same calculations as those of hypothesis tests or p-values, they describe both the size of the effect and the uncertainty surrounding it.

- Fallacy of the transposed conditional, aka prosecutor's fallacy: criticisms arise because the hypothesis testing approach forces one hypothesis (the null hypothesis) to be favored, since what is being evaluated is probability of the observed result given the null hypothesis and not probability of the null hypothesis given the observed result. An alternative to this approach is offered by Bayesian inference, although it requires establishing a prior probability.[18]

- Rejecting the null hypothesis does not automatically prove the alternative hypothesis.

- As everything in inferential statistics it relies on sample size, and therefore under fat tails p-values may be seriously mis-computed.

29.5.6   Examples

Some well-known statistical tests and procedures are:

- Analysis of variance (ANOVA)

- Chi-squared test

- Correlation

- Factor analysis

- Mann–Whitney U

- Mean square weighted deviation (MSWD)

- Pearson product-moment correlation coefficient

- Regression analysis

- Spearman's rank correlation coefficient

- Student's t-test

- Time series analysis

- Conjoint Analysis

29.6   Misuse of statistics

Main article: Misuse of statistics

Misuse of statistics can produce subtle, but serious errors in description and interpretation—subtle in the sense that even experienced professionals make such errors, and serious in the sense that they can lead to devastating decision errors. For instance, social policy, medical practice, and the reliability of structures like bridges all rely on the proper use of statistics.

Even when statistical techniques are correctly applied, the results can be difficult to interpret for those lacking expertise. The statistical significance of a trend in the data—which measures the extent to which a trend could be caused by random variation in the sample—may or may not agree with an intuitive sense of its significance. The set of basic statistical skills (and skepticism) that people need to deal with information in their everyday lives properly is referred to as statistical literacy.

There is a general perception that statistical knowledge is all-too-frequently intentionally misused by finding ways to interpret only the data that are favorable to the presenter.[19] A mistrust and misunderstanding of statistics is associated with the quotation, "There are three kinds of lies: lies, damned lies, and statistics". Misuse of statistics can be both inadvertent and intentional, and the book *How to Lie with Statistics*[19] outlines a range of considerations. In an attempt to shed light on the use and misuse of statistics, reviews of statistical techniques used in particular fields are conducted (e.g. Warne, Lazo, Ramos, and Ritter (2012)).[20]

Ways to avoid misuse of statistics include using proper diagrams and avoiding bias.[21] Misuse can occur when conclusions are overgeneralized and claimed to be representative of more than they really are, often by either deliberately or unconsciously overlooking sampling bias.[22] Bar graphs are arguably the easiest diagrams to use and understand, and they can be made either by hand or with simple computer programs.[21] Unfortunately, most people do not look for bias or errors, so they are not noticed. Thus, people may often believe that something is true even if it is not well represented.[22] To make data gathered from statistics believable and accurate, the sample taken must be representative of the whole.[23] According to Huff, "The dependability of a sample can be destroyed by [bias]... allow yourself some degree of skepticism."[24]

To assist in the understanding of statistics Huff proposed a series of questions to be asked in each case:[24]

- Who says so? (Does he/she have an axe to grind?)

- How does he/she know? (Does he/she have the resources to know the facts?)

- What's missing? (Does he/she give us a complete picture?)

- Did someone change the subject? (Does he/she offer us the right answer to the wrong problem?)

- Does it make sense? (Is his/her conclusion logical and consistent with what we already know?)

29.6.1   Misinterpretation: correlation

The concept of correlation is particularly noteworthy for the potential confusion it can cause. Statistical analysis of a data set often reveals that two variables (properties) of the population under consideration tend to vary together, as if they were connected. For example, a study of annual income that also looks at age of death might find that poor people tend to have shorter lives than affluent people. The two variables are said to be correlated; however, they may or may not be the cause of one another. The correlation phenomena could be caused by a third, previously unconsidered phenomenon, called a lurking variable or confounding variable. For this reason, there is no way to immediately infer the existence of a causal relationship between the two variables. (See Correlation does not imply causation.)

29.7   History of statistical science

Main articles: History of statistics and Founders of statistics

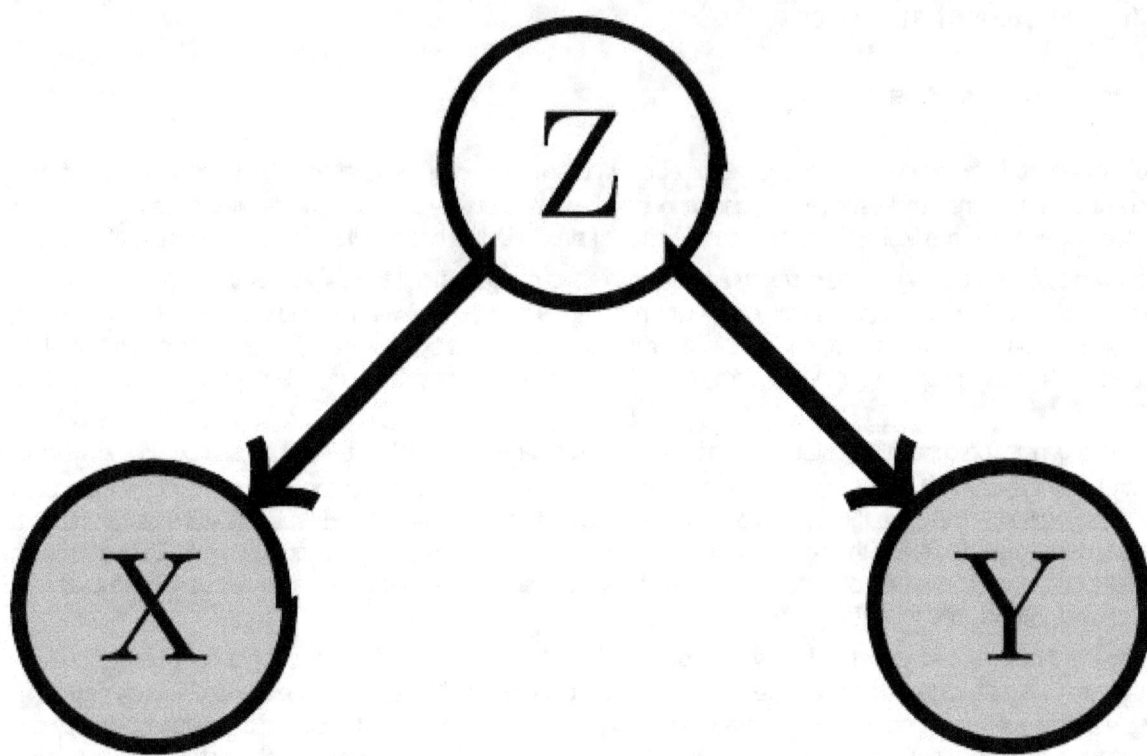

The confounding variable problem: X and Y may be correlated, not because there is causal relationship between them, but because both depend on a third variable Z. Z is called a confounding factor.

Statistical methods date back at least to the 5th century BC.

Some scholars pinpoint the origin of statistics to 1663, with the publication of *Natural and Political Observations upon the Bills of Mortality* by John Graunt.[25] Early applications of statistical thinking revolved around the needs of states to base policy on demographic and economic data, hence its *stat-* etymology. The scope of the discipline of statistics broadened in the early 19th century to include the collection and analysis of data in general. Today, statistics is widely employed in government, business, and natural and social sciences.

Its mathematical foundations were laid in the 17th century with the development of the probability theory by Gerolamo Cardano, Blaise Pascal and Pierre de Fermat. Mathematical probability theory arose from the study of games of chance, although the concept of probability was already examined in medieval law and by philosophers such as Juan Caramuel.[26] The method of least squares was first described by Adrien-Marie Legendre in 1805.

The modern field of statistics emerged in the late 19th and early 20th century in three stages.[27] The first wave, at the turn of the century, was led by the work of Sir Francis Galton and Karl Pearson, who transformed statistics into a rigorous mathematical discipline used for analysis, not just in science, but in industry and politics as well. Galton's contributions to the field included introducing the concepts of standard deviation, correlation, regression and the application of these methods to the study of the variety of human characteristics – height, weight, eyelash length among others.[28] Pearson developed the Correlation coefficient, defined as a product-moment,[29] the method of moments for the fitting of distributions to samples and the Pearson's system of continuous curves, among many other things.[30] Galton and Pearson founded *Biometrika* as the first journal of mathematical statistics and biometry, and the latter founded the world's first university statistics department at University College London.[31]

The second wave of the 1910s and 20s was initiated by William Gosset, and reached its culmination in the insights of Sir Ronald Fisher, who wrote the textbooks that were to define the academic discipline in universities around the world. Fisher's most important publications were his 1916 seminal paper *The Correlation between Relatives on the Supposition of Mendelian Inheritance* and his classic 1925 work *Statistical Methods for Research Workers*. His paper was the first to use the statistical term, variance. He developed rigorous experimental models and also originated the concepts of sufficiency,

Gerolamo Cardano, the earliest pioneer on the mathematics of probability.

ancillary statistics, Fisher's linear discriminator and Fisher information.[32]

The final wave, which mainly saw the refinement and expansion of earlier developments, emerged from the collaborative

Karl Pearson, a founder of mathematical statistics.

work between Egon Pearson and Jerzy Neyman in the 1930s. They introduced the concepts of "Type II" error, power of a test and confidence intervals. Jerzy Neyman in 1934 showed that stratified random sampling was in general a better method of estimation than purposive (quota) sampling.[33]

Today, statistical methods are applied in all fields that involve decision making, for making accurate inferences from a collated body of data and for making decisions in the face of uncertainty based on statistical methodology. The use of modern computers has expedited large-scale statistical computations, and has also made possible new methods that are impractical to perform manually. Statistics continues to be an area of active research, for example on the problem of how to analyze Big data.[34]

29.8   Applications

29.8.1   Applied statistics, theoretical statistics and mathematical statistics

"Applied statistics" comprises descriptive statistics and the application of inferential statistics.[35][36] *Theoretical statistics* concerns both the logical arguments underlying justification of approaches to statistical inference, as well encompassing *mathematical statistics*. Mathematical statistics includes not only the manipulation of probability distributions necessary for deriving results related to methods of estimation and inference, but also various aspects of computational statistics and the design of experiments.

Ronald Fisher coined the term "null hypothesis".

29.8.2 Machine learning and data mining

There are two applications for machine learning and data mining: data management and data analysis. Statistics tools are necessary for the data analysis.

Anil Kumar Gain, famous Indian statistician

29.8.3 Statistics in society

Statistics is applicable to a wide variety of academic disciplines, including natural and social sciences, government, and business. Statistical consultants can help organizations and companies that don't have in-house expertise relevant to their particular questions.

29.8.4 Statistical computing

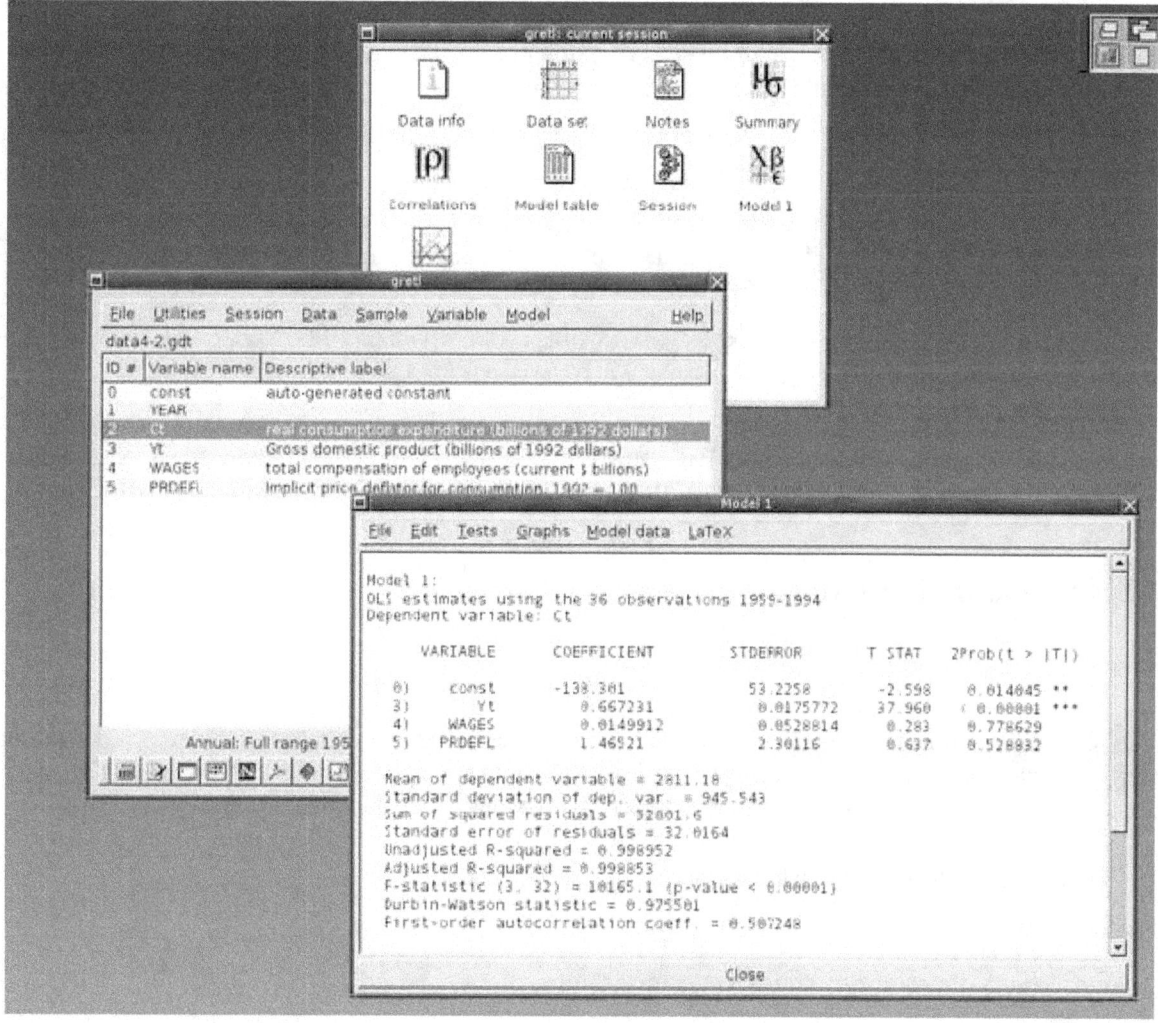

gretl, an example of an open source statistical package

Main article: Computational statistics

The rapid and sustained increases in computing power starting from the second half of the 20th century have had a substantial impact on the practice of statistical science. Early statistical models were almost always from the class of linear models, but powerful computers, coupled with suitable numerical algorithms, caused an increased interest in nonlinear models (such as neural networks) as well as the creation of new types, such as generalized linear models and multilevel models.

Increased computing power has also led to the growing popularity of computationally intensive methods based on resampling, such as permutation tests and the bootstrap, while techniques such as Gibbs sampling have made use of Bayesian models more feasible. The computer revolution has implications for the future of statistics with new emphasis on "experimental" and "empirical" statistics. A large number of both general and special purpose statistical software are now available.

29.8.5   Statistics applied to mathematics or the arts

Traditionally, statistics was concerned with drawing inferences using a semi-standardized methodology that was "required learning" in most sciences. This has changed with use of statistics in non-inferential contexts. What was once considered a dry subject, taken in many fields as a degree-requirement, is now viewed enthusiastically. Initially derided by some mathematical purists, it is now considered essential methodology in certain areas.

- In number theory, scatter plots of data generated by a distribution function may be transformed with familiar tools used in statistics to reveal underlying patterns, which may then lead to hypotheses.

- Methods of statistics including predictive methods in forecasting are combined with chaos theory and fractal geometry to create video works that are considered to have great beauty.

- The process art of Jackson Pollock relied on artistic experiments whereby underlying distributions in nature were artistically revealed. With the advent of computers, statistical methods were applied to formalize such distribution-driven natural processes to make and analyze moving video art.

- Methods of statistics may be used predicatively in performance art, as in a card trick based on a Markov process that only works some of the time, the occasion of which can be predicted using statistical methodology.

- Statistics can be used to predicatively create art, as in the statistical or stochastic music invented by Iannis Xenakis, where the music is performance-specific. Though this type of artistry does not always come out as expected, it does behave in ways that are predictable and tunable using statistics.

29.9   Specialized disciplines

Main article: List of fields of application of statistics

Statistical techniques are used in a wide range of types of scientific and social research, including: biostatistics, computational biology, computational sociology, network biology, social science, sociology and social research. Some fields of inquiry use applied statistics so extensively that they have specialized terminology. These disciplines include:

- Actuarial science (assesses risk in the insurance and finance industries)

- Applied information economics

- Astrostatistics (statistical evaluation of astronomical data)

- Biostatistics

- Business statistics

- Chemometrics (for analysis of data from chemistry)

- Data mining (applying statistics and pattern recognition to discover knowledge from data)

- Demography

- Econometrics (statistical analysis of economic data)

- Energy statistics

- Engineering statistics

- Epidemiology (statistical analysis of disease)

- Geography and Geographic Information Systems, specifically in Spatial analysis

- Image processing

- Medical Statistics

- Psychological statistics

- Reliability engineering

- Social statistics

- Statistical Mechanics

In addition, there are particular types of statistical analysis that have also developed their own specialised terminology and methodology:

- Bootstrap / Jackknife resampling

- Multivariate statistics

- Statistical classification

- Structured data analysis (statistics)

- Structural equation modelling

- Survey methodology

- Survival analysis

- Statistics in various sports, particularly baseball - known as 'Sabremetrics' - and cricket

Statistics form a key basis tool in business and manufacturing as well. It is used to understand measurement systems variability, control processes (as in statistical process control or SPC), for summarizing data, and to make data-driven decisions. In these roles, it is a key tool, and perhaps the only reliable tool.

29.10   See also

Main article: Outline of statistics

- Abundance estimation

- Glossary of probability and statistics

- List of academic statistical associations

- List of important publications in statistics

- List of national and international statistical services

- List of statistical packages (software)

- List of statistics articles

- List of university statistical consulting centers

- Notation in probability and statistics

Foundations and major areas of statistics

- Foundations of statistics
- List of statisticians
- Official statistics
- Multivariate analysis of variance

29.11   References

[1] Dodge, Y. (2006) *The Oxford Dictionary of Statistical Terms*, OUP. ISBN 0-19-920613-9

[2] Lund Research Ltd. "Descriptive and Inferential Statistics". statistics.laerd.com. Retrieved 2014-03-23.

[3] Moses, Lincoln E. (1986) *Think and Explain with Statistics*, Addison-Wesley, ISBN 978-0-201-15619-5 . pp. 1–3

[4] Hays, William Lee, (1973) *Statistics for the Social Sciences*, Holt, Rinehart and Winston, p.xii, ISBN 978-0-03-077945-9

[5] Moore, David (1992). "Teaching Statistics as a Respectable Subject". In F. Gordon and S. Gordon. *Statistics for the Twenty-First Century.* Washington, DC: The Mathematical Association of America. pp. 14–25. ISBN 978-0-88385-078-7.

[6] Chance, Beth L.; Rossman, Allan J. (2005). "Preface". *Investigating Statistical Concepts, Applications, and Methods* (PDF). Duxbury Press. ISBN 978-0-495-05064-3.

[7] Lakshmikantham,, ed. by D. Kannan,... V. (2002). *Handbook of stochastic analysis and applications.* New York: M. Dekker. ISBN 0824706609.

[8] Schervish, Mark J. (1995). *Theory of statistics* (Corr. 2nd print. ed.). New York: Springer. ISBN 0387945466.

[9] Freedman, D.A. (2005) *Statistical Models: Theory and Practice*, Cambridge University Press. ISBN 978-0-521-67105-7

[10] McCarney R, Warner J, Iliffe S, van Haselen R, Griffin M, Fisher P (2007). "The Hawthorne Effect: a randomised, controlled trial". *BMC Med Res Methodol* 7 (1): 30. doi:10.1186/1471-2288-7-30. PMC 1936999. PMID 17608932.

[11] Mosteller, F., & Tukey, J. W. (1977). *Data analysis and regression.* Boston: Addison-Wesley.

[12] Nelder, J. A. (1990). The knowledge needed to computerise the analysis and interpretation of statistical information. In *Expert systems and artificial intelligence: the need for information about data.* Library Association Report, London, March, 23–27.

[13] Chrisman, Nicholas R (1998). "Rethinking Levels of Measurement for Cartography". *Cartography and Geographic Information Science* 25 (4): 231–242. doi:10.1559/152304098782383043.

[14] van den Berg, G. (1991). *Choosing an analysis method.* Leiden: DSWO Press

[15] Hand, D. J. (2004). *Measurement theory and practice: The world through quantification.* London, UK: Arnold.

[16] Piazza Elio, Probabilità e Statistica, Esculapio 2007

[17] Rubin, Donald B.; Little, Roderick J. A.,Statistical analysis with missing data, New York: Wiley 2002

[18] Ioannidis, J. P. A. (2005). "Why Most Published Research Findings Are False". *PLoS Medicine* 2 (8): e124. doi:10.24. PMC 1182327. PMID 16060722.

[19] Huff, Darrell (1954) *How to Lie with Statistics*, WW Norton & Company, Inc. New York, NY. ISBN 0-393-31072-8

[20] Warne, R. Lazo; Ramos, T.; Ritter, N. (2012). "Statistical Methods Used in Gifted Education Journals, 2006–2010". *Gifted Child Quarterly* 56 (3): 134–149. doi:10.1177/0016986212444122.

[21] Drennan, Robert D. (2008). "Statistics in archaeology". In Pearsall, Deborah M. *Encyclopedia of Archaeology*. Elsevier Inc. pp. 2093–2100. ISBN 978-0-12-373962-9.

[22] Cohen, Jerome B. (December 1938). "Misuse of Statistics". *Journal of the American Statistical Association* (JSTOR) 33 (204): 657–674. doi:10.1080/01621459.1938.10502344.

[23] Freund, J. E. (1988). "Modern Elementary Statistics". *Credo Reference*.

[24] Huff, Darrell; Irving Geis (1954). *How to Lie with Statistics*. New York: Norton. The dependability of a sample can be destroyed by [bias]... allow yourself some degree of skepticism.

[25] Willcox, Walter (1938) "The Founder of Statistics". *Review of the International Statistical Institute* 5(4):321–328. JSTOR 1400906

[26] J. Franklin, The Science of Conjecture: Evidence and Probability before Pascal,Johns Hopkins Univ Pr 2002

[27] Helen Mary Walker (1975). *Studies in the history of statistical method*. Arno Press.

[28] Galton, F (1877). "Typical laws of heredity". *Nature* 15: 492–553. doi:10.1038/015492a0.

[29] Stigler, S. M. (1989). "Francis Galton's Account of the Invention of Correlation". *Statistical Science* 4(2): 73–79. doi:10.1214/ss/1177012580.

[30] Pearson, K. (1900). "On the Criterion that a given System of Deviations from the Probable in the Case of a Correlated System of Variables is such that it can be reasonably supposed to have arisen from Random Sampling". *Philosophical Magazine Series 5* 50 (302): 157–175. doi:10.1080/14786440009463897.

[31] "Karl Pearson (1857–1936)". Department of Statistical Science – University College London.

[32] Agresti, Alan; David B. Hichcock (2005). "Bayesian Inference for Categorical Data Analysis" (PDF). *Statistical Methods & Applications* 14 (14): 298. doi:10.1007/s10260-005-0121-y.

[33] Neyman, J (1934). "On the two different aspects of the representative method: The method of stratified sampling and the method of purposive selection". *Journal of the Royal Statistical Society* 97 (4): 557–625. JSTOR 2342192.

[34] "Science in a Complex World - Big Data: Opportunity or Threat?". *Santa Fe Institute*.

[35] Nikoletseas, M. M. (2014) "Statistics: Concepts and Examples." ISBN 978-1500815684

[36] Anderson, D.R.; Sweeney, D.J.; Williams, T.A. (1994) *Introduction to Statistics: Concepts and Applications*, pp. 5–9. West Group. ISBN 978-0-314-03309-3

Chapter 30

Numerical analysis

Numerical analysis is the study of algorithms that use numerical approximation (as opposed to general symbolic manipulations) for the problems of mathematical analysis (as distinguished from discrete mathematics).

One of the earliest mathematical writings is a Babylonian tablet from the Yale Babylonian Collection (YBC 7289), which gives a sexagesimal numerical approximation of $\sqrt{2}$, the length of the diagonal in a unit square. Being able to compute the sides of a triangle (and hence, being able to compute square roots) is extremely important, for instance, in astronomy, carpentry and construction.[2]

Numerical analysis continues this long tradition of practical mathematical calculations. Much like the Babylonian approximation of $\sqrt{2}$, modern numerical analysis does not seek exact answers, because exact answers are often impossible to obtain in practice. Instead, much of numerical analysis is concerned with obtaining approximate solutions while maintaining reasonable bounds on errors.

Numerical analysis naturally finds applications in all fields of engineering and the physical sciences, but in the 21st century also the life sciences and even the arts have adopted elements of scientific computations. Ordinary differential equations appear in celestial mechanics (planets, stars and galaxies); numerical linear algebra is important for data analysis; stochastic differential equations and Markov chains are essential in simulating living cells for medicine and biology.

Before the advent of modern computers numerical methods often depended on hand interpolation in large printed tables. Since the mid 20th century, computers calculate the required functions instead. These same interpolation formulas nevertheless continue to be used as part of the software algorithms for solving differential equations.

30.1   General introduction

The overall goal of the field of numerical analysis is the design and analysis of techniques to give approximate but accurate solutions to hard problems, the variety of which is suggested by the following:

- Advanced numerical methods are essential in making numerical weather prediction feasible.

- Computing the trajectory of a spacecraft requires the accurate numerical solution of a system of ordinary differential equations.

- Car companies can improve the crash safety of their vehicles by using computer simulations of car crashes. Such simulations essentially consist of solving partial differential equations numerically.

- Hedge funds (private investment funds) use tools from all fields of numerical analysis to attempt to calculate the value of stocks and derivatives more precisely than other market participants.

- Airlines use sophisticated optimization algorithms to decide ticket prices, airplane and crew assignments and fuel needs. Historically, such algorithms were developed within the overlapping field of operations research.

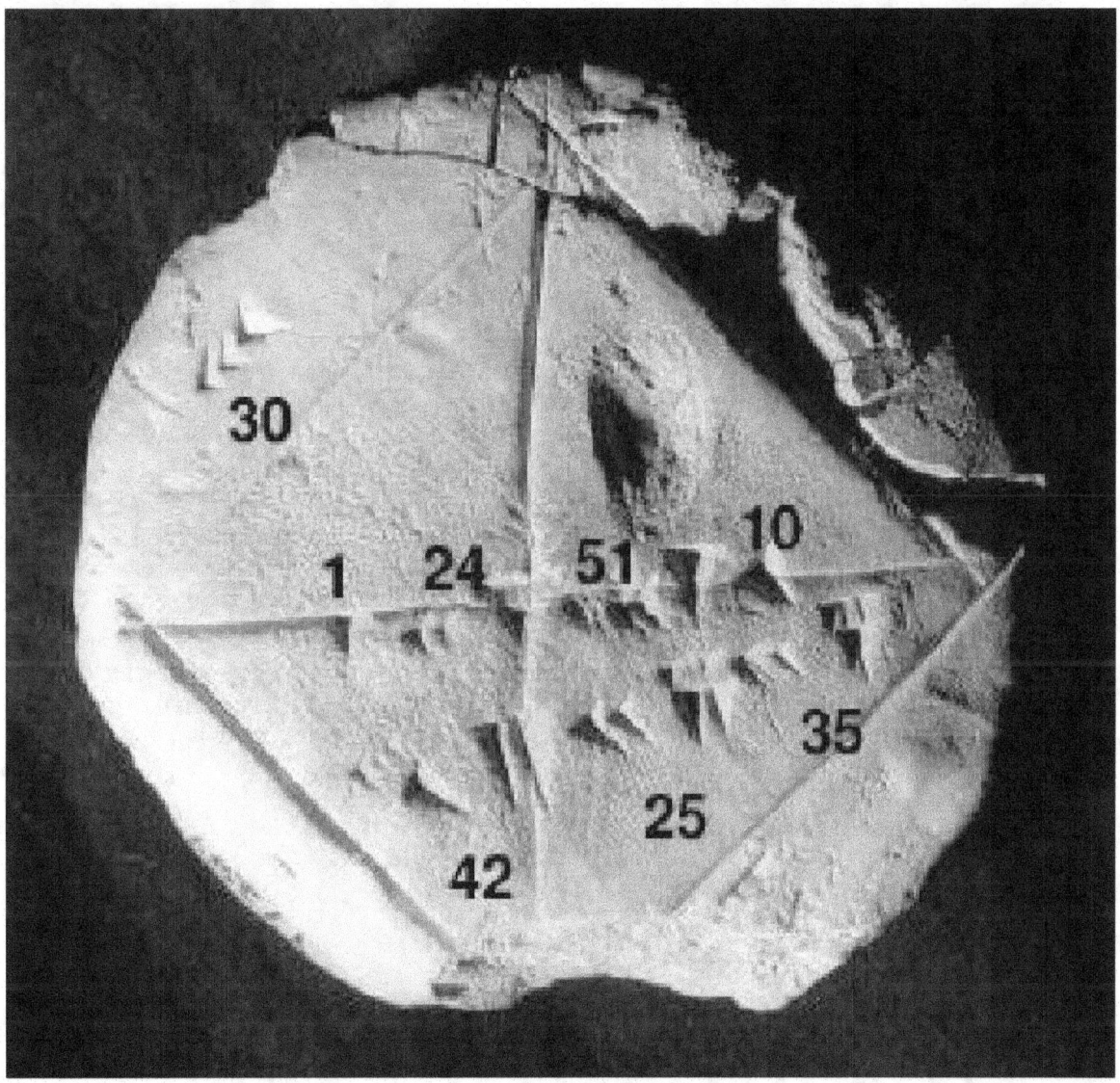

Babylonian clay tablet YBC 7289 (c. 1800–1600 BC) with annotations. The approximation of the square root of 2 is four sexagesimal figures, which is about six decimal figures. $1 + 24/60 + 51/60^2 + 10/60^3 = 1.41421296...$[1]

- Insurance companies use numerical programs for actuarial analysis.

The rest of this section outlines several important themes of numerical analysis.

30.1.1 History

The field of numerical analysis predates the invention of modern computers by many centuries. Linear interpolation was already in use more than 2000 years ago. Many great mathematicians of the past were preoccupied by numerical analysis, as is obvious from the names of important algorithms like Newton's method, Lagrange interpolation polynomial, Gaussian elimination, or Euler's method.

To facilitate computations by hand, large books were produced with formulas and tables of data such as interpolation points and function coefficients. Using these tables, often calculated out to 16 decimal places or more for some functions, one could look up values to plug into the formulas given and achieve very good numerical estimates of some functions. The canonical work in the field is the NIST publication edited by Abramowitz and Stegun, a 1000-plus page book of a very large number of commonly used formulas and functions and their values at many points. The function values are no longer very useful when a computer is available, but the large listing of formulas can still be very handy.

The mechanical calculator was also developed as a tool for hand computation. These calculators evolved into electronic computers in the 1940s, and it was then found that these computers were also useful for administrative purposes. But the invention of the computer also influenced the field of numerical analysis, since now longer and more complicated calculations could be done.

30.1.2   Direct and iterative methods

Direct methods compute the solution to a problem in a finite number of steps. These methods would give the precise answer if they were performed in infinite precision arithmetic. Examples include Gaussian elimination, the QR factorization method for solving systems of linear equations, and the simplex method of linear programming. In practice, finite precision is used and the result is an approximation of the true solution (assuming stability).

In contrast to direct methods, iterative methods are not expected to terminate in a finite number of steps. Starting from an initial guess, iterative methods form successive approximations that converge to the exact solution only in the limit. A convergence test, often involving the residual, is specified in order to decide when a sufficiently accurate solution has (hopefully) been found. Even using infinite precision arithmetic these methods would not reach the solution within a finite number of steps (in general). Examples include Newton's method, the bisection method, and Jacobi iteration. In computational matrix algebra, iterative methods are generally needed for large problems.

Iterative methods are more common than direct methods in numerical analysis. Some methods are direct in principle but are usually used as though they were not, e.g. GMRES and the conjugate gradient method. For these methods the number of steps needed to obtain the exact solution is so large that an approximation is accepted in the same manner as for an iterative method.

30.1.3   Discretization

Furthermore, continuous problems must sometimes be replaced by a discrete problem whose solution is known to approximate that of the continuous problem; this process is called *discretization*. For example, the solution of a differential equation is a function. This function must be represented by a finite amount of data, for instance by its value at a finite number of points at its domain, even though this domain is a continuum.

30.2   Generation and propagation of errors

The study of errors forms an important part of numerical analysis. There are several ways in which error can be introduced in the solution of the problem.

30.2.1   Round-off

Round-off errors arise because it is impossible to represent all real numbers exactly on a machine with finite memory (which is what all practical digital computers are).

30.2.2   Truncation and discretization error

Truncation errors are committed when an iterative method is terminated or a mathematical procedure is approximated, and the approximate solution differs from the exact solution. Similarly, discretization induces a discretization error because the solution of the discrete problem does not coincide with the solution of the continuous problem. For instance, in the iteration in the sidebar to compute the solution of $3x^3 + 4 = 28$", after 10 or so iterations, we conclude that the root is roughly 1.99 (for example). We therefore have a truncation error of 0.01.

Once an error is generated, it will generally propagate through the calculation. For instance, we have already noted that the operation + on a calculator (or a computer) is inexact. It follows that a calculation of the type $a+b+c+d+e$ is even more inexact.

What does it mean when we say that the truncation error is created when we approximate a mathematical procedure? We know that to integrate a function exactly requires one to find the sum of infinite trapezoids. But numerically one can find the sum of only finite trapezoids, and hence the approximation of the mathematical procedure. Similarly, to differentiate a function, the differential element approaches to zero but numerically we can only choose a finite value of the differential element.

30.2.3 Numerical stability and well-posed problems

Numerical stability is an important notion in numerical analysis. An algorithm is called *numerically stable* if an error, whatever its cause, does not grow to be much larger during the calculation. This happens if the problem is *well-conditioned*, meaning that the solution changes by only a small amount if the problem data are changed by a small amount. To the contrary, if a problem is *ill-conditioned*, then any small error in the data will grow to be a large error.

Both the original problem and the algorithm used to solve that problem can be *well-conditioned* and/or *ill-conditioned*, and any combination is possible.

So an algorithm that solves a well-conditioned problem may be either numerically stable or numerically unstable. An art of numerical analysis is to find a stable algorithm for solving a well-posed mathematical problem. For instance, computing the square root of 2 (which is roughly 1.41421) is a well-posed problem. Many algorithms solve this problem by starting with an initial approximation x_1 to $\sqrt{2}$, for instance $x_1 = 1.4$, and then computing improved guesses x_2, x_3, etc.. One such method is the famous Babylonian method, which is given by $x_{k+1} = x_k/2 + 1/x_k$. Another iteration, which we will call Method X, is given by $x_{k+1} = (x_k^2 - 2)^2 + x_k$.[3] We have calculated a few iterations of each scheme in table form below, with initial guesses $x_1 = 1.4$ and $x_1 = 1.42$.

Observe that the Babylonian method converges fast regardless of the initial guess, whereas Method X converges extremely slowly with initial guess 1.4 and diverges for initial guess 1.42. Hence, the Babylonian method is numerically stable, while Method X is numerically unstable.

Numerical stability is affected by the number of the significant digits the machine keeps on, if we use a machine that keeps on the first four floating-point digits, a good example on loss of significance is given by these two equivalent functions

$$\frac{}{(\sqrt{}} \pi \quad \text{and } g(x) = \frac{x}{\sqrt{x+1} + \sqrt{x}}$$

If we compare the results of

$$f(500) = 500 \sqrt{(\sqrt{2}} \quad \sqrt{500} = 500\,(22.3830 - 22.3607) = 500(0.0223) = 11.1500$$

and

$$g(500) = \sqrt{\frac{500}{\sqrt{500}}}$$
$$= \frac{500}{22.3830 + 22.3607}$$
$$= \frac{500}{44.7437} = 11.1748$$

by looking to the two results above, we realize that **loss of significance** which is also called **Subtractive Cancelation** has a huge effect on the results, even though both functions are equivalent; to show that they are equivalent simply we need to start by f(x) and end with g(x), and so

$$f(x) = x\left(\sqrt{x+1} - \sqrt{x}\right)$$

$$= x\sqrt{\frac{x+1-x}{x}}$$

$$= x\sqrt{\frac{1}{x}}$$

$$= \sqrt{\frac{x}{x}}$$

The true value for the result is 11.174755..., which is exactly $g(500) = 11.1748$ after rounding the result to 4 decimal digits.

Now imagine that lots of terms like these functions are used in the program; the error will increase as one proceeds in the program, unless one uses the suitable formula of the two functions each time one evaluates either $f(x)$, or $g(x)$; the choice is dependent on the parity of x.

- The example is taken from Mathew; Numerical methods using matlab, 3rd ed.

30.3   Areas of study

The field of numerical analysis includes many sub-disciplines. Some of the major ones are:

30.3.1   Computing values of functions

One of the simplest problems is the evaluation of a function at a given point. The most straightforward approach, of just plugging in the number in the formula is sometimes not very efficient. For polynomials, a better approach is using the Horner scheme, since it reduces the necessary number of multiplications and additions. Generally, it is important to estimate and control round-off errors arising from the use of floating point arithmetic.

30.3.2   Interpolation, extrapolation, and regression

Interpolation solves the following problem: given the value of some unknown function at a number of points, what value does that function have at some other point between the given points?

Extrapolation is very similar to interpolation, except that now we want to find the value of the unknown function at a point which is outside the given points.

Regression is also similar, but it takes into account that the data is imprecise. Given some points, and a measurement of the value of some function at these points (with an error), we want to determine the unknown function. The least squares-method is one popular way to achieve this.

30.3.3   Solving equations and systems of equations

Another fundamental problem is computing the solution of some given equation. Two cases are commonly distinguished, depending on whether the equation is linear or not. For instance, the equation $2x + 5 = 3$ is linear while $2x^2 + 5 = 3$ is not.

Much effort has been put in the development of methods for solving systems of linear equations. Standard direct methods, i.e., methods that use some matrix decomposition are Gaussian elimination, LU decomposition, Cholesky decomposition for symmetric (or hermitian) and positive-definite matrix, and QR decomposition for non-square matrices. Iterative methods such as the Jacobi method, Gauss–Seidel method, successive over-relaxation and conjugate gradient method are usually preferred for large systems. General iterative methods can be developed using a matrix splitting.

Root-finding algorithms are used to solve nonlinear equations (they are so named since a root of a function is an argument for which the function yields zero). If the function is differentiable and the derivative is known, then Newton's method is a popular choice. Linearization is another technique for solving nonlinear equations.

30.3.4 Solving eigenvalue or singular value problems

Several important problems can be phrased in terms of eigenvalue decompositions or singular value decompositions. For instance, the spectral image compression algorithm[4] is based on the singular value decomposition. The corresponding tool in statistics is called principal component analysis.

30.3.5 Optimization

Main article: Mathematical optimization

Optimization problems ask for the point at which a given function is maximized (or minimized). Often, the point also has to satisfy some constraints.

The field of optimization is further split in several subfields, depending on the form of the objective function and the constraint. For instance, linear programming deals with the case that both the objective function and the constraints are linear. A famous method in linear programming is the simplex method.

The method of Lagrange multipliers can be used to reduce optimization problems with constraints to unconstrained optimization problems.

30.3.6 Evaluating integrals

Main article: Numerical integration

Numerical integration, in some instances also known as numerical quadrature, asks for the value of a definite integral. Popular methods use one of the Newton–Cotes formulas (like the midpoint rule or Simpson's rule) or Gaussian quadrature. These methods rely on a "divide and conquer" strategy, whereby an integral on a relatively large set is broken down into integrals on smaller sets. In higher dimensions, where these methods become prohibitively expensive in terms of computational effort, one may use Monte Carlo or quasi-Monte Carlo methods (see Monte Carlo integration), or, in modestly large dimensions, the method of sparse grids.

30.3.7 Diff erential equations

Main articles: Numerical ordinary differential equations and Numerical partial differential equations

Numerical analysis is also concerned with computing (in an approximate way) the solution of differential equations, both ordinary differential equations and partial differential equations.

Partial differential equations are solved by first discretizing the equation, bringing it into a finite-dimensional subspace. This can be done by a finite element method, a finite difference method, or (particularly in engineering) a finite volume method. The theoretical justification of these methods often involves theorems from functional analysis. This reduces the problem to the solution of an algebraic equation.

30.4 Software

Main articles: List of numerical analysis software and Comparison of numerical analysis software

Since the late twentieth century, most algorithms are implemented in a variety of programming languages. The Netlib repository contains various collections of software routines for numerical problems, mostly in Fortran and C. Commercial products implementing many different numerical algorithms include the IMSL and NAG libraries; a free alternative is the GNU Scientific Library.

There are several popular numerical computing applications such as MATLAB, TK Solver, S-PLUS, LabVIEW, and IDL as well as free and open source alternatives such as FreeMat, Scilab, GNU Octave (similar to Matlab), IT++ (a C++ library), R (similar to S-PLUS) and certain variants of Python. Performance varies widely: while vector and matrix operations are usually fast, scalar loops may vary in speed by more than an order of magnitude.[5][6]

Many computer algebra systems such as Mathematica also benefit from the availability of arbitrary precision arithmetic which can provide more accurate results.

Also, any spreadsheet software can be used to solve simple problems relating to numerical analysis.

30.5 See also

- Analysis of algorithms
- Computational science
- List of numerical analysis topics
- Numerical differentiation
- *Numerical Recipes*
- Symbolic-numeric computation

30.6 Notes

[1] Photograph, illustration, and description of the *root(2)* tablet from the Yale Babylonian Collection

[2] The New Zealand Qualification authority specifically mentions this skill in document 13004 version 2, dated 17 October 2003 titled CARPENTRY THEORY: Demonstrate knowledge of setting out a building

[3] This is a fixed point iteration for the equation $x = (x^2 - 2)^2 + x = f(x)$, whose solutions include $\sqrt{2}$. The iterates always move to the right since $f(x) \geq x$. Hence $x_1 = 1.4 < \sqrt{2}$ converges and $x_1 = 1.42 > \sqrt{2}$ diverges.

[4] The Singular Value Decomposition and Its Applications in Image Compression

[5] Speed comparison of various number crunching packages

[6] Comparison of mathematical programs for data analysis Stefan Steinhaus, ScientificWeb.com

30.7 References

- Golub, Gene H. and Charles F. Van Loan (1986). *Matrix Computations, Third Edition (Johns Hopkins University Press, ISBN 0-8018-5413-X)*.
- Higham, Nicholas J. (1996). *Accuracy and Stability of Numerical Algorithms (Society for Industrial and Applied Mathematics, ISBN 0-89871-355-2)*.

- Hildebrand, F. B. (1974). *Introduction to Numerical Analysis* (2nd edition ed.). McGraw-Hill. ISBN 0-07-028761-9.

- Leader, Jeffery J. (2004). *Numerical Analysis and Scientific Computation*. Addison Wesley. ISBN 0-201-73499-0.

- Wilkinson, J.H. (1965). *The Algebraic Eigenvalue Problem (Clarendon Press)*.

- Kahan, W. (1972). ""A survey of error-analysis," in Info. Processing 71 (Proc. IFIP Congress 71 in Ljubljana), vol. 2, pp. 1214–39, North-Holland Publishing, Amsterdam". (examples of the importance of accurate arithmetic).

- Trefethen, Lloyd N. (2006). "Numerical analysis", 20 pages. In: Timothy Gowers and June Barrow-Green (editors), *Princeton Companion of Mathematics*, Princeton University Press.

30.8   External links

Journals

- Numerische Mathematik, volumes 1-66, Springer, 1959-1994 (searchable; pages are images). (English) (German)

- Numerische Mathematik at SpringerLink, volumes 1-112, Springer, 1959–2009

- SIAM Journal on Numerical Analysis, volumes 1-47, SIAM, 1964–2009

Online texts

- Hazewinkel, Michiel, ed. (2001), "Numerical analysis", *Encyclopedia of Mathematics*, Springer, ISBN 978-1-55608-010-4

- *Numerical Recipes*, William H. Press (free, downloadable previous editions)

- *First Steps in Numerical Analysis* (archived), R.J.Hosking, S.Joe, D.C.Joyce, and J.C.Turner

- *CSEP* (Computational Science Education Project), U.S. Department of Energy

Online course material

- Numerical Methods, Stuart Dalziel University of Cambridge

- Lectures on Numerical Analysis, Dennis Deturck and Herbert S. Wilf University of Pennsylvania

- Numerical methods, John D. Fenton University of Karlsruhe

- Numerical Methods for Science, Technology, Engineering and Mathematics, Autar Kaw University of South Florida

- Numerical Analysis Project, John H. Mathews California State University, Fullerton

- Numerical Methods - Online Course, Aaron Naiman Jerusalem College of Technology

- Numerical Methods for Physicists, Anthony O'Hare Oxford University

- Lectures in Numerical Analysis (archived), R. Radok Mahidol University

- Introduction to Numerical Analysis for Engineering, Henrik Schmidt Massachusetts Institute of Technology

- Numerical Methods for time-dependent Partial Differential Equations, J.W. Haverkort, based on a course by P.A. Zegeling Utrecht_University

- *Numerical Analysis for Engineering*, D. W. Harder University of Waterloo

Chapter 31

Symbolic computation

Not to be confused with symbolic execution.

In mathematics and computer science, **computer algebra**, also called **symbolic computation** or **algebraic computation** is a scientific area that refers to the study and development of algorithms and software for manipulating mathematical expressions and other mathematical objects. Although, properly speaking, computer algebra should be a subfield of scientific computing, they are generally considered as distinct fields because scientific computing is usually based on numerical computation with approximate floating point numbers, while symbolic computation emphasizes *exact* computation with expressions containing variables that have not any given value and are thus manipulated as symbols (therefore the name of *symbolic computation*).

Software applications that perform symbolic calculations are called *computer algebra systems*, with the term *system* alluding to the complexity of the main applications that include, at least, a method to represent mathematical data in a computer, a user programming language (usually different from the language used for the implementation), a dedicated memory manager, a user interface for the input/output of mathematical expressions, a large set of routines to perform usual operations, like simplification of expressions, differentiation using chain rule, polynomial factorization, indefinite integration, etc.

At the beginning of computer algebra, circa 1970, when the long-known algorithms were first put on computers, they turned out to be highly inefficient.[1] Therefore, a large part of the work of the researchers in the field consisted in revisiting classical algebra in order to make it effective and to discover efficient algorithms to implement this effectiveness. A typical example of this kind of work is the computation of polynomial greatest common divisors, which is required to simplify fractions. Surprisingly, the classical Euclid's algorithm turned out to be inefficient for polynomials over infinite fields, and thus new algorithms needed to be developed. The same was also true for the classical algorithms from linear algebra.

Computer algebra is widely used to experiment in mathematics and to design the formulas that are used in numerical programs. It is also used for complete scientific computations, when purely numerical methods fail, like in public key cryptography or for some non-linear problems.

31.1   Terminology

Some authors distinguish *computer algebra* from *symbolic computation* using the latter name to refer to kinds of symbolic computation other than the computation with mathematical formulas. Some authors use *symbolic computation* for the computer science aspect of the subject and "computer algebra" for the mathematical aspect.[2] In some languages the name of the field is not a direct translation of its English name. Typically, it is called *calcul formel* in French, which means "formal computation".

Symbolic computation has also been referred to, in the past, as *symbolic manipulation, algebraic manipulation, symbolic processing, symbolic mathematics*, or *symbolic algebra*, but these terms, which also refer to non-computational manipulation, are no more in use for referring to computer algebra.

31.2 Scientific community

There is no learned society that is specific to computer algebra, but this function is assumed by the special interest group of the Association for Computing Machinery named SIGSAM (Special Interest Group on Symbolic and Algebraic Manipulation).[3]

There are several annual conferences on computer algebra, the premier being ISSAC (International Symposium on Symbolic and Algebraic Computation), which is regularly sponsored by SIGSAM.[4]

There are several journals specializing in computer algebra, the top one being Journal of Symbolic Computation founded in 1985 by Bruno Buchberger.[5] There are also several other journals that regularly publish articles in computer algebra.[6]

31.3 Computer science aspects

31.3.1 Data representation

As numerical software are highly efficient for approximate numerical computation, it is common, in computer algebra, to emphasize on *exact* computation with exactly represented data. Such an exact representation implies that, even when the size of the output is small, the intermediate data generated during a computation may grow in an unpredictable way. This behavior is called *expression swell*. To obviate this problem, various methods are used in the representation of the data, as well as in the algorithms that manipulate them.

Numbers

The usual numbers systems used in numerical computation are either the floating point numbers and the integers of a fixed bounded size, that are improperly called *integers* by most programming languages. None is convenient for computer algebra, because of the expression swell.

Therefore, the basic numbers used in computer algebra are the integers of the mathematicians, commonly represented by an unbounded signed sequence of digits in some base of numeration, usually the largest base allowed by the machine word. These integers allow to define the rational numbers, which are irreducible fractions of two integers.

Programming an efficient implementation of the arithmetic operations is a hard task. Therefore, most free computer algebra systems and some commercial ones, like Maple (software), use the GMP library, which is thus a *de facto* standard.

Expressions

Except for numbers and variables, every mathematical expression may be viewed as the symbol of an operator followed by a sequence of operands. In computer algebra software, the expressions are usually represented in this way. This representation is very flexible, and many things, that seem not to be mathematical expressions at first glance, may be represented and manipulated as such. For example, an equation is an expression with "=" as an operator, a matrix may be represented as an expression with "matrix" as an operator and its rows as operands.

Even programs may be considered and represented as expressions with operator "procedure" and, at least, two operands, the list of parameters and the body, which is itself an expression with "body" as an operator and a sequence of instructions as operands. Conversely, any mathematical expression may be viewed as a program. For example, the expression $a + b$ may be viewed as a program for the addition, with a and b as parameters. Executing this program consists in *evaluating* the expression for given values of a and b; if they do not have any value—that is they are indeterminates—, the result of the evaluation is simply its input.

This process of delayed evaluation is fundamental in computer algebra. For example, the operator "=" of the equations is also, in most computer algebra systems, the name of the program of the equality test: normally, the evaluation of an equation results in an equation, but, when an equality test is needed,—either explicitly asked by the user through an "evaluation to a Boolean" command, or automatically started by the system in the case of a test inside a program—then the evaluation to a boolean 0 or 1 is executed.

As the size of the operands of an expression is unpredictable and may change during a working session, the sequence of the operands is usually represented as a sequence of either pointers (like in Macsyma) or entries in a hash table (like in Maple).

31.3.2   Simplification

The raw application of the basic rules of differentiation with respect to x on the expression a^x gives the result $x \cdot a^{x-1} \cdot 0 + a^x \cdot 1 \cdot \log a + x \cdot \frac{0}{a}$. Such an awful expression is clearly not acceptable, and a procedure of simplification is needed as soon as one works with general expressions.

This simplification is normally done through rewriting rules. There are several classes of rewriting rules that have to be considered. The simplest consists in the rewriting rules that always reduce the size of the expression, like $E - E \to 0$ or $\sin(0) \to 0$. They are systematically applied in the computer algebra systems.

The first difficulty occurs with associative operations like addition and multiplication. The standard way to deal with associativity is to consider that addition and multiplication have an arbitrary number of operands, that is that $a + b + c$ is represented as "+"(a, b, c). Thus $a + (b + c)$ and $(a + b) + c$ are both simplified to "+"(a, b, c), which is displayed $a + b + c$. What about $a - b + c$? To deal with this problem, the simplest way is to rewrite systematically $-E, E - F, E/F$ as, respectively, $(-1) \cdot E, E + (-1) \cdot F, EF^{-1}$. In other words, in the internal representation of the expressions, there is no subtraction nor division nor unary minus, outside the representation of the numbers.

A second difficulty occurs with the commutativity of addition and multiplication. The problem is to recognize quickly the like terms in order to combine or canceling them. In fact, the method for finding like terms, consisting of testing every pair of terms, is too costly for being practicable with very long sums and products. For solving this problem, Macsyma sorts the operands of sums and products with a function of comparison that is designed in order that like terms are in consecutive places, and thus easily detected. In Maple, the hash function is designed for generating collisions when like terms are entered, allowing to combine them as soon as they are introduced. This design of the hash function allows also to recognize immediately the expressions or subexpressions that appear several times in a computation and to store them only once. This allows not only to save some memory space, but also to speed up computation, by avoiding to repeat the same operations on several identical expressions.

Some rewriting rules sometimes increase and sometimes decrease the size of the expressions to which they are applied. This is the case of distributivity or trigonometric identities. For example the distributivity law allows rewriting $(x+1)^4 \to x^4 + 4x^3 + 6x^2 + 4x + 1$ and $(x-1)(x^4 + x^3 + x^2 + x + 1) \to x^5 - 1$. As there is no way to make a good general choice of applying or not such a rewriting rule, such rewritings are done only when explicitly asked by the user. For the distributivity, the computer function that apply this rewriting rule is generally called "expand". The reverse rewriting rule, called "factor", requires a non-trivial algorithm, which is thus a key function in computer algebra systems (see Polynomial factorization).

31.4   Mathematical aspects

In this section we consider some fundamental mathematical questions that arise as soon as one want to manipulate mathematical expressions in a computer. We consider mainly the case of the multivariate rational fractions. This is not a real restriction, because, as soon as the irrational functions appearing in an expression are simplified, they are usually considered as new indeterminates. For example $(\sin(x + y)^2 + \log(z^2 - 5))^3$ is viewed as a polynomial in $\sin(x + y)$ and $\log(z^2 - 5)$

31.4.1 Equality

There are two notions of equality for mathematical expressions. The *syntactic equality* is the equality of the expressions which means that they are written (or represented in a computer) in the same way. As trivial, it is rarely considered by mathematicians, but it is the only equality that is easy to test with a program. The *semantic equality* is when two expressions represent the same mathematical object, like in $(x + y)^2 = x^2 + 2xy + y^2$.

It is known that there may not exist an algorithm that decides if two expressions representing numbers are semantically equal, if exponentials and logarithms are allowed in the expressions. Therefore (semantical) equality may be tested only on some classes of expressions such as the polynomials and the rational fractions.

To test the equality of two expressions, instead to design a specific algorithm, it is usual to put them in some *canonical form* or to put their difference in a *normal form* and to test the syntactic equality of the result.

Unlike in usual mathematics, "canonical form" and "normal form" are not synonymous in computer algebra. A *canonical form* is such that two expressions in canonical form are semantically equal if and only if they are syntactically equal, while a *normal form* is such that an expression in normal form is semantically zero only if it is syntactically zero. In other words zero has a unique representation by expressions in normal form.

Normal forms are usually preferred in computer algebra for several reasons. Firstly, canonical forms may be more costly to compute than normal forms. For example, to put a polynomial in canonical form, one has to expand by distributivity every product, while it is not necessary with a normal form (see below). Secondly, It may be the case, like for expressions involving radicals, that a canonical form, if it exists, depends on some arbitrary choices and that these choices may be different for two expressions that have been computed independently. This may make impracticable the use of a canonical form.

31.5 See also

- Automated theorem prover
- Computer-assisted proof
- Computer algebra system
- Proof checker
- Model checker
- Symbolic-numeric computation
- Symbolic simulation

31.6 References

[1] Kaltofen, Erich (1982), "Factorization of polynomials", in Buchberger, B.; Loos, R.; Collins, G., *Computer Algebra*, Springer Verlag, CiteSeerX: 10.1.1.39.7916

[2] Watt, Stephen M. (2006). *Making Computer Algebra More Symbolic (Invited)* (PDF). Proc. Transgressive Computing 2006: A conference in honor or Jean Della Dora, (TC 2006). pp. 43–49.

[3] SIGSAM official site

[4] SIGSAM list of conferences

[5] Cohen, Joel S. (2003). *Computer Algebra and Symbolic Computation: Mathematical Methods*. A K Peters, Ltd. p. 14. ISBN 978-1-56881-159-8.

[6] SIGSAM list of journals

31.7   Further reading

For a detailed definition of the subject:

- Symbolic Computation (An Editorial), Bruno Buchberger, Journal of Symbolic Computation (1985) 1, pp. 1–6.

For textbooks devoted to the subject:

- Davenport, James H.; Siret, Yvon; Tournier, Èvelyne (1988). *Computer algebra: systems and algorithms for algebraic computation*. Translated from the French by A. Davenport and J.H. Davenport. Academic Press. ISBN 978-0-12-204230-0.

- von zur Gathen, Joachim; Gerhard, Jürgen (2003). *Modern computer algebra* (second ed.). Cambridge University Press. ISBN 0-521-82646-2.

- Geddes, K. O.; Czapor, S. R.; Labahn, G. (1992). "Algorithms for Computer Algebra". doi:10.1007/b102438. ISBN 978-0-7923-9259-0.

- Buchberger, Bruno; Collins, George Edwin; Loos, Rüdiger; Albrecht, Rudolf, eds. (1983). "Computer Algebra". Computing Supplementa 4. doi:10.1007/978-3-7091-7551-4. ISBN 978-3-211-81776-6.

Chapter 32

Mechanics

This article is about an area of scientific study. For other uses, see Mechanic (disambiguation).

Mechanics (Greek μηχανική) is an area of science concerned with the behavior of physical bodies when subjected to forces or displacements, and the subsequent effects of the bodies on their environment. The scientific discipline has its origins in Ancient Greece with the writings of Aristotle and Archimedes[1][2][3] (see History of classical mechanics and Timeline of classical mechanics). During the early modern period, scientists such as Galileo, Kepler, and especially Newton, laid the foundation for what is now known as classical mechanics. It is a branch of classical physics that deals with particles that are either at rest or are moving with velocities significantly less than the speed of light. It can also be defined as a branch of science which deals with the motion of and forces on objects.

32.1 Classical versus quantum

The major division of the mechanics discipline separates classical mechanics from quantum mechanics.

Historically, classical mechanics came first, while quantum mechanics is a comparatively recent invention. Classical mechanics originated with Isaac Newton's laws of motion in *Principia Mathematica*; Quantum Mechanics was discovered in 1925. Both are commonly held to constitute the most certain knowledge that exists about physical nature. Classical mechanics has especially often been viewed as a model for other so-called exact sciences. Essential in this respect is the relentless use of mathematics in theories, as well as the decisive role played by experiment in generating and testing them.

Quantum mechanics is of a wider scope, as it encompasses classical mechanics as a sub-discipline which applies under certain restricted circumstances. According to the correspondence principle, there is no contradiction or conflict between the two subjects, each simply pertains to specific situations. The correspondence principle states that the behavior of systems described by quantum theories reproduces classical physics in the limit of large quantum numbers. Quantum mechanics has superseded classical mechanics at the foundational level and is indispensable for the explanation and pre-diction of processes at molecular and (sub)atomic level. However, for macroscopic processes classical mechanics is able to solve problems which are unmanageably difficult in quantum mechanics and hence remains useful and well used. Modern descriptions of such behavior begin with a careful definition of such quantities as displacement (distance moved), time, velocity, acceleration, mass, and force. Until about 400 years ago, however, motion was explained from a very different point of view. For example, following the ideas of Greek philosopher and scientist Aristotle, scientists reasoned that a cannonball falls down because its natural position is in the Earth; the sun, the moon, and the stars travel in circles around the earth because it is the nature of heavenly objects to travel in perfect circles.

The Italian physicist and astronomer Galileo brought together the ideas of other great thinkers of his time and began to analyze motion in terms of distance traveled from some starting position and the time that it took. He showed that the speed of falling objects increases steadily during the time of their fall. This acceleration is the same for heavy objects as for light ones, provided air friction (air resistance) is discounted. The English mathematician and physicist Isaac Newton improved this analysis by defining force and mass and relating these to acceleration. For objects traveling at speeds close to the speed of light, Newton's laws were superseded by Albert Einstein's theory of relativity. For atomic and subatomic particles, Newton's laws were superseded by quantum theory. For everyday phenomena, however, Newton's three laws of motion remain the cornerstone of dynamics, which is the study of what causes motion.

32.2   Relativistic versus Newtonian mechanics

In analogy to the distinction between quantum and classical mechanics, Einstein's general and special theories of relativity have expanded the scope of Newton and Galileo's formulation of mechanics. The differences between relativistic and Newtonian mechanics become significant and even dominant as the velocity of a massive body approaches the speed of light. For instance, in Newtonian mechanics, Newton's laws of motion specify that $F = ma$, whereas in Relativistic mechanics and Lorentz transformations, which were first discovered by Hendrik Lorentz, $F = \gamma ma$ (γ is the Lorentz factor, which is almost equal to 1 for low speeds).

32.3   General relativistic versus quantum

Relativistic corrections are also needed for quantum mechanics, although general relativity has not been integrated. The two theories remain incompatible, a hurdle which must be overcome in developing a theory of everything.

32.4   History

Main articles: History of classical mechanics and History of quantum mechanics

32.4.1   Antiquity

Main article: Aristotelian mechanics

The main theory of mechanics in antiquity was Aristotelian mechanics.[4] A later developer in this tradition is Hipparchus.[5]

32.4.2   Medieval age

Main article: Theory of impetus

In the Middle Ages, Aristotle's theories were criticized and modified by a number of figures, beginning with John Philoponus in the 6th century. A central problem was that of projectile motion, which was discussed by Hipparchus and Philoponus. This led to the development of the theory of impetus by 14th century French Jean Buridan, which developed into the modern theories of inertia, velocity, acceleration and momentum. This work and others was developed in 14th century England by the Oxford Calculators such as Thomas Bradwardine, who studied and formulated various laws regarding falling bodies.

On the question of a body subject to a constant (uniform) force, the 12th century Jewish-Arab Nathanel (Iraqi, of Baghdad) stated that constant force imparts constant acceleration, while the main properties are uniformly accelerated motion (as of falling bodies) was worked out by the 14th century Oxford Calculators.

32.4.3   Early modern age

Two central figures in the early modern age are Galileo Galilei and Isaac Newton. Galileo's final statement of his mechanics, particularly of falling bodies, is his *Two New Sciences* (1638). Newton's 1687 *Philosophiæ Naturalis Principia*

Arabic Machine Manuscript. Unknown date (at a guess: 16th to 19th centuries).

Mathematica provided a detailed mathematical account of mechanics, using the newly developed mathematics of calculus and providing the basis of Newtonian mechanics.[5]

There is some dispute over priority of various ideas: Newton's *Principia* is certainly the seminal work and has been tremendously influential, and the systematic mathematics therein did not and could not have been stated earlier because calculus had not been developed. However, many of the ideas, particularly as pertain to inertia (impetus) and falling bodies had been developed and stated by earlier researchers, both the then-recent Galileo and the less-known medieval predecessors. Precise credit is at times difficult or contentious because scientific language and standards of proof changed, so whether medieval statements are *equivalent* to modern statements or *sufficient* proof, or instead *similar* to modern statements and *hypotheses* is often debatable.

32.4.4   Modern age

Two main modern developments in mechanics are general relativity of Einstein, and quantum mechanics, both developed in the 20th century based in part on earlier 19th century ideas.

32.5   Types of mechanical bodies

The often-used term **body** needs to stand for a wide assortment of objects, including particles, projectiles, spacecraft, stars, parts of machinery, parts of solids, parts of fluids (gases and liquids), etc.

Other distinctions between the various sub-disciplines of mechanics, concern the nature of the bodies being described. Particles are bodies with little (known) internal structure, treated as mathematical points in classical mechanics. Rigid bodies have size and shape, but retain a simplicity close to that of the particle, adding just a few so-called degrees of freedom, such as orientation in space.

Otherwise, bodies may be semi-rigid, i.e. elastic, or non-rigid, i.e. fluid. These subjects have both classical and quantum divisions of study.

For instance, the motion of a spacecraft, regarding its orbit and attitude (rotation), is described by the relativistic theory of classical mechanics, while the analogous movements of an atomic nucleus are described by quantum mechanics.

32.6   Sub-disciplines in mechanics

The following are two lists of various subjects that are studied in mechanics.

Note that there is also the "theory of fields" which constitutes a separate discipline in physics, formally treated as distinct from mechanics, whether classical fields or quantum fields. But in actual practice, subjects belonging to mechanics and fields are closely interwoven. Thus, for instance, forces that act on particles are frequently derived from fields (electromagnetic or gravitational), and particles generate fields by acting as sources. In fact, in quantum mechanics, particles themselves are fields, as described theoretically by the wave function.

32.6.1   Classical mechanics

The following are described as forming classical mechanics:

- Newtonian mechanics, the original theory of motion (kinematics) and forces (dynamics).

- Analytical mechanics is a reformulation of Newtonian mechanics with an emphasis on system energy, rather than on forces. There are two main branches of analytical mechanics:

  - Hamiltonian mechanics, a theoretical formalism, based on the principle of conservation of energy.

  - Lagrangian mechanics, another theoretical formalism, based on the principle of the least action.

Prof. Walter Lewin explains Newton's law of gravitation in MIT course 8.01[6]

- Classical statistical mechanics generalizes ordinary classical mechanics to consider systems in an unknown state; often used to derive thermodynamic properties.

- Celestial mechanics, the motion of bodies in space: planets, comets, stars, galaxies, etc.

- Astrodynamics, spacecraft navigation, etc.

- Solid mechanics, elasticity, the properties of deformable bodies.

- Fracture mechanics

- Acoustics, sound (= density variation propagation) in solids, fluids and gases.

- Statics, semi-rigid bodies in mechanical equilibrium

- Fluid mechanics, the motion of fluids

- Soil mechanics, mechanical behavior of soils

- Continuum mechanics, mechanics of continua (both solid and fluid)

- Hydraulics, mechanical properties of liquids

- Fluid statics, liquids in equilibrium

- Applied mechanics, or Engineering mechanics

- Biomechanics, solids, fluids, etc. in biology

- Biophysics, physical processes in living organisms

- Relativistic or Einsteinian mechanics, universal gravitation.

32.6.2   Quantum mechanics

The following are categorized as being part of quantum mechanics:

- Schrödinger wave mechanics, used to describe the motion of the wavefunction of a single particle.

- Matrix mechanics is an alternative formulation that allows considering systems with a finite-dimensional state space.

- Quantum statistical mechanics generalizes ordinary quantum mechanics to consider systems in an unknown state; often used to derive thermodynamic properties.

- Particle physics, the motion, structure, and reactions of particles

- Nuclear physics, the motion, structure, and reactions of nuclei

- Condensed matter physics, quantum gases, solids, liquids, etc.

32.7   Professional organizations

- Applied Mechanics Division, American Society of Mechanical Engineers

- Fluid Dynamics Division, American Physical Society

- Society for Experimental Mechanics

- Institution of Mechanical Engineers is the United Kingdom's qualifying body for Mechanical Engineers and has been the home of Mechanical Engineers for over 150 years.

- International Union of Theoretical and Applied Mechanics

32.8   See also

- Applied mechanics

- Dynamics

- Engineering

- Index of engineering science and mechanics articles

- Kinematics

- Kinetics

- Non-autonomous mechanics

- Statics

- Wiesen Test of Mechanical Aptitude (WTMA)

32.9 References

[1] Dugas, Rene. A History of Classical Mechanics. New York, NY: Dover Publications Inc, 1988, pg 19.

[2] Rana, N.C., and Joag, P.S. Classical Mechanics. West Petal Nagar, New Delhi. Tata McGraw-Hill, 1991, pg 6.

[3] Renn, J., Damerow, P., and McLaughlin, P. Aristotle, Archimedes, Euclid, and the Origin of Mechanics: The Perspective of Historical Epistemology. Berlin: Max Planck Institute for the History of Science, 2010, pg 1-2.

[4] "*A history of mechanics*". René Dugas (1988). p.19. ISBN 0-486-65632-2

[5] "A Tiny Taste of the History of Mechanics". The University of Texas at Austin.

[6] Walter Lewin (October 4, 1999). *Work, Energy, and Universal Gravitation. MIT Course 8.01: Classical Mechanics, Lecture 11.* (OGG) (videotape). Cambridge, MA USA: MIT OCW. Event occurs at 1:21-10:10. Retrieved December 23, 2010.

32.10 Further reading

- Robert Stawell Ball (1871) Experimental Mechanics from Google books.
- Landau, L. D.; Lifshitz, E. M. (1972). *Mechanics and Electrodynamics, Vol. 1.* Franklin Book Company, Inc. ISBN 0-08-016739-X.

32.11 External links

- iMechanica: the web of mechanics and mechanicians
- Mechanics Blog by a Purdue University Professor
- The Mechanics program at Virginia Tech
- Physclips: Mechanics with animations and video clips from the University of New South Wales
- U.S. National Committee on Theoretical and Applied Mechanics
- Interactive learning resources for teaching Mechanics
- The Archimedes Project

Chapter 33

Fluid mechanics

Fluid mechanics is the branch of physics which involves the study of fluids (liquids, gases, and plasmas) and the forces on them. Fluid mechanics can be divided into fluid statics, the study of fluids at rest; and fluid dynamics, the study of the effect of forces on fluid motion. It is a branch of continuum mechanics, a subject which models matter without using the information that it is made out of atoms; that is, it models matter from a *macroscopic* viewpoint rather than from *microscopic*. Fluid mechanics, especially fluid dynamics, is an active field of research with many problems that are partly or wholly unsolved. Fluid mechanics can be mathematically complex, and can best be solved by numerical methods, typically using computers. A modern discipline, called computational fluid dynamics (CFD), is devoted to this approach to solving fluid mechanics problems. Particle image velocimetry, an experimental method for visualizing and analyzing fluid flow, also takes advantage of the highly visual nature of fluid flow.

33.1   Brief history

Main article: History of fluid mechanics

The study of fluid mechanics goes back at least to the days of ancient Greece, when Archimedes investigated fluid statics and buoyancy and formulated his famous law known now as the Archimedes' principle, which was published in his work *On Floating Bodies* – generally considered to be the first major work on fluid mechanics. Rapid advancement in fluid mechanics began with Leonardo da Vinci (observations and experiments), Evangelista Torricelli (invented the barometer), Isaac Newton (investigated viscosity) and Blaise Pascal (researched hydrostatics, formulated Pascal's law), and was continued by Daniel Bernoulli with the introduction of mathematical fluid dynamics in *Hydrodynamica* (1738).

Inviscid flow was further analyzed by various mathematicians (Leonhard Euler, Jean le Rond d'Alembert, Joseph Louis Lagrange, Pierre-Simon Laplace, Siméon Denis Poisson) and viscous flow was explored by a multitude of engineers including Jean Léonard Marie Poiseuille and Gotthilf Hagen. Further mathematical justification was provided by Claude-Louis Navier and George Gabriel Stokes in the Navier–Stokes equations, and boundary layers were investigated (Ludwig Prandtl, Theodore von Kármán), while various scientists such as Osborne Reynolds, Andrey Kolmogorov, and Geoffrey Ingram Taylor advanced the understanding of fluid viscosity and turbulence.

33.2   Main branches

33.2.1   Fluid statics

Main article: Fluid statics

Fluid statics or **hydrostatics** is the branch of fluid mechanics that studies fluids at rest. It embraces the study of the conditions under which fluids are at rest in stable equilibrium; and is contrasted with fluid dynamics, the study of fluids in motion.

Hydrostatics is fundamental to hydraulics, the engineering of equipment for storing, transporting and using fluids. It is also relevant to geophysics and astrophysics (for example, in understanding plate tectonics and the anomalies of the Earth's gravitational field), to meteorology, to medicine (in the context of blood pressure), and many other fields.

Hydrostatics offers physical explanations for many phenomena of everyday life, such as why atmospheric pressure changes with altitude, why wood and oil float on water, and why the surface of water is always flat and horizontal whatever the shape of its container.

33.2.2   Fluid dynamics

Main article: Fluid dynamics

Fluid dynamics is a subdiscipline of fluid mechanics that deals with **fluid flow**—the natural science of fluids (liquids and gases) in motion. It has several subdisciplines itself, including **aerodynamics** (the study of air and other gases in motion) and **hydrodynamics** (the study of liquids in motion). Fluid dynamics has a wide range of applications, including calculating forces and moments on aircraft, determining the mass flow rate of petroleum through pipelines, predicting weather patterns, understanding nebulae in interstellar space and modelling fission weapon detonation. Some of its principles are even used in traffic engineering, where traffic is treated as a continuous fluid, and crowd dynamics.

Fluid dynamics offers a systematic structure—which underlies these practical disciplines—that embraces empirical and semi-empirical laws derived from flow measurement and used to solve practical problems. The solution to a fluid dynamics problem typically involves calculating various properties of the fluid, such as velocity, pressure, density, and temperature, as functions of space and time.

33.3   Relationship to continuum mechanics

Fluid mechanics is a subdiscipline of continuum mechanics, as illustrated in the following table.

In a mechanical view, a fluid is a substance that does not support shear stress; that is why a fluid at rest has the shape of its containing vessel. A fluid at rest has no shear stress.

33.4   Assumptions

Like any mathematical model of the real world, fluid mechanics makes some basic assumptions about the materials being studied. These assumptions are turned into equations that must be satisfied if the assumptions are to be held true.

For example, consider a fluid in three dimensions. The assumption that mass is conserved means that for any fixed control volume (for example a sphere) – enclosed by a control surface – the rate of change of the mass contained is equal to the rate at which mass is passing from *outside* to *inside* through the surface, minus the rate at which mass is passing the other way, from *inside* to *outside*. (A special case would be when the mass *inside* and the mass *outside* remain constant). This can be turned into an equation in integral form over the control volume.[1]

Fluid mechanics assumes that every fluid obeys the following:

- Conservation of mass

- Conservation of energy

- Conservation of momentum

Rate of Change of Property, N for a System

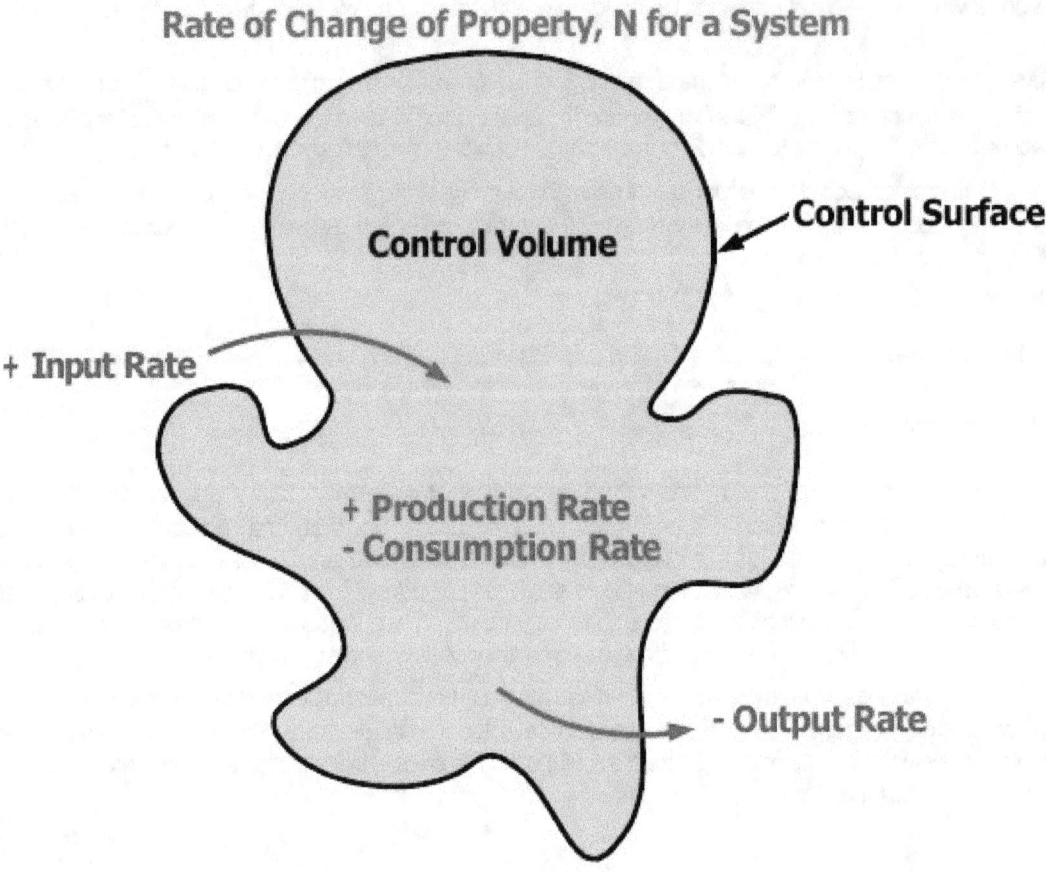

Balance for some integrated fluid quantity in a control volume enclosed by a control surface.

- The *continuum hypothesis*, detailed below.

Further, it is often useful (at subsonic conditions) to assume a fluid is incompressible – that is, the density of the fluid does not change.

Similarly, it can sometimes be assumed that the viscosity of the fluid is zero (the fluid is *inviscid*). Gases can often be assumed to be inviscid. If a fluid is viscous, and its flow contained in some way (e.g. in a pipe), then the flow at the boundary must have zero velocity. For a viscous fluid, if the boundary is not porous, the shear forces between the fluid and the boundary results also in a zero velocity for the fluid at the boundary. This is called the no-slip condition. For a porous media otherwise, in the frontier of the containing vessel, the slip condition is not zero velocity, and the fluid has a discontinuous velocity field between the free fluid and the fluid in the porous media (this is related to the Beavers and Joseph condition).

33.4.1   Continuum hypothesis

Main article: Continuum mechanics

Fluids are composed of molecules that collide with one another and solid objects. The continuum assumption, however, considers fluids to be continuous. That is, properties such as density, pressure, temperature, and velocity are taken to be well-defined at "infinitely" small points, defining a REV (Reference Element of Volume), at the geometric order of the distance between two adjacent molecules of fluid. Properties are assumed to vary continuously from one point to another, and are averaged values in the REV. The fact that the fluid is made up of discrete molecules is ignored.

The continuum hypothesis is basically an approximation, in the same way planets are approximated by point particles when dealing with celestial mechanics, and therefore results in approximate solutions. Consequently, assumption of the continuum hypothesis can lead to results which are not of desired accuracy. However, under the right circumstances, the continuum hypothesis produces extremely accurate results.

Those problems for which the continuum hypothesis does not allow solutions of desired accuracy are solved using statistical mechanics. To determine whether or not to use conventional fluid dynamics or statistical mechanics, the Knudsen number is evaluated for the problem. The Knudsen number is defined as the ratio of the molecular mean free path length to a certain representative physical length scale. This length scale could be, for example, the radius of a body in a fluid. (More simply, the Knudsen number is how many times its own diameter a particle will travel on average before hitting another particle). Problems with Knudsen numbers at or above one are best evaluated using statistical mechanics for reliable solutions.,,,,,,,,

33.5 Navier–Stokes equations

Main article: Navier–Stokes equations

The **Navier–Stokes equations** (named after Claude-Louis Navier and George Gabriel Stokes) are the set of equations that describe the motion of fluid substances such as liquids and gases. These equations state that changes in momentum (force) of fluid particles depend only on the external pressure and internal viscous forces (similar to friction) acting on the fluid. Thus, the Navier–Stokes equations describe the balance of forces acting at any given region of the fluid.

The Navier–Stokes equations are differential equations which describe the motion of a fluid. Such equations establish relations among the rates of change of the variables of interest. For example, the Navier–Stokes equations for an ideal fluid with zero viscosity states that acceleration (the rate of change of velocity) is proportional to the derivative of internal pressure.

This means that solutions of the Navier–Stokes equations for a given physical problem must be sought with the help of calculus. In practical terms only the simplest cases can be solved exactly in this way. These cases generally involve non-turbulent, steady flow (flow does not change with time) in which the Reynolds number is small.

For more complex situations, involving turbulence, such as global weather systems, aerodynamics, hydrodynamics and many more, solutions of the Navier–Stokes equations can currently only be found with the help of computers. This branch of science is called computational fluid dynamics.

33.5.1 General form of the equation

The general form of the Cauchy momentum equation is:

$$\rho \frac{D\mathbf{u}}{}$$

where

- ρ is the fluid density,
- D
- $\mathbf{u}$ is the flow velocity vector,
- $\mathbf{f}$ is the specific body force vector, and

- $\boldsymbol{\sigma}$ is the stress tensor.

Unless the fluid is made up of spinning degrees of freedom like vortices, $\boldsymbol{\sigma}$ is a symmetric tensor. In Navier Stokes equations the stress tensor can be decomposed as

$$\sigma_{ij} = -p\delta_{ij} + \tau_{ij}$$

where $-p\delta_{ij}$ is a static isotropic stress state (that would exist if the fluid were at rest), and τ_{ij} is the deviatoric stress tensor, corresponding to the part of the stress due to the fluid motion. Generally, the scalar p can be taken as the thermodynamic pressure, whereas τ_{ij} is called the viscous stress tensor. Furthermore, the diagonal components of tensor τ are called normal stresses and the off-diagonal components are called shear stresses.

The vectorial Cauchy equation above can be written then as

$$\rho \frac{D\mathbf{u}}{}$$

This is actually a set of three equations, one per dimension. By themselves, these equations are not sufficient to produce a solution. However, adding other conservation laws and appropriate boundary conditions to the system of equations produces a solvable set of equations. The conservation of mass provides another equation relating the density and the flow velocity:

$$\frac{\partial \rho}{}$$

On the other hand, the identification of p with the thermodynamic pressure is usually possible (unless the fluid is not in thermodynamic equilibrium; such situation is however rare [e.g. shock waves]). Therefore, a thermodynamic equation of state must be used to connect the pressure with the density and another state property, such as temperature or enthalpy. This in turn brings another unknown to the problem so that an equation for conservation of thermal energy must also be solved along with momentum and mass conservations.

In the case of an incompressible fluid there is no relationship between the pressure and the density. The Navier–Stokes equations and mass conservation are then sufficient to determine the solution to a fluid mechanics problem. Actually, the absolute pressure in an incompressible fluid is indeterminate, and only its gradient is relevant for the equations of motion. Taking the divergence of the Navier–Stokes equation and using the mass conservation equation to simplify the result gives a Poisson equation for the pressure.

Additionally, in order to close the system of equations a constitutive equation relating the viscous stress tensor to the velocity field must be introduced. This constitutive model, which depends on the nature of the fluid, is the basis for the distinction between Newtonian and non-Newtonian fluids.

33.6   Newtonian versus non-Newtonian fluids

A **Newtonian fluid** (named after Isaac Newton) is defined to be a fluid whose shear stress is linearly proportional to the velocity gradient in the direction perpendicular to the plane of shear. This definition means regardless of the forces acting on a fluid, it *continues to flow*. For example, water is a Newtonian fluid, because it continues to display fluid properties no matter how much it is stirred or mixed. A slightly less rigorous definition is that the drag of a small object being moved slowly through the fluid is proportional to the force applied to the object. (Compare friction). Important fluids, like water as well as most gases, behave – to good approximation – as a Newtonian fluid under normal conditions on Earth.[2]

By contrast, stirring a non-Newtonian fluid can leave a "hole" behind. This will gradually fill up over time – this behaviour is seen in materials such as pudding, oobleck, or sand (although sand isn't strictly a fluid). Alternatively, stirring a non-Newtonian fluid can cause the viscosity to decrease, so the fluid appears "thinner" (this is seen in non-drip paints). There are many types of non-Newtonian fluids, as they are defined to be something that fails to obey a particular property – for example, most fluids with long molecular chains can react in a non-Newtonian manner.[2]

33.6.1   Equations for a Newtonian fluid

Main article: Newtonian fluid

The constant of proportionality between the viscous stress tensor and the velocity gradient is known as the viscosity. A simple equation to describe incompressible Newtonian fluid behaviour is

$$\tau = -\mu \frac{uv}{dy}$$

where

> τ is the shear stress exerted by the fluid ("drag")
>
> μ is the fluid viscosity – a constant of proportionality
>
> $\frac{dv}{}$

For a Newtonian fluid, the viscosity, by definition, depends only on temperature and pressure, not on the forces acting upon it. If the fluid is incompressible the equation governing the viscous stress (in Cartesian coordinates) is

$$\tau_{ij} = \mu \left(\frac{\partial v_i}{\partial x_j} + \frac{\partial v_j}{\partial x_i} \right)$$

where

> τ_{ij} is the shear stress on the ith face of a fluid element in the jth direction
>
> v_i is the velocity in the ith direction
>
> x_j is the jth direction coordinate.

If the fluid is not incompressible the general form for the viscous stress in a Newtonian fluid is

$$\tau_{ij} = \mu \left(\frac{\partial v_i}{\partial x_j} + \frac{\partial v_j}{} \frac{2}{3} \right)$$

where κ is the second viscosity coefficient (or bulk viscosity). If a fluid does not obey this relation, it is termed a non-Newtonian fluid, of which there are several types. Non-Newtonian fluids can be either plastic, Bingham plastic, pseudoplastic, dilatant, thixotropic, rheopectic, viscoelastic.

In some applications another rough broad division among fluids is made: ideal and non-ideal fluids. An Ideal fluid is non-viscous and offers no resistance whatsoever to a shearing force. An ideal fluid really does not exist, but in some calculations, the assumption is justifiable. One example of this is the flow far from solid surfaces. In many cases the viscous effects are concentrated near the solid boundaries (such as in boundary layers) while in regions of the flow field far away from the boundaries the viscous effects can be neglected and the fluid there is treated as it were inviscid (ideal flow). When the viscosity is neglected, the term containing the viscous stress tensor τ in the Navier–Stokes equation vanishes. The equation reduced in this form is called the Euler equation.

33.7   See also

- Aerodynamics

- Applied mechanics
- Bernoulli's principle
- Communicating vessels
- Secondary flow
- Different types of boundary conditions in fluid dynamics

33.8   Notes

[1] Batchelor (1967), p. 74.

[2] Batchelor (1967), p. 145.

33.9   References

- Batchelor, George K. (1967), *An Introduction to Fluid Dynamics*, Cambridge University Press, ISBN 0-521-66396-2

33.10   Further reading

- Falkovich, Gregory (2011), *Fluid Mechanics (A short course for physicists)*, Cambridge University Press, ISBN 978-1-107-00575-4

- Kundu, Pijush K.; Cohen, Ira M. (2008), *Fluid Mechanics* (4th revised ed.), Academic Press, ISBN 978-0-12-373735-9

- Currie, I. G. (1974), *Fundamental Mechanics of Fluids*, McGraw-Hill, Inc., ISBN 0-07-015000-1

- Massey, B.; Ward-Smith, J. (2005), *Mechanics of Fluids* (8th ed.), Taylor & Francis, ISBN 978-0-415-36206-1

- White, Frank M. (2003), *Fluid Mechanics*, McGraw–Hill, ISBN 0-07-240217-2

- Nazarenko, Sergey (2014), *Fluid Dynamics via Examples and Solutions*, CRC Press (Taylor & Francis group), ISBN 978-1-43-988882-7

33.11   External links

- Free Fluid Mechanics books
- Annual Review of Fluid Mechanics
- CFDWiki – the Computational Fluid Dynamics reference wiki.
- Educational Particle Image Velocimetry – resources and demonstrations

Chapter 34

Operations research

For the academic journal, see Operations Research.

Operations research, or **operational research** in British usage, is a discipline that deals with the application of advanced analytical methods to help make better decisions.[1] It is often considered to be a sub-field of mathematics.[2] The terms management science and decision science are sometimes used as synonyms.[3]

Employing techniques from other mathematical sciences, such as mathematical modeling, statistical analysis, and mathematical optimization, operations research arrives at optimal or near-optimal solutions to complex decision-making problems. Because of its emphasis on human-technology interaction and because of its focus on practical applications, operations research has overlap with other disciplines, notably industrial engineering and operations management, and draws on psychology and organization science. Operations research is often concerned with determining the maximum (of profit, performance, or yield) or minimum (of loss, risk, or cost) of some real-world objective. Originating in military efforts before World War II, its techniques have grown to concern problems in a variety of industries.[4]

34.1 Overview

Operational research (OR) encompasses a wide range of problem-solving techniques and methods applied in the pursuit of improved decision-making and efficiency, such as simulation, mathematical optimization, queueing theory and other stochastic-process models, Markov decision processes, econometric methods, data envelopment analysis, neural networks, expert systems, decision analysis, and the analytic hierarchy process.[5] Nearly all of these techniques involve the construction of mathematical models that attempt to describe the system. Because of the computational and statistical nature of most of these fields, OR also has strong ties to computer science and analytics. Operational researchers faced with a new problem must determine which of these techniques are most appropriate given the nature of the system, the goals for improvement, and constraints on time and computing power.

The major subdisciplines in modern operational research, as identified by the journal *Operations Research*,[6] are:

- Computing and information technologies
-
- Financial engineering
- Manufacturing, service sciences, and supply chain management
- Marketing Engineering[7]
- Policy modeling and public sector work
- Revenue management

- Simulation

- Stochastic models

- Transportation

34.2   History

As a formal discipline, operational research originated in the efforts of military planners during World War II. In the decades after the war, the techniques were more widely applied to problems in business, industry and society. Since that time, operational research has expanded into a field widely used in industries ranging from petrochemicals to airlines, finance, logistics, and government, moving to a focus on the development of mathematical models that can be used to analyse and optimize complex systems, and has become an area of active academic and industrial research.[4]

34.2.1   Historical origins

Early work in operational research was carried out by individuals such as Charles Babbage. His research into the cost of transportation and sorting of mail led to England's universal "Penny Post" in 1840, and studies into the dynamical behaviour of railway vehicles in defence of the GWR's broad gauge.[8] Percy Bridgman brought operational research to bear on problems in physics in the 1920s and would later attempt to extend these to the social sciences.[9]

Modern operational research originated at the Bawdsey Research Station in the UK in 1937 and was the result of an initiative of the station's superintendent, A. P. Rowe. Rowe conceived the idea as a means to analyse and improve the working of the UK's early warning radar system, Chain Home (CH). Initially, he analysed the operating of the radar equipment and its communication networks, expanding later to include the operating personnel's behaviour. This revealed unappreciated limitations of the CH network and allowed remedial action to be taken.[10]

Scientists in the United Kingdom including Patrick Blackett (later Lord Blackett OM PRS), Cecil Gordon, Solly Zuckerman, (later Baron Zuckerman OM, KCB, FRS), C. H. Waddington, Owen Wansbrough-Jones, Frank Yates, Jacob Bronowski and Freeman Dyson, and in the United States with George Dantzig looked for ways to make better decisions in such areas as logistics and training schedules

34.2.2   Second World War

The modern field of operational research arose during World War II. In the World War II era, operational research was defined as "a scientific method of providing executive departments with a quantitative basis for decisions regarding the operations under their control."[11] Other names for it included operational analysis (UK Ministry of Defence from 1962)[12] and quantitative management.[13]

During the Second World War close to 1,000 men and women in Britain were engaged in operational research. About 200 operational research scientists worked for the British Army.[14]

Patrick Blackett worked for several different organizations during the war. Early in the war while working for the Royal Aircraft Establishment (RAE) he set up a team known as the "Circus" which helped to reduce the number of anti-aircraft artillery rounds needed to shoot down an enemy aircraft from an average of over 20,000 at the start of the Battle of Britain to 4,000 in 1941.[15]

In 1941 Blackett moved from the RAE to the Navy, after first working with RAF Coastal Command, in 1941 and then early in 1942 to the Admiralty.[16] Blackett's team at Coastal Command's Operational Research Section (CC-ORS) included two future Nobel prize winners and many other people who went on to be pre-eminent in their fields.[17] They undertook a number of crucial analyses that aided the war effort. Britain introduced the convoy system to reduce shipping losses, but while the principle of using warships to accompany merchant ships was generally accepted, it was unclear whether it was better for convoys to be small or large. Convoys travel at the speed of the slowest member, so small convoys can travel faster. It was also argued that small convoys would be harder for German U-boats to detect. On the other hand, large convoys could deploy more warships against an attacker. Blackett's staff showed that the losses suffered by convoys depended largely on the number of escort vessels present, rather than the size of the convoy. Their conclusion was that a few large convoys are more defensible than many small ones.[18]

A Liberator in standard RAF green/dark earth/black night bomber finish as originally used by Coastal Command

While performing an analysis of the methods used by RAF Coastal Command to hunt and destroy submarines, one of the analysts asked what colour the aircraft were. As most of them were from Bomber Command they were painted black for night-time operations. At the suggestion of CC-ORS a test was run to see if that was the best colour to camouflage the aircraft for daytime operations in the grey North Atlantic skies. Tests showed that aircraft painted white were on average not spotted until they were 20% closer than those painted black. This change indicated that 30% more submarines would be attacked and sunk for the same number of sightings.[19] As a result of these findings Coastal Command changed their aircraft to using white undersurfaces.

Other work by the CC-ORS indicated that on average if the trigger depth of aerial-delivered depth charges (DCs) were changed from 100 feet to 25 feet, the kill ratios would go up. The reason was that if a U-boat saw an aircraft only shortly before it arrived over the target then at 100 feet the charges would do no damage (because the U-boat wouldn't have had time to descend as far as 100 feet), and if it saw the aircraft a long way from the target it had time to alter course under water so the chances of it being within the 20-foot kill zone of the charges was small. It was more efficient to attack those submarines close to the surface when the targets' locations were better known than to attempt their destruction at greater depths when their positions could only be guessed. Before the change of settings from 100 feet to 25 feet, 1% of submerged U-boats were sunk and 14% damaged. After the change, 7% were sunk and 11% damaged. (If submarines were caught on the surface, even if attacked shortly after submerging, the numbers rose to 11% sunk and 15% damaged). Blackett observed "there can be few cases where such a great operational gain had been obtained by such a small and simple change of tactics".[20]

Bomber Command's Operational Research Section (BC-ORS), analysed a report of a survey carried out by RAF Bomber Command. For the survey, Bomber Command inspected all bombers returning from bombing raids over Germany over a particular period. All damage inflicted by German air defences was noted and the recommendation was given that armour be added in the most heavily damaged areas. This recommendation was not adopted because the fact that the aircraft returned with these areas damaged indicated these areas were NOT vital, and adding armour to non-vital areas where

A Warwick in the revised RAF Coastal Command green/dark grey/white finish

damage is acceptable negatively affects aircraft performance. Their suggestion to remove some of the crew so that an aircraft loss would result in fewer personnel losses, was also rejected by RAF command. Blackett's team made the logical recommendation that the armour be placed in the areas which were completely untouched by damage in the bombers which returned. They reasoned that the survey was biased, since it only included aircraft that returned to Britain. The untouched areas of returning aircraft were probably vital areas, which, if hit, would result in the loss of the aircraft.[21]

When Germany organised its air defences into the Kammhuber Line, it was realised by the British that if the RAF bombers were to fly in a bomber stream they could overwhelm the night fighters who flew in individual cells directed to their targets by ground controllers. It was then a matter of calculating the statistical loss from collisions against the statistical loss from night fighters to calculate how close the bombers should fly to minimise RAF losses.[22]

The "exchange rate" ratio of output to input was a characteristic feature of operational research. By comparing the number of flying hours put in by Allied aircraft to the number of U-boat sightings in a given area, it was possible to redistribute aircraft to more productive patrol areas. Comparison of exchange rates established "eff ectiveness ratios" useful in planning. The ratio of 60 mines laid per ship sunk was common to several campaigns: German mines in British ports, British mines on German routes, and United States mines in Japanese routes.[23]

Operational research doubled the on-target bomb rate of B-29s bombing Japan from the Marianas Islands by increasing the training ratio from 4 to 10 percent of flying hours; revealed that wolf-packs of three United States submarines were the most effective number to enable all members of the pack to engage targets discovered on their individual patrol stations; revealed that glossy enamel paint was more effective camouflage for night fighters than traditional dull camouflage paint finish, and the smooth paint finish increased airspeed by reducing skin friction.[23]

On land, the operational research sections of the Army Operational Research Group (AORG) of the Ministry of Supply (MoS) were landed in Normandy in 1944, and they followed British forces in the advance across Europe. They analysed, among other topics, the effectiveness of artillery, aerial bombing and anti-tank shooting.

34.2.3 After World War II

With expanded techniques and growing awareness of the field at the close of the war, operational research was no longer limited to only operational, but was extended to encompass equipment procurement, training, logistics and infrastructure. Operations Research also grew in many areas other than the military once scientists learned to apply its principles to the civilian sector. With the development of the simplex algorithm for Linear Programming in 1947 [24] and the development of computers over the next three decades, Operations Research can now "solve problems with hundreds of thousands of variables and constraints. Moreover, the large volumes of data required for such problems can be stored and manipulated very efficiently." [24]

34.3   Problems addressed

- Critical path analysis or project planning: identifying those processes in a complex project which affect the overall duration of the project

- Floorplanning: designing the layout of equipment in a factory or components on a computer chip to reduce manufacturing time (therefore reducing cost)

- Network optimization: for instance, setup of telecommunications networks to maintain quality of service during outages

- Allocation problems

- Facility location

- Assignment Problems:

    - Assignment problem
    - Generalized assignment problem
    - Quadratic assignment problem
    - Weapon target assignment problem

- Bayesian search theory : looking for a target

- Optimal search

- Routing, such as determining the routes of buses so that as few buses are needed as possible

- Supply chain management: managing the flow of raw materials and products based on uncertain demand for the finished products

- Efficient messaging and customer response tactics

- Automation: automating or integrating robotic systems in human-driven operations processes

- Globalization: globalizing operations processes in order to take advantage of cheaper materials, labor, land or other productivity inputs

- Transportation: managing freight transportation and delivery systems (Examples: LTL Shipping, intermodal freight transport, travelling salesman problem)

- Scheduling:

    - Personnel staffing
    - Manufacturing steps
    - Project tasks

- Network data traffic: these are known as queueing models or queueing systems.
  - Sports events and their television coverage
- Blending of raw materials in oil refineries
- Determining optimal prices, in many retail and B2B settings, within the disciplines of pricing science

Operational research is also used extensively in government where evidence-based policy is used.

34.4   Management science

Main article: Management science

In 1967 Stafford Beer characterized the field of management science as "the business use of operations research".[25] However, in modern times the term management science may also be used to refer to the separate fields of organizational studies or corporate strategy. Like operational research itself, management science (MS) is an interdisciplinary branch of applied mathematics devoted to optimal decision planning, with strong links with economics, business, engineering, and other sciences. It uses various scientific research-based principles, strategies, and analytical methods including mathematical modeling, statistics and numerical algorithms to improve an organization's ability to enact rational and meaningful management decisions by arriving at optimal or near optimal solutions to complex decision problems. In short, management sciences help businesses to achieve their goals using the scientific methods of operational research.

The management scientist's mandate is to use rational, systematic, science-based techniques to inform and improve decisions of all kinds. Of course, the techniques of management science are not restricted to business applications but may be applied to military, medical, public administration, charitable groups, political groups or community groups.

Management science is concerned with developing and applying models and concepts that may prove useful in helping to illuminate management issues and solve managerial problems, as well as designing and developing new and better models of organizational excellence.[26]

The application of these models within the corporate sector became known as management science.[27]

34.4.1   Related fields

Some of the fields that have considerable overlap with Operations Research and Management Science include:

34.4.2   Applications

Applications of management science is abundant in industry as airlines, manufacturing companies, service organizations, military branches, and in government. The range of problems and issues to which management science has contributed insights and solutions is vast. It includes:[26]

- scheduling airlines, including both planes and crew,
- deciding the appropriate place to site new facilities such as a warehouse, factory or fire station,
- managing the flow of water from reservoirs,
- identifying possible future development paths for parts of the telecommunications industry,
- establishing the information needs and appropriate systems to supply them within the health service, and
- identifying and understanding the strategies adopted by companies for their information systems

Management science is also concerned with so-called "soft-operational analysis", which concerns methods for strategic planning, strategic decision support, and Problem Structuring Methods (PSM). In dealing with these sorts of challenges mathematical modeling and simulation are not appropriate or will not suffice. Therefore, during the past 30 years, a number of non-quantified modeling methods have been developed. These include:

- stakeholder based approaches including metagame analysis and drama theory

- morphological analysis and various forms of influence diagrams.

- approaches using cognitive mapping

- the Strategic Choice Approach

- robustness analysis

34.5 Societies and journals

Societies

The International Federation of Operational Research Societies (IFORS)[28] is an umbrella organization for operational research societies worldwide, representing approximately 50 national societies including those in the US,[29] UK,[30] France,[31]Germany, Canada,[32]Australia,[33]New Zealand,[34]Philippines,[35]India,[36]Japan and South Africa (ORSSA). [37] The constituent members of IFORS form regional groups, such as that inEurope.[38]Other important operational research organizations are Simulation Interoperability Standards Organization (SISO) [39] and Interservice/Industry Training, Simulation and Education Conference (I/ITSEC).[40]

In 2004 the US-based organization INFORMS began an initiative to market the OR profession better, including a website entitled *The Science of Better*[41] which provides an introduction to OR and examples of successful applications of OR to industrial problems. This initiative has been adopted by the Operational Research Society in the UK, including a website entitled *Learn about OR.*[42]

Journals

The Institute for Operations Research and the Management Sciences (INFORMS) publishes thirteen scholarly journals about operations research, including the top two journals in their class, according to 2005 Journal Citation Reports.[43] They are:

- Decision Analysis

- Information Systems Research

- INFORMS Journal on Computing

- *INFORMS Transactions on Education*[44] (an open access journal)

- *Interfaces: An International Journal of the Institute for Operations Research and the Management Sciences*

- *Management Science: A Journal of the Institute for Operations Research and the Management Sciences*

- *Manufacturing & Service Operations Management*

- *Marketing Science*

- *Mathematics of Operations Research*

- *Operations Research: A Journal of the Institute for Operations Research and the Management Sciences*

- Organization Science
- *Service Science*
- *Transportation Science.*

Other journals

- *4OR-A Quarterly Journal of Operations Research:* jointly published the Belgian, French and Italian Operations Research Societies (Springer);
- *Decision Sciences* published by Wiley-Blackwell on behalf of the Decision Sciences Institute
- *European Journal of Operational Research (EJOR):* Founded in 1975 and is presently by far the largest operational research journal in the world, with its around 9,000 pages of published papers per year. In 2004, its total number of citations was the second largest amongst Operational Research and Management Science journals;
- *INFOR Journal:* published and sponsored by the Canadian Operational Research Society;
- *International Journal of Operations Research and Information Systems (IJORIS)":* an official publication of the Information Resources Management Association, published quarterly by IGI Global;[45]
- *Journal of Defense Modeling and Simulation (JDMS): Applications, Methodology, Technology:* a quarterly journal devoted to advancing the science of modeling and simulation as it relates to the military and defense.[46]
- *Journal of the Operational Research Society (JORS):* an official journal of The OR Society; this is the oldest continuously published journal of OR in the world, published by Palgrave;[47]
- *Journal of Simulation (JOS):* an official journal of The OR Society, published by Palgrave;[47]
- *Mathematical Methods of Operations Research (MMOR):* the journal of the German and Dutch OR Societies, published by Springer;[48]
- *Military Operations Research (MOR):* published by the Military Operations Research Society;
- *Opsearch:* official journal of the Operational Research Society of India;
- *OR Insight:* a quarterly journal of The OR Society, published by Palgrave;[47]
- *Production and Operations Management,* the official journal of the Production and Operations Management Society
- *TOP:* the official journal of the Spanish Society of Statistics and Operations Research.[49]

34.6   See also

34.7   References

[1] "About Operations Research". INFORMS.org. Retrieved 7 January 2012.

[2] "Mathematics Subject Classification". American Mathematical Society. 23 May 2011. Retrieved 7 January 2012.

[3] Wetherbe, James C. (1979), *Systems analysis for computer-based information systems,* West series in data processing and information systems, West Pub. Co., ISBN 9780829902280, A systems analyst who contributes in the area of DSS must be skilled in such areas as management science (synonymous with decision science and operations research), modeling, simulation, and advanced statistics.

[4] "What is OR". HSOR.org. Retrieved 13 November 2011.

[5] "Operations Research Analysts". Bls.gov. Retrieved 27 January 2012.

[6] "OR / Pubs / IOL Home". INFORMS.org. 2 January 2009. Retrieved 13 November 2011.

[7] "DecisionPro, Inc. – Makers of Marketing Engineering Software – Home Page". Decisionpro.biz. Retrieved 13 November 2011.

[8]M.S. Sodhi, "What about the 'O' in O.R.?" OR/MS Today, December, 2007, p. 12,http://www.lionhrtpub.com/orms/orms-12-0 7/ frqed.html

[9] P. W. Bridgman, The Logic of Modern Physics, The MacMillan Company, New York, 1927

[10] "operations research (industrial engineering) :: History – Britannica Online Encyclopedia". Britannica.com. Retrieved 13 November 2011.

[11]"Operational Research in the British Army 1939–1945, October 1947, Report C67/3/4/48, UK National Archives file WO291/13 01Quoted on the dust-jacket of: Morse, Philip M, and Kimball, George E, *Methods of Operations Research*, 1st Edition Revised,pu b MIT Press & J Wiley, 5th printing, 1954.

[12] UK National Archives Catalogue for WO291 lists a War Office organisation called Army Operational Research Group (AORG) that existed from 1946 to 1962. "In January 1962 the name was changed to Army Operational Research Establishment (AORE). Following the creation of a unified Ministry of Defence, a tri-service operational research organisation was established: the Defence Operational Research Establishment (DOAE) which was formed in 1965, and it the Army Operational Research Establishment based at West Byfleet."

[13] http://brochure.unisa.ac.za/myunisa/data/subjects/Quantitative%20Management.pdf

[14] Kirby, p. 117

[15] Kirby, pp. 91–94

[16] Kirby, p. 96,109

[17] Kirby, p. 96

[18] ""Numbers are Essential": Victory in the North Atlantic Reconsidered, March–May 1943". Familyheritage.ca. 24 May 1943. Retrieved 13 November 2011.

[19] Kirby, p. 101

[20] (Kirby, pp. 102,103)

[21] James F. Dunnigan (1999). *Dirty Little Secrets of the Twentieth Century*. Harper Paperbacks. pp. 215–217.

[22] "RAF History – Bomber Command 60th Anniversary". Raf.mod.uk. Retrieved 13 November 2011.

[23] Milkman, Raymond H. (May 1968). "Operations Research in World War II". United States Naval Institute Proceedings.

[24] http://www.pitt.edu/~{}jrclass/or/or-intro.html#history

[25] Stafford Beer (1967) *Management Science: The Business Use of Operations Research*

[26] What is Management Science? Lancaster University, 2008. Retrieved 5 June 2008.

[27] What is Management Science? The University of Tennessee, 2006. Retrieved 5 June 2008.

[28] "IFORS". IFORS. Retrieved 13 November 2011.

[29] Leszczynski, Mary (8 November 2011). "Informs". Informs. Retrieved 13 November 2011.

[30] "The OR Society". Orsoc.org.uk. Retrieved 13 November 2011.

[31] "Société française de Recherche Opérationnelle et d'Aide à la Décision". ROADEF. Retrieved 13 November 2011.

[32] www.cors.ca. "CORS". Cors.ca. Retrieved 13 November 2011.

[33] "ASOR". ASOR. 1 January 1972. Retrieved 13 November 2011.

[34] "ORSNZ". ORSNZ. Retrieved 13 November 2011.

[35] "ORSP". ORSP. Retrieved 13 November 2011.

[36] "ORSI". Orsi.in. Retrieved 13 November 2011.

[37] "ORSSA". ORSSA. 23 September 2011. Retrieved 13 November 2011.

[38] "EURO (EURO)". Euro-online.org. Retrieved 13 November 2011.

[39] "SISO". Sisostds.org. Retrieved 13 November 2011.

[40] "I/Itsec". I/Itsec. Retrieved 13 November 2011.

[41] "The Science of Better". The Science of Better. Retrieved 13 November 2011.

[42] "Learn about OR". Learn about OR. Retrieved 13 November 2011.

[43] "INFORMS Journals". Informs.org. Retrieved 13 November 2011.

[44] "INFORMS Transactions on Education". Informs.org. Retrieved 19 March 2015.

[45] "International Journal of Operations Research and Information Systems (IJORIS) (1947–9328)(1947–9336): John Wang: Journals". IGI Global. Retrieved 13 November 2011.

[46] The Society for Modeling & Simulation International. "JDMS". Scs.org. Retrieved 13 November 2011.

[47] The OR Society;

[48] "Mathematical Methods of Operations Research website". Springer.com. Retrieved 13 November 2011.

[49] "TOP". Springer.com. Retrieved 13 November 2011.

34.8   Notes

- Kirby, M. W. (Operational Research Society (Great Britain)). Operational Research in War and Peace: The British Experience from the 1930s to 1970, Imperial College Press, 2003. ISBN 1-86094-366-7, ISBN 978-1-86094-366-9

34.9   Further reading

- C. West Churchman, Russell L. Ackoff & E. L. Arnoff, *Introduction to Operations Research*, New York: J. Wiley and Sons, 1957
- Joseph G. Ecker & Michael Kupferschmid, *Introduction to Operations Research*, Krieger Publishing Co.
- Frederick S. Hillier & Gerald J. Lieberman, *Introduction to Operations Research*, McGraw-Hill: Boston MA; 8th. (International) Edition, 2005
- Michael Pidd, *Tools for Thinking: Modelling in Management Science*, J. Wiley & Sons Ltd., Chichester; 2nd. Edition, 2003
- Hamdy A. Taha, *Operations Research: An Introduction*, Prentice Hall; 9th. Edition, 2011
- Wayne Winston, *Operations Research: Applications and Algorithms*, Duxbury Press; 4th. Edition, 2003
- Kenneth R. Baker, Dean H. Kropp (1985). *Management Science: An Introduction to the Use of Decision Models*
- David Charles Heinze (1982). *Management Science: Introductory Concepts and Applications*
- Lee J. Krajewski, Howard E. Thompson (1981). "Management Science: Quantitative Methods in Context"
- Thomas W. Knowles (1989). *Management science: Building and Using Models*

- Kamlesh Mathur, Daniel Solow (1994). *Management Science: The Art of Decision Making*

- Laurence J. Moore, Sang M. Lee, Bernard W. Taylor (1993). *Management Science*

- William Thomas Morris (1968). *Management Science: A Bayesian Introduction.*

- William E. Pinney, Donald B. McWilliams (1987). *Management Science: An Introduction to Quantitative Analysis for Management*

- Shuchman, Abraham. *Scientific decision making in business.* New York: Holt, Rinehart and Winston (1963)

- Shrader, Charles R. (2006). *History of Operations Research in the United States Army, Volume 1:1942–1962.* Washington, D.C.: United States Army Center of Military History. CMH Pub 70-102-1.

- Gerald E. Thompson (1982). *Management Science: An Introduction to Modern Quantitative Analysis and Decision Making. New York : McGraw-Hill Publishing Co.*

- Saul I. Gass & Arjang A. Assad (2005). *An Annotated Timeline of Operations Research: An Informal History. New York : Kluwer Academic Publishers.*

- C. H. Waddington. "O. R. in World War 2: Operational Research against the U-boat", Elek Science, London, 1973.

34.10 External links

- What is Operations Research?

- International Federation of Operational Research Societies

- The Institute for Operations Research and the Management Sciences (INFORMS)

- Occupational Outlook Handbook, U.S. Department of Labor Bureau of Labor Statistics

- "Operation Everything: It Stocks Your Grocery Store, Schedules Your Favorite Team's Games, and Helps Plan Your Vacation. The Most Influential Academic Discipline You've Never Heard Of." Boston Globe, 27 June 2004

- "Optimal Results: IT-powered advances in operations research can enhance business processes and boost the corporate bottom line." Computerworld, 20 November 2000

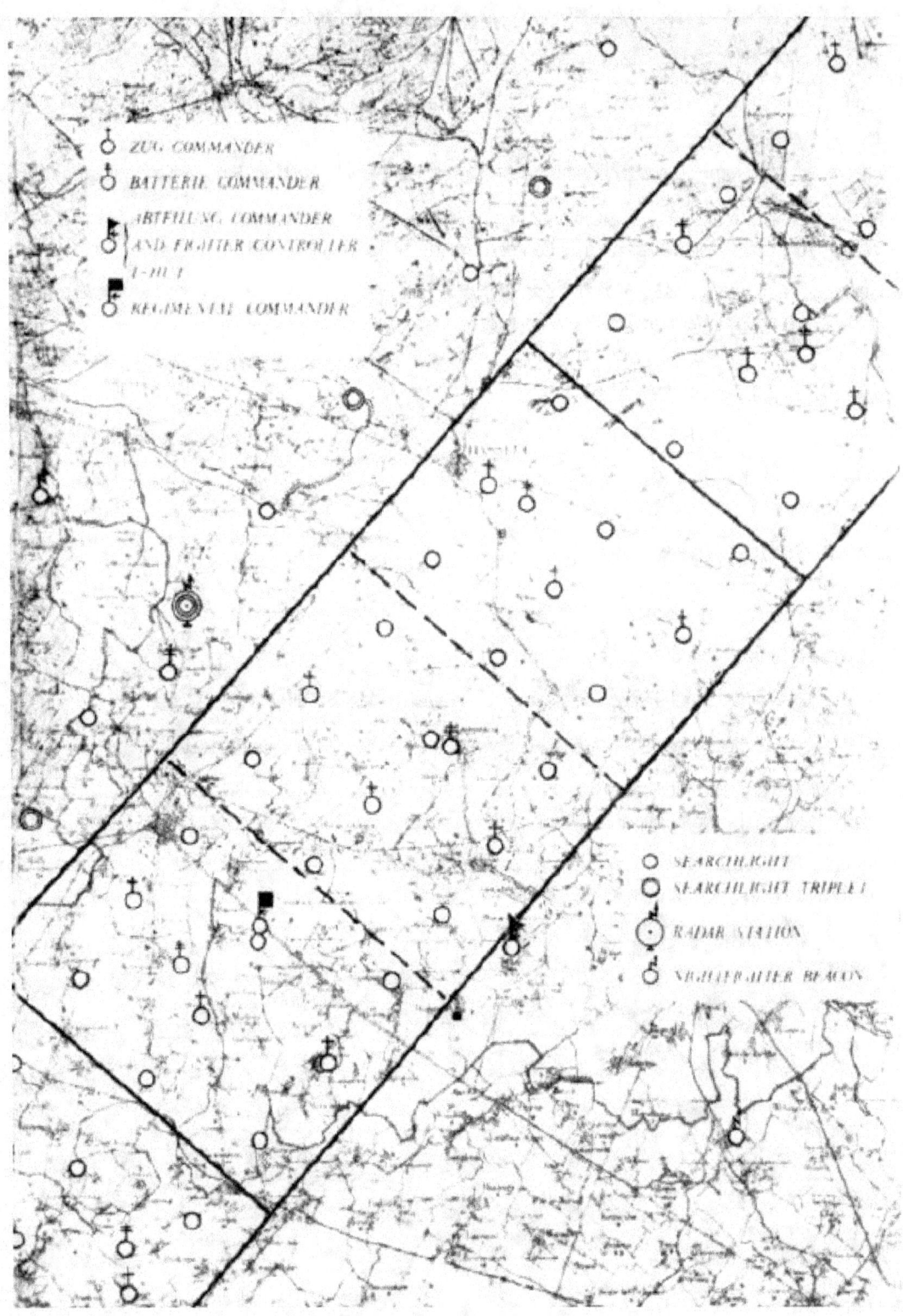

Map of Kammhuber Line

Chapter 35

Mathematical optimization

"Optimization" and "Optimum" redirect here. For other uses, see Optimization (disambiguation) and Optimum (disambiguation).
"Mathematical programming" redirects here. For the peer-reviewed journal, see Mathematical Programming.
In mathematics, computer science, operations research, **mathematical optimization** (alternatively, **optimization** or

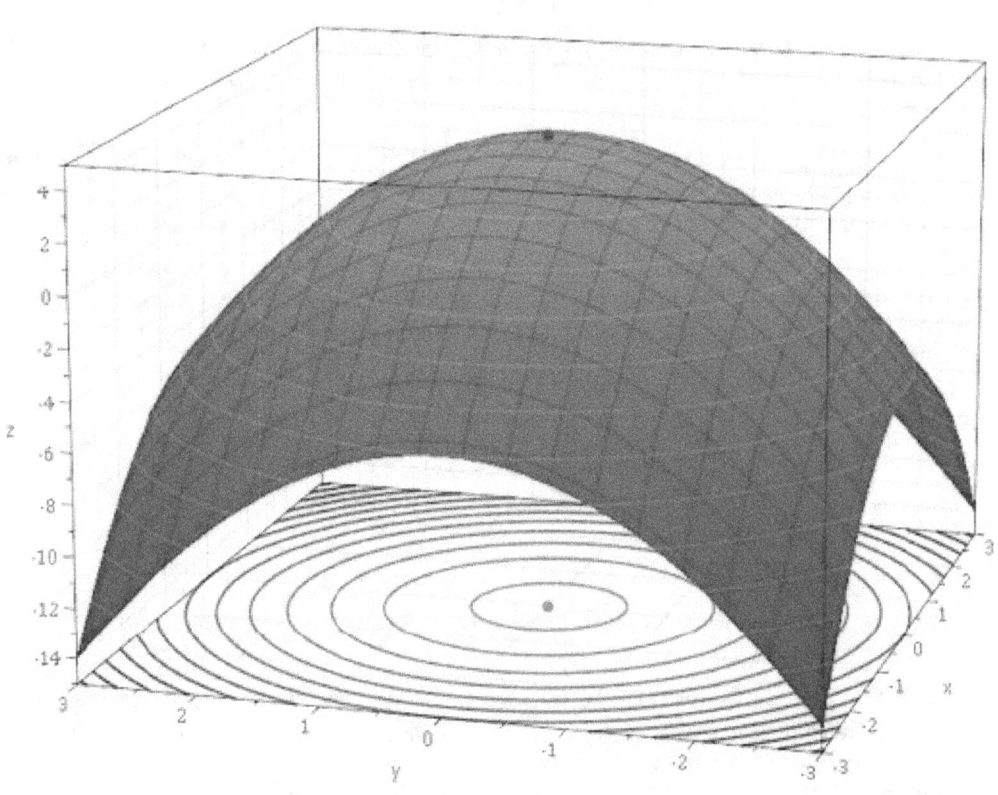

Graph of a paraboloid given by $f(x, y) = -(x^2 + y^2) + 4$. The global maximum at $(0, 0, 4)$ is indicated by a red dot.

mathematical programming) is the selection of a best element (with regard to some criteria) from some set of available alternatives.[1]

In the simplest case, an optimization problem consists of maximizing or minimizing a real function by systematically choosing input values from within an allowed set and computing the value of the function. The generalization of optimization theory and techniques to other formulations comprises a large area of applied mathematics. More generally, optimization includes finding "best available" values of some objective function given a defined domain (or a set of constraints), including a variety of different types of objective functions and different types of domains.

35.1 Optimization problems

Main article: Optimization problem

An optimization problem can be represented in the following way:

> *Given:* a function $f : A \to \mathbf{R}$ from some set A to the real numbers
>
> *Sought:* an element x_0 in A such that $f(x_0) \le f(x)$ for all x in A ("minimization") or such that $f(x_0) \ge f(x)$ for all x in A ("maximization").

Such a formulation is called an **optimization problem** or a **mathematical programming problem** (a term not directly related to computer programming, but still in use for example in linear programming – see History below). Many real-world and theoretical problems may be modeled in this general framework. Problems formulated using this technique in the fields of physics and computer vision may refer to the technique as **energy minimization**, speaking of the value of the function f as representing the energy of the system being modeled.

Typically, A is some subset of the Euclidean space $\mathbf{R}^n$, often specified by a set of *constraints*, equalities or inequalities that the members of A have to satisfy. The domain A of f is called the *search space* or the *choice set*, while the elements of A are called *candidate solutions* or *feasible solutions*.

The function f is called, variously, an **objective function**, a **loss function** or **cost function** (minimization),[2] a **utility function** or **fitness function** (maximization), or, in certain fields, an **energy function** or **energy functional**. A feasible solution that minimizes (or maximizes, if that is the goal) the objective function is called an *optimal solution*.

In mathematics, by convention optimization problems are usually stated in terms of minimization. Generally, unless both the objective function and the feasible region are convex in a minimization problem, there may be several local minima, where a *local minimum* x^* is defined as a point for which there exists some $\delta > 0$ so that for all x such that

$$|\mathbf{x} - \mathbf{x}'| \le \delta,$$

the expression

$$f(\mathbf{x}') \le f(\mathbf{x})$$

holds; that is to say, on some region around x^* all of the function values are greater than or equal to the value at that point. Local maxima are defined similarly.

A large number of algorithms proposed for solving non-convex problems – including the majority of commercially available solvers – are not capable of making a distinction between local optimal solutions and rigorous optimal solutions, and will treat the former as actual solutions to the original problem. The branch of applied mathematics and numerical analysis that is concerned with the development of deterministic algorithms that are capable of guaranteeing convergence in finite time to the actual optimal solution of a non-convex problem is called global optimization.

35.2 Notation

Optimization problems are often expressed with special notation. Here are some examples.

35.2.1 Minimum and maximum value of a function

Consider the following notation:

$$\min_{x \in R} (x^2 + 1)$$

This denotes the minimum value of the objective function $x^2 + 1$, when choosing x from the set of real numbers R. The minimum value in this case is 1, occurring at $x = 0$.

Similarly, the notation

$$\max_{x \in R} 2x$$

asks for the maximum value of the objective function $2x$, where x may be any real number. In this case, there is no such maximum as the objective function is unbounded, so the answer is "infinity" or "undefined".

35.2.2 Optimal input arguments

Main article: Arg max

Consider the following notation:

$$\underset{x \in (-\infty, -1]}{\arg\min} \; x^2 + 1,$$

or equivalently

$$\underset{x}{\arg\min} \; x^2 + 1, \text{ to: subject } x \in (-\infty, -1].$$

This represents the value (or values) of the argument x in the interval $(-\infty, -1]$ that minimizes (or minimize) the objective function $x^2 + 1$ (the actual minimum value of that function is not what the problem asks for). In this case, the answer is $x = -1$, since $x = 0$ is infeasible, i.e. does not belong to the feasible set.

Similarly,

$$\underset{x \in [-5,5], \; y \in R}{\arg\max} \; x\cos(y),$$

or equivalently

$$\underset{x, y}{\arg\max} \; x\cos(y), \text{ to: subject } x \in [-5,5], \; y \in R,$$

represents the (x, y) pair (or pairs) that maximizes (or maximize) the value of the objective function $x\cos(y)$, with the added constraint that x lie in the interval $[-5, 5]$ (again, the actual maximum value of the expression does not matter). In this case, the solutions are the pairs of the form $(5, 2k\pi)$ and $(-5,(2k+1)\pi)$, where k ranges over all integers.

arg min and **arg max** are sometimes also written **argmin** and **argmax**, and stand for **argument of the minimum** and **argument of the maximum**.

35.3   History

Fermat and Lagrange found calculus-based formulas for identifying optima, while Newton and Gauss proposed iterative methods for moving towards an optimum.

The term "linear programming" for certain optimization cases was due to George B. Dantzig, although much of the theory had been introduced by Leonid Kantorovich in 1939. (*Programming* in this context does not refer to computer programming, but from the use of *program* by the United States military to refer to proposed training and logistics schedules, which were the problems Dantzig studied at that time.) Dantzig published the Simplex algorithm in 1947, and John von Neumann developed the theory of duality in the same year.

Other major researchers in mathematical optimization include the following:

35.4   Major subfields

- Convex programming studies the case when the objective function is convex (minimization) or concave (maximization) and the constraint set is convex. This can be viewed as a particular case of nonlinear programming or as generalization of linear or convex quadratic programming.

    - Linear programming (LP), a type of convex programming, studies the case in which the objective function f is linear and the set of constraints is specified using only linear equalities and inequalities. Such a set is called a polyhedron or a polytope if it is bounded.
    - Second order cone programming (SOCP) is a convex program, and includes certain types of quadratic programs.
    - Semidefinite programming (SDP) is a subfield of convex optimization where the underlying variables are semidefinite matrices. It is generalization of linear and convex quadratic programming.
    - Conic programming is a general form of convex programming. LP, SOCP and SDP can all be viewed as conic programs with the appropriate type of cone.
    - Geometric programming is a technique whereby objective and inequality constraints expressed as posynomials and equality constraints as monomials can be transformed into a convex program.

- Integer programming studies linear programs in which some or all variables are constrained to take on integer values. This is not convex, and in general much more difficult than regular linear programming.

- Quadratic programming allows the objective function to have quadratic terms, while the feasible set must be specified with linear equalities and inequalities. For specific forms of the quadratic term, this is a type of convex programming.

- Fractional programming studies optimization of ratios of two nonlinear functions. The special class of concave fractional programs can be transformed to a convex optimization problem.

- Nonlinear programming studies the general case in which the objective function or the constraints or both contain nonlinear parts. This may or may not be a convex program. In general, whether the program is convex affects the difficulty of solving it.

- Stochastic programming studies the case in which some of the constraints or parameters depend on random variables.

- Robust programming is, like stochastic programming, an attempt to capture uncertainty in the data underlying the optimization problem. Robust optimization targets to find solutions that are valid under all possible realizations of the uncertainties.

- Combinatorial optimization is concerned with problems where the set of feasible solutions is discrete or can be reduced to a discrete one.

- Stochastic optimization for use with random (noisy) function measurements or random inputs in the search process.

- Infinite-dimensional optimization studies the case when the set of feasible solutions is a subset of an infinite-dimensional space, such as a space of functions.

- Heuristics and metaheuristics make few or no assumptions about the problem being optimized. Usually, heuristics do not guarantee that any optimal solution need be found. On the other hand, heuristics are used to find approximate solutions for many complicated optimization problems.

- Constraint satisfaction studies the case in which the objective function f is constant (this is used in artificial intelligence, particularly in automated reasoning).

    - Constraint programming.

- Disjunctive programming is used where at least one constraint must be satisfied but not all. It is of particular use in scheduling.

In a number of subfields, the techniques are designed primarily for optimization in dynamic contexts (that is, decision making over time):

- Calculus of variations seeks to optimize an objective defined over many points in time, by considering how the objective function changes if there is a small change in the choice path.

- Optimal control theory is a generalization of the calculus of variations.

- Dynamic programming studies the case in which the optimization strategy is based on splitting the problem into smaller subproblems. The equation that describes the relationship between these subproblems is called the Bellman equation.

- Mathematical programming with equilibrium constraints is where the constraints include variational inequalities or complementarities.

35.4.1 Multi-objective optimization

Main article: Multi-objective optimization

Adding more than one objective to an optimization problem adds complexity. For example, to optimize a structural design, one would want a design that is both light and rigid. Because these two objectives conflict, a trade-off exists. There will be one lightest design, one stiffest design, and an infinite number of designs that are some compromise of weight and stiffness. The set of trade-off designs that cannot be improved upon according to one criterion without hurting another criterion is known as the Pareto set. The curve created plotting weight against stiffness of the best designs is known as the Pareto frontier.

A design is judged to be "Pareto optimal" (equivalently, "Pareto efficient" or in the Pareto set) if it is not dominated by any other design: If it is worse than another design in some respects and no better in any respect, then it is dominated and is not Pareto optimal.

The choice among "Pareto optimal" solutions to determine the "favorite solution" is delegated to the decision maker. In other words, defining the problem as multiobjective optimization signals that some information is missing: desirable objectives are given but not their detailed combination. In some cases, the missing information can be derived by interactive sessions with the decision maker.

Multi-objective optimization problems have been generalized further to vector optimization problems where the (partial) ordering is no longer given by the Pareto ordering.

35.4.2   Multi-modal optimization

Optimization problems are often multi-modal; that is, they possess multiple good solutions. They could all be globally good (same cost function value) or there could be a mix of globally good and locally good solutions. Obtaining all (or at least some of) the multiple solutions is the goal of a multi-modal optimizer.

Classical optimization techniques due to their iterative approach do not perform satisfactorily when they are used to obtain multiple solutions, since it is not guaranteed that different solutions will be obtained even with different starting points in multiple runs of the algorithm. Evolutionary algorithms are however a very popular approach to obtain multiple solutions in a multi-modal optimization task.

35.5   Classification of critical points and extrema

35.5.1   Feasibility problem

The satisfiability problem, also called the feasibility problem, is just the problem of finding any feasible solution at all without regard to objective value. This can be regarded as the special case of mathematical optimization where the objective value is the same for every solution, and thus any solution is optimal.

Many optimization algorithms need to start from a feasible point. One way to obtain such a point is to relax the feasibility conditions using a slack variable; with enough slack, any starting point is feasible. Then, minimize that slack variable until slack is null or negative.

35.5.2   Existence

The extreme value theorem of Karl Weierstrass states that a continuous real-valued function on a compact set attains its maximum and minimum value. More generally, a lower semi-continuous function on a compact set attains its minimum; an upper semi-continuous function on a compact set attains its maximum.

35.5.3   Necessary conditions for optimality

One of Fermat's theorems states that optima of unconstrained problems are found at stationary points, where the first derivative or the gradient of the objective function is zero (see first derivative test). More generally, they may be found at critical points, where the first derivative or gradient of the objective function is zero or is undefined, or on the boundary of the choice set. An equation (or set of equations) stating that the first derivative(s) equal(s) zero at an interior optimum is called a 'first-order condition' or a set of first-order conditions.

Optima of equality-constrained problems can be found by the Lagrange multiplier method. The optima of problems with equality and/or inequality constraints can be found using the 'Karush–Kuhn–Tucker conditions'.

35.5.4   Sufficient conditions for optimality

While the first derivative test identifies points that might be extrema, this test does not distinguish a point that is a minimum from one that is a maximum or one that is neither. When the objective function is twice differentiable, these cases can be distinguished by checking the second derivative or the matrix of second derivatives (called the Hessian matrix) in unconstrained problems, or the matrix of second derivatives of the objective function and the constraints called the bordered Hessian in constrained problems. The conditions that distinguish maxima, or minima, from other stationary points are called 'second-order conditions' (see 'Second derivative test'). If a candidate solution satisfies the first-order conditions, then satisfaction of the second-order conditions as well is sufficient to establish at least local optimality.

35.5.5 Sensitivity and continuity of optima

The envelope theorem describes how the value of an optimal solution changes when an underlying parameter changes. The process of computing this change is called comparative statics.

The maximum theorem of Claude Berge (1963) describes the continuity of an optimal solution as a function of underlying parameters.

35.5.6 Calculus of optimization

Main article: Karush–Kuhn–Tucker conditions
See also: Critical point (mathematics), Differential calculus, Gradient, Hessian matrix, Positive definite matrix, Lipschitz continuity, Rademacher's theorem, Convex function and Convex analysis

For unconstrained problems with twice-differentiable functions, some critical points can be found by finding the points where the gradient of the objective function is zero (that is, the stationary points). More generally, a zero subgradient certifies that a local minimum has been found for minimization problems with convex functions and other locally Lipschitz functions.

Further, critical points can be classified using the definiteness of the Hessian matrix: If the Hessian is *positive* definite at a critical point, then the point is a local minimum; if the Hessian matrix is negative definite, then the point is a local maximum; finally, if indefinite, then the point is some kind of saddle point.

Constrained problems can often be transformed into unconstrained problems with the help of Lagrange multipliers. Lagrangian relaxation can also provide approximate solutions to difficult constrained problems.

When the objective function is convex, then any local minimum will also be a global minimum. There exist efficient numerical techniques for minimizing convex functions, such as interior-point methods.

35.6 Computational optimization techniques

To solve problems, researchers may use algorithms that terminate in a finite number of steps, or iterative methods that converge to a solution (on some specified class of problems), or heuristics that may provide approximate solutions to some problems (although their iterates need not converge).

35.6.1 Optimization algorithms

- Simplex algorithm of George Dantzig, designed for linear programming.

- Extensions of the simplex algorithm, designed for quadratic programming and for linear-fractional programming.

- Variants of the simplex algorithm that are especially suited for network optimization.

- Combinatorial algorithms

35.6.2 Iterative methods

Main article: Iterative method
See also: Newton's method, Quasi-Newton method, Finite difference, Approximation theory and Numerical analysis

The iterative methods used to solve problems of nonlinear programming differ according to whether they evaluate Hessians, gradients, or only function values. While evaluating Hessians (H) and gradients (G) improves the rate of convergence, for functions for which these quantities exist and vary sufficiently smoothly, such evaluations increase the computational complexity (or computational cost) of each iteration. In some cases, the computational complexity may be excessively high.

One major criterion for optimizers is just the number of required function evaluations as this often is already a large computational effort, usually much more effort than within the optimizer itself, which mainly has to operate over the N variables. The derivatives provide detailed information for such optimizers, but are even harder to calculate, e.g. approximating the gradient takes at least $N+1$ function evaluations. For approximations of the 2nd derivatives (collected in the Hessian matrix) the number of function evaluations is in the order of N^2. Newton's method requires the 2nd order derivates, so for each iteration the number of function calls is in the order of N^2, but for a simpler pure gradient optimizer it is only N. However, gradient optimizers need usually more iterations than Newton's algorithm. Which one is best with respect to the number of function calls depends on the problem itself.

- Methods that evaluate Hessians (or approximate Hessians, using finite differences):

    - Newton's method

        - Sequential quadratic programming: A Newton-based method for small-medium scale *constrained* problems. Some versions can handle large-dimensional problems.

- Methods that evaluate gradients or approximate gradients using finite differences (or even subgradients):

    - Quasi-Newton methods: Iterative methods for medium-large problems (e.g. N<1000).

    - Conjugate gradient methods: Iterative methods for large problems. (In theory, these methods terminate in a finite number of steps with quadratic objective functions, but this finite termination is not observed in practice on finite-precision computers.)

    - Interior point methods: This is a large class of methods for constrained optimization. Some interior-point methods use only (sub)gradient information, and others of which require the evaluation of Hessians.

    - Gradient descent (alternatively, "steepest descent" or "steepest ascent"): A (slow) method of historical and theoretical interest, which has had renewed interest for finding approximate solutions of enormous problems.

    - Subgradient methods - An iterative method for large locally Lipschitz functions using generalized gradients. Following Boris T. Polyak, subgradient-projection methods are similar to conjugate-gradient methods.

    - Bundle method of descent: An iterative method for small-medium-sized problems with locally Lipschitz functions, particularly for convex minimization problems. (Similar to conjugate gradient methods)

    - Ellipsoid method: An iterative method for small problems with quasiconvex objective functions and of great theoretical interest, particularly in establishing the polynomial time complexity of some combinatorial optimization problems. It has similarities with Quasi-Newton methods.

    - Reduced gradient method (Frank-Wolfe) for approximate minimization of specially structured problems with linear constraints, especially with traffic networks. For general unconstrained problems, this method reduces to the gradient method, which is regarded as obsolete (for almost all problems).

    - Simultaneous perturbation stochastic approximation (SPSA) method for stochastic optimization; uses random (efficient) gradient approximation.

- Methods that evaluate only function values: If a problem is continuously differentiable, then gradients can be approximated using finite differences, in which case a gradient-based method can be used.

    - Interpolation methods

    - Pattern search methods, which have better convergence properties than the Nelder-Mead heuristic (with simplices), which is listed below.

35.6.3   Global convergence

More generally, if the objective function is not a quadratic function, then many optimization methods use other methods to ensure that some subsequence of iterations converges to an optimal solution. The first and still popular method for ensuring convergence relies on line searches, which optimize a function along one dimension. A second and increasingly popular method for ensuring convergence uses trust regions. Both line searches and trust regions are used in modern methods of non-differentiable optimization. Usually a global optimizer is much slower than advanced local optimizers (such as BFGS), so often an efficient global optimizer can be constructed by starting the local optimizer from different starting points.

35.6.4   Heuristics

Main article: Heuristic algorithm

Besides (finitely terminating) algorithms and (convergent) iterative methods, there are heuristics that can provide approximate solutions to some optimization problems:

- Memetic algorithm

- Differential evolution

- Evolutionary algorithms

- Dynamic relaxation

- Genetic algorithms

- Hill climbing with random restart

- Nelder-Mead simplicial heuristic: A popular heuristic for approximate minimization (without calling gradients)

- Particle swarm optimization

- Artificial bee colony optimization

- Simulated annealing

- Tabu search

- Reactive Search Optimization (RSO)[3] implemented in LIONsolver

35.7   Applications

35.7.1   Mechanics and engineering

Problems in rigid body dynamics (in particular articulated rigid body dynamics) often require mathematical programming techniques, since you can view rigid body dynamics as attempting to solve an ordinary differential equation on a constraint manifold; the constraints are various nonlinear geometric constraints such as "these two points must always coincide", "this surface must not penetrate any other", or "this point must always lie somewhere on this curve". Also, the problem of computing contact forces can be done by solving a linear complementarity problem, which can also be viewed as a QP (quadratic programming) problem.

Many design problems can also be expressed as optimization programs. This application is called design optimization. One subset is the engineering optimization, and another recent and growing subset of this field is multidisciplinary design optimization, which, while useful in many problems, has in particular been applied to aerospace engineering problems.

35.7.2   Economics

Economics is closely enough linked to optimization of agents that an influential definition relatedly describes economics *qua* science as the "study of human behavior as a relationship between ends and scarce means" with alternative uses.[4] Modern optimization theory includes traditional optimization theory but also overlaps with game theory and the study of economic equilibria. The *Journal of Economic Literature* codes classify mathematical programming, optimization techniques, and related topics under JEL:C61-C63.

In microeconomics, the utility maximization problem and its dual problem, the expenditure minimization problem, are economic optimization problems. Insofar as they behave consistently, consumers are assumed to maximize their utility, while firms are usually assumed to maximize their profit. Also, agents are often modeled as being risk-averse, thereby preferring to avoid risk. Asset prices are also modeled using optimization theory, though the underlying mathematics relies on optimizing stochastic processes rather than on static optimization. Trade theory also uses optimization to explain trade patterns between nations. The optimization of market portfolios is an example of multi-objective optimization in economics.

Since the 1970s, economists have modeled dynamic decisions over time using control theory. For example, microeconomists use dynamic search models to study labor-market behavior.[5] A crucial distinction is between deterministic and stochastic models.[6] Macroeconomists build dynamic stochastic general equilibrium (DSGE) models that describe the dynamics of the whole economy as the result of the interdependent optimizing decisions of workers, consumers, investors, and governments.[7][8]

35.7.3   Operations research

Another field that uses optimization techniques extensively is operations research.[9] Operations research also uses stochastic modeling and simulation to support improved decision-making. Increasingly, operations research uses stochastic programming to model dynamic decisions that adapt to events; such problems can be solved with large-scale optimization and stochastic optimization methods.

35.7.4   Control engineering

Mathematical optimization is used in much modern controller design. High-level controllers such as Model predictive control (MPC) or Real-Time Optimization (RTO) employ mathematical optimization. These algorithms run online and repeatedly determine values for decision variables, such as choke openings in a process plant, by iteratively solving a mathematical optimization problem including constraints and a model of the system to be controlled.

35.7.5   Petroleum engineering

Nonlinear optimization methods are used to construct computational models of oil reservoirs.[10]

35.7.6   Molecular modeling

Main article: Molecular modeling

Nonlinear optimization methods are widely used in conformational analysis.

35.8   Solvers

Main article: List of optimization software

35.9   See also

35.10   Notes

[1] "The Nature of Mathematical Programming," *Mathematical Programming Glossary*, INFORMS Computing Society.

[2] W. Erwin Diewert (2008). "cost functions," *The New Palgrave Dictionary of Economics*, 2nd Edition Contents.

[3] Battiti, Roberto; Mauro Brunato; Franco Mascia (2008). *Reactive Search and Intelligent Optimization*. Springer Verlag. ISBN 978-0-387-09623-0.

[4] Lionel Robbins (1935, 2nd ed.) *An Essay on the Nature and Significance of Economic Science*, Macmillan, p. 16.

[5] A. K. Dixit ([1976] 1990). *Optimization in Economic Theory*, 2nd ed., Oxford. Description and contents preview.

[6] A.G. Malliaris (2008). "stochastic optimal control," *The New Palgrave Dictionary of Economics*, 2nd Edition. Abstract.

[7] Julio Rotemberg and Michael Woodford (1997), "An Optimization-based Econometric Framework for the Evaluation of Monetary Policy. *NBER Macroeconomics Annual*, 12, pp. 297-346.

[8] From *The New Palgrave Dictionary of Economics* (2008), 2nd Edition with Abstract links:
   • "numerical optimization methods in economics" by Karl Schmedders
   • "convex programming" by Lawrence E. Blume
   • "Arrow–Debreu model of general equilibrium" by John Geanakoplos.

[9] "New force on the political scene: the Seophonisten". http://www.seophonist-wahl.de. Retrieved 14 September 2013.

[10] "History matching production data and uncertainty assessment with an efficient TSVD parameterization algorithm". *Journal of Petroleum Science and Engineering* 113: 54–71. doi:10.1016/j.petrol.2013.11.025.

35.11   Further reading

35.11.1   Comprehensive

Undergraduate level

• Bradley, S.; Hax, A.; Magnanti, T. (1977). *Applied mathematical programming*. Addison Wesley.

• Rardin, Ronald L. (1997). *Optimization in operations research*. Prentice Hall. p. 919. ISBN 0-02-398415-5. copyright: 1998

• Strang, Gilbert (1986). *Introduction to applied mathematics*. Wellesley, MA: Wellesley-Cambridge Press (Strang's publishing company). pp. xii+758. ISBN 0-9614088-0-4. MR 870634.

Graduate level

• Magnanti, Thomas L. (1989). "Twenty years of mathematical programming". In Cornet, Bernard; Tulkens, Henry. *Contributions to Operations Research and Economics: The twentieth anniversary of CORE (Papers from the symposium held in Louvain-la-Neuve, January 1987)*. Cambridge, MA: MIT Press. pp. 163–227. ISBN 0-262-03149-3. MR 1104662.

• Minoux, M. (1986). *Mathematical programming: Theory and algorithms*. Egon Balas foreword) (Translated by Steven Vajda from the (1983 Paris: Dunod) French ed.). Chichester: A Wiley-Interscience Publication. John Wiley & Sons, Ltd. pp. xxviii+489. ISBN 0-471-90170-9. MR 2571910. (2008 Second ed., in French: *Programmation mathématique: Théorie et algorithmes*. Editions Tec & Doc, Paris, 2008. xxx+711 pp. ISBN 978-2-7430-1000-3.

- Nemhauser, G. L.; Rinnooy Kan, A. H. G.; Todd, M. J., eds. (1989). *Optimization*. Handbooks in Operations Research and Management Science 1. Amsterdam: North-Holland Publishing Co. pp. xiv+709. ISBN 0-444-87284-1. MR 1105099.

    - J. E. Dennis, Jr. and Robert B. Schnabel, A view of unconstrained optimization (pp. 1–72);
    - Donald Goldfarb and Michael J. Todd, Linear programming (pp. 73–170);
    - Philip E. Gill, Walter Murray, Michael A. Saunders, and Margaret H. Wright, Constrained nonlinear programming (pp. 171–210);
    - Ravindra K. Ahuja, Thomas L. Magnanti, and James B. Orlin, Network flows (pp. 211–369);
    - W. R. Pulleyblank, Polyhedral combinatorics (pp. 371–446);
    - George L. Nemhauser and Laurence A. Wolsey, Integer programming (pp. 447–527);
    - Claude Lemaréchal, Nondifferentiable optimization (pp. 529–572);
    - Roger J-B Wets, Stochastic programming (pp. 573–629);
    - A. H. G. Rinnooy Kan and G. T. Timmer, Global optimization (pp. 631–662);
    - P. L. Yu, Multiple criteria decision making: five basic concepts (pp. 663–699).

- Shapiro, Jeremy F. (1979). *Mathematical programming: Structures and algorithms*. New York: Wiley-Interscience [John Wiley & Sons]. pp. xvi+388. ISBN 0-471-77886-9. MR 544669.

- Spall, J. C. (2003), *Introduction to Stochastic Search and Optimization: Estimation, Simulation, and Control*, Wiley, Hoboken, NJ.

35.11.2   Continuous optimization

- Roger Fletcher (2000). *Practical methods of optimization*. Wiley. ISBN 978-0-471-49463-8.

- Mordecai Avriel (2003). *Nonlinear Programming: Analysis and Methods*. Dover Publishing. ISBN 0-486-43227-0.

- P. E. Gill, W. Murray and M. H. Wright (1982). *Practical Optimization*. Emerald Publishing. ISBN 978-0122839528.

- Xin-She Yang (2010). *Engineering Optimization: An Introduction with Metaheuristic Applications*. Wiley. ISBN 978-0470582466.

- Bonnans, J. Frédéric; Gilbert, J. Charles; Lemaréchal, Claude; Sagastizábal, Claudia A. (2006). *Numerical optimization: Theoretical and practical aspects*. Universitext (Second revised ed. of translation of 1997 French ed.). Berlin: Springer-Verlag. pp. xiv+490. doi:10.1007/978-3-540-35447-5. ISBN 3-540-35445-X. MR 2265882.

- Bonnans, J. Frédéric; Shapiro, Alexander (2000). *Perturbation analysis of optimization problems*. Springer Series in Operations Research. New York: Springer-Verlag. pp. xviii+601. ISBN 0-387-98705-3. MR 1756264.

- Boyd, Stephen P.; Vandenberghe, Lieven (2004). *Convex Optimization* (PDF). Cambridge University Press. ISBN 978-0-521-83378-3. Retrieved October 15, 2011.

- Jorge Nocedal and Stephen J. Wright (2006). *Numerical Optimization*. Springer. ISBN 0-387-30303-0.

- Ruszczyński, Andrzej (2006). *Nonlinear Optimization*. Princeton, NJ: Princeton University Press. pp. xii+454. ISBN 978-0691119151. MR 2199043.

- Robert J. Vanderbei (2013). *Linear Programming: Foundations and Extensions, 4th Edition*. Springer. ISBN 978-1461476290.

35.11.3 Combinatorial optimization

- R. K. Ahuja, Thomas L. Magnanti, and James B. Orlin (1993). *Network Flows: Theory, Algorithms, and Applications*. Prentice-Hall, Inc. ISBN 0-13-617549-X.

- William J. Cook, William H. Cunningham, William R. Pulleyblank, Alexander Schrijver; *Combinatorial Optimization*; John Wiley & Sons; 1 edition (November 12, 1997); ISBN 0-471-55894-X.

- Gondran, Michel; Minoux, Michel (1984). *Graphs and algorithms*. Wiley-Interscience Series in Discrete Mathematics (Translated by Steven Vajda from the second (*Collection de la Direction des Études et Recherches d'Électricité de France* [Collection of the Department of Studies and Research of Électricité de France], v. 37. Paris: Éditions Eyrolles 1985. xxviii+545 pp. MR 868083) French ed.). Chichester: John Wiley & Sons, Ltd. pp. xix+650. ISBN 978-2-7430-1035-5. MR 2552933. (Fourth ed. Collection EDF R&D. Paris: Editions Tec & Doc 2009. xxxii+784 pp.

- Eugene Lawler (2001). *Combinatorial Optimization: Networks and Matroids*. Dover. ISBN 0-486-41453-1.

- Lawler, E. L.; Lenstra, J. K.; Rinnooy Kan, A. H. G.; Shmoys, D. B. (1985), *The traveling salesman problem: A guided tour of combinatorial optimization*, John Wiley & Sons, ISBN 0-471-90413-9.

- Jon Lee; *A First Course in Combinatorial Optimization*; Cambridge University Press; 2004; ISBN 0-521-01012-8.

- Christos H. Papadimitriou and Kenneth Steiglitz *Combinatorial Optimization : Algorithms and Complexity*; Dover Pubns; (paperback, Unabridged edition, July 1998) ISBN 0-486-40258-4.

35.11.4 Relaxation (extension method)

Methods to obtain suitable (in some sense) natural extensions of optimization problems that otherwise lack of existence or stability of solutions to obtain problems with guaranteed existence of solutions and their stability in some sense (typically under various perturbation of data) are in general called relaxation. Solutions of such extended (=relaxed) problems in some sense characterizes (at least certain features) of the original problems, e.g. as far as their optimizing sequences concerns. Relaxed problems may also possesses their own natural linear structure that may yield specific optimality conditions different from optimality conditions for the original problems.

- H. O. Fattorini: Infinite Dimensional Optimization and Control Theory. Cambridge Univ. Press, 1999.

- P. Pedregal: Parametrized Measures and Variational Principles. Birkhäuser, Basel, 1997

- T. Roubicek: "Relaxation in Optimization Theory and Variational Calculus". W. de Gruyter, Berlin, 1997. ISBN 3-11-014542-1.

- J. Warga: Optimal control of differential and functional equations. Academic Press, 1972.

35.12 Journals

- *Computational Optimization and Applications*

- Journal of Computational *Optimization in Economics and Finance*

- *Journal of Economic Dynamics and Control*

- *SIAM Journal on Optimization* (SIOPT) and Editorial Policy

- *SIAM Journal on Control and Optimization* (SICON) and Editorial Policy

35.13 External links

- COIN-OR—Computational Infrastructure for Operations Research

- Decision Tree for Optimization Software Links to optimization source codes

- Global optimization

- Mathematical Programming Glossary

- Mathematical Programming Society

- NEOS Guide currently being replaced by the NEOS Wiki

- Optimization Online A repository for optimization e-prints

- Optimization Related Links

- Convex Optimization I EE364a: Course from Stanford University

- Convex Optimization – Boyd and Vandenberghe Book on Convex Optimization

- Book and Course on Optimization Methods for Engineering Design

35.14 Text and image sources, contributors, and licenses

35.14.1 Text

Roland Deschain, Kevlar992, Iridescent, K, Kencf0618, Zootsuits, Onestone, Nilamdoc, C. Lee, CzarB, Polymerbringer, Joseph Solis in Australia, Newone, White wolf753, Muéro, David Little, Igoldste, Amakuru, Marysunshine, Maelor, Masshaj, Jatrius, Experiment123, Tawkerbot2, Daniel5127, Joshuagross, Emote, Pikminiman, Heyheyhey99, JForget, Smkumar0, Sakowski, Wolfdog, Sleeping123, CRGreathouse, Wafulz, Sir Vicious, Triage, Iced Kola, CBM, Page Up, Jester-Tester, Taylorhewitt, Nczempin, GHe, Green caterpillar, Phanu9000, Yarnalgo, Thomasmeeks, McVities, Requestion, FlyingToaster, MarsRover, Tac-Tics, Some P. Erson, Tim1988, Tuluat, Alaymehta, MrFish, Oo7565, Gregbard, Captmog, El3m3nt09, Antiwiki~enwiki, Cydebot, Meznaric, Cantras, Funwithbig, MC10, Meno25, Gogo Dodo, DVokes, ST47, Srinath555, Pascal.Tesson, Goldencako, Benjiboi, Andrewm1986, Michael C Price, Tawkerbot4, Dragomiloff. Juansempere, M a s, Chrislk02, Brotown3, Mamounjo, 5300abc, Roccorossi, Abtract, Daven200520, Omicronpersei8, Vanished User jdksfajlasd, Daniel Olsen, Ventifact, TAU710, Aditya Kabir, BetacommandBot, Thijs!bot, Epbr123, Bezking, Jpark3591, Daemen, TheEmaciatedStilson, MCrawford, Opabinia regalis, Mattyboy500, Kilva, Daniel, Loudsox, Ucanlookitup, Hazmat2, Wootwootwoot, Brian G. Wilson, Timo3, Mojo Hand, Djfeldman, Pjvpjv, West Brom 4ever, John254, Alientraveller, Mnemeson, Ollyrobotham, BadKarma14, Sethdoe92, Dfrg.msc, RobHar, CharlotteWebb, Dawnseeker2000, RoboServien, Escarbot, Itsfrankie1221, Thomaswgc, Thadius856, Sidasta, AntiVandalBot, Ais523, RobotG, Gioto, Luna Santin, Dark Load, DarkAudit, Ringleader1489, Dylan Lake, Doktor Who, Chill doubt, AxiomShell, Abc30, Matheor, Archmagusrm, Falconleaf, Labongo, Spacefarer, Chocolatepizza, JAnDbot, Kaobear, MyNamesLogan, MER-C, The Transhumanist, Db099221, AussieOzborn au, Thenub314, Mosesroses, Hut 8.5, Kipholbeck, Xact, Twospoonfuls, anacondabot, Yahel Guhan, Bencherlite, Yurei-eggtart, Bongwarrior, VoABot II, JamesBWatson, Swpb, EdwardLockhart, SineWave, Charlielee111, Cic, Ryeterrell, Caesarjbsquitti, Wikiwhat?, Bubba hotep, KConWiki, Meb43, Faustnh, Hiplibrarianship, Johnbibby, Seberle, MetsBot, Pawl Kennedy, 28421u2232nfenfcenc, Systemlover, Bmeguru, Hotmedal, Just James, EstebanF, Glen, Rajpaj, Memorymentor, TheRanger, Calltech, Gun Powder Ma, Welshleprechaun, Robin S, Seba5618, SquidSK, 0612, J0equ1nn, Riccardobot, Jtir, Hdt83, MartinBot, Vladimir m, Arjun01, Quanticle, Nocklas, Rettetast, Fuzzyhair2, R'n'B, Pbroks13, Cmurphy au, Snozzer, Ben2then, PrestonH, Crazybobson, Thefutureschannel, RockMFR, Hrishikesh.24889, J.delanoy, Nev1, Unlockitall, Phoenix1177, Numbo3, Sp3000, Maurice Carbonaro, Nigholith, Hellonicole, -jmac-, Boris Allen, 2boobies, Jerry, TheSeven, NerdyNSK, Syphertext, Yadar677, Taop, G. Campbell, Wayp123, Keesiewonder, Matt1314, Ksucemfof, Gzkn, Ivehaps, Smeira, DarkFalls, Thomas Larsen, Vishi-vie, Washington8785, Xyzaxis, Arkuski, JDQuimby, Batmanfan77, Alphapeta, Trd89, HiLo48, The Transhumanist (AWB), NewEnglandYankee, RANDP, MKoltnow, MhordeXsnipa, Milogardner, Nacrha, Balaam42, Mviergujerghs89fhsdifds, Cfrehr, Elvisfan2095, Tiyoringo, Juliancolton, Cometstyles, DavidCBryant, SlightlyMad, Jamesontai, Remember the dot, Ilya Voyager, Huzefahamid, Dandy mandy, Andreas2001, Ishap, Sarregouset, CANUTELOOL2, CANUTELOOL3, Devonboy69, Jeyarathan, Death blaze, Emo kid you?, Thedudester, Samlyn.josfyn, Mother69, Vinsfan368, Cartiod, Helldude99, Sternkampf, Steel1943, CardinalDan, RJASE1, Idioma-bot, Remi0o, Lights, Tamillimat, Bandaidboy, C.lettingaAV, VolkovBot, Somebodyreallycool, Pleasantville, Jeff G., JohnBlackburne, Hhjk, The Catcher in The Rye D:, Alexandria, AlnoktaBOT, Dboerstl, NikolaiLobachevsky, Bangvang, 62 (number), Tseay11, Soliloquial, Headforaheadeyeforaneye, Barneca, Sześćsetsześćdziesiątsześć, Zeuron, Yoyoyo9, Trehansiddharth, TXiKiBoT, Katoa, Jacob Lundberg, Candy-Panda, Chickenclucker, Antoni Barau, Walor, Anonymous Dissident, Qxz, Nukemason4, Retiono Virginian, Ocolon, Savagepine, DennyColt, Digby Tantrum, JhsBot, Leafyplant, Beanai, 20em89.01, Cremepuff222, Geometry guy, Canyonsupreme, Natural Philosopher, Teller33, Mathsmad, Unknown 987, Tarten5, Nickmuller, Robomonster, Wolfrock, Jacob501, Kreemy, Synthebot, Tomaxer, Careercornerstone, Enviroboy, Rurik3, Sardonicone, Evanbrown326, Alliashax, Sylent, Rubentimothy, SMIE SMIE, Gamahucher, Braindamage3, Animalalley12895, Moohahaha, Thanatos666, Dillydumdum, AlleborgoBot, Voicework, Symane, Katzmik, Monkeynuts27, Demmy, Cam275, GoonerDP, SieBot, Mikemoral, James Banogon, BotMultichill, Timgregg96, Triwbe, 5150pacer, Soler97, Andersmusician, Anubhav29, Keilana, Tiptoety, Arbor to SJ, Undead Herle King, Richardcraig, Paolo.dL, Boogster, Oxymoron83, Henry Delforn (old), Avnjay, MiNombreDeGuerra, RW Marloe, SH84, Deejaye6, Musse-kloge, Jorgen W, Kumioko, Correogsk, MadmanBot, Nomoneynotime, Nickm4c, Darkmyst932, Anchor Link Bot, Jacob.jose, Randomblue, Melcombe, CaptainIron555, Yhkhoo, Dabomb87, Jat99, Pinkadelica, Francvs, Athenean, Ooswesthoesbes, ClueBot, Volcom5347, Gladysamuel, GPdB, Bwfrank, DFRussia, PipepBot, Foxj, Dobermanji, C1932, Remus John Lupin, Chocoforfriends, Smithpith, ArdClose, IceUnshattered, Plastikspork, Lawrence Cohen, Gawaxay, Nnemo, Ukabia, Michael.Urban, Niceguyedc, Xenon54, Mspraveen, DragonBot, Isaac25, 4pario, Donkeyboya, Excirial, CBOrgatrope, Bedsandbellies, Soccermaster3112, Alexbot, TonyBallioni, Pjb14, 0na01der, Andy pyro, Wikibobspider, BrentLeah, Eeekster, Anonymous1324354657687980897867564534231, Mycatiscool, Greenjuice, Chance Jeong, Arunta007, Greenjuice3.0, Greenjuice4, AnimeFan7, MacedonianBoy, ZuluPapa5, NuclearWarfare, JoelDick, Honeyspots3121, Blondeychck7, Faty148, Jotterbot, RC-0722, Wulfric1, Thingg, Franklin.vp, Aitias, DerBorg, Versus22, Hwalee76, SoxBot III, Apparition11, Mofeed.sawan, Slayerteez, XLinkBot, Marc van Leeuwen, Moocow444, Joejil67~enwiki, Little Mountain 5, Drumbeatsofeden, SilvonenBot, Planb 89, Alexius08, Vianello, MystBot, Zodon, RyanCross, Aetherealize, Zoltan808, T.M.M. Dowd, Aceleo, Jetsboy101, Willking1979, Mattguzy, 3Nigma, DOI bot, Cdt laurence, Fgnievinski, Yobmod, Aaronthegr8, CanadianLinuxUser, Potatoscrub, Download, Protonk, Chamal N, CarsracBot, Favonian, LinkFA-Bot, ViskonBot, Barak Sh, Aldermalhir, Jubeidono, PRL42, Lightbot, Ann Logsdon, Floccinocin123, Matěj Grabovský, Fivexthethird, TeH nOmInAtOr, Jarble, Herve1729, Sitehut, Ptbotgourou, Senator Palpatine, TaBOTzerem, Legobot II, Kan8eDie, Nirvana888, Gugtup, Washbummav, Mikeedla, THEN WHO WAS PHONE?, Skyeliam, MeatJustice, Wierdox, AnomieBOT, Nastor, ThaddeusB, Connectonline, Taskualads, Themantheman, Galoubet, Neko85, Noahschultz, JackieBot, Commander Shepard, Chingchangriceball, Piano non troppo, Supersmashballs123, Agroose, Pm11189, Riekuh, Hamletö, Deverenn, Frank2710, Chief Heath, Easton12, Codycash33, Archaeopteryx, Citation bot, Merlissimo, ArthurBot, Tatarian, MauritsBot, Xqbot, TinucherianBot II, Sketchmoose, Timir2, Capricorn42, Johnferrer, Jmundo, Locos epraix, Br77rino, Isheden, Inferno, Lord of Penguins, Uarrin, LevenBoy, Quixotex, GrouchoBot, Resident Mario, ProtectionTaggingBot, Omnipaedista, Point-set topologist, Gott wisst, RibotBOT, Charvest, KrazyKosbyKidz, MarilynCP, Gingeninja12, Caleb7693, Deathiscomin90919, VictorPorton, Grg222, Daryl7569, Petes2176, GhalyBot, ThibautLienart, Prozo3190, Family400005, Bupsiij, Aaron Kauppi, Har56, Dr. Klim, Velblod, CES1596, GliderMaven, Thomasjackson, FrescoBot, RTFVerterra, Triwikanto, Tobby72, Mark Renier, Onefive15, VS6507, Alpboyraz, ParaDoxus, Slawomir Biały, Xefer, Zhentmdfan, Tzurvah MeRabannan, Citation bot 1, Amplitude101, Tkuvho, Rotje66, Kiefer.Wolfowitz, AwesomeHersh, ElNuevoEinstein, Gamewizard71, FoxBot, TobeBot, DixonDBot, Burritoburritoburrito, Fama Clamosa, Lotje, Dinamik-bot, Raiden09, Mrjames99, DJTrickyM, Stephen MUFC, Tbhotch, RjwilmsiBot, TjBot, Ripchip Bot, Galois fu, Alphanumeric Sheep Pig, BertSeghers, Mr magnolias, DarkLightA, LibertyDodzo, EmausBot, PrisonerOfIce, Nima1024, WikitanvirBot, Surlyduff50, AThornyKoanz, Mehdiirfani, Legajoe, Wham Bam Rock II, Bethnim, ZéroBot, John Cline, Joswe05a, Leafiest of Futures, Battoe19, Anmol9999, Scythia, Brandmeister, Vanished user fijti34toksdcknqrjn54yoimascj, Ain92, Agatecat2700, Herk1955, Teapeat, Mjbmrbot, Liuthar, ClueBot NG, Incompetence, Wcherowi, Movses-bot, Kindyin, LJosil, SilentResident, Braincricket, Rbellini, Zackaback, MillingMachine, Helpful Pixie Bot, Thisthat2011, Curb Chain, AnandVivekSatpathi, Nashhinton, EmilyREditor, Ariel C.M.K., Fraqtive42, AvocatoBot, Davidiad, Ropestring, Edward Gordon Gey, EliteforceMMA, Karthickraj007, VirusKA, MYustin, Brad7777, Idresjafary, Nbrothers, IkamusumeFan, Kavy32, Sklange, Blevintron, BlevintronBot, Sulphuric Glue, Dexbot, Rezonansowy, Mudcap, Augustus Leonhardus Cartesius, Pankaj Jyoti Mahanta, Ybidzian, TycoonSaad, Jarash, Chern038, FireflySixtySeven, Kind Tennis Fan, Justin86789, 12visakhva, Dodi 8238, Rcehy, Vanisheduser00348374562342, 115ash, AdditionSubtraction, Mario Castelán Castro, Arvind asia, Rctillinghast, KasparBot, Kafishabbir, Evropariver and Anonymous: 1222

- **Definitions of mathematics** *Source:* https://en.wikipedia.org/wiki/Definitions_of_mathematics?oldid=670591940 *Contributors:* BenKovitz, Tobias Bergemann, Bcameron54, Danski14, Woohookitty, DVdm, Bgwhite, Bhny, Rick Norwood, Hmains, CBM, Conquistador2k6, KConWiki, Hiplibrarianship, Seberle, David Eppstein, Goustien, Niceguyedc, SF007, Zodon, Addbot, LilHelpa, Acebulf, Charvest, Ravendrop, Citation bot 1, Kiefer.Wolfowitz, ClueBot NG, Wcherowi, Helpful Pixie Bot, Brad7777, Federicoaolivieri, Abhilakshay Singh Pathania, Panpog1, TerryAlex, Fractalcows, Xxx Dingez sWagloRd xxx and Anonymous: 25

- **Mathematical beauty** *Source:* https://en.wikipedia.org/wiki/Mathematical_beauty?oldid=665571253 *Contributors:* Tarquin, Miguel~enwiki, Michael Hardy, Nixdorf, MartinHarper, GTBacchus, Karada, Ahoerstemeier, DavidWBrooks, Jimfbleak, Angela, Nikai, Nikola Smolenski, Revolver, Dysprosia, KRS, Ann O'nyme, Jmartinezot, Jose Ramos, Bevo, Aleph4, Gandalf61, Hadal, Randomness~enwiki, Ancheta Wis, Giftlite, Dbenbenn, Lupin, Bfinn, WHEELER, Dav4is, JRR Trollkien, Andycjp, Alexf, Elroch, Sam Hocevar, Gscshoyru, ChaTo, Thorwald, Paul August, Billymac00, C S, Andrewbadr, IonNerd, Orimosenzon, Critical, Linas, Mindmatrix, LOL, Ruud Koot, Tedneeman, Ryan Reich, Lawrence King, Grammarbot, Hack-Man, Salix alba, Slac, R.e.b., Nihiltres, Jersey Devil, DVdm, YurikBot, Wolfmankurd, Wikinick~enwiki, NawlinWiki, Arichnad, Raven4x4x, EEMIV, Sandstein, Cullinane, Gesslein, Allens, SmackBot, Stepa, PeterSymonds, Kithburd, SMP, Silly rabbit, Nbarth, DHN-bot~enwiki, Scalene, Rrelf, Armend, Jon Awbrey, DMacks, Xiutwel, Lambiam, Howdoesthiswo, Amenzix, RandomCritic, Waggers, Mets501, E-Kartoffel, BranStark, Aeternus, Sakurambo, CRGreathouse, Wafulz, A civilian, WeggeBot, Gregbard, Nauticashades, Tsenapathy, MC10, M a s, Int3gr4te, Thijs!bot, Kilva, Liquid-aim-bot, Mhatham.shammaa, Narssarssuaq, HarmonicFeather, Xeno, David Eppstein, Ineffable3000, Maurice Carbonaro, Nigholith, Rnest2002, DadaNeem, Ogranut, Funandtrvl, Hammersoft, Chitownmack, SteveStrummer, Broadbot, Lambyte, Psyche825, Everything counts, Mouse is back, Billinghurst, Adam.J.W.C., Euryalus, Anchor Link Bot, Tautologist, ClueBot, KarenSutherland, MonoBot, XLinkBot, Addbot, Idbelange, Guffydrawers, Tassedethe, Yobot, Fleabox, AnomieBOT, Citation bot, ProtectionTaggingBot, Kurosuke88, Ron Aharoni, Citation bot 1, Daclyff, EdEveridge, PPdd, WikitanvirBot, Kiatdd, ZéroBot, Thewhyman, Mixedberries17, ZeroCool4ta, Fuzzy artist, Miegoreng, Rmashhadi, ClueBot NG, Frietjes, Delusion23, Reify-tech, Luckimg, Helpful Pixie Bot, BG19bot, Brad7777, Pankaj Jyoti Mahanta, Mathbeauty, Ashorocetus, Nigellwh, Lamaballa, LZNQBD, Hampton11235, Blue-Continent and Anonymous: 116

- **Mathematical notation** *Source:* https://en.wikipedia.org/wiki/Mathematical_notation?oldid=646730035 *Contributors:* Patrick, Michael Hardy, Fred Bauder, Dominus, MartinHarper, Smack, Pizza Puzzle, Charles Matthews, Jitse Niesen, Dtgm, David Shay, AndrewKepert, Donarreiskoffer, Robbot, Fredrik, RedWolf, Gandalf61, MathMartin, Rorro, Ancheta Wis, Giftlite, COMPATT, Frencheigh, DO'Neil, Langec, Skla~enwiki, Neilc, Beland, Oskar Sigvardsson, Brianjd, CALR, Leibniz, Luqui, Paul August, Flammifer, Atlant, Kazvorpal, Oleg Alexandrov, Katyare, Camw, Ruud Koot, WadeSimMiser, Mpatel, Sholtar, Wikiklrsc, Skoban, BD2412, OMouse, Dpv, Jshadias, Josh Parris, Jwmcleod, Yamamoto Ichiro, Mathbot, YurikBot, Wavelength, Shell Kinney, Trovatore, Hakeem.gadi, Tetracube, Rwxrwxrwx, Arthur Rubin, SmackBot, Melchoir, Jagged 85, Eskimbot, Gilliam, Chris the speller, Iain.dalton, MalafayaBot, Spellchecker, SundarBot, Matherson, BryanG, Lambiam, Alpha Omicron, Thedaydreamer1, Mets501, Courcelles, CRGreathouse, CBM, HenningThielemann, Gregbard, AndrewHowse, Julian Mendez, Retired user 0002, Jaerik, Thijs!bot, Epbr123, Al Lemos, SomeHuman, AnAj, Danny lost, P.L.A.R., Jeff560, Aflaksp, David Eppstein, R'n'B, Maurice Carbonaro, Tygrrr, Philip Trueman, Marcosaedro, Aaron Rotenberg, Brianga, SieBot, MiNombreDeGuerra, ClueBot, Brews ohare, Thingg, Qwfp, SoxBot III, DumZiBoT, Addbot, Fgnievinski, Morriswa, Zahd, Delaszk, Barak Sh, Tide rolls, Pastaburritoboy92, Yobot, AnomieBOT, Galoubet, Orhanghazi, Calmer Waters, Abc518, Gizmoitai3, Bomazi, John17890, ClueBot NG, Brad7777, Filing Flunky, Bideford1, Yardimsever, 1rir4er and Anonymous: 75

- **Areas of mathematics** *Source:* https://en.wikipedia.org/wiki/Areas_of_mathematics?oldid=665701158 *Contributors:* Michael Hardy, Dominus, Charles Matthews, Fredrik, Robinh, Andycjp, Beland, Profvk, PhotoBox, Hippojazz, Paul August, Tompw, Mwanner, Stephen Bain, Anthony Appleyard, Sligocki, PaePae, Pontus, Oleg Alexandrov, Natalya, Woohookitty, Linas, Jeff3000, Btyner, Porcher, Chenxlee, Salix alba, Juan Marquez, Mathbot, Nihiltres, DVdm, Wavelength, RussBot, Michael Slone, PaulGarner, Geraschenko, Tribaal, Fram, Sardanaphalus, SmackBot, RDBury, Jagged 85, Commander Keane bot, ChuckHG, Snori, Kevin Ryde, Hongooi, G716, John Reid, Mets501, Dlohcierekim, CRGreathouse, Gregbard, AndrewHowse, Cydebot, Dr.enh, Barticus88, AntiVandalBot, Thenub314, EagleFan, HEL, Adammarklenny, DavidCBryant, Sapphic, Yintan, Paolo.dL, Lisatwo, Garyzx, DragonBot, Hans Adler, Addbot, AkhtaBot, Mootros, Aboctok, Humbugde, Numbo3-bot, Lightbot, Yobot, KamikazeBot, Materialscientist, Twri, Xqbot, Kiefer.Wolfowitz, Gamewizard71, D.Lazard, Tolly4bolly, Kwalju, ClueBot NG, Joel B. Lewis, Helpful Pixie Bot, Mistory, Brad7777, Pankaj Jyoti Mahanta, RC711 and Anonymous: 38

- **Philosophy of mathematics** *Source:* https://en.wikipedia.org/wiki/Philosophy_of_mathematics?oldid=671116977 *Contributors:* Damian Yerrick, AxelBoldt, Matthew Woodcraft, Derek Ross, LC~enwiki, Bryan Derksen, Zundark, The Anome, Hhanke, SimonP, Zadcat, Ryguasu, Cwitty, IanS, The hanged man, Michael Hardy, Nixdorf, BoNoMoJo (old), Gabbe, Chinju, GTBacchus, Dori, Eric119, Ahoerstemeier, Snoyes, Darkwind, Cyan, Tim Retout, Rotem Dan, Andres, EdH, Schneelocke, Renamed user 4, Charles Matthews, RickK, Ww, Dtgm, Markhurd, Maximus Rex, Unknown, Robbot, Romanm, Gandalf61, Tim Ivorson, MathMartin, OmegaMan, Hadal, JohannesMarat, Aetheling, Tobias Bergemann, Adam78, Giftlite, Lee J Haywood, Lethe, Monedula, Everyking, Esap, WHEELER, Just Another Dan, JRR Trollkien, Stevietheman, Alberto da Calvairate~enwiki, Karol Langner, Rdsmith4, Pmanderson, Robin klein, Random account 47, Shahab, D6, Splatty, Rich Farmbrough, Wclark, YUL89YYZ, Paul August, Pban92, Elwikipedista~enwiki, Syp, Lycurgus, Rgdboer, Episcopo, Pokereth, Root4(one), Pearle, IonNerd, Jumbuck, Vesal, Droob, Eaglearn, Ossiemanners, Sligocki, Mr. Hyde~enwiki, Hu, Pernest2002, Bookandcoffee, Euphrosyne, Angr, Boothy443, Mel Etitis, Woohookitty, Linas, Barrylb, Jok2000, Al E., Dfranke, DaveApter, BD2412, Qwertyus, Gigapixel, Rjwilmsi, Salix alba, Mathbot, Nihiltres, Echeneida, RexNL, Mark J, David H Braun (1964), JamesLee, YurikBot, Wavelength, Shonk, Hairy Dude, Conscious, Hydrargyrum, Gaius Cornelius, Chaos, KSchutte, Aldux, Hakeem.gadi, Tomisti, Igiffin, Arthur Rubin, Skullfission, LeonardoRob0t, Canadianism, Meegs, Infinity0, Brentt, Sardanaphalus, JJL, SmackBot, Reedy, Tom Lougheed, Pokipsy76, Eskimbot, Gilliam, Ghosts&empties, Chris the speller, Bluebot, Trebor, JMSwtlk, Helder Ribeiro, Go for it!, Lesnail, Cybercobra, John wesley, Jon Awbrey, Just plain Bill, Bidabadi~enwiki, Vina-iwbot~enwiki, FlyHigh, Clicketyclack, Byelf2007, Igrant, Dbtfz, JHunterJ, Mets501, Ryulong, Iridescent, K, Aeternus, Rhetth, Tawkerbot2, 8754865, CRGreathouse, CmdrObot, CBM, Pan Camel, Gregbard, Logicus, Danman3459, Peterdjones, M a s, JamesAM, Old port, Thijs!bot, W3asal, 271828182, Headbomb, Marek69, Nick Number, Klausness, Vodello, Rjmars97, GeePriest, Wayiran, Leuqarte, Tayl1257, Ophion, Huphelmeyer, Andrewthomas10, Wlod, Lucaas, Revery~enwiki, Seberle, David Eppstein, Martynas Patasius, Nowietsgo, Ynotds, CommonsDelinker, Filll, Altes, Maurice Carbonaro, NerdyNSK, Dispenser, Mikael Häggström, Rnest2002, Infarom,

Milogardner, AlnoktaBOT, Jimmaths, TXiKiBoT, Aleph42, The Tetrast, Philogo, Broadbot, Geometry guy, Popopp, Finalfantasy2012, Wenli, Tomaxer, Sapphic, Dmcq, Newbyguesses, SieBot, Darrell Wheeler, Lightmouse, Roran659, DesolateReality, Classicalecon, ClueBot, DFRussia, Rockfang, CohesionBot, Azadeh.a, Sun Creator, Vegetator, JKeck, Marc van Leeuwen, Gerhardvalentin, Zodon, Addbot, Fyrael, With goodness in mind, MrOllie, Imtg5102, ProfessorThunderlips, Drdonzi, Nallimbot, Synchronism, AnomieBOT, Materialscientist, Citation bot, Xqbot, Capricorn42, Crzer07, Uarrin, J04n, Freddyfirre, Omnipaedista, Taekwandean, Aaron Kauppi, Constructive editor, Mr fabs, Hugetim, FrescoBot, Mark Renier, EricAndrewWallace, BrideOfKripkenstein, Alboran, Steve Quinn, CESSMASTER, Machine Elf 1735, Mary rose arias, Tkuvho, Pinethicket, Pollinosisss, Jdapayne, Vrenator, Hueyha, Jowa fan, Merehap, EmausBot, John of Reading, ZéroBot, Amacfiew. RaptureBot, HarmoniousMembrane, BartlebytheScrivener, RockMagnetist, Logicalgregory, Anita5192, E. Fokker, ClueBot NG, Alexander E Ross, Deer*lake, Helpful Pixie Bot, Beaumont877, Brian Tomasik, Harizotoh9, Brad7777, Gibbja, Dexbot, Jochen Burghardt, Mark viking, Stara729, Workstern, Kyle1009, Waluigigod, Cnbr15 and Anonymous: 225

- **Glossary of areas of mathematics** *Source:* https://en.wikipedia.org/wiki/Glossary_of_areas_of_mathematics?oldid=642557384 *Contributors:* Zundark, Michael Hardy, Topbanana, Sodin, Wavelength, Joel7687, Xaxafrad, John, Myasuda, Cydebot, JaGa, Davidmanheim, R'n'B, Niceguyedc, Arjayay, SchreiberBike, Offbeatcinema, Ozob, Yobot, LilHelpa, FrescoBot, MegaSloth, John of Reading, D.Lazard, SporkBot, DPL bot, Killikalli, Brad7777, The1337gamer, LegacyOfValor and Anonymous: 3

- **Arithmetic** *Source:* https://en.wikipedia.org/wiki/Arithmetic?oldid=662878715 *Contributors:* LC~enwiki, Tarquin, Christian List, Toby Bartels, Nonenmac, Juuitchan, TeunSpaans, Michael Hardy, JakeVortex, Meekohi, Ixfd64, Dcljr, Dori, J-Wiki, Angela, Julesd, AugPi, Rotem Dan, Mikue, Pizza Puzzle, Hashar, Revolver, Charles Matthews, Dysprosia, Jitse Niesen, Selket, Markhurd, Grendelkhan, Robbot, RedWolf, Gandalf61, Henrygb, OmegaMan, Rebrane, Wikibot, Fuelbottle, Mandel, Guy Peters, Mfc, Centrx, Giftlite, Gene Ward Smith, Herbee, Monedula, Michael Devore, Siroxo, Bobblewik, Chowbok, Alexf, Beland, Smallstraw~enwiki, Rdsmith4, APH, Histrion, Iantresman, Joyous!, Frenchwhale, Jh51681, CliffordEW, Zondor, Trevor MacInnis, Freakofnurture, Skal, Rich Farmbrough, Guanabot, Gadykozma, Rama, Paul August, ESkog, Andrejj, Blotwell, Jojit fb, Haham hanuka, Mdd, Ogress, Jumbuck, Msh210, Arthena, Diego Moya, Hippophaë~enwiki, Mrholybrain, Caesura, Velella, Computerjoe, Freyr, Kenyon, Oleg Alexandrov, Mindmatrix, Kzollman, ^demon, MFH, Amikeco, KingsleyIdehen, Waldir, Palica, BD2412, Josh Parris, Sdoman, Salix alba, Brighterorange, FlavrSavr, Yamamoto Ichiro, FlaBot, Mathbot, Greg321, RexNL, Gurch, Ayla, Brendan Moody, BradBeattie, Chobot, YurikBot, SpikeJones, RobotE, Red Slash, KSmrq, RadioFan, NawlinWiki, Rick Norwood, DryaUnda, Jhinman, Brisvegas, Lt-wiki-bot, Xaxafrad, DmitriyV, Sardanaphalus, SmackBot, YellowMonkey, Melchoir, Jagged 85, BiT, Timotheus Canens, CrypticBacon, Gilliam, G O T R, IMacWin95, Bluebot, Keegan, Pimrietbroek, Oli Filth, Adam Lewis, Frap, SundarBot, Khoikhoi, Acepectif, Jiddisch~enwiki, StephenMacmanus, Ruwanraj, Kalathalan, Bpeel, Via strass, SashatoBot, Mksword, Goodnightmush, Asdfv, 16@r, Mets501, Citicat, Robertwb, Captainj, Vanisaac, Stifynsemons, DWarrior, Mikeliuk, Patchouli, Ninetyone, Timichal, MarsRover, Nilfanion, Equendil, Cydebot, Richardguk, Flowerpotman, Dr.enh, Tdvance, PlanetCoder, Doug Weller, M a s, Daven200520, Emmett5, Kansas Sam, Thijs!bot, Pmagyar, Headbomb, AndresV, Marek69, Missvain, JustAGal, Dfrg.msc, Escarbot, Quintote, Edokter, Doktor Who, Danger, JAnDbot, Sangwinc, MER-C, The Transhumanist, 100110100, VoABot II, Mclay1, Akgupta, Redaktor, Schwarzbichler, CountingPine, Grape Soda, Ryeterrell, Seberle, David Eppstein, Spellmaster, JaGa, Khalid Mahmood, Misibacsi, MartinBot, Nono64, Roelvandijk~enwiki, Uncle Dick, Katalaveno, Mathwhizx2, Merceris, The Transhumanist (AWB), Bobianite, Milogardner, Jazzbruce, Treisijs, Idioma-bot, Ottershrew, VolkovBot, JohnBlackburne, Am Fiosaigear~enwiki, DoorsAjar, TXiKiBoT, Anonymous Dissident, Ocolon, Yk Yk Yk, Meters, Synthebot, Enviroboy, Dmcq, AlleborgoBot, Symane, NHRHS2010, SieBot, Gerakibot, Keilana, Flyer22, LDCutter, Oxymoron83, Janfri, Nic bor, UKe-CH, ClueBot, PipepBot, Justin W Smith, The Thing That Should Not Be, Cliff. VsBot, Ignorance is strength, JRD RockS, Rejka, Dekanfari, Muro Bot, Workman63, Workman64, Workman65, Workman100, DumZiBoT, Fastily, Pichpich, WikHead, Addbot, Jojhutton, Olli Niemitalo, Istvánka, Aboctok, NjardarBot, MrOllie, LaaknorBot, CarsracBot, AndersBot, Tassedethe, VASANTH S.N., Tide rolls, Lightbot, Jarble, Luckas-bot, Yobot, II MusLiM HyBRiD II, AnomieBOT, Jim1138, NickK, Materialscientist, Bob Burkhardt, LilHelpa, Xqbot, Timir2, Capricorn42, Johnferrer, Isheden, Almabot, Novonium, GrouchoBot, Omnipaedista, Charvest, Aashaa, Schekinov Alexey Victorovich, Aaron Kauppi, FrescoBot, Tobby72, VS6507, MacMed, Pinethicket, Niaz632, Martinvl, Shiva Khanal, MastiBot, SpaceFlight89, FoxBot, TobeBot, Yunshui, Lotje, Zvn, January, Joodeak, Reach Out to the Truth, RjwilmsiBot, Slon02, Mr. Anon515, EmausBot, John of Reading, WikitanvirBot, Lipsio, Dewritech, Gfuy, Fayimora, Savh, AvicBot, PBS-AWB, Fæ, Josve05a, Simulations, Skipper per, Bamyers99, D.Lazard, Lightbeamrider55, ChuispastonBot, Buriedundergound, Jordibuma, Strangely Real, ClueBot NG, Wcherowi, AnthonyNotes, Movses-bot, Quantamflux, Yster76, Frietjes, Braincricket, Lincoln Josh, Helpful Pixie Bot, Jerrydeanrsmith, Vagobot, Mrjohncummings, MusikAnimal, Furkaocean, DrTechDaddy, Brad7777, Cky2250, Helloimjustintimefordinner, Arcandam, Dexbot, Saehry, Lugia2453, Frosty, Konstantin.bay, Kevin12xd, JustAMuggle, NHCLS, 1canuckbuck, Ramanujan srinivasa, LateralMoraine, Galois2718, Suelru, Whoppabang, Luram, Kylieh10, FourViolas, KasparBot and Anonymous: 268

- **Order theory** *Source:* https://en.wikipedia.org/wiki/Order_theory?oldid=666554129 *Contributors:* Bryan Derksen, Toby Bartels, Michael Hardy, Dineshjk, Ehn, Charles Matthews, Dcoetzee, Jitse Niesen, Wik, Natevw, VeryVerily, Populus, Topbanana, Robbot, Henrygb, ElBenevolente, Tobias Bergemann, Giftlite, Markus Krötzsch, Elias, DefLog~enwiki, Yamover, APH, SimonLyall, Xrchz, Abar, Pjacobi, Paul August, Tompw, Nickj, Themusicgod1, Lysdexia, Arthena, Joriki, Linas, Jeff3000. Josh Parris, Salix alba, Mathbot, Hairy Dude, Dmharvey, Trovatore, Ott2, Arthur Rubin, Modify, Netrapt, That Guy, From That Show!, SmackBot, KnowledgeOfSelf, Mhss, Bluebot, RDBrown, Cybercobra, Kntrabssi, Dreadstar, JohnI, 16@r, Loadmaster, Dicklyon, Landonproctor, Levineps, Dreftymac, Majora4, CRGreathouse, CBM, Sam Staton, Skittleys, Ankit mcgill, MER-C, VoABot II, David Eppstein, Gwern, Maurice Carbonaro, Inquam, Daniel5Ko, Jorfer, JohnBlackburne, Trondarild, Philip Trueman, GcSwRhIc, The Tetrast, Magmi, Geometry guy, Tomaxer, StevenJohnston, AS, Anchor Link Bot, Randomblue, ClueBot, Justin W Smith, Hans Adler, Wikidsp, Addbot, Barak Sh, Badou517, Legobot, Buenasdiaz, Smallman12q, Tuetschek, FrescoBot, Mark Renier, Orhanghazi, Chenopodiaceous, Gamewizard71, Genezistan, John of Reading, Dadaist6174, ClueBot NG, Syamino, MerlIwBot, Helpful Pixie Bot, Knwlgc, VolunBute, Brad7777, Nicuchalan1, K401sTL3, JaconaFrere, Srlgator, Pheello87, KasparBot and Anonymous: 59

- **Algebraic structure** *Source:* https://en.wikipedia.org/wiki/Algebraic_structure?oldid=666895712 *Contributors:* AxelBoldt, Zundark, Danny, Toby Bartels, Edward, Marvinfreeman, Michael Hardy, Wshun, Ideyal, Revolver, Charles Matthews, Dysprosia, Aleph4, Robbot, RedWolf, Naddy, Wikibot, Tobias Bergemann, Giftlite, Lethe, Fropuff, Waltpohl, Tristanreid, Guanabot, ArnoldReinhold, HeikoEvermann, Paul August, Tompw, Rgdboer, Szquirrel, Obradovic Goran, Varuna, Kuratowski's Ghost, Msh210, Eric Kvaalen, Mysdaao, Kusma, Galaxiaad, Linas, BD2412, Icey, Jshadias, Josh Parris, Bgohla, Zinoviev, Staecker, Salix alba, Michal.burda, RexNL, Don Gosiewski, Chobot, Algebraist, YurikBot, Spiderboy, Dmharvey, SoroSuub1, Grubber, Grafen, Trovatore, Moe Epsilon, Crasshopper, Bota47, Reyk, Modify, Netrapt, SoberEmu, SmackBot, Incnis Mrsi, Melchoir, PJTraill, MaxSem, G716, FlyHigh, Byelf2007, Cronholm144, IronGargoyle, Mets501, Rschwieb, Simon12, Adriatikus, Johnfuhrmann, CRGreathouse, Alexey Feldgendler, Myasuda, DustinBernard, Goldencako, Thijs!bot, Berria, Icep, Escarbot, Olaf,

bomb, Albmont, Ling.Nut, Jakob.scholbach, Pomte, Ambrose H. Field, Robert Stanforth, Hesam7, Ferengi, Joeldl, Philmac, Arcfrk, SieBot, Oxymoron83, Bob1960evens, Ideal gas equation, Harpreetgr, Niceguyedc, Ncsinger, Addbot, Prostarplayer999, Calle, AnomieBOT, Xqbot, Depassp, CBoeckle, SassoBot, Thehelpfulbot, FrescoBot, Citation bot 1, Foobarnix, Vovchyck, EmausBot, Chharvey, Quondum, Brad7777, BattyBot, Deltahedron, CsDix, Hamoudafg, Danneks, GeoffreyT2000, KasparBot and Anonymous: 27

- **Commutative algebra** *Source:* https://en.wikipedia.org/wiki/Commutative_algebra?oldid=667008180 *Contributors:* AxelBoldt, Zundark, Charles Matthews, Phys, Altenmann, MathMartin, Giftlite, Gene Ward Smith, Schopenhauer, Fropuff, Berjoh, Waltpohl, Gauss, Paul August, Obradovic Goran, R.e.b., FlaBot, Chobot, YurikBot, Grafen, MalafayaBot, RyanEberhart, Cicero, Gleuschk, Noodlez84, Valoem, CR-Greathouse, Martín Oregón, Myasuda, HStel, Turgidson, RebelRobot, Magioladitis, STBot, JohnBlackburne, WarddrBOT, Joeldl, Arcfrk, SieBot, ClueBot, Addbot, Luckas-bot, D'ohBot, Zero Thrust, Gamewizard71, TobeBot, Miracle Pen, E.songhori, EmausBot, WikitanvirBot, Tonyxty, ZéroBot, Midas02, Quondum, D.Lazard, ChuispastonBot, LJosil, Daviddwd, Brad7777, ChrisGualtieri, Mogism, Brirush, Limit-theorem, Mark viking and Anonymous: 20

- **Mathematical analysis** *Source:* https://en.wikipedia.org/wiki/Mathematical_analysis?oldid=668076634 *Contributors:* AxelBoldt, Lee Daniel Crocker, Tarquin, Miguel~enwiki, Peterlin~enwiki, Ben-Zin~enwiki, Youandme, Michael Hardy, Wshun, Norm, Iulianu, Snoyes, Andrewa, Cyan, Charles Matthews, Dino, Dysprosia, Tpbradbury, Traroth, Robbot, Fredrik, Romanm, Gandalf61, MathMartin, Fuelbottle, Tobias Bergemann, Snobot, Giftlite, Lethe, Dratman, Sam Hocevar, PhotoBox, D6, HedgeHog, Urvabara, Paul August, Bender235, Tompw, Art LaPella, Nk, Mdd, Msh210, Alansohn, Dallashan~enwiki, Sligocki, Olegalexandrov, Almafeta, Oleg Alexandrov, Linas, Igny, Mandarax, Rjwilmsi, Mayumashu, Koavf, MarSch, Juan Marquez, FlaBot, JYOuyang, Otets, Malhonen, Chobot, DVdm, Borgx, Spacepotato, Hairy Dude, Deeptrivia, RussBot, KSmrq, Chaos, Amplimax, MaNeMeBasat, Pred, Ilmari Karonen, Lunch, Finell, Sardanaphalus, SmackBot, Self-worm, Lestrade, InverseHypercube, Melchoir, Bomac, Jagged 85, Alsandro, Grokmoo, SMP, Darth Panda, Vanished User 0001, SundarBot, LkNsngth, Stefano85, Bidabadi~enwiki, SashatoBot, Lambiam, Ckatz, Daphne A, Aetemus, CRGreathouse, CBM, Thomasmeeks, FilipeS, Rifleman 82, M a s, Omicronpersei8, Cj67, Urdutext, Escarbot, AntiVandalBot, Luna Santin, JAnDbot, MER-C, Thenub314, Yill577, Hurmata, Kuyabribri, Khalid Mahmood, Jtir, R'n'B, ZRV, J.delanoy, Maurice Carbonaro, Jonathanzung, Jwuthe2, KCinDC, Juliancolton, DavidCBryant, Treisijs, GregWoodhouse, Useight, Funandtrvl, VolkovBot, JohnBlackburne, Greclevoir, Altruism, Rei-bot, BotKung, Falcon8765, Thric3, Symane, SieBot, Neworder1, Lagrange613, Zedlik, DesolateReality, Altzinn, Smithpith, Cenarium, BOTarate, Kruusamägi, XLinkBot, SilvonenBot, JinJian, ElMeBot, D.M. from Ukraine, Leonini, Addbot, Friginator, LaaknorBot, CarsracBot, Ozob, Legobot, Luckas-bot, Yobot, Ht686rg90, Quangbao, 9258fahsflkh917fas, Wrelwser43, ArthurBot, Xqbot, Dowjgyta, Txebixev, Psyoptix, Gaussy, GrouchoBot, Point-set topologist, RibotBOT, Charvest, Geoffreybernardo, Tkuvho, Hard Sin, RedBot, Allen 6666, FoxBot, EmausBot, WikitanvirBot, Slawekb, Bethnim, QuentinUK, Wayne Slam, Lorem Ip, Herebo, ClueBot NG, Rurik the Varangian, Helpful Pixie Bot, Bcapetta, Alelbre, BG19bot, Mistory, Artem Karimov, Huntingg, Hillcrest98, Brad7777, Christian Glodzinski, Webclient101, Jcardazzi, Brirush, Dave Bowman - Discovery Won, Limit-theorem, K401sTL3, Monkbot, Raghav statistics jaipur, Degenerate prodigy, KasparBot and Anonymous: 117

- **Cryptography** *Source:* https://en.wikipedia.org/wiki/Cryptography?oldid=671227688 *Contributors:* AxelBoldt, WojPob, LC~enwiki, Brion VIBBER, Mav, Uriyan, Zundark, The Anome, Taw, Ap, Tao~enwiki, Ted Longstaffe, Dachshund, Arvindn, Gianfranco, PierreAbbat, Ortolan88, Roadrunner, Boleslav Bobcik, Maury Markowitz, Imran, Graft, Heron, Sfdan, Stevertigo, Nevilley, Patrick, Chas zzz brown, Michael Hardy, GABaker, Dante Alighieri, Liftarn, Ixfd64, Cyde, TakuyaMurata, Karada, Dori, (, Goatasaur, Card~enwiki, Ahoerstemeier, DavidW-Brooks, ZoeB, Theresa knott, Cferrero, Jdforrester, Julesd, Glenn, Kylet, Nikai, Andres, Cimon Avaro, Evercat, Delifisek, Dgreen34, Schnee-locke, Norwikian, Revolver, Novum, Htaccess, Timwi, Wikiborg, Dmsar, Ww, Dysprosia, Jitse Niesen, Phr, The Anomebot, Greenrd, Dtgm, Tpbradbury, GimmeFuel, K1Bond007, Tempshill, Ed g2s, Raul654, Rbellin, Pakaran, Jeffq, Ckape, Robbot, Fredrik, Chris 73, RedWolf, Donreed, Altenmann, Kuszi, Securiger, Georg Muntingh, MathMartin, Jsdeancoearthlink.net, Academic Challenger, Meelar, Timrollpickering, Rasmus Faber, Cyrius, Mattflaschen, Ludraman, Tobias Bergemann, Dave6, Snobot, Giftlite, Dbenbenn, Jacoplane, HippoMan, Wolfkeeper, Netoholic, Farnik, Peruvianllama, Michael Devore, Yekrats, Per Honor et Gloria, Sietse, Mboverload, Ferdinand Pienaar, Matt Crypto, Mobius, Neilc, Gubbubu, Geni, CryptoDerk, Antandrus, Beland, Vanished user 1234567890, Pale blue dot, Rdsmith4, APH, Mzajac, Euphoria, SimonLyall, Oiarbovnb, TiMike, Ta bu shi da yu, Freakofnurture, Monkeyman, Blokhead, Heryu~enwiki, Mark Zinthefer, Moverton, Discospinster, Rich Farmbrough, Guanabot, MaxMad, ArnoldReinhold, YUL89YYZ, Ivan Bajlo, Paul August, DcoetzeeBot~enwiki, Bender235, TerraFrost, Surachit, JRM, Prsephone1674, Bobo192, Stesmo, Harley peters, AnyFile, John Vandenberg, Myria, Jericho4.0, Davidgothberg, Slipperyweasel, Wrs1864, ClementSeveillac, M5, Stephen G. Brown, LoganK, Msh210, Wereldburger758, Alansohn, JYolkowski, Dhar, Mo0, Fg, Seamusandrosy, Complex01, ABCD, Logologist, InShanee, Avenue, Snowolf, Super-Magician, Saga City, Zyarb, Daedelus, Egg, H2g2bob, Vadim Makarov, Richwales, Oleg Alexandrov, Zntrip, Woohookitty, Mindmatrix, Justinlebar, Deeahbz, Jacobolus, Madchester, E=MC^2, Brentdax, Duncan.france, Nfearnley, Shmitra, Jok2000, Wikiklrsc, Mangojuice, SDC, Plrk, DarkBard, Cedrus-Libani, Stefanomione, Turnstep, Jimgawn, Tslocum, Graham87, Abach, FreplySpang, Vyse, JIP, Sinar~enwiki, Jorunn, Sjakkalle, Ner102, Rjwilmsi, Demian12358, Adjusting, MarSch, Mike Segal, Edggar, Miserlou, HappyCamper, Brighterorange, The wub, DoubleBlue, Volfy, CBR1kboy, Vuong Ngan Ha, RobertG, Mathbot, Gouldja, PleaseSendMoneyToWikipedia, Crazycomputers, Jameshfisher, RobyWayne, KFP, King of Hearts, Chobot, Manscher, Roboto de Ajvol, Siddhant, Wavelength, Laurentius, Ayongcheemeng, Mukkakukaku, RussBot, Lpmusix, Pigman, Manop, The1physicist, Gaius Cornelius, Chaos, Zeno of Elea, NawlinWiki, Welsh, Joel7687, Exir Kamalabadi, Proidiot, ONEder Boy, Schlafly, DavidJablon, Thiseye, Dhollm, Peter Delmonte, Misza13, Grafikm fr, Xompanthy, Deckiller, BOT-Superzerocool, Jeremy Visser, FF2010, 21655, Papergrl, Closedmouth, Nemu, CharlesHBennett, Aeon1006, Peyna, Bernd Paysan, Echartre, Anclation~enwiki, Wbrameld, Who-is-me, MagneticFlux, Crazyquesadilla, EndymiOn, Dr1819, DVD R W, ChemGardener, Yakudza, A bit iffy, SmackBot, Sean.nobles, Mmemex, Nihonjoe, 1dragon, Impaciente, Uncle Lemon, Jacek Kendysz, Jagged 85, Jrockley, David G Brault, BiT, JohnMac777, Mauls, Peter Isotalo, Gilliam, Ohnoitsjamie, Hmains, Skizzik, Chaojoker, Lakshmin, Chris the speller, Ciacchi, Agateller, Hibbleton, Thumperward, Delfeye, Snori, Alan smithee, PrimeHunter, Iago4096, NYKevin, DevSolar, Vkareh, ZachPruckowski, DrDnar, Wes!, Rashad9607, Alieseraj, Kazov, Wonderstruck, Maxt, DRLB, OutRIAAge, Sovietmah, Bidabadi~enwiki, Chungc, Andrewrabbott, Harryboyles, Dr. Sunglasses, Molerat, Fatespeaks, Ksn, Sidmow, JoshuaZ, Minna Sora no Shita, ManiF, Michael miceli, Jacopo, Ryanwammons, Slayemin, Chrisd87, Eltzermay, Meco, TastyPoutine, Dhp1080, Serlin, DeathLoofah, Drink666, Hectorian, DouglasCalvert, RudyB, Judgesurreal777, Pegasus1138, Detach, Shenron, Nightswatch, Gilabrand, Tawkerbot2, Chetvorno, Jafet, Powerslide, Sansbras, CRGreathouse, Hermitage17, Crownjewel82, BeenAroundAWhile, Thehockeydude44, CWY2190, Saoirse11, Raghunath88, Blackvault, Grandexandi, Cydebot, Ntsimp, Mblumber, John Yesberg, Gogo Dodo, Corpx, Tawkerbot4, XP105, Kozuch, Brad101, Omicronpersei8, Robertsteadman, Antura, Pallas44, Saber Cherry, Oerjan, Mojo Hand, Lotte Monz, Dgies, DPdH, Scircle, AntiVandalBot, Luna Santin, Jj137, Dylan Lake, Oddity-, G Rose, JAnDbot, Monkeymonkey11, Komponisto, WPIsFlawed, Hut 8.5, GurchBot, SCCC, Jahoe, Richard Burr, Acroterion, KooIkirby, Calcton, Hong ton po, Mol-

Bob1960evens, Thegeneralguy, PixelBot, Bracton, Thingg, Heyzeuss, Erodium, RMFan1, Vadimvadim, XLinkBot, Marc van Leeuwen, SilvonenBot, Willking1979, Betterusername, Ron B. Thomson, Tomthecool, NjardarBot, MrOllie, Delaszk, Nassrat, Zorrobot, Jarble, Noumenon, Legobot, Luckas-bot, Yobot, Kilom691, KamikazeBot, AnomieBOT, Ciphers, , Citation bot, Timir2, Thore Husfeldt, GrouchoBot, Arid Zkwelty, Charvest, BoomerAB, Alexisastupidnoob, Tobby72, HRoestBot, RedBot, Fparnon, TjBot, EmausBot, Jmencisom, Bethnim, Phoenixthebird, Sourabh Katagade, Ahughes6, Devanshuhpandey, OnePt618, Agatecat2700, ClueBot NG, Wcherowi, CocuBot, Joel B. Lewis, Widr, Helpful Pixie Bot, Leonxlin, Ijgt, Piguy101, Rajathsbhat, Will Gladstone, Brad7777, Dtotoo, JYBot, RichardMarioFratini, Mark viking, Purnendu Karmakar, SakeUPenn, Dough34, Jorgelin10, Theodora.ser, Pts46 and Anonymous: 231

- **Geometry** *Source:* https://en.wikipedia.org/wiki/Geometry?oldid=670750286 *Contributors:* William Avery, Michael Hardy, Ixfd64, Seav, Tregoweth, Ahoerstemeier, Charles Matthews, Dcoetzee, Dino, Tpbradbury, Robbot, Lowellian, Gandalf61, Raeky, Dina, Tobias Bergemann, Giftlite, DocWatson42, NeoJustin, Michael Devore, Python eggs, Ran, Oneiros, Tomruen, Shotwell, Grstain, Shahab, Discospinster, Rich Farmbrough, YUL89YYZ, Paul August, Bender235, ESkog, Tompw, El C, Shrike, RoyBoy, Bobo192, C S, Mdd, Jumbuck, Alansohn, Gary, Iothiania, Lord Pistachio, Mlm42, Bart133, Wtmitchell, Velella, Harej, Agutie, Blaxthos, Oleg Alexandrov, Angr, Starblind, Woohookitty, Mindmatrix, Igny, LOL, WadeSimMiser, KingsleyIdehen, Noetica, Mandarax, Rjwilmsi, Salix alba, The wub, Flam2006, Latka, RexNL, Malhonen, BradBeattie, King of Hearts, Chobot, DVdm, Algebraist, Siddhant, Wavelength, KSmrq, Wimt, NawlinWiki, Semperf, BOT-Superzerocool, Nescio, Dv82matt, Ninly, Closedmouth, Arthur Rubin, Kier07, KNHaw, Fatih Kurt, DVD R W, Luk, Yvwv, Sardanaphalus, A13ean, SmackBot, MattieTK, YellowMonkey, Selfworm, Jagged 85, Hardyplants, Srnec, Yamaguchi

Taric25, Valley2city, Full Shunyata, MalafayaBot, Silly rabbit, SchfiftyThree, PureRED, Viewfinder, John Reaves, Can't sleep, clown will eat me, Emrrans, Tamfang, The Placebo Effect, Rrburke, Addshore, Stevenmitchell, Aldaron, Flyguy649, Jaimie Henry, Syncopator, Jiddisch~enwiki, Kukini, Tesseran, Ged UK, SashatoBot, Lambiam, Xdamr, ArglebargleIV, UberCryxic, Cronholm144, Dumelow, Jim.belk, Majorclanger, Mr Stephen, Waggers, Mets501, Doczilla, Asyndeton, Hu12, Rschwieb, Iridescent, Nilamdoc, Elharo, Courcelles, JRSpriggs, JForget, CRGreathouse, Bigred625, CBM, KyraVixen, JohnCD, NickW557, MarsRover, Moreschi, Cydebot, Kanags, Metanoid, Gogo Dodo, JFreeman, Scott14, Julian Mendez, Doug Weller, Christian75, Juansempere, DumbBOT, Bookgrrl, Lee, JamesAM, Thijs!bot, Epbr123, Mojo Hand, Marek69, NorwegianBlue, Cool Blue, Sinn, RFerreira, Escarbot, Mentifisto, Hmrox, Sidasta, AntiVandalBot, Luna Santin, Seaphoto, Shirt58, TimVickers, Malcolm, The man stephen, Steelpillow, JAnDbot, DuncanHill, MER-C, The Transhumanist, Ophion, Bongwarrior, VoABot II, Avjoska, JNW, Rich257, Seberle, 28421u2232nfenfcenc, David Eppstein, DerHexer, Edward321, Esanchez7587, MartinBot, PAK Man, Arjun01, Burnedthru, R'n'B, CommonsDelinker, Whale plane, Tgeairn, Ssolbergj, J.delanoy, Trusilver, Rgoodermote, Bogey97, Jonpro, Eliz81, Lantonov, DarkFalls, Gargiapama, Ryan Postlethwaite, The Transhumanist (AWB), Happy138, Darrenderg, NewEnglandYankee, SJP, Touch Of Light, Piggywiggy23, Jackaranga, Cometstyles, Vindicated123, Xiahou, Idioma-bot, Wikieditor06, VolkovBot, CWii, Macedonian, Jeff G., JohnBlackburne, AlnoktaBOT, NikolaiLobachevsky, Am Fiosaigear~enwiki, Philip Trueman, DoorsAjar, TXiKiBoT, Gaj8, Z.E.R.O., Anonymous Dissident, Sankalpdravid, Qxz, Voorlandt, Anna Lincoln, Ocolon, Blahblahpoopyface, Speedy2alex, JhsBot, Globe-Gores, Slysplace, Jackfork, LeaveSleaves, Frankmoon, Prb4, Val001, Mr.Kennedy1, Sploonie, Wolfrock, Synthebot, Enviroboy, Insanity Incarnate, Dmcq, Arcfrk, Twooars, AlleborgoBot, Symane, Logan, ZBrannigan, PeridesofAthens, NHRHS2010, EmxBot, Rybu, D. Recorder, SieBot, Zmanq, Frans Fowler, Tiddly Tom, Chriscorbell, Caltas, Matthew Yeager, DBishop1984, Triwbe, Zsniew, Yintan, GlassCobra, Cmb71129, Savagemania, Chromaticity, Tiptoety, HannibalofCarthage, Solar clathrate, B123e, Arbor to SJ, Momo san, Wilson44691, CutOffTies, Prestonmag, Oxymoron83, Faradayplank, MiNombreDeGuerra, Xeltran, Nikkislayer7, N96, StaticGull, Gon56, Pinkadelica, Escape Orbit, LarRan, Athenean, Jamesfranklingresham, Atif.t2, Deavenger, ClueBot, PipepBot, Snigbrook, Justin W Smith, The Thing That Should Not Be, TrigWorks, Razimantv, Mild Bill Hiccup, CounterVandalismBot, Lbertolotti, Chrispaypwnsyou, Excirial, Alexbot, Jusdafax, Eeekster, Teds4 life, Gtstricky, Rejka, Brews ohare, NuclearWarfare, Cenarium, Snicoulaud, Razorflame, Deskuntil, Thingg, Franklin.vp, Zig-Zag Zig-Zag, Grant12345, Egmontaz, Goodvac, Vanished user uih38riiw4hjlsd, Bücherwürmlein, Skunkboy74, XLinkBot, BodhisattvaBot, Rror, Gerhardvalentin, Bert Carpenter, Mitch Ames, SilvonenBot, Joebob797979, Badgernet, HarlandQPitt, Thatguyflint, HexaChord, Tusharkrsna, Addbot, Proofreader77, Ronhjones, TutterMouse, Fluffernutter, MrOllie, Glass Sword, Favonian, Doniago, LinkFA-Bot, Numbo3-bot, Tide rolls, BrianKnez, Solid State, Romanskolduns, MuZemike, TeH nOmInAtOr, Vistalover, Bopha, Sitehut, Luckas-bot, Ajpj999, Allemandtando, 2D, Newportm, II MuSLiM HyBRiD II, Jgmoxness, Pcap, QueenCake, South Bay, MacTire02, Magog the Ogre, AnomieBOT, Cch0404, Floquenbeam, Jim1138, Dwayne, Piano non troppo, Kingpin13, Law, NickK, RandomAct, Materialscientist, La comadreja, ArthurBot, LilHelpa, Gsmgm, The Firewall, Andrewmc123, Xqbot, Cah says down, TinucherianBot II, Timir2, Cureden, Capricorn42, Nasnema, TheWeakWilled, Jsharpminor, Grim23, ReubenQuickel24477, Laramie Huggins, Foomanchu1234, Ksagittariusr, Ched, Ruy Pugliesi, J04n, GrouchoBot, Omnipaedista, Prunesqualer, RibotBOT, Mathonius, Seeleschneider, Guitarist1897, Doulos Christos, Geometryfan, Prari, Djcam, Tobby72, Zombolas, Cpretzelh, Romanlezner, Craig Pemberton, Machine Elf 1735, PrBeacon, Tkuvho, Thesavagenation12345, Pinethicket, HRoestBot, Heiwa.peace167, Rameshngbot, A8UDI, Martken 95, Olivier Bommel, Mjs1991, TobeBot, Lionslayer, Lotje, Reaper Eternal, Dizbemyname, Diannaa, JV Smithy, Reach Out to the Truth, Bobby122, DARTH SIDIOUS 2, Mean as custard, Viennaiswaiting, TjBot, TheArguer, Midhart90, Jowa fan, DASHBot, EmausBot, John of Reading, Domesticenginerd, WikitanvirBot, Stryn, Chlgrg, Cetinozyurt, GoingBatty, Minimac's Clone, RenamedUser01302013, KHamsun, Otyler27, Wikipelli, JSquish, Chasesd16eg, Ldboer, Fae, Stuartisgreat, 1Veertje, Wayne Slam, OnePt618, Mrvitiamwater, Arman Cagle, Alvalv69, L Kensington, Ready, ChuispastonBot, Jordibuma, LikeLakers2, Weimer, DASHBotAV, Rocketrod1960, Thuytnguyen48, ClueBot NG, Smtchahal, Stephenou, Wcherowi, MelbourneStar, Zg001, Gilderien, Deepvalley, SusikMkr, Nepal.sudeep123, Oleovc, Widr, Swiffer11. Helpful Pixie Bot, 299x333=99567, Thisthat2011, Monkeyfirends, Subhammishra, Ariel C.M.K., Grimr117, AvocatoBot, Dan653, Dazakip, Whyubhatin123, Solomon7968, Xosé Antonio, Himanway12, FertileCatfish. Sparklegirl007me, Potiroivsw, Brad7777, Rocky2222, TejasDiscipulus2, Kiilidiplomus, Anbu121, Diddy711, Darylgolden, Ak16899, Iloveyou711, Pratyya Ghosh, Stigmatella aurantiaca, ChrisGualtieri, Sewlin, Brade222, DeathStar696969696969, Dexbot, Magentic Manifestations, Ty000000, Dr.Arrieta, Lugia2453, Falcon Galdo, Jt24, Josophie, Melissa Bennett, Wywin, Vishrocks, Lauraotero136, Atomsk1321, Corkwii, Eyesnore, Lasmbcflr, Cloudyjbg27512, Kharkiv07, Username314159, Ugog Nizdast, Ginsuloft, Lakshay7124, Nikki MJ, GRIMM SOUL666, Skr15081997, Newnewisjazz, IamForCereal, Vieque, Chuluojun, Sandwich8849, FriendlyCaribou, JacobParham, Cowlord12, BabyChastie, Ptucktheduck, JellyPatotie, I love Kellin Quinn, Hishamthn, SimpleFloater555, KasparBot, 236benderavenue and Anonymous: 722

- **Convex geometry** *Source:* https://en.wikipedia.org/wiki/Convex_geometry?oldid=661180288 *Contributors:* Fred Bauder, Altenmann, Giftlite, Zaslav, RJFJR, Oleg Alexandrov, Sodin, Melchoir, Eserikto, Kopeikins, Riedel~enwiki, Van helsing, OrenBochman, David Eppstein, Yannledu, Xeltifon, JohnBlackburne, Val001, Dattorro, JP.Martin-Flatin, Addbot, Xario, Yobot, Ciphers, DrilBot, EmausBot, Brad7777 and Anonymous: 17

nasc, Rybu, YohanN7, SieBot, Ivan Štambuk, Meldor, Triwbe, Aristolaos, Nicinic, Pendlehaven, Daniarmo, MiNombreDeGuerra, Jorgen W, Kumioko, Valeria.depaiva, Vituzzu, Stfg, Laurentseries, Jludwig, ClueBot, The Thing That Should Not Be, Stevanspringer, TheSmuel, Monty42, SapphireJay, SchreiberBike, Triathematician, Manatee331, Robertabrams, Novjunulo, Fastily, Pi.C.Noizecehx, ErickOrtiz, Tilmanbauer, MystBot, Addbot, Some jerk on the Internet, Yobmod, Fieldday-sunday, MrOllie, CarsracBot, Bazza1971, LinkFA-Bot, Jasper Deng, K-topology, Tide rolls, Lightbot, OlEnglish, Zorrobot, TeH nOmInAtOr, Jarble, Sammtamm, Legobot, Cote d'Azur, Luckas-bot, 2D, Deputyduck, AnomieBOT, 1exec1, JackieBot, AdjustShift, Materialscientist, Citation bot, Xqbot, Ekwos, J04n, GrouchoBot, Point-set topologist, RibotBOT, Charvest, Contraverse, Divisbyzero, Orhanghazi, VI, Anilkumarphysics, Commit charge, Pinethicket, Tom.Reding, CrowzRSA, PoincaresChild, TobeBot, Jws401, Enthdegree, Integrals4life, Unbitwise, Jesse V., EmausBot, Fly by Night, Slawekb, ZéroBot, Chimpdmunk, The Nut, Caspertheghost, QEDK, Staszek Lem, Lorem Ip, ProteoPhenom, Anita5192, ResearchRave, ClueBot NG, Wcherowi, O.Koslowski, Widr, Helpful Pixie Bot, Daheadhunter, BG19bot, TCN7JM, Bigdon128, Wimvdam, Brad7777, Charismaa, Waleed.598, Sboosali, JYBot, MrBubbleFace, Dexbot, Paulo Henrique Macedo, King jakob c, Brirush, Mark viking, Ayesh2788, I am One of Many, TJLaher123, SakeUPenn, K401sTL3, Lizia7, Sesamo12, Btomoiaga, Chuluojun, Je.est.un.autre, Betapictoris, SoSivr, Jainmskip, Hriton, KasparBot, Pollock137 and Anonymous: 348

• **General topology** *Source:* https://en.wikipedia.org/wiki/General_topology?oldid=670342233 *Contributors:* AxelBoldt, Eloquence, Toby Bartels, AugPi, Revolver, Charles Matthews, Dysprosia, Jitse Niesen, Robbot, MathMartin, Tosha, Giftlite, Maximaximax, Rgdboer, Remuel, Tsirel, Msh210, Dallashan~enwiki, Sligocki, Brookie, Linas, BD2412, R.e.b., Mathbot, Dnwoodbury, Masnevets, YurikBot, Trovatore, Number 57, Googl, Reyk, Sardanaphalus, Bluebot, Nbarth, CBM, Chadnash, MER-C, DGG, VolkovBot, JohnBlackburne, YohanN7, JackSchmidt, Hans Adler, M.K.R.S. VEERA KUMAR, Addbot, Yobot, Sławomir Biały, Trappist the monk, TjBot, EmausBot, Fly by Night, ZéroBot, Anita5192, Oleg Viro, Brad7777, Majesty of Knowledge, ChrisGualtieri, APerson, Mogism, Aymankamelwiki, Brirush, Mark viking, Steynberg, Purnendu Karmakar, Nigellwh, K401sTL3 and Anonymous: 22

• **Algebraic topology** *Source:* https://en.wikipedia.org/wiki/Algebraic_topology?oldid=663311337 *Contributors:* AxelBoldt, Zundark, Youandme, Chas zzz brown, Michael Hardy, Alodyne, TakuyaMurata, Revolver, Charles Matthews, Timwi, Dysprosia, Jitse Niesen, Phys, MathMartin, Wikibot, Fuelbottle, Giftlite, Gtrmp, Lethe, Lupin, Fropuff, APH, Icairns, Smimram, Rich Farmbrough, Dave Foley, Gauge, Momotaro, Cyc~enwiki, Obradovic Goran, Marc van Woerkom, Msh210, Alansohn, Oleg Alexandrov, Linas, Rjwilmsi, FlaBot, YurikBot, Wavelength, Michael Slone, Banus, RonnieBrown, Sardanaphalus, SmackBot, Commander Keane bot, Yamaguchio, Akriasas, Archgoon, Loadmaster, ChazYork, Newone, CRGreathouse, Anonymi, Sam Staton, Mojo Hand, AntiVandalBot, David Eppstein, Policron, Mjg0, Bl4ck54bb4th, JohnBlackburne, Father Christmastime, Ambrose H. Field, Aaeamdar, Piclark, Spinningspark, Agüeybaná, Katzmik, Haiviet~enwiki, SieBot, OKBot, LarRan, Polyrhythm, Matt Hellige, Horoball, TimothyRias, MystBot, Addbot, TutterMouse, De-laszk, Debresser, Verbal, RobertHannah89, Luckas-bot, Tinyde Evenstar, Citation bot, Bci2, ArthurBot, FrescoBot, Hexagon70, Citation bot 1, D stankov, GustavLa, Dangerous wasp, EmausBot, Fly by Night, Ebrambot, D.Lazard, AManWithNoPlan, Chewings72, Anita5192, ClueBot NG, Frietjes, Helpful Pixie Bot, Brad7777, Makecat-bot, Brirush, Mark viking, SakeUPenn, CalTechMD, Eric Corbett, Stamptrader, K9re11, KasparBot and Anonymous: 65

• **Manifold** *Source:* https://en.wikipedia.org/wiki/Manifold?oldid=669627064 *Contributors:* Zundark, The Anome, XJaM, Edemaine, Michael Hardy, Alodyne, Ciphergoth, Rmilson, Charles Matthews, Timwi, Joshuabowman, Jitse Niesen, Mtcv, Banno, Catskul, Fredrik, Sverdrup, Timrollpickering, Tobias Bergemann, Tosha, Giftlite, BenFrantzDale, Lethe, Anville, Jason Quinn, Python eggs, Bobblewik, Melikamp, Pmanderson, Elroch, Tzanko Matev, Smimram, Cacycle, Sperling, Paul August, Ben Standeven, Gauge, El C, Szquirrel, Bobo192, Kevin Lamoreau, Quasicharacter, Obradovic Goran, Haham hanuka, Crust, Varuna, Schissel, Alansohn, Anthony Appleyard, Cdc, Oleg Alexandrov, Optimusnauta, Joriki, Linas, Daniel Case, Ruud Koot, Joke137, Mandarax, BD2412, Chun-hian, NatusRoma, OneWeirdDude, MarSch, Salix alba, NeonMerlin, Juan Marquez, R.e.b., Quuxplusone, DaGizza, Krishnavedala, Bgwhite, Algebraist, Wavelength, Gene.arboit, Loom91, Markus Schmaus, KSmrq, Rick Norwood, Tong~enwiki, Dethomas, Vb, Voidxor, Hv, Acer, Pred, Marlasdad, Jaysbro, Zvika, SmackBot, Nihonjoe, Incnis Mrsi, Slashme, Gcdart, Wisygig, Jjalexand, Silly rabbit, StrangerInParadise, SEIBasaurus, Nbarth, Sisodia, AdamSmithee, Foxjwill, Chlewbot, Berland, QFT, HLwiKi, Leland McInnes, Jdlambert, Jon Awbrey, Acdx, ArglebargleIV, Loodog, Loadmaster, Adriferr, EmreDuran, Newone, James pic, Ranicki, Gregbard, Vanished user fj0390923roktg4tlkm2pkd, Ntsimp, Corpx, Madmonk325, Invisible Capybara, Pascal.Tesson, Amitushtush, Msnicki, Xtv, Gimmetrow, Bioguy, Thijs!bot, AntiVandalBot, Exteray, Dougher, JAnDbot, Michael Tiemann, Norailyain, Y K Times, JamesBWatson, Jakob.scholbach, Maniwar, David Eppstein, Martynas Patasius, Renetus, R'n'B, Pbroks13, J.delanoy, Maurice Carbonaro, Lantonov, Sormani, Policron, Vanished user 47736712, LeighvsOptimvsMaximvs, Sarregouset, Camm86, JohnBlackburne, LokiClock, Bovineboy2008, Aesopos, Greclevoir, Anonymous Dissident, Appoose, Geometry guy, Sploonie, Arcfrk, Bikasuishin, Katzmik, Rybu, GirasoleDE, Stca74, WereSpielChequers, Domination989, Squelle, Soler97, JackSchmidt, Stfg, Randomblue, Jludwig, Sidiropo, LarRan, Mr. Granger, Loren.wilton, Gopalkrishnan83, ClueBot, Plastikspork, JuPitEer, Sun Creator, Invive, Brews ohare, Eloifigueiredo, AnonyScientist, XLinkBot, Bradv, Luca Antonelli, CàlcuIntegral, Addbot, Topology Expert, MrVanBot, LinkFA-Bot, Lightbot, عبد, Yobot, Ht686rg90, AnomieBOT, VanishedUser sdu9aya9fasdsopa, Citation bot, Gsmgm, Expooz, TinucherianBot II, Tasudrty, Adccon, Philipp Kuehl, Cpryby, Br77rino, Point-set topologist, FrescoBot, AllCluesKey, Sławomir Biały, Lost-n-translation, Tkuvho, Pmokeefe, RedBot, Rausch, Gryllida, FoxBot, ActuallyRationalThinker, Bj norge, Jowa fan, WildBot, EmausBot, Slawekb, ZéroBot, Moorechen, Quondum, D.Lazard, Patatas101, Ali Dadsetan, Anita5192, Mgvongoeden, Helpful Pixie Bot, Tub82911, AvocatoBot, Paweł Ziemian, Agent Swipe, Dexbot, BeaumontTaz, Mark viking, Purnendu Karmakar, Samreid94, Pwm86, K9re11, Rcehy, BethNaught, Karandodia, UtherSB, KasparBot and Anonymous: 161

• **Probability theory** *Source:* https://en.wikipedia.org/wiki/Probability_theory?oldid=667950379 *Contributors:* Lee Daniel Crocker, Bryan Derksen, Zundark, The Anome, Miguel~enwiki, Boleslav Bobcik, Patrick, Michael Hardy, Goatasaur, Den fjättrade ankan~enwiki, Jonik, Bjcairns, Charles Matthews, Dysprosia, Gutsul, Johannes Hüsing, MH~enwiki, Robbot, MathMartin, TMLutas, Hadal, Borislav, Tobias Bergemann, Weialawaga~enwiki, Tosha, Giftlite, Raymond Meredith, ShaunMacPherson, Lethe, Fastfission, Niteowlneils, ChicXulub, Utcursch, Knutux, Ynh~enwiki, Beland, APH, Maximaximax, Gbr3~enwiki, Vivacissamamente, Rich Farmbrough, Paul August, Bender235, RJHall, MisterSheik, El C, Zenohockey, AvidDismantler, Hayabusa future, Bobo192, Cretog8, Zwilson, Flammifer, Obradovic Goran, Cyrillic, Mdd, Tsirel, Msh210, Rgdegg, PAR, KingTT, Frankman, Sunena, Jheald, RainbowOfLight, Oleg Alexandrov, INic, Btyner, Graham87, Porcher, Rjwilmsi, Mayumashu, Tizio, Salix alba, FutureNJGov, Nguyen Thanh Quang, Mathbot, RexNL, Krun, Malhonen, Chobot, Wavelength, Sceptre, Spudbeach, Michael Slone, Lenthe, TheMandarin, ENeville, Robertetaylor, Trovatore, Srinivasasha, User27091, Wknight94, Hirak 99, Pb30, JoanneB, Kungfuadam, Sardanaphalus, SmackBot, Snielsen, Arjay369, Aastrup, PJTraill, Chris the speller, Bluebot, Jprg1966, MalafayaBot, DHN-bot~enwiki, Wynand.winterbach, Gala.martin, Decltype, Skiminki, Andeggs, Ncmathsadist, Lambiam, Dbtfz, Bjankuloski06en~enwiki, Levineps, Spebudmak, Tawkerbot2, Kurtan~enwiki, Devourer09, Sleeping123, CRGreathouse, Unionhawk, Ali Obeid,

tats, Brougham96, Mhmolitor, AnomieBOT, DemocraticLuntz, VX, Cavarrone, Galoubet, Dwayne, Piano non troppo, Youkbam, Template-hater, Walter Grassroot, Htim, Materialscientist, The High Fin Sperm Whale, Citation bot, OllieFury, Markmagdy, Sweeraha, GB fan, Apollo, Neurolysis, ArthurBot, Herreradavid33, LilHelpa, Xqbot, TinucherianBot II, Class ruiner, Kenz0402, Drilnoth, Fishiface, Locos epraix, Spetz-znaz, AbigailAbernathy, Clear range, Coretheapple, GrouchoBot, Ute in DC, SassoBot, Loizbec, 78.26, Rstabx, Stynyr, Doulos Christos, Chen-Pan Liao, N.j.hansen, Shadowjams, Joaquin008, Brennan41292, FrescoBot, Tobby72, Hallway916, Shadowpsi, HJ Mitchell, Winterswift, Cita-tion bot 1, PrBeacon, Boxplot, Yuanfangdelang, Pinethicket, Kiefer.Wolfowitz, Stpasha, Brian Everlasting, Île flottante, Bwana2009, Dee539, Florendobe, White Shadows, Gamewizard71, FoxBot, Mjs1991, Ruzihm, TobeBot, LAUD, Arfgab, Decstop, MrX, Spegali, Keepitup.sid, Sourishdas, Tbhotch, Drivi86, Sandman888, DARTH SIDIOUS 2, Chrisrayner, Whisky drinker, Mean as custard, Updatehelper, TjBot, Kastchei, Karlheinz037, Becritical, Elitropia, Jordan.brayanov, EmausBot, Orphan Wiki, Gfoley4, Racerx11, Hiamy, Tommy2010, Kellylautt, Tuxedo junction, Bae88, Daonguyen95, Fæ, Josve05a, Bollyjeff, Tastewrong1234, WeijiBaikeBianji, Cbratsas, JA(000)Davidson, Access Denied, Dylthaavatar, Kgwet, SporkBot, Jorjulio, GrindtXX, Makecat, Sak11sl, Future ahead, Anglais1, Sunur7, Mr. Kenan Bek, Noodleki, Donner60, Agatecat2700, NTox, DemonicPartyHat, 28bot, Petrb, ClueBot NG, MelbourneStar, This lousy T-shirt, Chrisminter, Dvsbmx, Bar-relProof, Bped1985, Andreas.Persson, Shawnluft, Cntras, Braincricket, ScottSteiner, Widr, Hikenstuff, Ryan Vesey, Amircrypto, Helpful Pixie Bot, Xandrox, Mishnadar, Ldownss00, Calabe1992, KLBot2, Lowercase sigmabot, BG19bot, WikiTryHardDieHard, Juro2351, Northamer-ica1000, Absconded Northerner, Muhehej1000, MusikAnimal, Marcocapelle, Stalve, EmadIV, Rm1271, Htrkaya, Omiswiki, Manoguru, Kit-tipatv, Meclee, Brad7777, Glacialfox, Roleren, Anbu121, Europeancentralbank, Bsutradhar, Ca3tki, Kodiologist, Codeh, Gr khan veroana kharal, Markk waugh, Illia Connell, SelmanRepišti, Dexbot, Ubertook, Mogism, Wikignome1213, Princessandthepi, Lugia2453, Brownstat, Norazoey, Speakel, 069952497a, Faizan, RG57, FallingGravity, AmericanLemming, Tentinator, Beasarah, Butter7938, Seppi333, Spuri-ousTwist, Ginsuloft, Sean4424, Sarwan khan, Adirlanz, AddWittyNameHere, Narasandraprabhakara, Science.philosophy.arts, Akuaku123, Mendisar Esarimar Desktrwaimar, Mconnolly17, Zib2542, Therealthings, MelaniePS, Monkbot, Poepkop, Soon Son Simps, Vieque, Ma-jormuesli, Trackteur, Andri Kuawko, Romelthomas, Umkan, Ybergner, Amortias, NQ, Morgantaschuk, VanishedUser sdu9aya9fs654654, Sumonratin, Charlotte Aryanne, Thebearedguy, Mj3322, Rainamagdalena, Lucky457, JohnDae123, Kreplach123, BabyChastie, SolidPhase, Amira Swedan, Isambard Kingdom, All-wikipro, Asyraf Afthanorhan, KasparBot, Hilopmip, Replypartyeuclides, Chonzom and Anonymous: 1217

• Numerical analysis *Source:* https://en.wikipedia.org/wiki/Numerical_analysis?oldid=666224558 *Contributors:* Tobias Hoevekamp, The Ano, Tar quin, Taw, Eijkhout, Michael Hardy, Dominus, Nixdorf, Zeno Gantner, Loisel, Stevan White, Jurgen~enwiki, Charles Matthews, Dys-prosia ,Jitse Niesen, Taxman, Topbanana, Ldo, Rogper~enwiki, Robbot, Jaredwf, Troworld, PedroPVZ, Blainster, Bkell, JesseW, Safor-rest, Reg e~enwiki, Robinh, Peter L, Decrypt3, Giftlite, BenFrantzDale, Lethe, Tom harrison, Fleminra, Guanaco, Matt Crypto, Gadfium,APH, Ama deusKlocker, Fintor, Kate, PhotoBox, Zowie, Rich Farmbrough, Yuval madar, Elwikipedista~enwiki, MisterSheik, Oyz, Dungo-dung, Greenle af~enwiki, Nk, Crust, Msh210, Alansohn, Arthena, Sligocki, Bugg, Aitter, Gene Nygaard, Bookandcoffee, Beliavsky, OlegAlexandrov, Lin as, DavidHaslam, Decrease789, KymFarnik, Graham87, Xask Linus, Koavf, Salix alba, TeemuN, Hlangeveld, Bubba73,Arnero, Mathbot, N ihiltres, RexNL, Carrionluggage, Windharp, NevilleDNZ, Chobot, Turidoth, Wavelength, RussBot, Hede2000, KSmrq,Chaos, Yahya Abdal-A ziz, Cat2020,Boivie, Zzuuzz, Ninly, Closedmouth, Josh3580, Dontaskme, Willtron, JLaTondre, Pred, Gesslein, Xiao-jeng~enwiki, Lunch, Yvwv , Sardanaphalus, A bit iffy, JJL, SmackBot, Tomchen~enwiki, Adam majewski, Jagged85, Wygk, Hongooi, TimothyClemans, Berland, Brianboon stra, Mlpkr, Drunken Pirate, The undertow, Lambiam, Sina2, ExamplePuzzle, IronGargoyle, Asyndeton, StephenB Streater, Ginkgo100, MystRi venExile, FatBastardInk, Audiosmurf, Paul Matthews, Didimos, CRGreathouse, CBM, Vitolis, Requestion,Bocianski, Ntsimp, Mattisse, Thij s!bot, Oliver202, Quantufinity, HussainAbbas, Urdutext, BigJohnHenry, Darklilac, VictorAnyakin, Niko-las Karalis, MER-C, EagleFan, Syst emlover, Khalid Mahmood, SamShearman, Gwern, Jtir, Leyo, Kawautar, Salih, Gombang, Nonphixion,JonMcLoone, Policron, Robertgreer, CMaes, Sheliak, JohnBlackburne, Redgecko, Stephenkirkup, Antoni Barau, Jmath666, Aither~enwiki,Wolfrock, SieBot, Un4v41l48l3, Light mouse, Svick, Amahoney, WikipedianMarlith, SlackerMom, Justin W Smith, EXTER7, Wikijens,Tomtzigt, ich, XLinkBot, Chrismacgr, JinJian, Addbot, Wordsoup, Manuel Trujillo Berges, Ekojekoj, Met-savend, Tide rolls, Loupeter, Legobot, Yobot , KamikazeBot, 1exec1, Kingpin13,Citation bot, Happyrabbit, Isheden, Rsmn, Charvest, Gtpjg,Bekus, CES1596, FrescoBot, Steve Quinn, Gaba p, Boulaur, RedBot, Meaghan, Jauhienij, David Binner, TjBot, DSP-user, EmausBot, Jmenci-som, Bethnim, Mirificium, Google Child, Christina Silverman, Darthhappyface, Shootji, Tot12, Ramiamro, Anita5192, ClueBot NG, Satelizer,Vacation9, Snotbot, GiantReliant, 3rdiw, Helpful Pixi e Bot, Enneth with a k, Brad7777, ChrisGualtieri, Chunliang Lyu, JYBot, Shtamy, Dra-jay1976, Purnendu Karmakar, Zhangzk, LaguerreLege ndre, Brtietz, Stormmeteo, Msalimi122, Msalimi1222, Linnet1979, Sasadd, MateuszKonieczny, KasparBot, Herobeaner and Anonymous: 188

• Symbolic computation *Source:* https://en.wikipedia.org/wiki/Symbolic_computation?oldid=663603063 *Contributors:* Camembert, CesarB, Charles Matthews, Jitse Niesen, Bevo, Phil Boswell, Stewartadcock, Saforrest, Lumingz, Tobias Bergemann, Discospinster, DonDiego, Liber-atus, Sciurinæ, Oleg Alexandrov, Tomhab, Rjwilmsi, Salix alba, Chobot, YurikBot, LeonardoRob0t, Gesslein, Sardanaphalus, Optikos, Bird of paradox, Cybercobra, MvH, Antonielly, Yaris678, Oerjan, Seaphoto, AllenDowney, Bender2k14, Mikaey, Addbot, Jarble, Luckas-bot, Amirobot, AnomieBOT, Charvest, RoyLeban, TobeBot, Vladislav Pogorelov, Alvaro Vidal-Abarca, Trevor hansen, WikitanvirBot, Jmenci-som, Bethnim, ArachanoxReal, D.Lazard, RockMagnetist, ClueBot NG, Ruddyscent, ChrisGualtieri, YFdyh-bot, Mogism, Bsalvy, HoboM-cJoe, Lizia7, Zieglerk, Wallnut tree, Papersandcoffee. Haosjaboeces and Anonymous: 33

• Mechanics *Source:* https://en.wikipedia.org/wiki/Mechanics?oldid=669837544 *Contributors:* Tarquin, Andre Engels, Youssefsan, Youandme, Stevertigo, Michael Hardy, GABaker, Ahoerstemeier, Mac, Docu, J-Wiki, Александър, Glenn, Mxn, Lommer, Charles Matthews, Reddi, Gre-glocock, Fruggo, Robbot, Altenmann, Rorro, Buster2058, Giftlite, Tom harrison, Lupin, Fropuff, Zhen Lin, Utcursch, Andycjp, Antandrus, Karol Langner, Icairns, Gscshoyru, Martpol, BBB~enwiki, CanisRufus, El C, Rgdboer, Joanjoc~enwiki, CDN99, Smalljim, Cmdrjameson, I9Q79oL78KiL0QTFHgyc, Alansohn, Comrade009, Natalya, Feezo, Jimbryho, Duncan.france, Mpatel, SCEhardt, Mandarax, Graham87, Magister Mathematicae, Nanite, Mayumashu, Erebus555, Bruce1ee, RexNL, Ewlyahoocom, Snailwalker, Chobot, Zath42, DVdm, Gwer-nol, Wavelength, Deeptrivia, RussBot, Fabartus, Epolk, Polyvios, Stephenb, The Merciful, Shakehandsman, Paul D. Anderson, Kungfuadam, ChemGardener, Sardanaphalus, SmackBot, Tom Lougheed, KnowledgeOfSelf, Pgk, Bomac, Dome359, Jagged 85, Pennywisdom2099, One-bravemonkey, Alsandro, Pzavon, Gilliam, Ohnoitsjamie, Bluebot, TimBentley, Dahn, Nbarth, Hongooi, Argyriou, CorbinSimpson, Fuhghet-taboutit, Zhigangsuo, Cybercobra, Jiddisch~enwiki, J.Wolfe@unsw.edu.au, Hgilbert, DMacks, Kotjze, John, Gobonobo, Peterlewis, MarkSut-ton, Special-T, Mets501, Hu12, Wizard191, Iridescent, Polymerbringer, Francl, JoannaSerah, Rchase, Irwangatot, Van helsing, Makeem-lighter, Qrc2006, Cydebot, Fnlayson, LouisBB, Jon Stockton, ST47, Quibik, Qwyrxian, LeeG, Mbell, Andyjsmith, Oerjan, Headbomb, John254, Thljcl, Dainis, AntiVandalBot, Yonatan, RDT2, Autocracy, Âme Errante, Altamel, JAnDbot, Samar, T L Miles, Acroterion, Pe-

Dor, Arthur Rubin, MaNeMeBasat, Nojhan, RandallZ, Sardanaphalus, SmackBot, InverseHypercube, Vermorel, Mcld, Misfeldt, Oli Filth, EncMstr, DHN-bot~enwiki, MaxSem, Jasonb05, Ggpauly, Hua001, Brianboonstra, Tbbooher~enwiki, Asm4, Twocs, G.de.Lange, Antonielly, JHunterJ, Xprime, Carlo.milanesi, Hu12, Riedel~enwiki, Polar Bear~enwiki, Lavaka, CRGreathouse, CmdrObot, Jonnat, Jackzhp, Van hels-ing, Thomasmeeks, Solomon Douglas, Stebulus, Schaber, Cydebot, Mikewax, Czenek~enwiki, Headbomb, Arnab das, KrakatoaKatie, AnAj, VictorAnyakin, JAnDbot, JPRBW, Ravelite, Roleplayer, LSpring, DRHagen, Smartcat, MartinDK, Sabamo, Mrbynum, Charlesreid1, Cic, Armehrabian, LisaTK, David Eppstein, Martynas Patasius, Bradgib, Erkan Yilmaz, Mange01, Stochastics, Salih, Burgaz, Nwbeeson, JonM-cLoone, Epistemenical, Bonadea, LastChanceToBe, VolkovBot, LokiClock, Philip Trueman, Anonymous Dissident, TPlantenga, Broadbot, Srinnath, Johngcarlsson, JFPuget, Sapphic, Struway, Mangogirl2, Wikibuki, Zwgeem, Truecobb, Gerakibot, Paolo.dL, CharlesGillingham, Iknowyourider, Dattorro, Rinconsoleao, Procellarum, PipepBot, Justin W Smith, PimBeers, Suegerman, Daveagp, Dmitrey, Auntof6, Lber-tolotti, DragonBot, Dianegarey, Doobliebop, DumZiBoT, Thoughtfire, Oğuz Ergin, Sliders06, Addbot, Ablevy, Щегол, Ekojnoekoj, MrOl-lie, Download, HiYoSilver01, Delaszk, Kiril Simeonovski, Zorrobot, Jarble, ConstantLearner, Luckas-bot, Ashburyjohn, Yobot, Diracula, Travis.a.buckingham, Pcap, Lylenorton, Deuxoursendormis, AnomieBOT, Erel Segal, Galoubet, Citation bot, ArthurBot, Pownuk, Quebec99, Obersachsebot, Xqbot, Carbaholic, P99am, Micemug, Wamanning, Isheden, Almabot, Knillinux, Georg Stillfried, Omnipaedista, RibotBOT, Klochkov.ivan, YuriyMikhaylovskiy, Amaury, Optiy, Mcmlxxxi, Schlitz4U, Eiro06, Ct529, Kamitsaha, FrescoBot, Nageh, X7q, Asadi1978, D'ohBot, Pschaus, Boxplot, HRoestBot, Kiefer.Wolfowitz, MaximizeMinimize, Jmc200, FoxBot, Mdwang, Dsz4, Trappist the monk, Dpbert, Rbdevore, Bpadmakumar, Duoduoduo, BlakeJRiley, Crisgh, H.ehsaan, RjwilmsiBot, Hosseininassab, Jowa fan, EmausBot, John of Read-ing, Nacopt, Carbo1200, Cfg1777, Metafun, Psogeek, Daryakav, Optimering, Krystofer, Bonnans, ZéroBot, The Nut, AManWithNoPlan, Katie O'Hare, Robbiemorrison, Edesigner, Ванко1, Zfeinst, Syst analytic, Brycehughes, ClueBot NG, Jean-Charles.Gilbert, Chester Markel, Robiminer, Leonardo61, Sahinidis, Harrycaterpillar, Helpful Pixie Bot, Dwassel, Rxnt, Thiscomments voice, MSchlueter, FiveColourMap, Heroestr, Albert triv, Osiris, Centathlete, Mdann52, ChrisGualtieri, Denmartines, Makecat-bot, Limit-theorem, Delafé, PiotrGregorczyk, SakeUPenn, Lazars53, Buzzzman, Cubism44, Thennicke, Βελτιστοποίησης, Mehr86, Mitsisko, Tyrneaeplith, Loraof, KasparBot and Anony-mous: 250

35.14.2   Images

- **File:10_DM_Serie4_Vorderseite.jpg** *Source:* https://upload.wikimedia.org/wikipedia/commons/0/0d/10_DM_Serie4_Vorderseite.jpg *Li-cense:* Public domain *Contributors:* http://www.bundesbank.de/Redaktion/DE/Standardartikel/Kerngeschaeftsfelder/Bargeld/dm_banknoten.html#doc18118bodyText2 *Original artist:* Deutsche Bundesbank, Frankfurt am Main, Germany

- **File:16th_century_French_cypher_machine_in_the_shape_of_a_book_with_arms_of_Henri_II.jpg** *Source:* https://upload.wikimedia.org/wikipedia/commons/a/a2/16th_century_French_cypher_machine_in_the_shape_of_a_book_with_arms_of_Henri_II.jpg *License:* CC BY-SA 3.0 *Contributors:* Own work, photographed at Musee d'Ecouen *Original artist:* Uploadalt

- **File:2008-09_Kaiserschloss_Kryptologen.JPG** *Source:* https://upload.wikimedia.org/wikipedia/commons/a/ad/2008-09_Kaiserschloss_Kryptologen.JPG *License:* CC BY-SA 3.0 *Contributors:* Own work *Original artist:* Ziko

- **File:Abacus_6.png** *Source:* https://upload.wikimedia.org/wikipedia/commons/a/af/Abacus_6.png *License:* Public domain *Contributors:*

- Article for "abacus", 9th edition Encyclopedia Britannica, volume 1 (1875); scanned and uploaded by Malcolm Farmer *Original artist:* Encyclopaedia Britannica

- **File:Al-kindi_cryptographic.png** *Source:* https://upload.wikimedia.org/wikipedia/commons/7/76/Al-kindi_cryptographic.png *License:* Pub-lic domain *Contributors:* en:Image:Al-kindi_cryptographic.gif *Original artist:* Al-Kindi

- **File:Alphabet_homeo.png** *Source:* https://upload.wikimedia.org/wikipedia/commons/7/74/Alphabet_homeo.png *License:* Public domain *Contributors:* Transferred from en.wikipedia; transferred to Commons by User:Ylebru using CommonsHelper.
Original artist: C S (talk). Original uploader was C S at en.wikipedia

- **File:Alphabet_homotopy.png** *Source:* https://upload.wikimedia.org/wikipedia/commons/7/7c/Alphabet_homotopy.png *License:* Public do-main *Contributors:* Own work (Original text: *I created this work entirely by myself.*) *Original artist:* C S (talk)

- **File:Anil_Kumar_Gain.png** *Source:* https://upload.wikimedia.org/wikipedia/en/d/d4/Anil_Kumar_Gain.png *License:* Public domain *Con-tributors:*
http://vidyasagar.ac.in/About/AKGayen.aspx *Original artist:*
Vidyasagar University

- **File:Arabic_machine_manuscript_-_Anonym_-_Ms._or._fol._3306_c.jpg** *Source:* https://upload.wikimedia.org/wikipedia/commons/9/93/Arabic_machine_manuscript_-_Anonym_-_Ms._or._fol._3306_c.jpg *License:* Public domain *Contributors:* Max Planck Digital Library *Original artist:* Anonymous

- **File:Arbitrary-gametree-solved.svg** *Source:* https://upload.wikimedia.org/wikipedia/commons/d/d7/Arbitrary-gametree-solved.svg *License:* Public domain *Contributors:*

- Arbitrary-gametree-solved.png *Original artist:*

- derivative work: Qef (talk)

- **File:Archimedes_pi.svg** *Source:* https://upload.wikimedia.org/wikipedia/commons/c/c9/Archimedes_pi.svg *License:* CC-BY-SA-3.0 *Con-tributors:* Own work *Original artist:* Leszek Krupinski (disputed, see File talk:Archimedes pi.svg)

- **File:B_24_in_raf_service_23_03_05.jpg** *Source:* https://upload.wikimedia.org/wikipedia/commons/a/a1/B_24_in_raf_service_23_03_05.jpg *License:* Public domain *Contributors:* Transferred from en.wikipedia to Commons. *Original artist:* The original uploader was Bzuk at English Wikipedia

- **File:BernoullisLawDerivationDiagram.svg** *Source:* https://upload.wikimedia.org/wikipedia/commons/2/20/BernoullisLawDerivationDiagram.svg *License:* CC-BY-SA-3.0 *Contributors:* Image:BernoullisLawDerivationDiagram.png *Original artist:* MannyMax (original)

35.14.3   Content license